Genetic Analyses of Plant and Molecular Marker-Assisted Breeding

Genetic Analyses of Plant and Molecular Marker-Assisted Breeding

Edited by Kiara Woods

STATES
ACADEMIC PRESS
www.statesacademicpress.com

States Academic Press,
109 South 5th Street,
Brooklyn, NY 11249, USA

Visit us on the World Wide Web at:
www.statesacademicpress.com

ISBN: 978-1-63989-748-3

Cataloging-in-Publication Data

Genetic analyses of plant and molecular marker-assisted breeding / edited by Kiara Woods.
p. cm.
Includes bibliographical references and index.
ISBN 978-1-63989-748-3
1. Plant molecular genetics. 2. Plant breeding. 3. Plants--Molecular aspects. I. Woods, Kiara.
QK981.4 .M65 2023
572.865 2--dc23

Table of Contents

Preface

The purpose of the book is to provide a glimpse into the dynamics and to present opinions and studies of some of the scientists engaged in the development of new ideas in the field from very different standpoints. This book will prove useful to students and researchers owing to its high content quality.

Plants have a complex gene pool that is capable of identifying invading pathogens and initiating effective defense mechanisms. Plant genetics refers to a specialized field of biology and botany that is concerned with the study of genes, genetic variation and heredity in plants. Molecular marker-assisted breeding (MAB) is a method for plant improvement that applies molecular biotechnologies to practical breeding and selection. MAB combines traditional plant breeding with the tools of molecular biology and genetics, prominently making use of molecular markers. MAB offers a number of benefits as compared with conventional breeding methods. It involves genetic manipulation performed at the DNA level for improving the traits of interest in plants. This book is a compilation of chapters that discuss the most vital concepts and emerging trends in the genetic analyses of plants and molecular marker-assisted breeding. It consists of contributions made by international experts. This book will serve as a reference to a broad spectrum of readers.

At the end, I would like to appreciate all the efforts made by the authors in completing their chapters professionally. I express my deepest gratitude to all of them for contributing to this book by sharing their valuable works. A special thanks to my family and friends for their constant support in this journey.

Editor

1

Comparative Transcriptomic Analysis of the Development of Sepal Morphology in Tomato (*Solanum Lycopersicum* L.)

Jingyi Liu [1,2], Meijing Shi [1,2], Jing Wang [1,2], Bo Zhang [1,2], Yushun Li [1,2], Jin Wang [1,2], Ahmed. H. El-Sappah [1,2,3] and Yan Liang [1,2,*]

1 College of Horticulture, Northwest A&F University, Shaanxi 712100, China; Liujingyi1987@nwsuaf.edu.cn (J.L.); shimeijing4@gmail.com (M.S.); wangjingwj518@gmail.com (J.W.); zhang-bo@nwafu.edu.cn (B.Z.); liyushun2016@nwafu.edu.cn (Y.L.); jw6127@nwafu.edu.cn (J.W.); Ahmed_elsappah2006@yahoo.com (A.H.E.-S.)
2 State Agriculture Ministry Laboratory of Northwest Horticultural Plant Germplasm Resources & Genetic Improvement, Northwest A&F University, Shaanxi 712100, China
3 Genetics Department, Faculty of Agriculture, Zagazig University, Zagazig 44511, Egypt
* Correspondence: liangyan@nwsuaf.edu.cn

Abstract: Sepal is an important component of the tomato flower and fruit that typically protects the flower in bud and functions as a support for petals and fruits. Moreover, sepal appearance influences the commercial property of tomato nowadays. However, the phenotype information and development mechanism of the natural variation of sepal morphology in the tomato is still largely unexplored. To study the developmental mechanism and to determine key genes related to downward sepal in the tomato, we compared the transcriptomes of sepals between *downward sepal* (*dsp*) mutation and the wild-type by RNA sequencing and found that the differentially expressed genes were dominantly related to cell expansion, auxin, gibberellins and cytokinin. *dsp* mutation affected cell size and auxin, and gibberellins and cytokinin contents in sepals. The results showed that cell enlargement or abnormal cell expansion in the adaxial part of sepals in *dsp*. As reported, auxin, gibberellins and cytokinin were important factors for cell expansion. Hence, *dsp* mutation regulated cell expansion to control sepal morphology, and auxin, gibberellins and cytokinin may mediate this process. One *ARF* gene and nine *SAUR* genes were dramatically upregulated in the sepal of the *dsp* mutant, whereas seven *AUX/IAA* genes were significantly downregulated in the sepal of *dsp* mutant. Further bioinformatic analyses implied that seven *AUX/IAA* genes might function as negative regulators, while one *ARF* gene and nine *SAUR* genes might serve as positive regulators of auxin signal transduction, thereby contributing to cell expansion in *dsp* sepal. Thus, our data suggest that 17 auxin-responsive genes are involved in downward sepal formation in the tomato. This study provides valuable information for dissecting the molecular mechanism of sepal morphology control in the tomato.

Keywords: tomato; sepal morphology; RNA-seq; differential expression; cell expansion; auxin; gibberellins; cytokinin

---

1. Introduction

The tomato (*Solanum lycopersicum* L.) is an important commercial crop and model for studying the floral organ development of angiosperms. After flowering is completed, tomato sepals are persistently protect young fruits and improve the quality of the appearance of mature fruits. However, as living standards increase, many people started to consider the quality and appearance of tomato sepals.

Healthy and flat sepals have become an important standard for measuring the quality of tomato fruits, enhancing visual esthetics and reflecting fruit freshness. Therefore, the molecular mechanism of the regulation of sepal morphology regulation should be investigated.

Sepals affect the flower development by coordinating cell division, cell differentiation and cell expansion with other parts of the flower whorl. The morphology and size of the sepals have been associated with the yield and quality of the fruit. The larger sepal size tightly associates with the protection of flower whorl and better fruit quality [1]. In *SlMBP21*-RNAi tomato, the sepals are longer and fruit sets are improved [2]. Among the green parts of the flower, sepal has the greatest ability to photosynthesis, follow by the receptacle [3]. The contents of Chl and the activity of ribulose-1, 5-bisphosphate carboxylase/oxygenase, the key photosynthetic enzyme, are both increased in longer sepals, and photosynthesis is enhanced in longer sepals, which may contribute by improving the fruit set [2]. Conclusively, sepal morphology is closely associated with fruit development.

Cell size affects sepal morphology. In *Arabidopsis* sepals, the loss of single-cell variability in an *ftsh4* mutant leads to the destruction of the entire sepal shape. This result indicates that changes in individual cell shape and size are important factors influencing the final organ morphology [4]. Other studies have shown that cell growth rate also results in differences in sepal morphology. The growth rate between cells in sepals significantly differs, the relative growth rate of cells in different regions of *Arabidopsis* sepals is about 0% to 5% (cell growth size/h) [5]. Although the cell growth rate during sepal development is variable, all cells achieve the maximum relative growth rate almost at the same time during development [5]. However, cell growth is not synchronized, and the time needed to reach the maximum growth rate varies between cells [5].

The morphological characteristics and growth of plant organs are regulated by hormones, including auxin, cytokinin, gibberellin, ethylene, abscisic acid (ABA), jasmonic acid (JA), brassinosteroids, stearolactone, and many peptides [6,7]. Auxin and gibberellin affect petal expansion and flowering [8,9]. The petal growth of *Arabidopsis thaliana* is regulated by *AUXIN RESPONSE FACTOR8* (*ARF8*), and the petals of *arf8* mutants are significantly larger than those of wild-type (WT) petals because of the increased number of cells and their expansion [8]. The *miRNA319a* mutant in *Arabidopsis* has narrow and short petals, and this trait is regulated by *TCP* transcription factors, which regulate auxin function [10–13]. A large number of signals (e.g., hormones, mechanical signals and polar fields) are distributed in various regions of plant organs, thereby coordinating the developmental behavior of multiple cells; as a result, different growth regions form. For example, *TCPs* promote the cessation of cell division and slow down the growth rate from the sepal apex to the base, resulting in a mechanical conflict that leads to sepal apex formation. The functional redundancy of these signals reinforces them, thereby enhancing the stability of organ morphology.

Despite a large diversity in flower morphology among *Solanum* species, studies on the genetic, basis of the variations in these organs are limited. In the tomato, *MACROCALYX* (*MC*), *TAGL1*, *SlFYFL* *SlMBP21* and, *SlCMB1*, are involved in the regulation of sepal development [14–19]. The T-DNA insertion of *Arlequin* (*Alq*) and the overexpression of the MADS-box gene *TAGL1* result in the conversion of sepals into fleshy, fruitlike organs [15]. The nucleotide sequences of *TAGL1* (syn. *ALQ*) show a high similarity to those of *Arabidopsis* D-class genes *SHATTERPROOF1* (*SHP1* and *AGL1*) [15]. The overexpression of another MADS-box gene, namely *SlFYFL*, gives rise to longer sepals [16]. The MADS-box gene *SlMBP21* regulates the development of sepals and the floral abscission zone [2,17,18]. Moreover, the SEPALLATA (E-function) MADS-box gene *SlCMB1* participates in the development of inflorescence architecture and also regulates sepal size in tomato plants [19].

In this study, high-throughput sequencing was conducted to detect differentially expressed genes (DEGs) during sepal development in the tomato. One wild-type (WT) accession with typical flat-spread sepal and *dsp* mutant with downward sepal were used to identify sepal shape-related candidate genes by comparing sepal transcriptomes at key stages, and to reveal the pathways and related genes possibly involved in sepal development. This work could offer valuable information as a basis for further studying the tomato flower shape. It would also provide insights into the molecular mechanism underlying flower organ development in the tomato for future breeding programs.

2. Results

2.1. Phenotypic Development Features and Diversity of Tomato Sepals

One sepal morphology mutant was verified. Although the WT tomato plant was characterized by typical flat-spread sepals, the *dsp* mutant, which was generated through natural mutation, exhibited an unusual "downward sepal" phenotype (Figure 1a). The morphological development of sepals from the flower bud stage to the fruit maturity stage in WT and *dsp* was subdivided into seven stages (Figure 1a). At stage 1, i.e., unopened stage, sepals combined with embracing the three inner whorls of floral organs. At stage 2, i.e., slightly opened stage, sepals began to separate, and an inflated corolla emerged at the top of the calyx. At stage 3, i.e., fully opened stage, sepals and petals were completely separated, and the calyx and corolla were fully expanded. At stage 4, i.e., reclosed stage, the calyx and corolla began to close again, and the corolla started to wilt. At stage 5, i.e., reopened stage, the calyx of WT began to open again and bore fruit, but the *dsp* calyx was not completely separated. At stage 6, i.e., morphological differentiation stage, the WT sepals were fully opened again, but the *dsp* sepals were still closed and wrapped around the fruit. At stage 7, i.e., final morphology, the morphological characteristics of sepals were fully developed, the sepals of WT were fully expanded and flat, whereas the *dsp* sepals were oriented downward. Notably, the differences in the sepal morphology between the WT and *dsp* mutant were apparent even at stage 4 of sepal development (Figure 1a,b). The upturned degree and rolling index of the sepals from >10 independent WT and *dsp* plants at different stages were measured. The upturned degree of the sepals of the *dsp* plants was significantly lower than that of the WT plants at stages 4 to 7. The rolling index of the sepals of the *dsp* plants was significantly greater than that of the WT plants at stages 5 to 7 (Figure 1b). The WT and *dsp* plants showed the same morphological differentiation of sepals from stages 1 to 3. Sepal morphology was significantly diverse during stages 4 to 7.

2.2. Identification of Differentially Expressed Genes in Sepals from WT and dsp Mutant Plants

Genome-wide expression analysis was conducted to compare the transcriptome profiles of the sepals between WT and *dsp* mutant plants through a differential gene expression (DGE) approach and to identify the genes involved in the downward sepals of tomato [20]. Phenotypic changes occurred at stage 4 (Figure 1a), so the sepals at stage 3 (0 days after flowering [DAF]; length of ~8 mm) and stage 4 (15 DAF; length of ~11 mm) were chosen for RNA-seq analyses. After the RNA was sequenced, 49.11 million and 48.89 million reads were obtained from the WT and *dsp* plants at stage 3, respectively; of these reads, 91.76% and 90.74% could be mapped to the annotated tomato genome, respectively (Figure 2a). Meanwhile, 55.56 million and 52.56 million reads were obtained from the WT and *dsp* plants at stage 4, respectively. Of these reads, 68.12% and 65.39% could be mapped to the annotated tomato genome, respectively (Figure 2a). The RNA data displayed good correlations between the two biological replicates and were used for further analysis (Figure S1).

After the transcriptomes were compared with those of WT plants, 3066 and 2459 DEGs were detected in the *dsp* plants at stages 3 and 4, respectively, and a total of 4729 DEGs were identified at two stages (Figurer 2b,e). Of the DEGs at stage 3, 1355 were upregulated and 1711 were downregulated (Figure 2b,f). However, 1236 DEGs were upregulated and 1223 DEGs were downregulated in the *dsp* plants at stage 4 (Figure 2b,g).

The top five genes in the upregulated DEGs at stage 3 were two unknown protein-coding genes (*Solyc08g044230.1*, 19065.64-fold; *Solyc05g010380.1*, 3690.21-fold), one wound-induced proteinase inhibitor gene (*Solyc09g084450.2*, 12106.3 9-fold), one 1-aminocyclopropane-1-carboxylate oxidase homolog gene (*Solyc09g089580.2*, 8600.12-fold) and one galactan beta-1,4-galactosyltransferase gene (*Solyc05g015790.1*, 5977.76) (Table S1). The top five significantly downregulated genes at stage 3 were two unknown protein-coding gene (*Solyc01g016460.2*, 2873.56-fold; *Solyc05g032670.1*, 2132.20-fold), one TMV resistance protein (*Solyc07g052790.1*, 2105.26-fold), one phosphoethanolamine N-methyltransferase (*Solyc06g068950.2*, 2083.33-fold) and one BTB/POZ domain-containing protein (*Solyc12g010080.1*, 1515.15-fold). At stage 4, the top five upregulated genes were three unknown protein-coding gene (*Solyc05g010380.1*, 2746.95-fold; *Solyc08g044230.1*, 1315.58-fold, *Solyc05g018060.1*, 1144.65-fold), one 1-aminocyclopropane-1-carboxylate oxidase homolog gene (*Solyc09g089580.2*, 12576.37-fold), and one serine/threonine-protein kinase (*Solyc06g005880.2*, 2158.385-fold). The top five downregulated genes were one ascorbate peroxidase protein (*Solyc06g005150.2*, 4878.05-fold), one unknown protein gene (*Solyc01g016460.2*, 1960.78-fold), one chlorophyll a-b binding protein-coding gene (*Solyc02g071030.1*, 1597.44-fold), one TMV resistance protein-coding gene (*Solyc07g052790.1*, 891.27-fold) and one BTB/POZ domain-containing protein-coding gene (*Solyc12g010080.1*, 780.03-fold). *Solyc09g089580.2* which encodes an ethylene synthesis-related protein, namely, ACO3, was significantly upregulated and remarkably changed in the sepals at both stages.

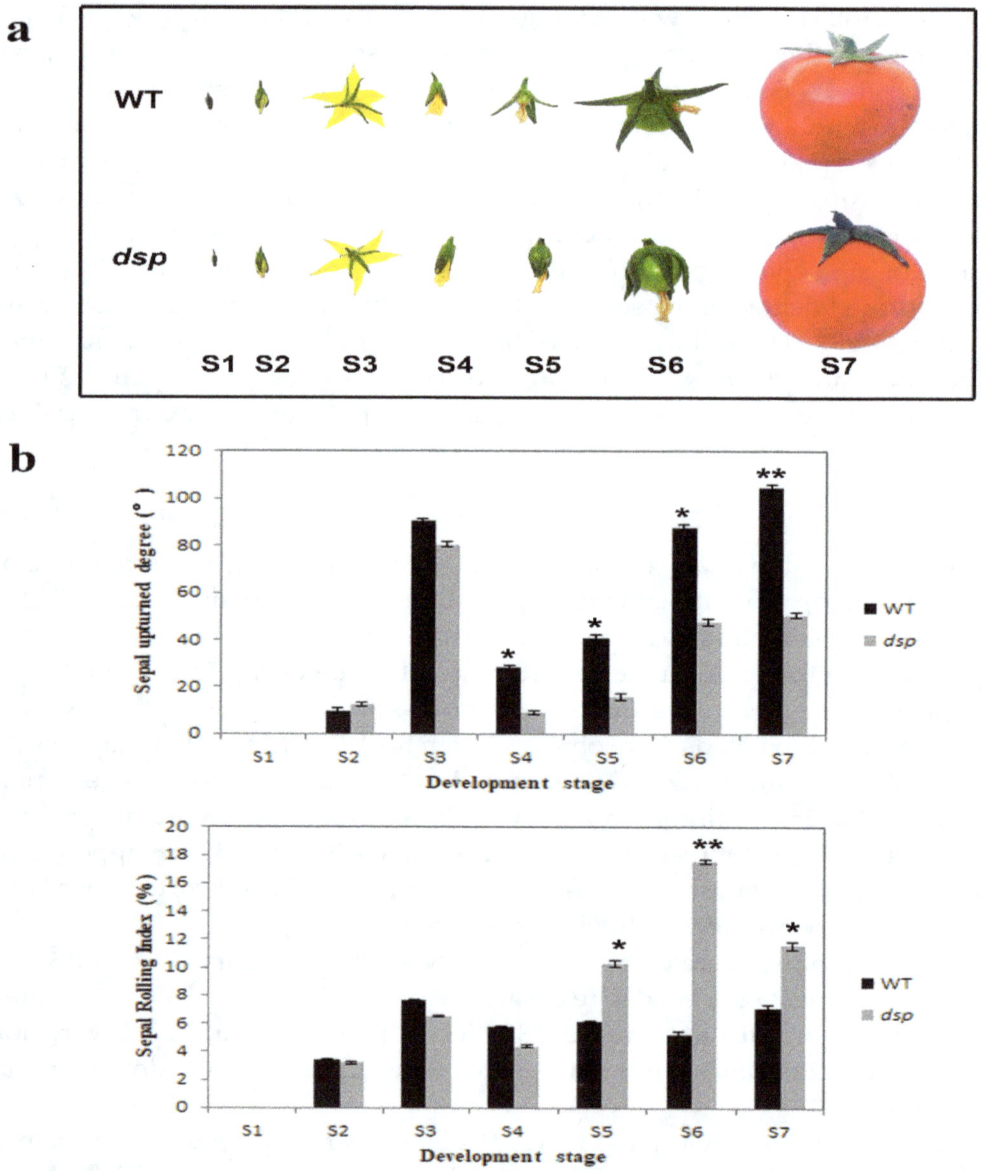

Figure 1. Phenotypes of WT and *dsp (downward sepal)* sepal at different developmental stages. (**a**) Sepal of the wild type and the *dsp* at different stages. (**b**) Sepal upturned degree and sepal rolling index at different stages. *p*-values were determined by *t*-test. * $p < 0.05$; ** $p < 0.01$ (*t*-test).

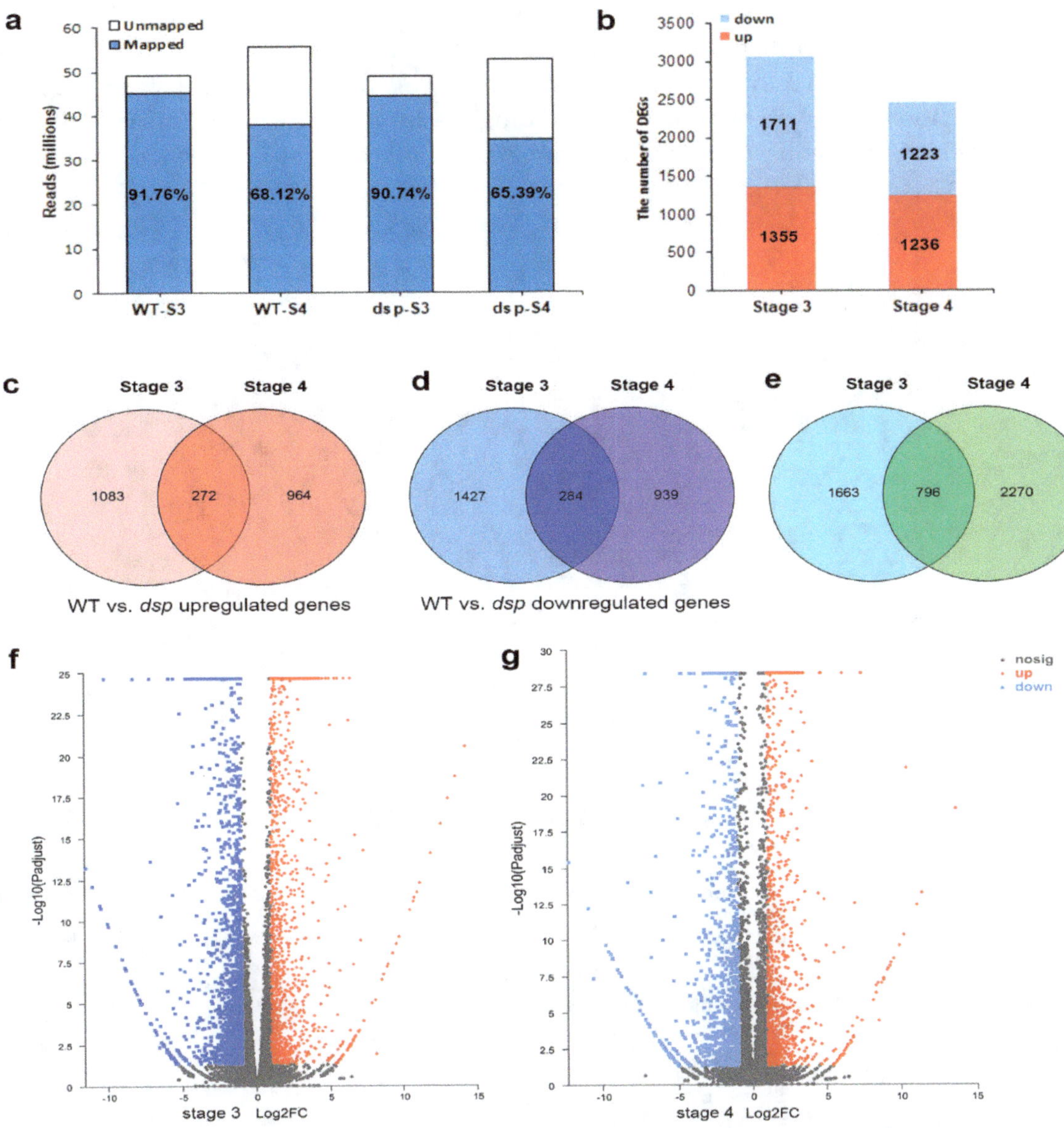

Figure 2. Analysis of transcriptomes from sepals of wild-type and *dsp* plants. (**a**) The number of clean reads obtained from the WT and *dsp* in stage 3 and stage 4, and the percentage of clean reads mapped to the genome. (**b**) The number of up- or down-regulated DEGs for WT vs. *dsp* at stage 3 and stage 4. (**c**) Venn diagram analysis of both upregulated genes of stage 3 and stage 4 groups. (**d**) Venn diagram analysis of both downregulated genes of stage 3 and stage 4 groups. (**e**) Venn diagram analysis of genes with both DEGs of stage 3 and stage 4 groups. (**f**,**g**) The volcano map of differentially expressed genes (DEGs) in stage 3 and stage 4 groups. Red dots indicate upregulated genes; blue dots indicate downregulated genes; grey dots represent no significant difference.

2.3. Annotation of DEGs in Sepals from WT and dsp Mutant Plants

A total of 4729 DEGs were classified into three categories based on gene ontology (GO) assignments: biological processes (BP), molecular functions (MF), and cellular components (CC). Further GO enrichment analysis resulted in the identification of 205 GO terms (114 BP, 23 CC, and 68 MF; false discovery rate [FDR] <0.05; Table S2). For MF, the top five enriched GO terms were "transferase activity" (738 genes), "oxidoreductase activity" (428 genes), "kinase activity" (307 genes) "phosphotransferase activity" (294 genes), and "transporter activity" (294 genes; Figure 3a; Table S2). For BP, the DEGs were primarily enriched in response to stimuli, lipid metabolism, and organic acid metabolism (Figure 3a; Table S2). Changes in CC occurred primarily in the membrane and cell wall

(Figure 3a; Table S2). MF analysis revealed that pectinesterase activity, pectinesterase inhibitor activity, and pectate lyase activity significantly changed, and pectate lyase activity significantly changed according to MF analysis, and they were related to cell wall modification. (Figure 3a; Table S2). The DEGs that function in cell wall organization and biogenesis were enriched. These results suggested that cell wall biogenesis and components in downward sepals were influenced.

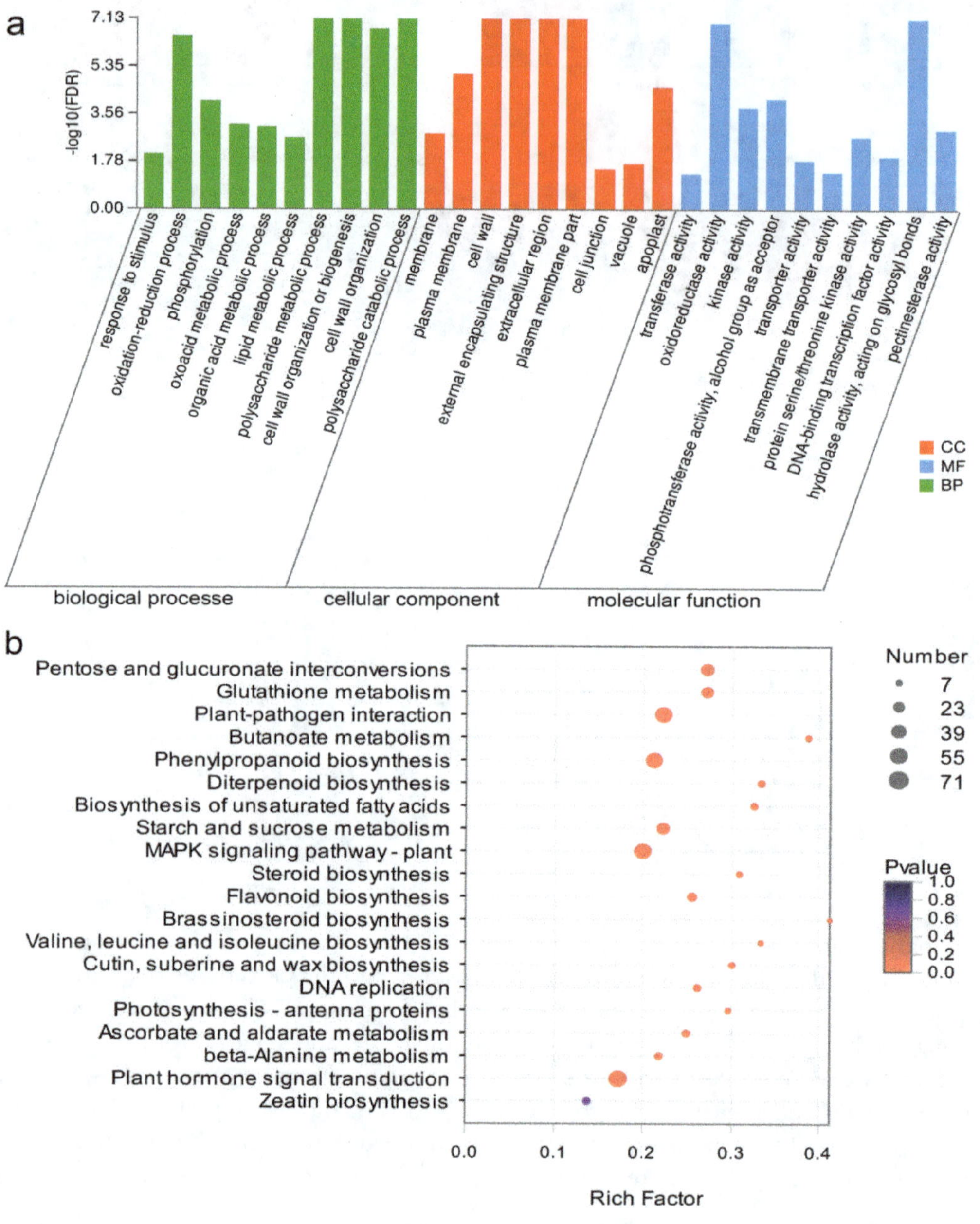

Figure 3. The Gene Ontology (GO) and (Kyoto Encyclopedia of Genes and Genomes) KEGG enrichment analysis of DEGs. (**a**) The GO enrichment analysis. The top ten enriched biological processes, molecular function and cellular component GO terms for DEGs. The x-axis represents GO term. The y-axis represents the significance level of enrichment ($-\log_{10}$ FDR—false discovery rate). (**b**) The KEGG enrichment scatter plot of DEGs. The y-axis represents the name of the pathway, and the x-axis represents the rich factor, the degree of KEGG pathway enrichment. Top 20 KEGG pathway enrichments with DEGs were showed. Dot size represents the number of genes and the color indicates the *p*-value.

Furthermore, 4729 DEGs were mapped into 177 Kyoto Encyclopedia of Genes and Genomes (KEGG) signaling pathways. Of these DEGs, 20 were significantly enriched ($p < 0.05$; Figure 3b; Table S3). The top five enriched KEGG pathways were "brassinosteroid biosynthesis (rich factor = 0.41)," "butanoate metabolism (rich factor = 0.39)", "taurine and hypotaurine metabolism (rich factor = 0.38)", "diterpenoid biosynthesis (rich factor = 0.33)", and "valine, leucine, and isoleucine biosynthesis (rich factor = 0.33)," as shown in Figure 3b and Table S3. KEGG analysis showed that most of the DEGs (71 genes) were enriched in "plant hormone signal transduction". A number of DEGs were enriched in the following pathways: 15 genes in "diterpenoid biosynthesis," 7 genes in "brassinosteroid biosynthesis," and 16 genes in "zeatin biosynthesis". In addition, many DEGs that were involved in the pathways of cell wall metabolism and "cutin, suberine, and wax biosynthesis" (12 genes) were altered in the *dsp* plants. These results indicated that *dsp* affected sepal morphology by controlling the genes related to cell wall modification and plant hormone regulation.

2.4. Coexpression Pattern of DEGs in Sepals from WT and dsp Mutant Plants

Hierarchical clustering was applied to 4729 DEGs between WT and *dsp* sepals at stages 3 and 4 to further examine the transcriptomic response to downward sepal. Five large clusters were obtained, and the DEGs were clustered into five clusters by using the *k*-means ($k = 5$) clustering algorithm. These clusters were then visualized with a heat map and a cluster trend chart (Figure 4a,b). The results revealed the general patterns of transcriptomic profiles during sepal development.

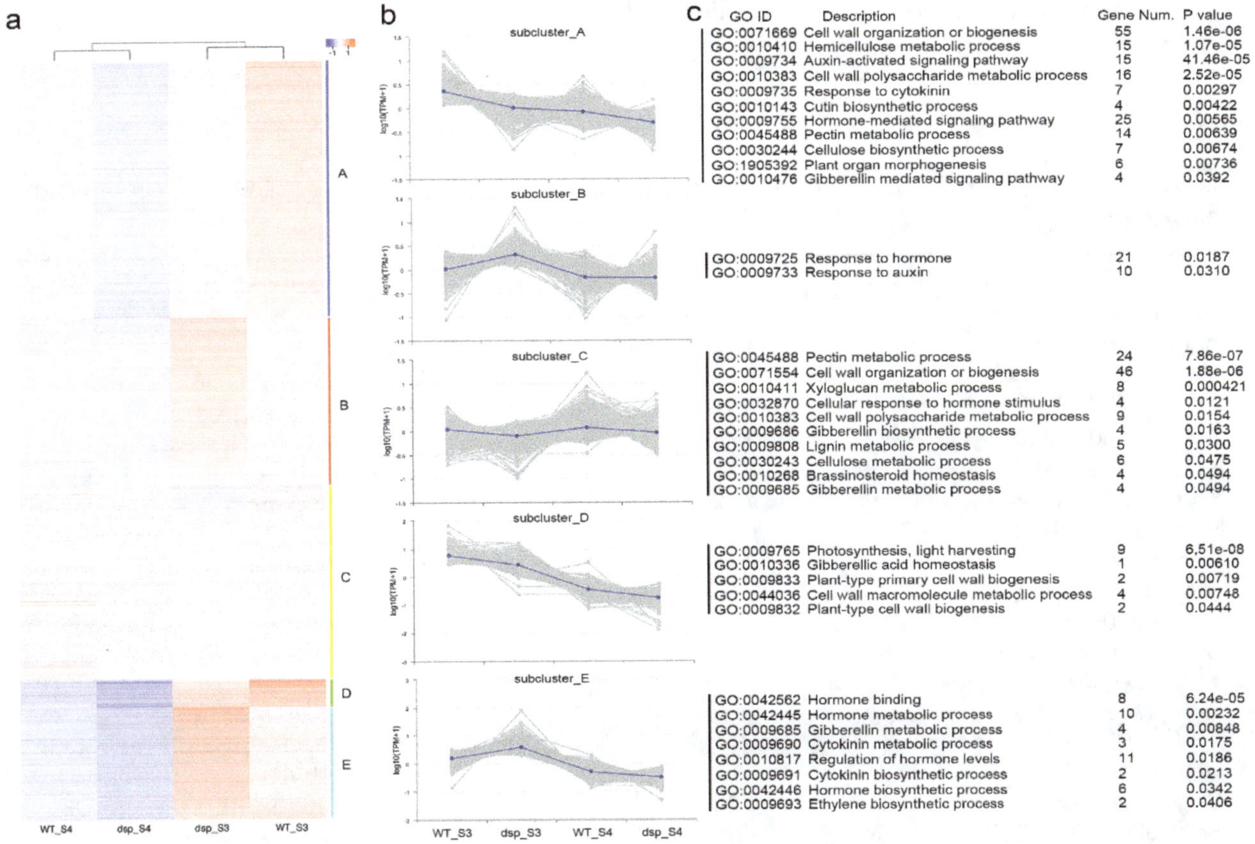

Figure 4. Cluster analysis of differentially expressed genes. (**a**) K-means clustering of DEGs in sepals at stage 3 and 4 of the WT and *dsp* plants. Red and blue in the heat maps represent up-regulated and down-regulated genes, respectively. (**b**) The trend chart of each subcluster. (**c**) The GO enrichment analysis of genes in each cluster.

The expression levels of the genes in clusters A and C of the *dsp* plants decreased at stage 3, whereas the expression levels of the genes in cluster D of the *dsp* plants continuously decreased from stage 3 to stage 4 (Figure 4b). The expression levels of the genes in cluster C did not significantly differ in the same plant sepals at different stages (Figure 4b). By contrast, the expression levels of the genes from clusters B and E of the *dsp* plants decreased at stage 3. The five clusters were subjected to GO annotation enrichment analysis to gain further insights into the transcriptomic changes in the sepals of the *dsp* plants. In clusters A and C, the genes involved in cell wall organization and biogenesis were enriched (Figure 4c), suggesting that changes in the cell wall were one of the reasons for the altered sepal morphology in the *dsp* plants. Conversely, the gene functioning in hormone metabolism and biogenesis were enriched in clusters B and E. This result indicated that the differences in the sepal morphology of the *dsp* plants were attributed to alterations in hormonal levels, particularly auxin, gibberellin and cytokinin.

2.5. Regulation of Cell Expansion by dsp to Control Sepal Morphology

According to the results of DEG analysis, the expression levels of genes related to cell expansion significantly changed in *dsp* sepals (Table 1; Table S4). Among the DEGs, 16 were *XTH* genes, 1 was xyloglucan galactosyltransferase gene, 24 were pectinesterase genes, 1 was pectin methyltransferase gene, 10 were pectin lyase protein genes, 5 were pectin acetylesterase genes, 4 were pectinesterase inhibitor genes, 10 were pectin methylesterase inhibitor genes, 14 were cellulose synthase genes, 1 was a glycine-rich cell wall structural protein 1 precursor gene, and 4 were cell wall protein genes; these genes were upregulated or downregulated from 287.36-fold to 2.00-fold (Table 1; Table S4). They were also involved in cell wall metabolism, which is important for cell expansion and cell growth rates. In addition, the expression of 14 expansion (-like) protein genes and 12 extension (-like) protein genes remarkably differed (from 196.85-fold to 2.16-fold; Table 1 and Table S4). These genes were required for cell expansion, cell size, and organ shape development.

Table 1. The top 10 DEGs related to cell expansion, auxin, gibberellins and cytokinin in *dsp* sepal at stage 3 and stage 4.

ID	logFC	*p* Value	Annotation
Cell expansion (stage 3)			
Solyc02g080220.2	1.61	4.34×10^{-53}	Pectinesterase
Solyc07g052980.2	−1.30	3.22×10^{-33}	Xyloglucan endotransglucosylase/hydrolase protein 9
Solyc08g005800.2	−1.71	2.64×10^{-36}	Pectin acetylesterase
Solyc07g043390.2	−1.79	1.11×10^{-90}	cellulose synthase-like protein G2
Solyc02g088100.2	−1.99	4.22×10^{-47}	Expansin-A5
Solyc02g078040.2	−2.16	6.28×10^{-60}	pistil-specific extensin-like protein
Solyc05g007830.2	−2.39	1.79×10^{-59}	Expansin-A15
Solyc05g014000.2	−2.78	2.60×10^{-180}	Pectate lyase
Solyc03g083770.1	−4.14	0	pectin methylesterase inhibitor
Solyc06g005560.2	−4.60	1.83×10^{-35}	Expansin9
Cell expansion (stage 4)			
Solyc01g106650.2	3.28	9.68×10^{-58}	Xyloglucan endotransglucosylase/hydrolase protein 10
Solyc06g034370.1	1.64	6.36×10^{-134}	pectin methylesterase inhibitor
Solyc03g097050.2	1.46	4.00×10^{-65}	cellulose synthase-like protein D3
Solyc07g017600.2	1.28	7.32×10^{-60}	Pectinesterase
Solyc04g074290.2	1.28	2.21×10^{-74}	pectin methyltransferase QUA2
Solyc03g083730.1	−1.08	1.76×10^{-59}	pectin methylesterase inhibitor
Solyc03g025600.2	−1.17	4.64×10^{-62}	Pectin acetylesterase
Solyc09g097770.2	−1.34	1.22×10^{-168}	Cell wall protein
Solyc08g077330.2	−2.44	2.38×10^{-107}	expansin-like B1
Solyc06g084620.1	−2.86	2.12×10^{-48}	Pectinesterase

Table 1. *Cont.*

ID	logFC	*p* Value	Annotation
Auxin (stage 3)			
Solyc05g025920.2	8.66	1.98×10^{-8}	Auxin-induced protein 15A
Solyc06g075690.2	2.23	1.51×10^{-91}	Auxin-regulated protein
Solyc09g075210.2	1.08	3.11×10^{-32}	indole-3-acetic acid-induced protein ARG2
Solyc02g077880.2	1.05	1.14×10^{-36}	Auxin repressed/dormancy associated protein
Solyc10g008520.2	−1.36	1.43×10^{-29}	Indole-3-acetic acid-amido synthetase GH3.10
Solyc11g013310.1	−1.91	2.97×10^{-37}	Auxin transporter-like protein 3
Solyc03g123410.1	−1.96	6.59×10^{-35}	auxin-binding protein ABP19a
Solyc11g069190.1	−2.00	2.29×10^{-35}	Auxin response factor
Solyc09g056390.1	−2.13	2.09×10^{-66}	auxin-induced in root cultures protein 12
Solyc07g041720.1	−2.99	2.45×10^{-162}	auxin-binding protein ABP19a-like
Auxin (stage 4)			
Solyc07g066560.1	2.29	5.62×10^{-30}	auxin-responsive protein SAUR71
Solyc06g075690.2	2.14	0	Auxin-regulated protein
Solyc12g014500.1	1.87	2.91×10^{-13}	indole-3-acetate O-methyltransferase 1-like
Solyc04g081240.2	1.07	1.16×10^{-6}	Auxin response factor 5
Solyc04g076850.2	−1.14	1.62×10^{-94}	Auxin-responsive protein
Solyc09g075210.2	−1.35	2.30×10^{-182}	indole-3-acetic acid-induced protein ARG2
Solyc02g079190.2	−1.64	5.46×10^{-122}	protein AUXIN SIGNALING F-BOX 2
Solyc06g063060.2	−2.06	3.94×10^{-21}	Auxin repressed protein
Solyc09g083290.2	−2.79	1.06×10^{-35}	Auxin-responsive protein
Solyc06g053840.2	−2.79	3.09×10^{-34}	Auxin-responsive protein
Gibberellins (stage 3)			
Solyc02g070430.2	6.31	2.08×10^{-24}	gibberellin 2-beta-dioxygenase 1
Solyc12g042500.1	2.63	3.77×10^{-08}	Gibberellin regulated protein
Solyc07g061730.2	2.32	1.09×10^{-20}	Gibberellin 2-oxidase
Solyc07g056670.2	1.93	1.71×10^{-15}	Gibberellin 2-oxidase 2
Solyc01g111080.2	1.59	2.12×10^{-43}	Gibberellin-regulated protein 1
Solyc03g113910.2	1.42	2.34×10^{-06}	gibberellin-regulated protein 10
Solyc06g008870.2	1.11	4.70×10^{-12}	gibberellin receptor GID1B-like
Solyc02g089350.2	−2.50	8.37×10^{-46}	Gibberellin regulated protein
Solyc03g119910.2	−3.83	7.14×10^{-15}	Gibberellin 3-beta-dioxygenase 1
Solyc03g116060.2	−5.80	5.10×10^{-4}	Gibberellin-regulated protein 4
Gibberellins (stage 4)			
Solyc12g042500.1	2.48	2.00×10^{-4}	Gibberellin regulated protein
Solyc04g078200.2	1.95	2.94×10^{-27}	gibberellin-regulated family protein precursor
Solyc03g116060.2	1.70	7.30×10^{-3}	Gibberellin-regulated protein 4
Solyc06g007890.2	1.52	1.56×10^{-2}	Gibberellin regulated protein
Solyc07g063940.1	1.28	5.31×10^{-62}	Chitin-inducible gibberellin-responsive protein 1
Solyc01g111080.2	1.15	2.47×10^{-198}	Gibberellin-regulated protein 1
Solyc07g061720.2	−1.02	8.70×10^{-4}	Gibberellin 2-oxidase
Solyc11g072310.1	−1.54	1.00×10^{-2}	Gibberellin 20-oxidase-3
Solyc07g061730.2	−3.96	7.97×10^{-18}	Gibberellin 2-oxidase
Cytokinin (stage3)			
Solyc04g014990.1	3.46	1.82×10^{-8}	zeatin O-xylosyltransferase-like
Solyc01g088160.2	2.71	3.54×10^{-94}	Cytokinin oxidase/dehydrogenase-like
Solyc10g079930.1	2.25	2.63×10^{-48}	zeatin O-glucosyltransferase-like
Solyc06g062330.1	2.05	5.77×10^{-19}	zeatin O-glucosyltransferase-like
Solyc04g081290.2	2.00	9.30×10^{-19}	Cytokinin riboside 5′-monophosphate phosphoribohydrolase
Solyc04g008310.1	1.67	5.45×10^{-38}	zeatin O-xylosyltransferase-like

Table 1. *Cont.*

ID	logFC	p Value	Annotation
Solyc08g062820.2	1.56	7.56×10^{-31}	Cytokinin riboside 5′-monophosphate phosphoribohydrolase
Solyc12g008900.1	1.74	2.16×10^{-8}	cytokinin dehydrogenase 3 isoform X1
Solyc05g053400.1	−1.95	1.05×10^{-18}	zeatin O-xylosyltransferase-like
Solyc11g066670.1	−2.16	1.66×10^{-56}	zeatin O-glucosyltransferase-like
Cytokinin (stage4)			
Solyc01g088160.2	1.79	5.20×10^{-18}	Cytokinin oxidase/dehydrogenase-like
Solyc06g075090.2	1.61	1.25×10^{-6}	Cytokinin riboside 5′-monophosphate phosphoribohydrolase
Solyc06g062330.1	−1.18	1.60×10^{-3}	zeatin O-glucosyltransferase-like

The cell expansion and cell growth rate determine the size and shape of organs. The DEGs related to cell expansion remarkably varied in *dsp* sepals (Table 1 and Table S4). Hence, the cells in the middle part of the sepal in *dsp* and WT plants were observed through histological analysis. In the sections from stages 4 to 7, the cell size in the adaxial part of the sepals of *dsp* plants was larger than that in the sepals of the WT plants (Figure 5A–K). The cell number per unit view clearly decreased, whereas the cell area significantly increased ($p < 0.05$) in the sepals of *dsp* plants at stages 4 and 5 (Figure 5B,C,F,G,K). The cell area in the sepals of the *dsp* plants at stages 6 and 7 significantly increased ($p < 0.01$; Figure 5D,H–K). Thus, cell enlargement or abnormal cell expansion in the adaxial part of sepals might contribute to downward sepals in *dsp* plants.

The cell expansion and cell growth rate determine the size and shape of organs. The DEGs related to cell expansion remarkably varied in *dsp* sepals (Table 1; Table S4). Hence, the cells in the middle part of the sepal in *dsp* and WT plants were observed through histological analysis. In the sections from stages 4 to 7, the cell size in the adaxial part of the sepals of *dsp* plants was larger than that in the sepals of the WT plants (Figure 5A–K). The cell number per unit view clearly decreased, whereas the cell area significantly increased ($p < 0.05$) in the sepals of *dsp* plants at stages 4 and 5 (Figure 5B,C,F,G,K). The cell area in the sepals of the *dsp* plants at stages 6 and 7 significantly increased ($p < 0.01$; Figure 5D,H–K). Thus, cell enlargement or abnormal cell expansion in the adaxial part of sepals might contribute to downward sepals in *dsp* plants.

Consequently, the alteration of the expression of these cell expansion-related genes induced cell enlargement, which might contribute to downward sepals in *dsp* plants. Therefore, *dsp* might regulate cell expansion to control sepal morphology.

2.6. *Effects of* dsp *on Auxin, Cytokinin, and Gibberellin Levels in Sepals*

Hormones serve as crucial regulators of organ morphology development. The duration and rate of cell proliferation are positively controlled by auxin and cytokinin. The transition from cell division to expansion is correlated with gibberellin. Phytohormones play a critical role in integrating developmental signals to control organ morphology. The DEGs related to auxin, cytokinin, and gibberellin remarkably differed in *dsp* sepals (Figure 6; Table 1). The majority of these genes were related to auxin, and some of them were associated with cytokinin and gibberellin (Figure 6).

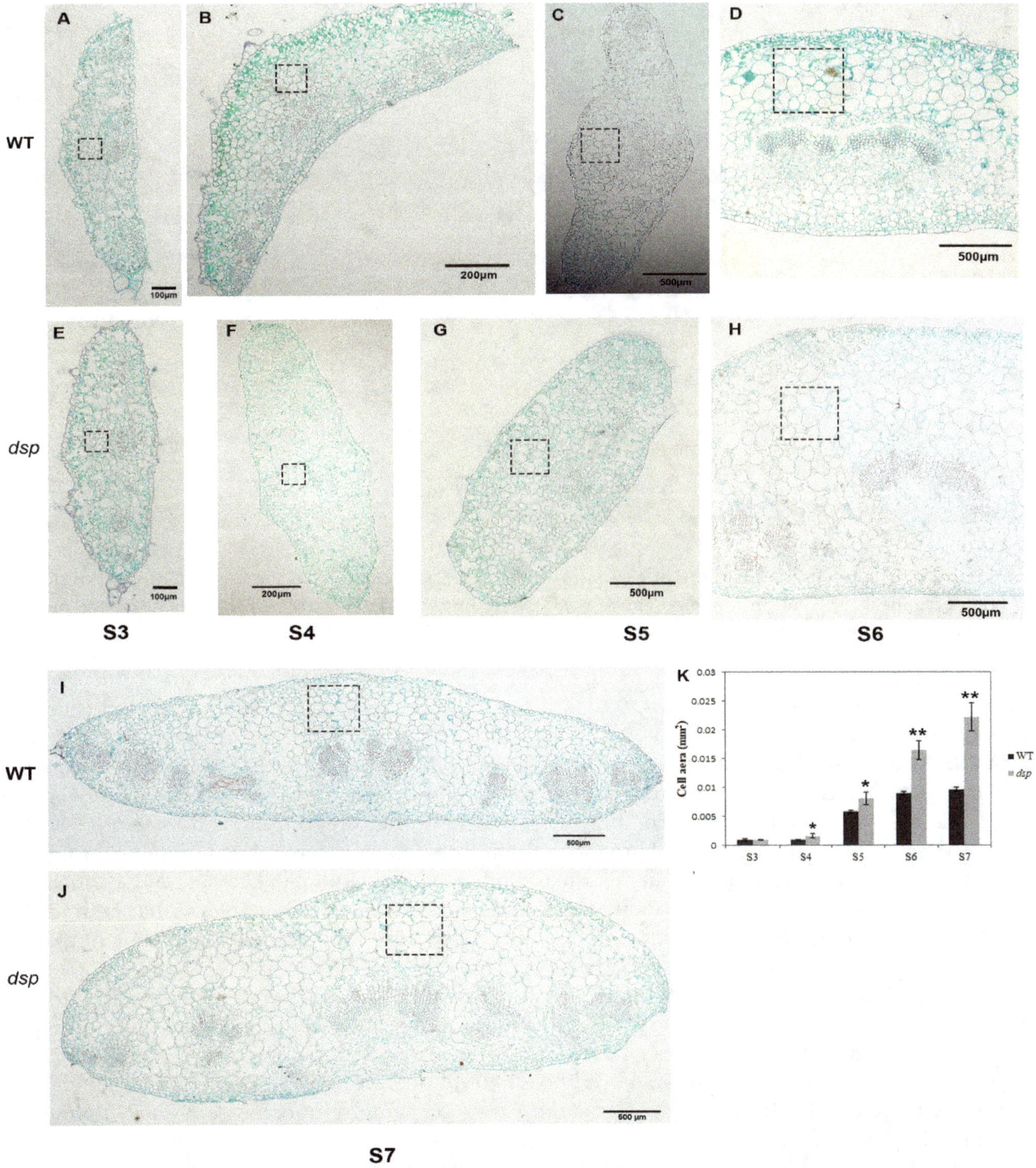

Figure 5. The paraffin cross-section of the sepal of WT and *dsp*. (**A**–**D** and **I**) Stage 3 to 7 of WT sepal. (**E**–**H** and **J**) Stage 3 to 7 of *dsp* sepal. (**K**) The average area of sepal cells in unit view. The area of unit view of A, B, E and F = 0.01mm^2; C and G = 0.09 mm^2; D, H–J = 0.25 mm^2. The cells in > 3 unit views were counted. * $p < 0.05$ (*t*-test), ** $p < 0.01$ (*t*-test).

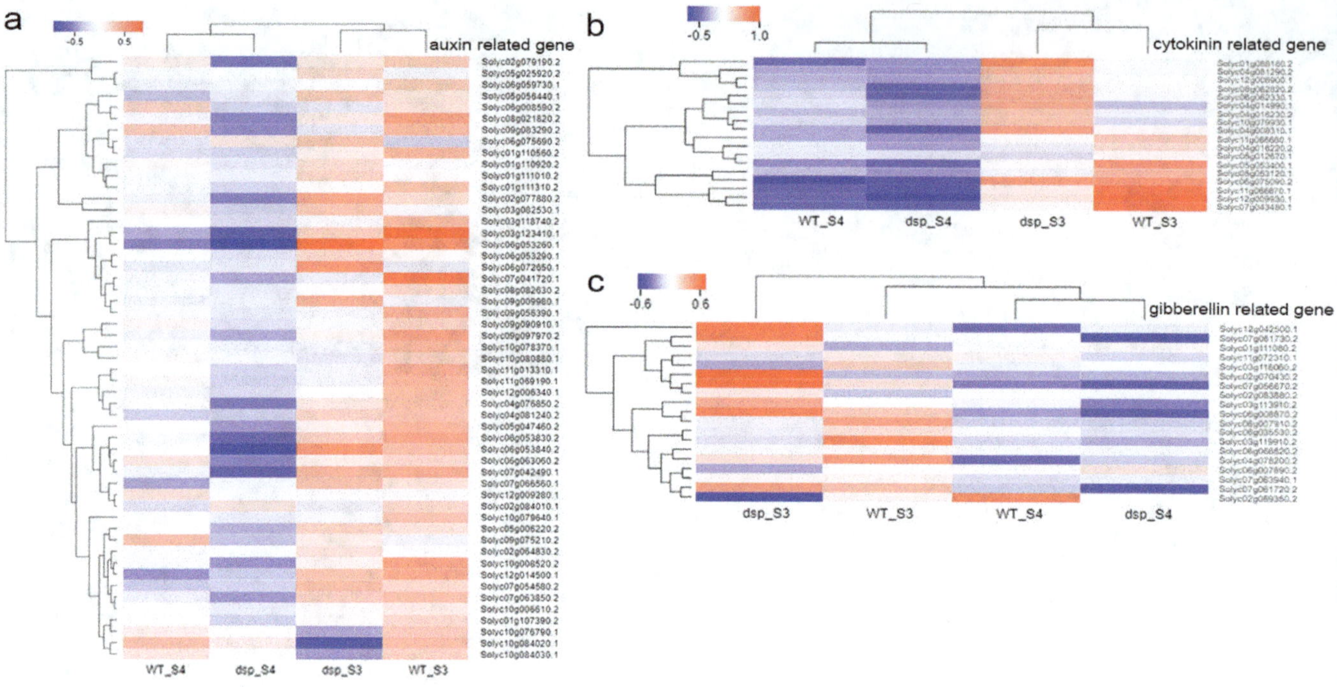

Figure 6. The result of clustering analysis for the differentially expressed genes (DEGs). (**a**) Auxin related genes. (**b**) Cytokinin related genes. (**c**) Gibberellin related genes. Blue and red colors indicate genes with higher expression and lower expression, respectively.

Among DEGs, 53 were linked to auxin, i.e., 13 of them were upregulated and 23 were downregulated at stage 3. At stage 4, 9 were upregulated and 16 were downregulated (Figure 6; Table 1; Table S4). The expression levels of the following genes were altered from 404.79-fold to 2.04-fold: 31 auxin-responsive protein-coding genes (*GH3* genes, *SAUR* genes, *ARF* genes, *IAA3*, etc.), four auxin efflux carrier protein-coding genes, nine auxin-induced protein-coding genes, three auxin transporter-like protein-coding genes, two auxin-binding protein-coding genes (ABP19a and ABP19a-like), one auxin-repressed protein-coding genes, one auxin-regulated protein-coding gene, and one small auxin-up protein 58 gene (Figure 6; Table S4). In addition, the content of IAA increased 1.69-fold and 1.44-fold in the *dsp* sepals at stages 3 and 4, respectively, compared with that in the WT sepals (Figure 7a). (Figure 7a). Consequently, *dsp* influenced the efflux, signaling, and content of auxin in sepals.

Among DEGs, 17 were cytokinin synthesis-related factors, and two were cytokinin degradation-related factors. Of these genes, nine were upregulated and nine were downregulated at stage 3; at stage 4, two were upregulated and one was downregulated (Figure 6; Table 1 and Table S4). Two cytokinin riboside 5″-monophosphate phosphoribohydrolase (*SlLOG*) genes were upregulated from 4.01-fold to 2.94-fold at stage 3, one isopentenyltransferase 2 (*SlIPT*) gene was downregulated up to 5.39-fold, and one *LOG* gene was upregulated up to 3.06-fold at stage 4. Two cytokinin oxidase/dehydrogenase (*SlCKX*) genes were upregulated from 6.54-fold to 3.33-fold at stages 3 and 4 (Table 1, Figure 6). The first step of CK biosynthesis in plants is the N-prenylation of adenosine 5′-phosphates via dimethylallyl diphosphate, resulting in the biosynthesis of CK nucleotides. [21,22]. This step is catalyzed by IPT. The next step is the phosphoribohydroxylation of CK nucleotides to synthesize biologically active CK nucleobases [23]. This step is catalyzed by CK riboside 5′-monophosphate phosphoribohydrolase (LOG). Cytokinin conjugation occurs mainly through cytokinin oxidase (CKX) enzymes. Furthermore, two zeatin O-glucosyltransferase genes were upregulated from 4.75-fold to 4.14-fold, whereas four zeatin O-glucosyltransferase genes were downregulated from 5.01-fold to 2.03-fold) (Table 1 and Table S4). Three zeatin O-xylosyltransferase genes were upregulated from 11.02-fold to 3.17-fold, whereas four zeatin O-xylosyltransferase genes were downregulated from 5.96-fold to 2.64-fold) (Table 1 and Table S4). Cytokinins with a hydroxyl group on the side chain can undergo O-glycosylations. Zeatin O-glucosyltransferase and zeatin O-xylosyltransferase, which are *zisZOG* genes, play important roles in regulating the levels of *cis*-zeatin and maintaining appropriate levels of active cytokinins because biosynthetic enzymes work slowly [24]. Zeatin O-glycosyl derivatives are resistant to the cytokinin degrading enzyme *CKX* and are considered to be storage forms because they can be cleaved by b-glucosidase [22,25]. In our study, the contents of cytokinin in the *dsp* sepals increased 1.16-, 2.35- and 1.34-fold at stages 3, 4, and 5, respectively, compared with those in the WT sepals (Figure 7). In our data, the upregulation of *SlCKX* might be responsible for the increase in the cytokinin levels in the sepals of *dsp* plants. The upregulation of *SlLOG* might accelerate cytokinin synthesis and result in an increase in cytokinin levels. These results indicated that the increased expression of cytokinin biosynthesis-related genes (*SlLOGs*), the increased expression of *SlCKXs*, and the increased cytokinin content contributed to the expansion of sepal cells in *dsp*.

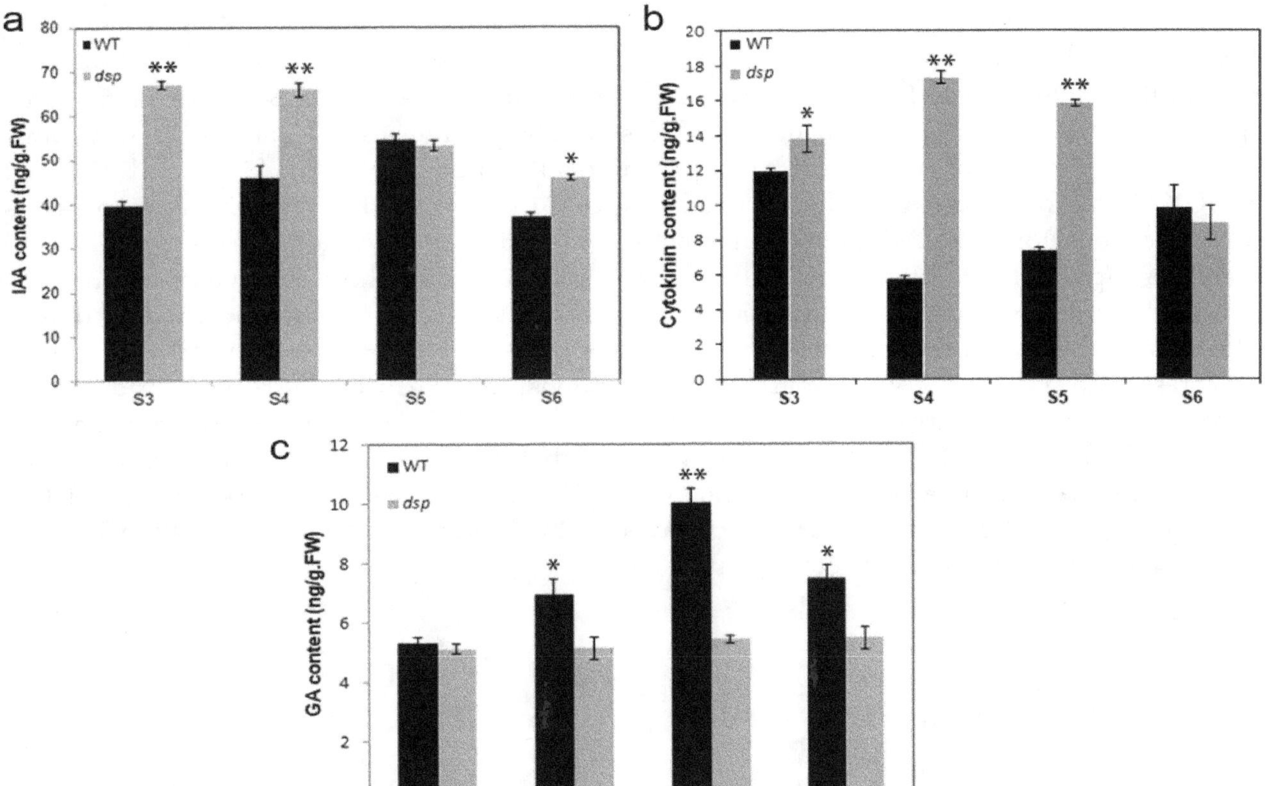

Figure 7. Analysis of hormones in sepals. (**a**) IAA (indole-3-acetic acid) content of sepals in the wild type and *dsp* from stage 3 to stage 6. (**b**) Cytokinin content of sepals in the wild type and *dsp* from stage 3 to stage 6. (**c**) Gibberellin content of sepals in the wild type and *dsp* from stage 3 to stage 6. * $p < 0.05$, ** $p < 0.01$ (*t*-test).

GAs are essential for cell elongation and other plant growth and developmental processes [26–28]. GA20oxs (GA20-oxidases) are key GA biosynthesis regulators that determine the GA content of plant species [29]. Among the DEGs in our study, seven were gibberellin synthesis-related genes and 11 were gibberellin response-related genes. Of these genes, eight were upregulated and seven were downregulated at stage 3 while six were upregulated and three were downregulated at stage 4 (Figure 6; Table 1 and Table S4). The two GA20ox genes in *dsp* were downregulated from 10.87-fold to 4.31-fold at stage 3. One of them was downregulated to 2.92-fold at stage 4. Two GA2ox genes, which are responsible for the deactivation of GAs, were significantly upregulated from 4.98-fold to 3.81-fold at stage 3 of the *dsp* plant (Table 1 and Table S4). Furthermore, the contents of gibberellin in the *dsp* sepals of the plant decreased by 1.35- and 1.85-fold at stages 4 and 5, respectively, compared with those in the WT sepals (Figure 7). These results indicated that the decreased expression of GA biosynthesis-related genes (GA20oxs) and the increased expression of GA2oxs contributed to a decrease in the GA content.

2.7. Possible Function of 17 Auxin-Responsive Genes as Regulators in Downward Sepal Formation in the Tomato

The detected DEGs were mapped to reference canonical pathways in the KEGG to further identify the key gene regulatory pathways responsible for the formation of downward sepals in the tomato [30]. Among the enriched KEGG pathways, the "auxin signal transduction" pathway, which was putatively associated with the downward sepal phenotype in the *dsp* mutant, was also enriched. Notably, one *ARF* gene (*Solyc04g081240.2*) and nine *SAUR* genes (*Solyc05g025920.2, Solyc05g056440.1, Solyc01g110920.2, Solyc03g082530.1, Solyc06g053290.1, Solyc06g072650.1, Solyc09g009980.1, Solyc07g066560.1, Solyc02g084010.1*) were dramatically upregulated in the *dsp* sepals of *dsp* mutant, whereas seven *AUX/IAA* genes (*Solyc06g008590.2, Solyc08g021820.2, Solyc09g083290.2, Solyc09g090910.1, Solyc04g076850.2, Solyc06g053830.2, Solyc06g053840.2*) were significantly downregulated in the *dsp* sepal mutant (Table 1 and Table S4). All of them were mapped to the "auxin signal transduction". Auxin signaling promotes cell expansion. In this pathway, once the concentration of auxin increases, it mediates the linkage of TIR1/AFBs with AUX/IAAs, and AUX /IAA proteins are degraded by proteasomes [31–33]. AUX /IAA proteins are repressors of auxin response factors (*ARFs*) that function as activators, and their degradation leads to the activation of the transcriptional regulation of *ARFs* (Figure 8; Table S4). Auxin induces cell expansion through the degradation of AUX/IAAs and the activation of *ARFs*. *ARFs* are transcription factors that bind to the promoters of auxin-responsive genes [31,34–36]. In Arabidopsis, *ARF7* positively regulates the expression of *EXP8* [37], which participates in extensive cell growth [38]. In addition, 9 *SAUR* genes stimulated the activity of H+-ATPase proton pumps in the plasma membrane to promote cell expansion (Figure 8) [39]. These observations indicated that the seven AUX/IAA genes might function as negative regulators, while one *ARF* gene and nine *SAUR* genes might serve as positive regulators of auxin signal transduction. These genes could play a positive role in cell expansion and the downward morphology of the sepals of *dsp* mutant plants.

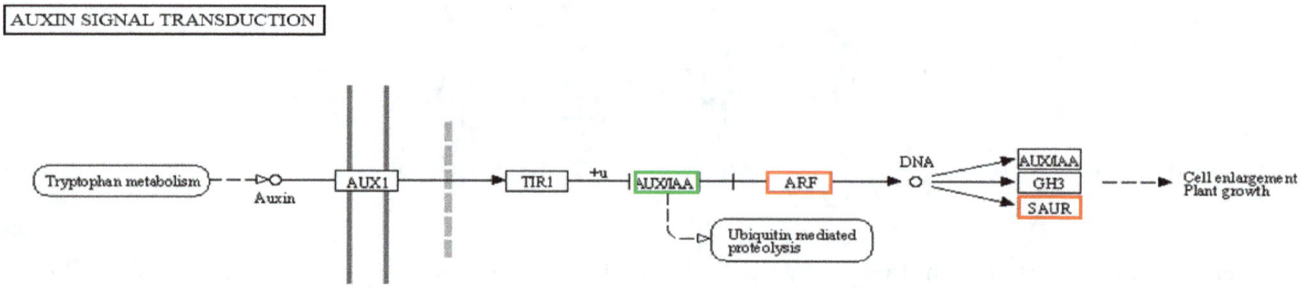

Figure 8. KEGG graph of auxin signal transduction pathway. Up-regulated, down-regulated and unchanged genes are shown in red, green and black boxes, respectively. "ARF" in the red box indicates the one *ARF* gene (*Solyc04g081240.2*). "AUX/IAA" in the green box represents the seven *AUX/IAA* genes (*Solyc06g008590.2, Solyc08g021820.2, Solyc09g083290.2, Solyc09g090910.1, Solyc04g076850.2, Solyc06g053830.2, Solyc06g053840.2*). "SAUR" in the red box indicates the nine *SAUR* genes (*Solyc05g025920.2, Solyc05g056440.1, Solyc01g110920.2, Solyc03g082530.1, Solyc06g053290.1, Solyc06g072650.1, Solyc09g009980.1, Solyc07g066560.1, Solyc02g084010.1*).

3. Discussion

The histological analysis showed that adaxial cells in the *dsp* sepals were larger than those in the WT sepals (Figure 6). Cell expansion is an important developmental force of organ morphology development at the cellular level. The transcriptome data indicated that cell expansion-related genes, such as *XTH*, pectinesterase genes, pectin lyase, cellulose synthases, expansins, and extensins, significantly changed (Table 1 and Table S4). Plant cell walls are complex structures composed of cellulose, xyloglucan, pectic polysaccharides, and structural proteins. To adapt to expansive forces and plant growth, plant cells selectively loosen their cell wall [40]. In the *dsp* sepals, seven *XTH* genes were upregulated at stage 4, and these genes were responsible for loosening the cell wall and promoting cell expansion and floral organ growth during flower opening [41,42]. Xyloglucan endotransglycosylase/hydrolases (XTHs) are encoded by a large multigene family, which cut xyloglucan and join the new reducing end to the non-reducing end of another xyloglucan (a transglucosylation) or to water (a hydrolysis) [43]. XTH itself cannot induce wall relaxation or creep, but can synergistically enhance wall extension, which might be termed an indirect or secondary loosening agent [43]. Moreover, 34 genes related to cell wall degradation were expressed differently in *dsp* sepals. For example, pectate lyase and pectinesterase break down pectins and participate in cell wall metabolism [44–46]. Ten pectin methylesterase (PME) inhibitor genes were expressed differently in *dsp* sepals, and they are implicated in the demethylation of pectin, resulting in the relaxation of the cell wall so that cells can enlarge [47]. Twelve DEGs related to cellulose synthesis were downregulated at stages 3 and 4 and involved in depositing cellulose to either primary or secondary walls in *Arabidopsis* [48]. In addition, 26 expansin/expansin-like and extensin/extensin- like genes were expressed differently in *dsp* sepals. They were considered two of the most important regulators of cell wall expansion and loosening during plant cell growth [49–51]. The α-expansins induced wall creep and wall relaxation, which mediated acid-induced extension of plant cell walls without mechanically weakening the cell walls [52–54]. In our study, 16 *XTHs*, 14 expansin genes and 11 PMEs/PME inhibitor genes were expressed differently in *dsp* sepals (Table S4). These results suggested that *dsp* positively regulated cell expansion through cell wall loosening, thus inducing the alteration of sepal morphology.

Organ development is regulated by different phytohormones, which manipulate appropriate cell division and cell expansion. As an important hormone in plant development, auxin plays important role in cell expansion [55–58]. In our work, the IAA content increased in *dsp* sepals, compared with those in WT sepals (Figure 7A). Moreover, the expression of auxin-related genes involved in response and efflux, especially *AUX/IAA* and *SAUR* family genes, significantly varied (Figure 8; Table S4). *SMALL AUXIN UP-RNA* (*SAUR*) genes constitute the largest family of auxin-induced genes, and 79 members are found in *Arabidopsis* [34]. The overexpression of stabilized SAUR19 proteins confers numerous phenotypes indicative of increased cell expansion, including increases in hypocotyl length and leaf size, and altered tropic growth [59–61] Similar findings were obtained with plants expressing SAUR63 proteins [62], suggesting that SAURs may positive effectors of cell expansion [59,62,63]. SAUR gene expression is upregulated by treatments/conditions that promote growth (e.g., IAA, brassinosteroids) and downregulated by factors that repress growth (e.g., abscisic acid, jasmonic acid, abiotic stress) [64,65]. Auxin is known to stimulate the activity of plasma membrane H^+-ATPase proton pumps, which promote proton efflux to acidify the apoplast and facilitate the uptake of solutes and water to drive plant cell expansion [39,66–71]. This process is regulated by auxin-inducible SAUR proteins [39,70]. The ensuing decrease in apoplastic pH alters the activity of cell wall-modifying proteins, including expansins [72], xyloglucan endotransglycosylase/hydrolases (XTHs) [73], and pectin methylesterases (PMEs) [74], leading to changes in wall extensibility. Auxin promotes the reorientation of microtubules from random to transverse, and suppresses the peroxidase activity in the cell wall. Consequently, cell wall extensibility is also promoted [75]. Therefore, *dsp* likely affected cell expansion by regulating auxin response, which induced the alteration of sepal morphology.

Transcriptomic analysis demonstrated that the genes related to cytokinin were expressed differentially; correspondingly, the cytokinin production level clearly increased in the *dsp* sepals compared with that in the WT sepals (Figure 7B). Furthermore, 19 DEGs were related to cytokinin: 17 biosynthesis-related genes (11.02-fold, the maximum fold change) and two degradation-related genes (6.54-fold, the maximum fold change; Figure 6C; Table S4). For example, two cytokinin oxidase/dehydrogenase (*CKX*) genes were upregulated 2.71-fold to 1.74-fold (Table S4). The overexpression of *CKX1* or *CKX2* in *Arabidopsis* and other species causes the elongation of the primary root and increases root branching [76–78], whereas the overexpression of *AtCKX7* results in an opposite phenotype [79]. Therefore, each *CKX* determined a specific developmental and physiological function. Cytokinin stimulated cell expansion in plants [80,81] and induced fourα-expansin subfamily members, namely, *EXPA1*, *EXPA10*, *EXPA14*, and *EXPA15*, in Arabidopsis root [82]. As a result, the cell wall loosened, and the cells expanded. In angiosperms, the ancient *tRNA-IPTs* and *CKXs* preferred the cZtype cytokinins as substrates, play a housekeeping role to maintain basic cellular functions. On the other hand, the nonancient ATP/ADP-IPTs and CKXs preferred the iP- and tZ-type cytokinins as bstrates, contribute more to the regulation of organ development and abiotic stress responses [83]. Five zeatin O-glucosyltransferase and zeatin O-xylosyltransferase (cisZOG) genes were upregulated 3.46 fold to 1.67-fold. cisZOG is involved in zeatin biosynthesis [24]. These results implied that cytokinin biosynthesis and signaling in *dsp* sepals were affected, and these processes might promote cell expansion and lead to downward sepals.

Gibberellin promotes cell expansion [84]. In *Petunia* and *Mimulus*, stamen removal leads to a reduced number of petals likely because stamens produce gibberellins [85,86]. In addition, GA-deficient *Arabidopsis* mutants exhibit a reduction in petal elongation, but gibberellins promote cell elongation by inhibiting the function of DELLA protein. The DELLA protein RGA, RGL1 and RGL2 in *Arabidopsis* inhibit the growth of petals in gibberellin-deficient plants [87]. Our results revealed that the content of gibberellins decreased, whereas the cell size increased in the *dsp* sepals, suggesting that gibberellins antagonized CK in a wide range of developmental events, including cell differentiation, shoot and root elongation, and meristem maintenance [88,89]. Active CK and recessive GA signals not only induce the expression of GA2ox, a GA-deactivating enzyme but also promote the expression of the cytokinin-biosynthesis gene *ISOPENTENYL TRANSFERASE7*. Consequently, cytokinin signaling is elevated [89,90]. Therefore, *dsp* promoted downward sepals through the regulation of adaxial cell size in sepals, and this process might be mediated by auxin, cytokinin and gibberellin.

4. Materials and Methods

4.1. Plant Materials

The *dsp* mutant of tomato was generated in the background of the inbred line TI1101 through natural mutation and stabilized via six generations of selfing prior to this study. The seeds of *dsp* mutant and WT were germinated on wet filter paper in a Petri dish at 28 °C in dark for 3 days. Then the resulting seedlings were grown in a growth chamber under a 16h/8h (light/dark) photoperiod with 25 °C/16 °C temperatures, respectively. Upon four true-leaf stage, plants were transferred to a greenhouse in the experimental field of the Northwest A&F University. Pest control and water management were carried out according to standard practices. All the materials were grown in a plastic greenhouse in the Northwest A&F University (Shaanxi, China).

4.2. Measurement of Sepal Morphology at Different Developmental Stages

Nine ripened fruits with whole calyx per accession were used as repeats in the phenotypic analysis. Each sepal was evaluated for 2 traits:

(1) Sepal Upturned Degree (SEUD, °) = $90° + \alpha = 90° + \arctan(H/L_1)$, degree of upwarp or down-wrap of sepal. Put a whole calyx on a L-square ruler, then keep the calyx base level on one side of L-square ruler and the sepal to be measured leaning against another side of L-square ruler. H represents the vertical distance between the furthest point from sepal to stalk in the horizontal line and the base line, L_1 is the longest distance from sepal to stalk (Figure 9).

(2) Sepal Rolling Index (SERI, %) = $(SEL - L_2)/SEL \times 100\%$, SEL (Sepal Length, mm) is the length of sepal in flat condition, L_2 is the distance from apex to base point (Figure 9).

Figure 9. Measurement of sepal traits. (**a**) Measurement of sepal length (SEL) and sepal width (SEW). (**b**) Measurement of Sepal Upturned Degree (SEUD) and Sepal Rolling Index (SERI). SEUD-Sepal Upturned Degree (α). $\alpha = \arctan(H/L1)$ (°). H represents the vertical distance between the furthest point from sepal to stalk in the horizontal line and the base line; L1 is the longest distance from sepal to stalk. SERI-Sepal Rolling Index = (SEL − L2)/SEL. SEL, Sepal Length; L2 is the distance from the sepal apex to base point.

4.3. DGE (Differentially Gene Expression) Library Construction and Sequencing

Sepal at stage 3 (0 DAF, length around 8mm) and stage 4 (15 DAF, length around 11 mm) were collected from WT and *dsp* mutant at the same time on the same day. Samples were immediately frozen in liquid nitrogen and stored at −80°C for RNA-Seq analyses. Total RNA was isolated using the TRIzol® Plant RNA Purification Reagent according to the manufacturer's instructions (Invitrogen, Carlsbad, CA, USA) and genomic DNA was removed using DNase I (TaKara, Dalian, China). RNA quality was determined by 2100 Bioanalyser (Agilent, Silicon Valley, CA, USA) and quantified using the ND-2000 (NanoDrop Technologies, Wilmington, DE, USA). Only high-quality RNA sample (OD260/280 = 1.8–2.2, OD260/230 ≥ 2.0, RIN ≥ 6.5, 28S:18S ≥ 1.0, >10 μg) was used to construct sequencing library.

RNA-seq transcriptome library was prepared following TruSeq™ RNA sample preparation Kit from Illumina (San Diego, CA, USA) using 5 μg of total RNA. Shortly, messenger RNA was isolated according to polyA selection method by oligo(dT) beads and then fragmented by fragmentation buffer firstly. Secondly double-stranded cDNA was synthesized using a SuperScript double-stranded cDNA synthesis kit (Invitrogen, Carlsbad, CA, USA) with random hexamer primers (Illumina, San Diego, CA, USA). Then the synthesized cDNA was subjected to end-repair, phosphorylation and "A" base addition according to Illumina's library construction protocol. Libraries were size selected for cDNA target fragments of 200–300 bp on 2% Low Range Ultra Agarose followed by PCR amplified using Phusion DNA polymerase (NEB) for 15 PCR cycles. After quantified by TBS380, the paired-end RNA-seq sequencing library was sequenced with the Illumina HiSeq xten (2 × 150 bp read length).

The raw paired end reads were trimmed and quality controlled by SeqPrep (https://github.com/jstjohn/SeqPrep) and Sickle (https://github.com/najoshi/sickle) with default parameters. Then clean reads were separately aligned to reference genome with orientation mode using TopHat (http://tophat.cbcb.umd.edu/,version2.0.0) [91] software. The mapping criteria of bowtie was as follows: sequencing reads should be uniquely matched to the genome allowing up to 2 mismatches, without insertions or deletions. Then, the region of the gene was expanded following depths of sites and the operon was obtained. In addition, the whole genome was split into multiple 15k bp windows that share 5k bp. New transcribed regions were defined as more than 2 consecutive windows without the overlapped region of the gene, where at least 2 reads mapped per window in the same orientation.

4.4. *Bioinformatics Analysis of DGE Data*

To identify DEGs (differential expression genes) between two different samples, the expression level of each transcript was calculated according to the fragments per kilobase of exon per million mapped reads (FRKM) method. RSEM (http://deweylab.biostat.wisc.edu/rsem/) [92] was used to quantify gene abundances. R statistical package software EdgeR (Empirical analysis of Digital Gene Expression in R, http://www.bioconductor.org/packages/2.12/bioc/html/edgeR.html) [93] was utilized for differential expression analysis. In addition, functional-enrichment analysis including GO and KEGG were performed to identify which DEGs were significantly enriched in GO terms and metabolic pathways at Bonferroni-corrected p-value ≤ 0.05 compared with the whole-transcriptome background. GO functional enrichment and KEGG pathway analysis were carried out by Goatools (https://github.com/tanghaibao/Goatools) and KOBAS (http://kobas.cbi.pku.edu.cn/home.do) [94].

4.5. *Paraffin Sectioning and Electron Microscopy Experiment*

The sepal cells were observed by paraffin sections stained with safranin-fast green. The sepals of WT and *dsp* at stage 3 to stage 7 were harvested and immediately fixed in formalin/glacial acetic acid (FAA). The samples were then dehydrated in a xylene and alcohol series (75, 85, 90, 95 and 100%) and then embedded in paraffin wax. Four-micrometer-thick sections were cut and stained with safranin (1%)-fast green (0.5%) for histological examination. Sepal cells underwent electron microscopy using a general laboratory biology electron microscope

4.6. *Extraction, Purification and Quantification of the Phytohormones in Sepals*

Stage 3, stage 4, stage 5 and, stage 6 sepals for phytohormones IAA, zeatin and GA quantification. The method for extraction, purification, and quantification of phytohormones was modified from the description of Wang [95]. ELISA kits used for estimation of the hormonal levels came from China Agricultural University (Beijing, China).

SPSS software was used for statistical analysis. The *dsp* mutants were compared with the WT using Tukey's test at $p < 0.05$.

5. Conclusions

Conclusively, in the first part of the study, we observed the seven stages of the sepal morphological development of *dsp* (downward sepal) mutant and WT. The WT and *dsp* plants showed to be significantly diverse during stages 4 to 7. To understand the molecular mechanisms of downward sepal development, a transcriptome analysis of *dsp* mutant and WT plants at stages 3 and 4 was performed. We observed that *dsp* affected sepal morphology by controlling genes related to cell wall biogenesis and modification and plant hormone regulation, particularly auxin, gibberellin, and cytokinin. Among these genes, seven AUX/IAA genes functioned as negative regulators, while one *ARF* gene and nine *SAUR* genes served as positive regulators of auxin signal transduction. According to cell morphology observation between *dsp* and WT, we found that cell enlargement or abnormal cell expansion in the adaxial part of sepals might contribute to downward sepals in *dsp*. Meanwhile, *dsp* mutant led to increase in auxin and cytokinin, and a decrease in gibberellin. In conclusion, *dsp* promoted downward sepals through the regulation of adaxial cell size in sepals, and this process might be mediated by auxin, cytokinin and gibberellin.

Author Contributions: Conceptualization, J.L. and Y.L. (Yan Liang); data curation, J.L.; Formal analysis, J.L., M.S., J.W. (Jing Wang) and B.Z.; investigation, J.L., M.S., B.Z., J.W. (Jin Wang) and Y.L. (Yushun Li); methodology, J.L.; resources, J.L. and Y.L. (Yan Liang); software, J.L.; validation, Y.L. (Yan Liang); visualization, J.L.; writing original draft, J.L.; writing review and editing, J.L., A.H.E.-S. and Y.L. (Yan Liang); supervision, Y.L. (Yan Liang); project administration, Y.L. (Yan Liang); funding acquisition, Y.L. (Yan Liang). All authors have read and agreed to the published version of the manuscript.

References

1. Zhong, F.L.; Zhou, X.Z.; Lin, Y.W.; Chen, X.; Xu, R.; Wang, S.B.; Lin, Y.Z.; Pang, J.; Wu, S. Sepal Morphology Affects Flower and Fruit Development in Luffa Cylindrica. *Fresenius. Environ. Bull.* **2018**, *27*, 4006–4013.
2. Li, N.; Huang, B.; Tang, N.; Jian, W.; Zou, J.; Chen, J.; Cao, H.; Habib, S.; Dong, X.; Wei, W.; et al. The MADS-Box Gene *SlMBP21* Regulates Sepal Size Mediated by Ethylene and Auxin in Tomato. *Plant Cell. Physiol.* **2017**, *58*, 2241–2256. [CrossRef] [PubMed]
3. Vemmos, S.N.; Goldwin, G.K. The Photosynthetic Activity of Cox Orange Pippin Apple Flowers in Relation to Fruit Setting. *Ann. Bot.* **1994**, *73*, 385–391. [CrossRef]
4. Hong, L.L.; Dumond, M.; Tsugawa, S.; Sapala, A.; Routier-Kierzkowska, A.L.; Zhou, Y.; Chen, C.; Kiss, A.; Zhu, M.Y.; Hamant, O.; et al. Variable Cell Growth Yields Reproducible Organ Development through Spatiotemporal Averaging. *Dev. Cell.* **2016**, *38*, 15–32. [CrossRef]
5. Tauriello, G.; Meyer, H.M.; Smith, R.S.; Koumoutsakos, P.; Roeder, A.H.K. Variability and Constancy in Cellular Growth of Arabidopsis Sepals. *Plant Physiol.* **2015**, *169*, 2342–2358. [CrossRef] [PubMed]
6. Santner, A.; Calderon-Villalobos, L.I.A.; Estelle, M. Plant hormones are versatile chemical regulators of plant growth. *Nat. Chem. Biol.* **2009**, *5*, 301–307. [CrossRef]
7. Wang, Y.H.; Irving, H.R. Developing a model of plant hormone interactions. *Plant Signal. Behav.* **2011**, *6*, 494–500. [CrossRef]
8. Varaud, E.; Brioudes, F.; Szécsi, J.; Leroux, J.; Brown, S.; Perrot-Rechenmann, C.; Bendahmane, M. *AUXIN RESPONSE FACTOR8* regulates Arabidopsis petal growth by interacting with the bHLH transcription factor BIGPETALp. *Plant Cell.* **2011**, *23*, 973–983. [CrossRef]
9. Bao, S.; Hua, C.; Shen, L.; Yu, H. New insights into gibberellin signaling in regulating flowering in Arabidopsis. *J. Integr. Plant Biol.* **2020**, *62*, 118–131. [CrossRef]
10. Danisman, S.; Van Dijk, A.D.J.; Bimbo, A.; Van der Wal, F.; Hennig, L.; De Folter, S.; Angenent, G.C.; Immink, R.G.H. Analysis of functional redundancies within the Arabidopsis TCP transcription factor family. *J. Exp. Bot.* **2013**, *64*, 5673–5685. [CrossRef]
11. Koyama, T.; Mitsuda, N.; Seki, M.; Shinozaki, K.; Ohme-Takagi, M. TCP Transcription Factors Regulate the Activities of *ASYMMETRIC LEAVES1* and *miR164*, as Well as the Auxin Response, during Differentiation of Leaves in Arabidopsis. *Plant Cell.* **2010**, *22*, 3574–3588. [CrossRef] [PubMed]
12. Nag, A.; King, S.; Jack, T. *miR319a* targeting of *TCP4* is critical for petal growth and development in Arabidopsis. *Proc. Natl. Acad. Sci. USA* **2009**, *106*, 22534–22539. [CrossRef] [PubMed]
13. Wang, M.Y.; Zhao, P.M.; Cheng, H.Q.; Han, L.B.; Wu, X.M.; Gao, P.; Wang, H.Y.; Yang, C.L.; Zhong, N.Q.; Zuo, J.R.; et al. The Cotton Transcription Factor TCP14 Functions in Auxin-Mediated Epidermal Cell Differentiation and Elongation. *Plant Physiol.* **2013**, *162*, 1669–1680. [CrossRef] [PubMed]
14. Vrebalov, J.; Ruezinsky, D.; Padmanabhan, V.; White, R.; Medrano, D.; Drake, R.; Schuch, W.; Giovannoni, J. A MADS-box gene necessary for fruit ripening at the tomato ripening-inhibitor (Rin) locus. *Science* **2002**, *296*, 343–346. [CrossRef] [PubMed]
15. Giménez, E.; Pineda, B.; Capel, J.; Teresa Antón, M.; Atarés, A.; Pérez-Martin, F.; Garcia-Sogo, B.; Angosto, T.; Moreno, V.; Lozano, R. Functional Analysis of the Arlequin Mutant Corroborates the Essential Role of the *ARLEQUIN/TAGL1* Gene during Reproductive Development of Tomato. *PLoS ONE* **2010**, *5*, e14427. [CrossRef] [PubMed]
16. Xie, Q.L.; Hu, Z.L.; Zhu, Z.G.; Dong, T.T.; Zhao, Z.P.; Cui, B.L.; Chen, G.P. Overexpression of a novel MADS-box gene *SlFYFL* delays senescence, fruit ripening and abscission in tomato. *Sci. Rep.* **2014**, *4*. [CrossRef] [PubMed]

17. Liu, D.; Wang, D.; Qin, Z.; Zhang, D.; Yin, L.; Wu, L.; Colasanti, J.; Li, A.; Mao, L. The SEPALLATA MADS-box protein *SLMBP21* forms protein complexes with *JOINTLESS* and *MACROCALYX* as a transcription activator for development of the tomato flower abscission zone. *Plant J.* **2014**, *77*, 284–296. [CrossRef]
18. Roldan, M.V.G.; Perilleux, C.; Morin, H.; Huerga-Fernandez, S.; Latrasse, D.; Benhamed, M.; Bendahmane, A. Natural and induced loss of function mutations in *SlMBP21* MADS-box gene led to *jointless-2* phenotype in tomato. *Sci. Rep.* **2017**, *7*. [CrossRef]
19. Zhang, J.l.; Hu, Z.l.; Wang, Y.s.; Yu, X.h.; Liao, C.g.; Zhu, M.k.; Chen, G.p. Suppression of a tomato SEPALLATA MADS-box gene, *SlCMB1*, generates altered inflorescence architecture and enlarged sepals. *Plant Sci.* **2018**, *272*, 75–87. [CrossRef]
20. Eveland, A.L.; Satoh-Nagasawa, N.; Goldshmidt, A.; Meyer, S.; Beatty, M.; Sakai, H.; Ware, D.; Jackson, D. Digital gene expression signatures for maize development. *Plant Physiol.* **2010**, *154*, 1024–1039. [CrossRef]
21. Zhang, Y.; Li, Z.G.; Tu, Y.; Cheng, W.J.; Yang, Y.W. Tomato (*Solarium lycopersicum*) *SIIPT4*, encoding an isopentenyltransferase, is involved in leaf senescence and lycopene biosynthesis during fruit ripening. *BMC Plant Biol.* **2018**. [CrossRef]
22. Wang, X.J.; Ding, J.; Lin, S.S.; Liu, D.C.; Gu, T.T.; Wu, H.; Trigiano, R.N.; McAvoy, R.; Huang, J.L.; Li, Y. Evolution and roles of cytokinin genes in angiosperms 2: Do ancient *CKXs* play housekeeping roles while non-ancient *CKXs* play regulatory roles? *Hortic. Res.-Engl.* **2020**. [CrossRef] [PubMed]
23. Kurakawa, T.; Ueda, N.; Maekawa, M.; Kobayashi, K.; Kojima, M.; Nagato, Y.; Sakakibara, H.; Kyozuka, J. Direct control of shoot meristem activity by a cytokinin-activating enzyme. *Nature* **2007**, *445*, 652–655. [CrossRef] [PubMed]
24. Veach, Y.K.; Martin, R.C.; Mok, D.W.; Malbeck, J.; Vankova, R.; Mok, M.C. O-glucosylation of cis-zeatin in maize. Characterization of genes, enzymes, and endogenous cytokinins. *Plant Physiol.* **2003**, *131*, 1374–1380. [CrossRef]
25. Frebort, I.; Kowalska, M.; Hluska, T.; Frebortova, J.; Galuszka, P. Evolution of cytokinin biosynthesis and degradation. *J. Exp. Bot.* **2011**, *62*, 2431–2452. [CrossRef]
26. De Lucas, M.; Davière, J.M.; Rodríguez-Falcón, M.; Pontin, M.; Iglesias-Pedraz, J.M.; Lorrain, S.; Fankhauser, C.; Blázquez, M.A.; Titarenko, E.; Prat, S. A molecular framework for light and gibberellin control of cell elongation. *Nature* **2008**, *451*, 480–484. [CrossRef]
27. Que, F.; Khadr, A.; Wang, G.L.; Li, T.; Wang, Y.H.; Xu, Z.S.; Xiong, A.S. Exogenous brassinosteroids altered cell length, gibberellin content, and cellulose deposition in promoting carrot petiole elongation. *Plant Sci.* **2018**. [CrossRef]
28. Richards, D.E.; King, K.E.; Ait-ali, T.; Harberd, N.P. How gibberellin regulates plant growth and development: A molecular genetic analysis of gibberellin signaling. *Annu. Rev. Plant Physiol. Plant Molec. Biol.* **2001**, *52*, 67–88. [CrossRef]
29. Yamaguchi, S. Gibberellin Metabolism and its Regulation. *Annu. Rev. Plant Biol.* **2008**, *59*, 225–251. [CrossRef]
30. Kanehisa, M.; Goto, S.; Kawashima, S.; Okuno, Y.; Hattori, M. The KEGG resource for deciphering the genome. *Nucleic. Acids. Res.* **2004**, *32*, D277–D280. [CrossRef]
31. Overvoorde, P.J.; Okushima, Y.; Alonso, J.M.; Chan, A.; Chang, C.; Ecker, J.R.; Hughes, B.; Liu, A.; Onodera, C.; Quach, H.; et al. Functional genomic analysis of the AUXIN/INDOLE-3-ACETIC ACID gene family members in *Arabidopsis thaliana*. *Plant Cell.* **2005**, *17*, 3282–3300. [CrossRef] [PubMed]
32. Ramos, J.A.; Zenser, N.; Leyser, O.; Callis, J. Rapid degradation of auxin/indoleacetic acid proteins requires conserved amino acids of domain II and is proteasome dependent. *Plant Cell.* **2001**, *13*, 2349–2360. [CrossRef] [PubMed]
33. Sato, A.; Yamamoto, K.T. Overexpression of the non-canonical Aux/IAA genes causes auxin-related aberrant phenotypes in Arabidopsis. *Physiol. Plant* **2008**, *133*, 397–405. [CrossRef] [PubMed]
34. Hagen, G.; Guilfoyle, T. Auxin-responsive gene expression: Genes, promoters and regulatory factors. *Plant Mol. Biol.* **2002**, *49*, 373–385. [CrossRef] [PubMed]
35. Tiwari, S.B.; Hagen, G.; Guilfoyle, T. The roles of auxin response factor domains in auxin-responsive transcription. *Plant Cell.* **2003**, *15*, 533–543. [CrossRef] [PubMed]
36. Chandler, J.W. Auxin response factors. *Plant Cell. Environ.* **2016**, *39*, 1014–1028. [CrossRef]
37. Esmon, C.A.; Tinsley, A.G.; Ljung, K.; Sandberg, G.; Hearne, L.B.; Liscum, E. A gradient of auxin and auxin-dependent transcription precedes tropic growth responses. *Proc. Natl. Acad. Sci. USA* **2006**, *103*, 236–241. [CrossRef]

38. Pelletier, S.; Van Orden, J.; Wolf, S.; Vissenberg, K.; Delacourt, J.; Ndong, Y.A.; Pelloux, J.; Bischoff, V.; Urbain, A.; Mouille, G.; et al. A role for pectin de-methylesterification in a developmentally regulated growth acceleration in dark-grown *Arabidopsis* hypocotyls. *New. Phytol.* **2010**, *188*, 726–739. [CrossRef]
39. Spartz, A.K.; Ren, H.; Park, M.Y.; Grandt, K.N.; Lee, S.H.; Murphy, A.S.; Sussman, M.R.; Overvoorde, P.J.; Gray, W.M. SAUR Inhibition of PP2C-D Phosphatases Activates Plasma Membrane H+-ATPases to Promote Cell Expansion in Arabidopsis. *Plant Cell.* **2014**, *26*, 2129–2142. [CrossRef]
40. Cosgrove, D.J. Plant expansins: Diversity and interactions with plant cell walls. *Curr. Opin. Plant Biol.* **2015**, *25*, 162–172. [CrossRef]
41. Harada, T.; Torii, Y.; Morita, S.; Masumura, T.; Satoh, S. Differential expression of genes identified by suppression subtractive hybridization in petals of opening carnation flowers. *J. Exp. Bot.* **2010**, *61*, 2345–2354. [CrossRef] [PubMed]
42. Hyodo, H.; Yamakawa, S.; Takeda, Y.; Tsuduki, M.; Yokota, A.; Nishitani, K.; Kohchi, T. Active gene expression of a xyloglucan endotransglucosylase/hydrolase gene, *XTH9*, in inflorescence apices is related to cell elongation in *Arabidopsis thaliana*. *Plant Mol. Biol.* **2003**, *52*, 473–482. [CrossRef] [PubMed]
43. Cosgrove, D.J. Catalysts of plant cell wall loosening. *F1000Res* **2016**. [CrossRef] [PubMed]
44. Hu, R.; Xu, Y.; Yu, C.; He, K.; Tang, Q.; Jia, C.; He, G.; Wang, X.; Kong, Y.; Zhou, G. Transcriptome analysis of genes involved in secondary cell wall biosynthesis in developing internodes of Miscanthus lutarioriparius. *Sci. Rep.* **2017**, *7*, 9034. [CrossRef] [PubMed]
45. Cosgrove, D.J. Growth of the plant cell wall. *Nat. Rev. Mol. Cell Biol.* **2005**, *6*, 850–861. [CrossRef]
46. Wolf, S.; Hématy, K.; Höfte, H. Growth control and cell wall signaling in plants. *Annu. Rev. Plant Biol.* **2012**, *63*, 381–407. [CrossRef]
47. Marzec-Schmidt, K.; Ludwików, A.; Wojciechowska, N.; Kasprowicz-Maluśki, A.; Mucha, J.; Bagniewska-Zadworna, A. Xylem Cell Wall Formation in Pioneer Roots and Stems of Populus trichocarpa (Torr. & Gray). *Front. Plant Sci.* **2019**, *10*, 1419. [CrossRef]
48. Somerville, C. Cellulose synthesis in higher plants. *Annu. Rev. Cell. Dev. Biol.* **2006**, *22*, 53–78. [CrossRef]
49. Cosgrove, D.J.; Li, L.C.; Cho, H.T.; Hoffmann-Benning, S.; Moore, R.C.; Blecker, D. The growing world of expansins. *Plant Cell. Physiol.* **2002**, *43*, 1436–1444. [CrossRef]
50. Cannon, M.C.; Terneus, K.; Hall, Q.; Tan, L.; Wang, Y.; Wegenhart, B.L.; Chen, L.; Lamport, D.T.; Chen, Y.; Kieliszewski, M.J. Self-assembly of the plant cell wall requires an extensin scaffold. *Proc. Natl. Acad. Sci. USA* **2008**, *105*, 2226–2231. [CrossRef]
51. Li, Y.; Jones, L.; McQueen-Mason, S. Expansins and cell growth. *Curr. Opin. Plant Biol.* **2003**, *6*, 603–610. [CrossRef] [PubMed]
52. McQueenmason, S.; Durachko, D.M.; Cosgrove, D.J. 2 Endogenous Proteins That Induce Cell-Wall Extension in Plants. *Plant Cell.* **1992**, *4*, 1425–1433. [CrossRef]
53. McQueen-Mason, S.J.; Cosgrove, D.J. Expansin mode of action on cell walls. Analysis of wall hydrolysis, stress relaxation, and binding. *Plant Physiol.* **1995**, *107*, 87–100. [CrossRef] [PubMed]
54. McQueen-Mason, S.J.; Fry, S.C.; Durachko, D.M.; Cosgrove, D.J. The relationship between xyloglucan endotransglycosylase and in-vitro cell wall extension in cucumber hypocotyls. *Planta* **1993**, *190*, 327–331. [CrossRef] [PubMed]
55. Braun, N.; Wyrzykowska, J.; Muller, P.; David, K.; Couch, D.; Perrot-Rechenmann, C.; Fleming, A.J. Conditional Repression of *AUXIN BINDING PROTEIN1* Reveals That It Coordinates Cell Division and Cell Expansion during Postembryonic Shoot Development in Arabidopsis and Tobacco. *Plant Cell.* **2008**, *20*, 2746–2762. [CrossRef]
56. Barbez, E.; Dünser, K.; Gaidora, A.; Lendl, T.; Busch, W. Auxin steers root cell expansion via apoplastic pH regulation in *Arabidopsis thaliana*. *Proc. Natl. Acad. Sci. USA* **2017**, *114*, E4884–E4893. [CrossRef]
57. Paque, S.; Mouille, G.; Grandont, L.; Alabadí, D.; Gaertner, C.; Goyallon, A.; Muller, P.; Primard-Brisset, C.; Sormani, R.; Blázquez, M.A.; et al. *AUXIN BINDING PROTEIN1* links cell wall remodeling, auxin signaling, and cell expansion in arabidopsis. *Plant Cell.* **2014**, *26*, 280–295. [CrossRef]
58. Schopfer, P.; Liszkay, A.; Bechtold, M.; Frahry, G.; Wagner, A. Evidence that hydroxyl radicals mediate auxin-induced extension growth. *Planta* **2002**, *214*, 821–828. [CrossRef]
59. Spartz, A.K.; Lee, S.H.; Wenger, J.P.; Gonzalez, N.; Itoh, H.; Inze, D.; Peer, W.A.; Murphy, A.S.; Overvoorde, P.J.; Gray, W.M. The SAUR19 subfamily of *SMALL AUXIN UP RNA* genes promote cell expansion. *Plant J.* **2012**, *70*, 978–990. [CrossRef]

60. Vanhaeren, H.; Gonzalez, N.; Coppens, F.; De Milde, L.; Van Daele, T.; Vermeersch, M.; Eloy, N.B.; Storme, V.; Inzé, D. Combining growth-promoting genes leads to positive epistasis in Arabidopsis thaliana. *eLife* **2014**, *3*, e02252. [CrossRef]
61. Vi, S.L.; Trost, G.; Lange, P.; Czesnick, H.; Rao, N.; Lieber, D.; Laux, T.; Gray, W.M.; Manley, J.L.; Groth, D.; et al. Target specificity among canonical nuclear poly(A) polymerases in plants modulates organ growth and pathogen response. *Proc. Natl. Acad. Sci. USA* **2013**, *110*, 13994–13999. [CrossRef] [PubMed]
62. Chae, K.; Isaacs, C.G.; Reeves, P.H.; Maloney, G.S.; Muday, G.K.; Nagpal, P.; Reed, J.W. Arabidopsis *SMALL AUXIN UP RNA63* promotes hypocotyl and stamen filament elongation. *Plant J.* **2012**, *71*, 684–697. [CrossRef] [PubMed]
63. Franklin, K.A.; Lee, S.H.; Patel, D.; Kumar, S.V.; Spartz, A.K.; Gu, C.; Ye, S.; Yu, P.; Breen, G.; Cohen, J.D.; et al. *PHYTOCHROME-INTERACTING FACTOR 4* (*PIF4*) regulates auxin biosynthesis at high temperature. *Proc. Natl. Acad. Sci. USA* **2011**, *108*, 20231–20235. [CrossRef] [PubMed]
64. Ren, H.; Gray, W.M. SAUR Proteins as Effectors of Hormonal and Environmental Signals in Plant Growth. *Mol. Plant* **2015**, *8*, 1153–1164. [CrossRef]
65. Stortenbeker, N.; Bemer, M. The SAUR gene family: The plant's toolbox for adaptation of growth and development. *J. Exp. Bot.* **2019**, *70*, 17–27. [CrossRef]
66. Arsuffi, G.; Braybrook, S.A. Acid growth: An ongoing trip. *J. Exp. Bot.* **2018**, *69*, 137–146. [CrossRef]
67. Majda, M.; Robert, S. The Role of Auxin in Cell Wall Expansion. *Int. J. Mol. Sci.* **2018**, 951. [CrossRef]
68. Takahashi, K.; Hayashi, K.-i.; Kinoshita, T. Auxin Activates the Plasma Membrane H+-ATPase by Phosphorylation during Hypocotyl Elongation in Arabidopsis. *Plant Physiol.* **2012**, *159*, 632–641. [CrossRef]
69. Inoue, S.I.; Takahashi, K.; Okumura-Noda, H.; Kinoshita, T. Auxin Influx Carrier AUX1 Confers Acid Resistance for Arabidopsis Root Elongation Through the Regulation of Plasma Membrane H+-ATPase. *Plant Cell. Physiol.* **2016**, *57*, 2194–2201. [CrossRef]
70. Farquharson, K.L. SAUR19 Links Auxin and Plasma Membrane H+-ATPases in Cell Expansion. *Plant Cell.* **2014**, *26*, 1835. [CrossRef]
71. Coenen, C.; Bierfreund, N.; Lüthen, H.; Neuhaus, G. Developmental regulation of H+-ATPase-dependent auxin responses in the diageotropica mutant of tomato (*Lycopersicon esculentum*). *Physiol. Plant* **2002**, *114*, 461–471. [CrossRef] [PubMed]
72. Lee, S.H. Recent research progress on acid-growth theory. *J. Exp. Bot.* **2016**, *43*, 405–410. [CrossRef]
73. Nishitani, K.; Vissenberg, K. Roles of the XTH Protein Family in the Expanding Cell. *Plant Cell. Monogr.* **2006**. [CrossRef]
74. Hocq, L.; Pelloux, J.; Lefebvre, V. Connecting Homogalacturonan-Type Pectin Remodeling to Acid Growth. *Trends Plant Sci.* **2017**, *22*, 20–29. [CrossRef] [PubMed]
75. Knöller, A.S.; Blakeslee, J.J.; Richards, E.L.; Peer, W.A.; Murphy, A.S. Brachytic2/ZmABCB1 functions in IAA export from intercalary meristems. *J. Exp. Bot.* **2010**, *61*, 3689–3696. [CrossRef]
76. Galuszka, P.; Frebortova, J.; Werner, T.; Yamada, M.; Strnad, M.; Schmulling, T.; Frebort, I. Cytokinin oxidase/dehydrogenase genes in barley and wheat - Cloning and heterologous expression. *Eur. J. Biochem.* **2004**, *271*, 3990–4002. [CrossRef]
77. Mrizova, K.; Jiskrova, E.; Vyroubalova, S.; Novak, O.; Ohnoutkova, L.; Pospisilova, H.; Frebort, I.; Harwood, W.A.; Galuszka, P. Overexpression of Cytokinin Dehydrogenase Genes in Barley (Hordeum vulgare cv. Golden Promise) Fundamentally Affects Morphology and Fertility. *PLoS ONE* **2013**, *8*, e79029. [CrossRef]
78. Werner, T.; Motyka, V.; Strnad, M.; Schmülling, T. Regulation of plant growth by cytokinin. *Proc. Natl. Acad. Sci. USA* **2001**, *98*, 10487–10492. [CrossRef]
79. Koellmer, I.; Novak, O.; Strnad, M.; Schmuelling, T.; Werner, T. Overexpression of the cytosolic cytokinin oxidase/dehydrogenase (*CKX7*) from *Arabidopsis* causes specific changes in root growth and xylem differentiation. *Plant J.* **2014**, *78*, 359–371. [CrossRef]
80. Efroni, I.; Blum, E.; Goldshmidt, A.; Eshed, Y. A Protracted and Dynamic Maturation Schedule Underlies Arabidopsis Leaf Development. *Plant Cell.* **2008**, *20*, 2293–2306. [CrossRef]
81. Skalak, J.; Vercruyssen, L.; Claeys, H.; Hradilova, J.; Cerny, M.; Novak, O.; Plackova, L.; Saiz-Fernandez, I.; Skalakova, P.; Coppens, F.; et al. Multifaceted activity of cytokinin in leaf development shapes its size and structure in *Arabidopsis*. *Plant J.* **2019**, *97*, 805–824. [CrossRef] [PubMed]
82. Bhargava, A.; Clabaugh, I.; To, J.P.; Maxwell, B.B.; Chiang, Y.-H.; Schaller, G.E.; Loraine, A.; Kieber, J.J. Identification of Cytokinin-Responsive Genes Using Microarray Meta-Analysis and RNA-Seq in *Arabidopsis*. *Plant Physiol.* **2013**, *162*, 272–294. [CrossRef] [PubMed]

83. Wang, X.; Lin, S.; Liu, D.; Gan, L.; McAvoy, R.; Ding, J.; Li, Y. Evolution and roles of cytokinin genes in angiosperms 1: Do ancient *IPTs* play housekeeping while non-ancient *IPTs* play regulatory roles? *Hortic. Res.* **2020**, *7*, 28. [CrossRef]
84. Gazzarrini, S.; McCourt, P. Cross-talk in plant hormone signalling: What Arabidopsis mutants are telling us. *Ann. Bot.* **2003**, *91*, 605–612. [CrossRef] [PubMed]
85. Barr, C.M.; Fishman, L. Cytoplasmic male sterility in Mimulus hybrids has pleiotropic effects on corolla and pistil traits. *Heredity* **2011**, *106*, 886–893. [CrossRef] [PubMed]
86. Weiss, D.; Halevy, A.H. Stamens and gibberellin in the regulation of corolla pigmentation and growth in *Petunia* hybrida. *Planta* **1989**, *179*, 89–96. [CrossRef]
87. Cheng, H.; Qin, L.J.; Lee, S.C.; Fu, X.D.; Richards, D.E.; Cao, D.N.; Luo, D.; Harberd, N.P.; Peng, J.R. Gibberellin regulates *Arabidopsis* floral development via suppression of DELLA protein function. *Development* **2004**, *131*, 1055–1064. [CrossRef]
88. Greenboim-Wainberg, Y.; Maymon, I.; Borochov, R.; Alvarez, J.; Olszewski, N.; Ori, N.; Eshed, Y.; Weiss, D. Cross talk between gibberellin and cytokinin: The Arabidopsis GA response inhibitor SPINDLY plays a positive role in cytokinin signaling. *Plant Cell.* **2005**, *17*, 92–102. [CrossRef]
89. Jasinski, S.; Piazza, P.; Craft, J.; Hay, A.; Woolley, L.; Rieu, I.; Phillips, A.; Hedden, P.; Tsiantis, M. KNOX action in Arabidopsis is mediated by coordinate regulation of cytokinin and gibberellin activities. *Curr. Biol.* **2005**, *15*, 1560–1565. [CrossRef]
90. Yanai, O.; Shani, E.; Dolezal, K.; Tarkowski, P.; Sablowski, R.; Sandberg, G.; Samach, A.; Ori, N. Arabidopsis KNOXI proteins activate cytokinin biosynthesis. *Curr. Biol.* **2005**, *15*, 1566–1571. [CrossRef]
91. Langmead, B.; Salzberg, S.L. Fast gapped-read alignment with Bowtie 2. *Nat. Methods* **2012**, *9*, 357–359. [CrossRef] [PubMed]
92. Li, B.; Dewey, C.N. RSEM: Accurate transcript quantification from RNA-Seq data with or without a reference genome. *BMC Bioinform.* **2011**, *12*, 323. [CrossRef] [PubMed]
93. Robinson, M.D.; McCarthy, D.J.; Smyth, G.K. edgeR: A Bioconductor package for differential expression analysis of digital gene expression data. *Bioinformatics* **2010**, *26*, 139–140. [CrossRef] [PubMed]
94. Xie, C.; Mao, X.; Huang, J.; Ding, Y.; Wu, J.; Dong, S.; Kong, L.; Gao, G.; Li, C.Y.; Wei, L. KOBAS 2.0: A web server for annotation and identification of enriched pathways and diseases. *Nucleic. Acids Res.* **2011**, *39*, W316–W322. [CrossRef] [PubMed]
95. Wang, Y.; Li, B.; Du, M.; Eneji, A.E.; Wang, B.; Duan, L.; Li, Z.; Tian, X. Mechanism of phytohormone involvement in feedback regulation of cotton leaf senescence induced by potassium deficiency. *J. Exp. Bot.* **2012**, *63*, 5887–5901. [CrossRef]

2

Phenotypic Divergence Analysis in Pigeonpea [*Cajanus cajan* (L.) Millspaugh] Germplasm Accessions

Esnart Nyirenda Yohane *, Hussein Shimelis, Mark Laing, Isack Mathew and Admire Shayanowako

African Centre for Crop Improvement, School of Agricultural, Earth and Environmental Sciences, University of KwaZulu-Natal, Private Bag X01, Scottsville 3201, Pietermaritzburg 3209, South Africa; Shimelish@ukzn.ac.za (H.S.); laing@ukzn.ac.za (M.L.); isackmathew@gmail.com (I.M.); Shayanowako@googlemail.com (A.S.)

* Correspondence: nyirendaesnart@yahoo.com

Abstract: Pigeonpea (*Cajanus cajan* (L.) Millspaugh) is an important source of grain protein for low-income countries such as Malawi. Knowledge of the genetic diversity in pigeonpea is essential for an effective breeding program. The study objective was to assess the genetic diversity among diverse pigeonpea accessions to select complementary and unique genotypes for breeding. Eighty-one pigeonpea accessions were evaluated in six environments in Malawi using a 9 × 9 alpha-lattice design with two replications. The cross-tabulation analysis revealed a significant genotype variation on plant growth, flower, and seed traits. The combined analysis of variance identified genotypes MWPLR 14, ICEAP 01170, ICEAP 871091, and ICEAP 01285 as early maturing varieties, while Kachangu, MWPLR 16, TZA 5582, No. 40, and MWPLR 14 were identified as high-yielding genotypes. The correlation analysis revealed a significant positive correlation between grain yield and a hundred seed weight (HSWT) (r = 0.50, $p < 0.01$), suggesting the usefulness of this trait for selection. The nonlinear principal component analysis identified grain yield (GDY), days to 50% flowering (DTF), days to 75% maturity (DTM), number of pods per plant (NPP), number of racemes per plant (NRP), 100 seed weight (HSWT), leaf hairiness (LH), and number of seeds per pod (NSP) as the most discriminated traits among the test genotypes. The cluster analysis using morphological traits delineated the accessions into three clusters. The selected high-yielding and early-maturing genotypes may be recommended as parental lines for breeding and grain yield improvement in Malawi or similar agro-ecologies.

Keywords: agronomic performance; correlation analysis; malawi; pigeonpea; yield stability

---

1. Introduction

Pigeonpea (*Cajanus cajan* (L.) Millspaugh, 2n = 2x = 22) is an essential cash and food crop in the tropical and subtropical regions of the world. It is a multi-purpose crop that is cultivated mainly for its edible grains that are high in dietary protein and essential amino acids such as leucine (16.48 g/kg), tyrosine (14.77 g/kg), and arginine (13.51 g/kg) [1].

Pigeonpea is an essential component of the agriculture systems in semi-arid ecologies due to its adaptation to growing with relatively low rainfall and with poor soil fertility. It has a deep root system and a unique ability to maintain optimal osmotic adjustment under limited water conditions [2]. Pigeonpea can fix atmospheric nitrogen in the soils through symbiosis with species of *Rhizobium* bacteria depositing up to 200 kg of nitrogen per hectare in agricultural lands [3,4]. Thus, pigeonpea has important roles in enhancing food security and livelihoods, especially during drought years, and providing ecosystem services through nitrogen fixation and soil health improvement.

Pigeonpea accounts for 5% of the world's pulse production [5]. India is the largest producer of pigeonpea, accounting for 25% of the world's production, followed by Myanmar and Malawi [6]. In Malawi, pigeonpea accounts for more than 22% of total legume production and ranks as the 3rd most important legume crop after groundnut and common beans. The grain productivity of pigeonpea in Malawi is low ($\approx$700 kg ha^{-1}) compared to its potential yield of 2500 kg ha^{-1} [7]. The yield gap is due to various constraints, including insect pests and diseases, drought stress, and a lack of improved cultivars. The breeding and deployment of improved cultivars can enhance pigeonpea production and productivity. The successful development of improved cultivars with the client and market-preferred traits depends on the availability of adequate genetic variation.

Reportedly, modern pigeonpea cultivars and varieties exhibit relatively low levels of genetic diversity [8]. The loss of genetic diversity is due to continuous artificial selection and breeding for a few targeted economic traits to meet the market requirements [9]. Hence, there is a need to initiate pre-breeding programs in the target production environments through divergence breeding involving modern and obsolete cultivars, landraces, and wild relatives that possess desirable traits. This will broaden the genetic diversity of pigeonpea through gene recombination and effective selection [9]. The International Crops Research Institute for the Semi-Arid Tropics (ICRISAT) and various national and regional improvement programs are actively involved in genetic improvement and conservation of the pigeonpea. Diverse pigeonpea collections are preserved globally, including by ICRISAT, the International Institute of Tropical Agriculture (IITA), and the Svalbard Global Seed Vault in Norway. These genetic resources can be used for pigeonpea improvement and breeding programs globally [10].

To date, only seven pigeonpea cultivars have been released in Malawi. These cultivars were introductions from ICRISAT [7] developed in Kenya with germplasm from eastern and southern Africa (ESA). The ESA region is recognized as a secondary center of genetic diversity for pigeonpea. The introduced cultivars are poorly adapted to local farming conditions in Malawi and lack farmer-preferred traits such as good cooking quality, resistance to pod borers, and high yield potential. Therefore, the development of high performance, locally adapted pigeonpea cultivars is an important target in Malawi. This requires a range of genetic resources and crosses to integrate adaptive and functional traits, according to the needs and preferences of farmers and the value chain. Introduced germplasm can provide useful genetic resources that can be introgressed into locally adapted germplasm to improve economic traits such as high yield, early maturity, and pest and disease resistance, among others [9]. Evaluating accessions maintained by the public and private breeding sectors within the ESA region provides an opportunity to identify stable and high-yielding genotypes for selection.

Many pigeonpea genotypes have been collected and maintained at the Department of Agricultural Services in Malawi for breeding purposes. The genotypes are adapted to the ESA region and possess valuable attributes including good cooking quality, insect pests, and disease resistance, but they are limited by their poor yield performance. The key traits present in the local and introduced germplasm should be assessed for pre-breeding and breeding purposes. Hence, the objectives of the study were to determine the genetic diversity among pigeonpea accessions in selected target production environments in Malawi to select complementary and unique genotypes for breeding.

2. Materials and Methods

2.1. Plant Materials

The study evaluated 81 pigeonpea genotypes comprising 28 landraces, 6 released cultivars, and 47 advanced elite lines (Table 1), which were sourced from the Department of Agricultural Research Services (DARS)/Malawi and the Tanzania Agriculture Research Institute (TARI) and the International Crop Research Institute for the Semi-Arid Tropics (ICRISAT)/Kenya. The landraces were included as checks for adaptation to local conditions and possessing farmers' traits, while the elite lines provide important genetic resources, since Tanzania and Kenya have more advanced pigeonpea-breeding programs. The released cultivars provided a benchmark against commercial standards that are currently in production.

Table 1. Description of the pigeonpea genotypes used in the study.

Code	Genotype Designation	Description	Source	Origin	Code	Genotype Designation	Description	Source	Origin
G1	ICEAP 0673/1	Breeding line	ICRISAT	Kenya	G42	ICEAP 87105	Cultivar	ICRISAT	Kenya
G2	ICEAP 00554	Breeding line	ICRISAT	Kenya	G43	MWPLR 16	Landrace	GENEBANK	Malawi
G3	ICEAP 01164/1	Breeding line	ICRISAT	Kenya	G44	TZA 2496	Landrace	TARI	Tanzania
G4	MWPLR 19	Landrace	GENEBANK	Malawi	G45	TZA 5582	Landrace	TARI	Tanzania
G5	MWPLR 22	Landrace	GENEBANK	Malawi	G46	TZA 5596	Landrace	TARI	Tanzania
G6	ICEAP 01170	Breeding line	ICRISAT	Kenya	G47	Chitedze Pigeonpea 2	Cultivar	DARS	Malawi
G7	ICEAP 01169	Breeding line	ICRISAT	Tanzania	G48	MWPLR 7	Landrace	GENEBANK	Malawi
G8	TZA 2439	Landrace	TARI	Tanzania	G49	Babati	Landrace	TARI	Tanzania
G9	MWPLR 9	Landrace	GENEBANK	Malawi	G50	TZA 5557	Landrace	TARI	Tanzania
G10	MWPLR 6	Landrace	GENEBANK	Malawi	G51	MWPLR 14	Landrace	ICRISAT	Kenya
G11	MWPLR 17	Landrace	GENEBANK	Malawi	G52	ICEAP 01101/1	Breeding line	ICRISAT	Kenya
G12	TZA 253	Landrace	TARI	Tanzania	G53	TZA 2456	Landrace	TARI	Tanzania
G13	MWPLR 1	Landrace	GENEBANK	Malawi	G54	TZA 5464	Landrace	TARI	Tanzania
G14	MWPLR 18	Landrace	GENEBANK	Malawi	G55	ICEAP 01101/2	Breeding line	ICRISAT	Kenya
G15	TZA 2464	Landrace	TARI	Tanzania	G56	ICEAP 01285	Breeding line	ICRISAT	Kenya
G16	ICEAP 00604	Breeding line	ICRISAT	Kenya	G57	MWPLR 25	Landrace	GENEBANK	Malawi
G17	TZA 2509	Landrace	GENEBANK	Malawi	G58	ICEAP 87091	Breeding line	ICRISAT	Kenya
G18	ICEAP 01146/1	Breeding line	ICRISAT	Kenya	G59	TZA 2692	Landrace	TARI	Tanzania
G19	MWPLR 11	Landrace	GENEBANK	Malawi	G60	TZA 2807	Landrace	TARI	Tanzania
G20	TZA 5555	Landrace	TARI	Tanzania	G61	ICEAP 00068	Breeding line	ICRISAT	Kenya
G21	No. 40	Landrace	TARI	Tanzania	G62	TZA 2785	Landrace	TARI	Tanzania
G22	ICEAP 01150	Breeding line	ICRISAT	Kenya	G63	MWPLR 10	Landrace	GENEBANK	Malawi
G23	MZ2/9	Breeding line	TARI	Tanzania	G64	ICEAP 00612	Breeding line	ICRISAT	Kenya
G24	ICEAP 01172/1	Breeding line	ICRISAT	Kenya	G65	MWPLR 21	Landrace	GENEBANK	Malawi
G25	ICEAP 01103/1	Breeding line	ICRISAT	Kenya	G66	TZA 2514	Landrace	TARI	Tanzania
G26	MWPLR 24	Landrace	GENEBANK	Malawi	G67	TZA 2466	Landrace	TARI	Tanzania
G27	ICEAP 01155	Breeding line	ICRISAT	Kenya	G68	ICEAP 01179	Breeding line	ICRISAT	Kenya
G28	ICEAP 01180/2	Breeding line	ICRISAT	Malawi	G69	MWPLR 13	Landrace	GENEBANK	Malawi
G29	MWPLR 4	Landrace	GENEBANK	Malawi	G70	MWPLR 2	Landrace	GENEBANK	Malawi
G30	Kachangu	Cultivar	DARS	Malawi	G71	TZA 250	Landrace	DARS	Malawi
G31	Mwayiwathualimi	Cultivar	DARS	Kenya	G72	MWPLR 3	Landrace	GENEBANK	Malawi
G32	MWPLR 8	Landrace	ICRISAT	Malawi	G73	TZA 5541	Landrace	TARI	Tanzania
G33	ICEAP 01154/2	Breeding line	ICRISAT	Kenya	G74	MWPLR 23	Landrace	GENEBANK	Malawi
G34	Chitedze Pigeonpea 1	Cultivar	DARS	Malawi	G75	ICEAP 00979/1	Breeding line	ICRISAT	Kenya
G35	ICEAP 01164	Breeding line	ICRISAT	Kenya	G76	TZA 197	Landrace	TARI	Tanzania
G36	Bangili	Landrace	TARI	Tanzania	G77	MWPLR 20	Landrace	GENEBANK	Malawi
G37	ICEAP 00053	Breeding line	ICRISAT	Kenya	G78	HOMBOLO	Landrace	TARI	Tanzania
G38	MWPLR 12	Landrace	GENEBANK	Malawi	G79	ICEAP 86012	Breeding line	ICRISAT	Kenya
G39	TZA5463	Landrace	TARI	Tanzania	G80	ICEAP 01106/1	Breeding line	ICRISAT	Kenya
G40	MWPLR 5	Landrace	GENEBANK	Malawi	G81	Sauma	Cultivar	DARS	Malawi
G41	MWPLR 15	Landrace	GENEBANK	Malawi					

ICRISAT = International Crops Research Institute for the Semi-Arid Tropics, DARS = Department of Agricultural Research Services, TARI = Tanzania Agricultural Research Institute.

2.2. Study Sites

Field experiments were conducted in Malawi at three sites, Bvumbwe, Chitedze, and Makoka Research Stations, during the 2017/18 and 2018/19 cropping seasons. The geographic location, altitude, weather, and soil characteristics of the study locations are presented in Table 2. Each season and site combination presented unique environmental conditions due to variations in temperature, rainfall, and agronomic practices. Therefore, due to site × season combinations, six environments were identified for evaluating the genotypes: Bvumbwe during 2017/18 (Environment 1), Bvumbwe in 2018/19 (Environment 2), Chitedze in 2017/18 (Environment 3), Chitedze in 2018/19 (Environment 4), Makoka in 2017/18 (Environment 5), and Makoka in 2018/19 (Environment 6).

Table 2. Physical and weather characteristics of the study locations.

Site	Latitude	Longitude	Altitude (Masl)	Soil Texture	Rainfall (mm)		Min Temp (°C)		Max Temp (°C)	
					2017/18	2018/19	2017/18	2018/19	2017/18	2018/19
Bvumbwe	15°55′ S	35°04′ E	1228	Sandy clay loam	975.2	1442	16.2	17.9	22.6	24.9
Chitedze	13°59′ S	33°38′ E	1146	Sandy clay	929.8	693.4	18.5	20.2	24.7	29.4
Makoka	15°32′ S	35°11′ E	1029	Sandy clay loam	566.6	1184.8	16.3	15.6	23.2	28.2

Masl = meters above sea level, mm = millimeters, min = minimum, max = maximum, temp = temperature, °C = degrees Celsius.

2.3. Experimental Design and Data Collection

The experiment at each site was laid out in an alpha-lattice design with two replications. Each genotype was planted on a plot consisting of two rows. The rows were 5 m in length and 0.90 m apart, giving a plot size of 4.5 m^2. Seeds were planted at 0.75 m apart within a row. Three seeds were planted per planting station and thinned to one plant two weeks after emergence. All agronomic practices were applied following standard practices for pigeonpea production in Malawi [7]. Both qualitative and quantitative phenotypic traits' data were collected as presented in Table 3 according to pigeonpea descriptors of the International Board for Plant Genetic Resource (IBPGR) and International Centre for Research Institute for Semi-Arid Tropics (ICRISAT) [11].

2.4. Statistical Analysis

Data collected on qualitative traits (Table 3) were subjected to frequency distribution and cross-tabulation analyses using SPSS for Windows 25.0 [12].

The quantitative data from each variable were tested for homogeneity of variances using Bartlett's test and data normality using the Shapiro–Wilkes test before the analysis of variance (ANOVA). Subsequently, the data were pooled across sites and subjected to a combined analysis of variance following the alpha lattice procedure in Genstat 18th edition [13]. The total variance was partitioned into genotype (σ^2g), environment (σ^2e), and genotype by environment (σ^2ge) components based on the mean squares derived from the partial analysis of variance adapted from [14]. Correlation and principal component analyses were performed using Genstat 18th edition [13] to determine influential components and trait relationships. A nonlinear principal component analysis was conducted in SPSS (SPSS 2016). The nominal variables (qualitative traits) were transformed using the categorical principal component analysis (CATPCA) procedure described by [15]. The nonlinear PCA can standardize both quantitative and qualitative data to deduce their associations and identified the most important components.

Table 3. Descriptors for the pigeonpea qualitative and quantitative traits.

Traits	Code	Description
		Qualitative Traits
Plant habit	PH	1 = Compact (erect), 2 = semi-spreading (semi-erect) or 3 = spreading
Flower streak pattern	FSP	0 = no streaks, 1 = sparse, 2 = medium and 3 = dense streaks, 4 = uniform coverage of second color
Flower base/main color	FBC	1 = ivory (green white), 2 = light yellow, 3 = yellow, 4 = orange, 5 = red, 6 = purple
Leaf shape	LS	1 = ovate, 2 = triangular, 3 = trullate
Leaf hairiness	LH	1 = hairy, 2 = non-hairy
Pod form	PF	1 = flat, 2 = cylindrical
Pod color	PC	1 = green, 2 = purple, 3 = mixed (green +purple) and 4 = dark purple
Seed color pattern	SCP	1 = plain, 2 = mottled, 3 = speckled, 4 = mottled and speckled, 5 = ringed
Seed main color	SMC	1 = white (yellow white), 2 = cream (gray white), 3 = orange, 4 = brown, 5 = grey, 6 = purple, 7 = black
Seed eye color	SEC	1 = purple, 2 = light brown, 3 = reddish brown, 4 = gray/dark, 5 = cream/white
Seed shape	SSH	1 = Oval, 2 = pea-shape, 3 = square/angular, 4 = elongate
		Quantitative Traits
Plant height	PH	Measured in cm from plant base to the tip of the main stem
Days to 50% flowering	DTF	Number of days from sowing until when 50% of the plants have at least one open flower
Primary branches	PBR	The average number of primary branches of 10 randomly selected and tagged plants
Secondary branches	NSB	The average number of secondary branches of 10 randomly selected and tagged plants
Days to 75% maturity	DTM	Number of days from sowing until when 75% of the pods in a plot turn brown
Number of seeds per pod	NSP	The average number of pods per plant from 10 randomly selected and tagged pods
Number of pods per plant	NPP	The average number of pods from 10 randomly selected and tagged plants
Number of racemes per plant	NRP	The average number of racemes from 10 randomly selected and tagged plants
Grain yield (t/ha)	GYD	Weight of the grain harvested in a plot extrapolated to t/ha
100 seed weight (g)	HSWT	Weight of a random sample of 100 grain

3. Results

3.1. Genotype Variation Based on Qualitative Traits

Significant variations were exhibited among genotypes for all assessed qualitative traits ($p < 0.001$) such as growth habit, flower main color, flower streak pattern, pod color, and seed traits (Table 4, Figure 1A–D). A large proportion of test genotypes (61.9%) were semi-spreading, followed by spreading (26.6%) and compact (11.5%) in growth habits. A majority of the test genotypes (64.9%) had yellow flower color (Table 4, Figure 1A), while 16.8% had purple flowers, 13.6% had ivory flowers, and 7.4%

had light yellow flowers (Table 4, Figure 1A). A large population of the genotypes (60.5%) had no flower streaks, and the rest of the genotypes had sparse, medium, dense, and uniform coverage streaks at 8.1%, 1.9%, 14.5%, and 15%, respectively (Table 4, Figure 1B). About 48.7% of the genotypes had a green pod color, while 33.9% had a mixed pod color and 7.1% had purple pods (Table 4, Figure 1C). A majority of the genotypes (76.8%) had a cream seed coat color, while 11% had a brown seed coat color and the rest had gray, orange, and purple seed coat colors (Table 4, Figure 1D). About 70.2% of the test genotypes had a brown seed eye, and 20.7% had a purple seed eye, while the remainder had gray or cream seed eyes. The most common seed shape was square or angular shapes, which were exhibited by 69.3% of the test genotypes.

Figure 1. Genetic variability for some qualitative traits in pigeonpea genotypes: (**A**) flower color: genotype Sauma (ivory), ICEAP 87105 (purple), TZA 5582 (yellow), Mwaiwathualimi (light yellow); (**B**) flower streak pattern: genotype MWPLR 14 (no streak), MWPLR 23 (medium streaks), ICEAP 00068 (dense streaks), MWPLR 16 (uniform coverage); (**C**) pod color: genotype MWPLR 16 (purple), ICEAP 01106/1 (green), ICEAP 01103/1 (mixed), MWPRL 22 (dark purple) and (**D**) seed coat color: genotype MWPLR 19 (orange), ICEAP 00612 (brown), No. 40 (cream), and TZA 5463 (purple).

3.2. Genotype and Environment Variances for Quantitative Traits

The quantitative agronomic data were pooled across sites after applying tests for homogeneity of variance and normality. The genotype × environment interaction effects were significant ($p < 0.001$) for grain yield (GYD), days to 50% flowering (DTF), days to 75% maturity (DTM), plant height (PH), number of primary branches (NPB), number of pods per plant (NPP), number of racemes per plant (NRP), 100 seed weight (HSWT), and number of secondary branches per plant (NSB) (Table 5). The genotype and environment had significant ($p < 0.001$) effects on all assessed traits except the NSP. The site × season × type interaction effects were significant ($p \leq 0.001$ and $p = 0.05$, respectively) for DTM, PH, and NSP (Table 6). The site × type interaction effects were only significant ($p = 0.01$ and $p = 0.05$) for DTM, NPP, NRP, and HSWT, respectively. However, season × type interaction effects were not significant for all the variables except for GYD, which was highly significant.

Table 4. Frequency distribution and significance tests among 81 pigeonpea genotypes assessed based on qualitative traits.

Trait	Description	Frequency (%)	DF	Chi-Square	Genotype Code [a]
Growth habit	Compact	11.5	160	304.52 **	G53, G2, G1, G27, G26
	Semi-spreading	61.9			G63, G50, G28, G70, G76, G80, G51, G78, G49, G32, G62, G39, G67, G5, G8, G13, G72, G24, G74, G3,32, G22, G4, G40, G30, G52, G56, G48, G79, G36, G23,G16, G77, G7, G71, G44, G67, G46, G69, G33, G54, G20, G43, G42, G71, G62, G65,G39, G69, G17, G18, G59
	Spreading	26.6			G45, G41, G29, G49, G56, G64, G37, G60, G15, G11, G65, G75, G81, G44, G67, G11, G46
Flower color	Ivory	13.6	240	910.08 ***	G78, G40, G36, G27, G33, G80, G51
	Light yellow	7.4			G13, G5, G31
	Yellow	64.9			G50, G45, G70, G53, G76, G72, G24, G74, G3, G22, G4, G58, G68, G18, G19, G17, G9, G62, G29, G32, G65, G21, G52, G1, G56, G37, G48, G79. G23, G16, G61, G77, G7, G71, G44, G15, G67, G11, G69, G65, G75, G20, G43, G26, G71, G44, G15, G67, G62, G11, G46, G65
	Purple	16.8			G63, G28, G41, G56, G60, G25, G46, G54, G26, G42
Flower streak pattern	No streaks	60.5	320	589.69 ***	G17, G53, G36, G12, G15, G37, G20, G60, G9, G54, G11, G66, G55, G80, G81, G71, G73, G23, G1, G65, G21, G18, G7, G13, G51, G62, G48, G49, G58, G14, G32, G16, G2, G27, G22, G6, G57, G10, G31, G8, G39, G30
	Sparse streaks	8.1			G49, G69, G42, G33, G28, G5, G70
	Medium sparse	1.9			G72, G74
	Dense streaks	14.5			G47, G61, G29, G60, G34, G40, G45, G67, G45, G68, G63, G77, G19
	Uniform coverage	15			G79, G50, G76, G59, G25, G46, G78, G38, G51, G75, G26, G35, G52, G56, G41, G43
Pod color	Green	48.7	240	647.43 ***	G73, G42, G1, G24, G74, G75, G52, G16, G65, G21, G18, G7, G13, G62, G17, G47, G61, G15, G20, G29, G44, G72, G60, G64, G9, G11, G66, G55, G80, G71, G58, G14, G27, G6, G57, G10, G8, G19
	Purple	7.1			G76, G45, G67, G38
	Mixed (green + purple)	33.9			G81, G70, G53, G36, G61, G43, G37, G34, G54, G79, G50, G40, G25, G33, G46, G42, G51, G4, G68, G26, G49, G3, G35, G32, G69, G2, G63, G22, G56, G77, G41, G30
	Dark purple	10.3			G31, G28, G39, G48, G59, G43

Table 4. *Cont.*

Trait	Description	Frequency (%)	DF	Chi-Square	Genotype Code [a]
Seed color pattern	Plain	56.6	240	841.57 ***	G59, G80, G5, G18, G6, G53, G65, G62, G35, G34, G67, GG4, G60, G66, G21, G70, G36, G42, G40, G14, G50, G66, G20, G79, G49, G2, G3, G69, G56, G81, G47, G72, G15, G44
	Mottled	15.3			G41, G25, G34, G48, G28, G78, G23, G31, G9, G37, G57
	Speckled	22.2			G75, G68, G43, G38, G10, G19, G52, G58, G51, G73, G59, G76, G16, G29, G13, GG3, G17, G8, G54, G1, G24, G7, G71, G27, G12, G22, G55, G77
	Mottled + speckled	5.9			G46, G33, G30, 632, G39, G45, G26
Seed main color	Cream	76.8	320	1049.31 ***	G75, G68, G59, G43, G5, G18, G6, G38, G10, G53, G65, G63, G35, G19, G34, 52, G72, G15, G44, G22, G55, G57, G77, G60, G58, G78, G32, G73, G51, G70, G36, G16, G29, G42, G40, G23, G14, G17, G8, G50, G66, G20, G49, G54, G2, G3, G69, G1, G24, G45, G7, G9, G71, G81, G12, G47
	Orange	3			G4, G46, G25
	Brown	11			G64, G76, G63, G30, G34, G48, G28, G31, G37, G26
	Gray	6.2			G80, G66, G67, G56
	Purple	3			G39, G33, G41
Seed shape	Oval	30.7	80	480.21 ***	G75, G22, G5, G25, G38, G53, G35, G34, G28, G73, G51, G70, G36, G29, G42, G40, G31, G8, G18, G49, G3, G45, G37, G28, G27, G12, G55, G57
	Square/angular	69.3			G15, G44, G22, G77, G68, G59, G43, G46, G80, G18, G33, G30, G41, G6, G10, G65, G62, G19, G34, G67, G4, G52, G48, G60, G58, G66, G32, G64, G76, G21, G16, G13, G23, G14, G63, G17, G39, G52, G66, G79, G54, G2, G69, G1, G24, G56, G7, G9, G71, G81
Seed eye color	Purple	20.7	240	848.32 ***	G68, G5, G34, G25, G60, G78, G51, G64, G76, G21, G16, G29, G42, G40, G31, G50, G49, G2, G69, G24, G81, G55, G57
	Light brown	70.2			G75, G59, G43, G46, G18, G33, G30, G41, G6, G10, G53, G65, G62, G35, G19, G34, G67, G52, G48, G58, G28, G66, G32, G73, G36, G23, G14, G17, G39, G74, G20, G79, G54, G1, G46, G45, G9, G71, G37, G27, G12, G47, G15, G44, G22
	Gray/dark	1.2			G25
	Cream	7.5			G80, G38, G63, G8, G7, G26

DF = degrees of freedom, ** and *** = significance at 0.01 and 0.001 levels, respectively; [a] see genotype codes (G1–G81) in Table 1.

Table 5. Mean squares and significant tests for grain yield and yield components measured in 81 pigeonpea genotypes across six environments in Malawi.

Source of Variation	DF	DTF	DTM	PH	NPB	NSB	NRP	NPP	NSP	GYD	HSWT
Location	2	9024.2 ***	8735.4 ***	54,965 ***	114.4 ***	93.7 *	226.9 ***	3236 **	22.5 ***	5,968,860 ***	1008.1 ***
Replication (Rep)	1	701.9 ns	289 ns	118 ns	1.2 ns	105.4 *	14,646 ns	9810 *	0.45 ns	1,663,232 *	9.5 ns
Block (Rep)	8	3168.5 ***	5703.4 ***	7710.9 ns	52.9 *	93.7 *	9099 *	6433.6 **	2.4 *	16,534,356.5 ***	72.2 **
Genotype (G)	80	879.2 ***	1234.9 ***	2137 ***	12.5 *	30.9 *	5004.9 *	1990.3 *	0.8 ns	351,745.3 *	16.8 *
Season (S)	1	3370.5 **	2945.3 *	447 ns	409.6 ***	650.1 ***	2,023,492 ***	437.5 ***	31.5 ***	30,308,789 ***	50.2 *
G × L	160	243 *	361.9 *	1106 *	18 *	35.6 *	6150.9 *	1916.1 *	0.9 *	360,816.9 *	20.7 **
G × S	80	3610.3 ns	606.9 ns	1198 ns	17.9 *	34.7 *	4642.7 ns	1060.3 *	0.9 ns	400,468.2 *	14.9 ns
G × L × S	160	330.6 ns	484.9 ns	744 ns	15.2*	34.5 *	6110.9 ns	1502.8 *	0.7 ns	919,105.3 ns	16.2 ns
Residual	469	345.4	585.8	1243.1	14.5	11.8	5822.9	5667.2	0.8	313,554	15.4

DF = degrees of freedom, Rep = replication, DTF = days to 50% flowering, DTM = days to 75% maturity, PH = plant height, NPB = number of primary branches, NSB = number of secondary branches per plant, NRP = number of racemes per plant, NPP = number of pods per plant, NSP = number of seeds per pod, GYD = grain yield, HSWT = 100 seed weight, *, ** and *** = significance at 0.05, 0.01 and 0.001 probability levels, respectively.

Table 6. Mean squares and significant tests for grain yield and yield components among the three categories of pigeonpea genotypes.

Source of Variation	DF	DTF	DTM	PH	NPB	NSB	NRP	NPP	NSP	GYD	HSWT
Site	2	9167 ***	8020 ***	55,114 ***	108.6 ***	93 *	1,309,332 ***	118,174 ***	22.54 ***	1.658 ***	80.16 ***
Rep	1	717	328	111	2.7	109	14184	9836	0.45	1.651	0.26
Rep (Block)	16	123	82	9769	33.3	71	14908	21,509 ***	1.35	0.206	11.78
Season	1	3797 **	3625 *	407	433.3 ***	4672 ***	2,043,617 ***	440,237 ***	31.51 ***	3.092 ***	0.04
Type	2	1629 *	4725 **	44,433 **	18.5	14	4686	5891 *	40.2 ***	2.087 ***	20.09
Site*Season	2	2523 **	700	55,081 ***	910.5 ***	253 ***	1,018,464 ***	149,039 ***	36.09 ***	6.38 ***	39.42 *
Site*Type	4	114	4385 **	1257	17	29	988,914 *	8167 ***	0.41	0.078	45.83 *
Season*Type	2	161	388	3023	9.9	2	642	380	1.94	2.006 ***	6.44
Site*Season*Type	4	676	7883 ***	65,810 ***	3.1	44	1771	2032	30.17 *	0.2	1.82
Residual	937	386	619	1177	15.5	31	5823	1682	0.79	0.149	12.45

DF = degrees of freedom, Rep = replication, DTF = days to 50% flowering, DTM = days to 75% maturity, PH = plant height, NPB = number of primary branches, NSB = number of secondary branches per plant, NRP = number of racemes per plant, NPP = number of pods per plant, GYD = grain yield, HSWT = 100 seed weight, *, ** and *** = significance at 0.05, 0.01 and 0.001 probability level.

3.3. *Mean Performance of Pigeonpea Genotypes across the Test Environments*

Tables 7–10 summarize the mean values and statistics for eight quantitative traits recorded from three locations in two seasons. The tables presents the best ten, and the bottom five genotypes on DTF, DTM, PH, NPB, NRP, NPP and HSWT ranked on grain yield response. The mean DTF and DTM were 112 and 157 days, respectively (Table 7). Genotype MWPLR 14 was the earliest to attain 50% flowering and maturity at 74 and 113 days, which was followed by ICEAP 01170 at 85 and 125 days, ICEAP 87091 at 85 and 132 days, ICEAP 01285 at 87 and 133 days, and ICEAP 01169 at 91 and 137 days, respectively. Sauma was among the latest genotypes to flower and mature at 145 and 205 days, respectively. There were marked genotype differences in plant height that varied from 125.3 to 202.4 cm (Table 8). The mean plant height of the test genotypes was 167.5 cm. The shortest genotype across the testing environments was ICEAP 87105. The tallest genotypes with plant heights exceeding 180 cm were Kachangu, No. 40, ICEAP 01106/3, ICEAP 00068, TZA 5596, MWPLR 6, Sauma, and ICEAP 00053. The mean number of the primary branch of the test genotypes was 15 (Table 8). The most productive genotypes with many primary branches per plant were MWPLR 12, MWPLR 20, ICEAP 01170, and MWPLR 23, with 19, 18, 17, and 17 primary branches per plant, in that order. The mean number of pods per plant varied from 67 to 144, with a grand mean of 94 pods per plant (Table 9). The highest number of pods per plant was 144, 134, 126, 124, and 123 observed on the genotypes Kachangu, MWPLR 16, TZA 5582, No. 40, and MWPLR 14, in that order. The number of seeds per pod exhibited non-significant differences among the assessed genotypes. The mean number of grains per pod was five. There was a wide genetic variation for grain yield ranging from 0.5 to 1.8 t ha^{-1} with a mean of 1.1 t ha^{-1} (Table 10). Accessions No. 40, MWPLR 14, and MWPLR 16 were the three best performing genotypes with mean yields of 1.8, 1.7, and 1.7 t ha^{-1}, respectively. The lowest grain yield response was 0.5 t/ha recorded for the genotypes ICEAP 00604 and ICEAP 01285. The 100 seed weight ranged from 11.0 to 17.3 g/100 seed (Table 10). Accessions MWPLR 22, TZA 5582, and MWPLR 14 expressed the highest HSWT ≥ 17 g/100 seed.

Table 7. Mean values for 10 quantitative traits among the ten top best and five bottom performing genotypes after evaluating 81 genotypes in six environments in Malawi.

Genotype	DTF							DTM						
	Y1			YII			Mean	YI			YII			Mean
	S1	S2	S3	S1	S2	S3		S1	S2	S3	S1	S2	S3	
Top Ten Genotypes														
21	129	131	141	124	131	132	131	173	191	211	158	176	176	181
43	125	105	119	117	105	105	113	177	166	172	156	161	154	164
51	63	65	64	87	67	98	74	95	105	102	127	116	132	113
30	100	97	118	128	116	118	113	133	150	164	159	159	164	155
45	107	96	91	128	101	124	108	143	158	146	170	153	165	156
81	163	127	155	132	165	130	145	215	201	254	171	211	178	205
17	147	120	125	109	120	106	121	182	167	174	156	160	147	164
66	120	95	115	116	108	116	111	155	151	170	154	158	161	158
74	118	78	123	113	115	118	110	163	145	166	153	165	163	159
20	116	120	129	122	120	127	122	143	163	175	156	160	172	161
Bottom Five Genotypes														
39	113	90	131	85	90	88	99	149	144	195	127	150	122	147
13	126	117	109	116	107	115	115	167	166	153	145	154	155	156
50	117	77	107	116	77	115	101	141	136	156	155	137	149	145
42	114	102	127	120	102	120	114	145	154	172	164	166	162	160
79	124	101	122	117	127	119	118	168	153	165	152	179	161	163
Mean	117.8	102.8	115.5	110.6	106.1	113.1	110.6	154.7	156.5	163.2	148.7	155.7	154.3	155.3
STD	17.9	18.2	15.1	13.0	16.9	12.3	10.5	22.0	22.0	21.1	13.7	18.4	14.9	11.9
SED±	2.0	2.0	1.7	1.4	1.9	1.4	1.2	2.4	2.4	2.3	1.5	2.0	1.7	1.3
CV (%)	15.2	17.7	13.1	11.8	15.9	10.8	9.5	14.2	14.0	12.9	9.2	11.8	9.6	7.7

STD = standard deviation, SED = standard error of difference, CV = coefficient of variation, S1 = site 1 (Bvumbwe), S2 = site 2 (Chitedze), S3 = site 3 (Makoka), Y1 = year 1 (2017/18), Y11 = year 2 (2018/19), DTF = days to flowering, DTM = days to 75% maturity, See genotype codes (G1–G81) in Table 1.

Table 8. Mean values for plant height and number of primary branches among the ten top best and five bottom performing genotypes after evaluating 81 genotypes in six environments in Malawi.

Genotype	PH							NPB						
	Y1			Y11			Mean	Y1			Y11			Mean
	S1	S2	S3	S1	S2	S3		S1	S2	S3	S1	S2	S3	
Top Ten Genotypes														
21	166.5	220.0	193.0	160.0	212.8	193.0	190.9	19	19	17	14	18	12	16
43	113.5	147.5	127.5	96.5	146.7	148.0	163.7	14	15	17	14	17	11	15
51	151.5	109.0	158.0	234.5	209.4	149.0	168.6	13	12	14	18	13	11	13
30	229.5	188.5	204.0	170.0	218.5	204.0	202.4	15	13	18	15	16	15	15
45	139.5	144.5	173.0	161.5	169.4	197.5	164.2	15	13	22	15	17	14	16
81	163.0	222.0	191.0	160.5	168.1	194.5	183.2	13	17	19	18	12	14	15
17	163.5	164.0	163.5	100.0	152.1	156.0	149.9	15	14	21	17	16	13	16
66	181.5	177.5	164.0	161.5	156.8	149.5	165.1	12	13	13	14	16	12	13
74	156.0	195.0	185.5	124.5	178.7	164.0	167.3	15	18	17	20	18	12	17
20	152.5	163.0	168.5	138.5	247.5	166.5	172.8	10	12	20	12	18	11	14
Bottom Five Genotypes														
39	203	154.5	174	157.5	200	151.5	173.4	16	18	17	15	12	12	15
13	169	171.5	134	134.5	203.3	156.5	161.5	18	12	18	15	10	15	14
50	119	101.5	149.5	130.5	218.5	166.5	147.6	18	13	14	15	17	13	15
42	140	153	175.5	104.5	207.7	120	125.3	14	9	16	14	13	13	13
79	174	165.5	167.5	120.5	201.4	148	162.8	11	18	23	13	14	13	15
Mean	168.0	166.7	166.2	143.4	195.5	166.1	167.3	14.6	13.6	18.0	14.9	14.6	12.8	14.5
STD	23.9	34.5	22.1	23.0	27.0	23.1	12.6	2.7	4.4	2.7	2.4	3.2	2.0	1.3
SED±	2.7	3.8	2.5	2.6	3.0	2.6	1.4	0.3	0.5	0.3	0.3	0.4	0.2	0.1
CV (%)	14.2	20.7	13.3	16.0	13.8	13.9	7.5	18.7	32.1	15.0	16.3	22.0	15.6	9.1

STD = standard deviation, SED = standard error of difference, CV = coefficient of variation, S1 = site 1 (Bvumbwe), S2 = site 2 (Chitedze), S3 = site 3 (Makoka), Y1 = year 1 (2017/18), Y11 = year 2 (2018/19), PH = plant height (cm), NPB = number of primary branches, see genotype codes (G1–G81) in Table 1.

Table 9. Mean values for number of racemes and number of pods per plant among the ten top best and five bottom performing genotypes after evaluating 81 genotypes in six environments in Malawi.

Genotype	NRP							NPP						
	Y1			Y11			Mean	Y1			Y11			Mean
	S1	S2	S3	S1	S2	S3		S1	S2	S3	S1	S2	S3	
Top Ten Genotypes														
21	214	402	71	130	61	47	154	157	270	66	61	92	98	124
43	138	173	97	117	95	58	113	119	315	98	72	110	90	134
51	260	155	146	113	80	51	134	167	231	109	65	76	90	123
30	178	430	134	132	73	52	166	127	362	95	97	83	101	144
45	191	647	160	151	88	83	220	96	261	106	81	92	122	126
81	200	536	85	89	69	40	170	140	240	69	61	70	82	110
17	184	258	96	139	94	61	139	102	158	65	35	112	89	93
66	148	168	108	119	76	49	111	69	186	82	26	78	64	84
74	196	414	98	125	84	81	166	128	112	64	46	40	94	81
20	126	259	106	148	130	73	140	115	177	78	38	157	45	101
Bottom Five Genotypes														
39	161	465	103	145	55	60	165	128	125	93	38	61	82	88
13	155	228	80	119	99	52	122	98	195	55	37	60	95	90
50	116	321	199	195	81	46	159	79	78	60	59	96	84	76
42	122	150	87	151	80	62	109	99	78	90	62	67	70	78
79	98	552	70	131	163	54	178	53	226	51	26	165	90	102
Mean	174.1	312.3	99.0	161.6	91.8	58.9	149.4	114.6	148.2	80.0	51.0	80.9	86.7	93.4
STD	43.9	146.5	27.7	39.8	30.0	12.1	26.2	30.5	56.7	22.1	16.1	33.4	19.7	14.1
SED±	4.9	16.3	3.1	4.4	3.3	1.3	2.9	3.4	6.3	2.5	1.8	3.7	2.2	1.6
CV (%)	25.2	46.9	28.0	24.7	32.7	20.6	17.5	26.6	38.2	27.7	31.5	41.3	22.8	15.1

STD = standard deviation, SED = standard error of difference, CV = coefficient of variation, S1 = site 1 (Bvumbwe), S2 = site 2 (Chitedze), S3 = site 3 (Makoka), Y1 = year 1 (2017/18), Y11 = year 2 (2018/19), NRP = number of racemes per plant, NPP = number of pods per plant, see genotype codes (G1–G81) in Table 1.

Table 10. Mean values for grain yield and hundred seed weight among the ten top best and five bottom performing genotypes after evaluating 81 genotypes in six environments in Malawi.

Genotype	GYD							HSWT						
	Y1			Y11			Mean	Y1			Y11			Mean
	S1	S2	S3	S1	S2	S3		S1	S2	S3	S1	S2	S3	
Top Ten Genotypes														
21	2.1	0.9	2.3	2.4	1.3	1.7	1.8	16.0	16.5	10.0	10.5	12.5	15.5	13.5
43	1.7	1.7	1.6	1.8	1.6	1.9	1.7	17.0	14.5	14.0	17.0	22.5	13.0	16.3
51	1.8	1.0	2.1	2.1	1.7	1.7	1.7	16.5	17.5	14.5	18.5	21.5	13.5	17.0
30	2.3	1.6	1.2	1.2	1.4	1.8	1.6	17.5	17.0	15.0	16.0	16.0	12.0	15.6
45	1.5	0.9	1.4	1.5	2.3	1.9	1.6	18.4	19.0	15.5	16.0	16.5	18.0	17.2
81	1.3	0.5	1.5	1.6	2.3	2.3	1.6	19.5	16.0	15.5	19.0	15.0	11.0	16.0
17	1.1	0.5	0.7	1.4	2.5	3.0	1.5	18.5	14.0	11.0	17.5	20.0	15.5	16.1
66	2.4	1.5	1.2	1.2	1.4	1.5	1.5	15.5	15.5	15.0	17.5	17.5	13.5	15.8
74	2.2	1.6	1.1	1.0	1.1	1.8	1.5	14.5	14.5	15.5	16.9	20.0	13.5	15.8
20	1.2	0.9	1.7	1.7	1.7	1.2	1.4	16.0	12.5	15.0	18.5	15.0	14.0	15.2
Bottom Five Genotypes														
39	0.4	0.4	1.1	1.1	1.2	0.9	0.8	15.5	14.5	14.5	16	15	16	15.3
13	0.8	0.2	0.5	1.4	0.5	0.3	0.6	12.5	15	14	15	16.5	16	14.8
50	0.9	0.5	0.4	0.5	0.4	0.7	0.6	13	10.5	17.5	21	19	14.5	15.9
42	0.6	0.4	0.9	0.5	0.4	0.4	0.5	12	12.5	14	19	20	14.5	15.3
79	0.8	0.3	0.4	0.5	0.3	0.5	0.5	13	16.5	14.5	17.5	17.5	14	15.5
Mean	1.1	0.6	1.3	1.3	1.5	1.3	1.2	15.9	13.9	13.5	17.6	12.9	14.2	14.7
STD	0.4	0.3	0.4	0.4	0.5	0.4	0.2	2.4	3.2	2.4	2.3	4.5	2.5	1.3
SED±	0.0	0.0	0.0	0.0	0.1	0.0	0.0	0.3	0.4	0.3	0.3	0.5	0.3	0.1
CV (%)	37.3	43.3	32.8	32.1	31.3	33.5	20.5	15.1	22.9	18.0	13.2	35.1	17.5	8.9

STD = standard deviation, SED = standard error of difference, CV = coefficient of variation, S1 = site 1 (Bvumbwe), S2 = site 2 (Chitedze), S3 = site 3 (Makoka), Y1 = year 1 (2017/18), Y11 = year 2 (2018/19), GYD = grain yield (t ha^{-1}), HSWT = 100 seed weight (g), see genotype codes (G1–G81) in Table 1.

3.4. Correlation Analysis among Phenotypic Traits

Assessed traits exhibited variable degrees of associations with grain yield (Table 11). Grain yield was moderately correlated with HSWT ($r = 0.50, p < 0.01$). A number of secondary traits exhibited variable pairwise correlations. DTF and DTM exhibited the strongest correlation ($r = 0.79, p < 0.01$). There were moderate correlations between DTF and PH ($r = 0.44, p < 0.01$), NPB and NSP ($r = 0.41, p < 0.01$), and DTM and PH ($r = 0.41, p < 0.01$). Relatively, HSWT exhibited weak correlations ($r < 0.30$) with NPB and NPP.

Table 11. Phenotypic correlation coefficients among the ten quantitative traits of 81 pigeonpea genotypes evaluated in six environments.

Trait	DTF	DTM	PH	NPB	NSB	NRP	NPP	NSP	GYD	HSWT
DTF	1	0.787 **	0.442 **	0.069	0.006	0.063	0.121	−0.134	0.232 *	−0.021
DTM		1	0.409 **	0.066	0.037	0.034	0.121	−0.020	0.131	0.023
PH			1	0.057	0.149	0.249 *	0.190	−0.123	0.123	0.021
NPB				1	0.044	0.261 *	0.145	0.406 **	0.174	0.350 **
NSB					1	0.024	0.152	−0.101	0.214	0.090
NRP						1	0.191	0.262 *	0.177	0.124
NPP							1	0.099	0.354 **	0.307 **
NSP								1	0.051	0.173
GYD									1	0.498 **
HSWT										1

**. Correlation is significant at the 0.01 level, *. Correlation is significant at the 0.05 level (2-tailed), DTF = days to 50% flowering, DTM = days to 75% maturity, PH = plant height, NPB = number of pods per plant, NSB = number of secondary branches per plant, NRP = number of racemes per plant, NPP = number of primary branches per plant, GYD = grain yield, HSWT = 100 seed weight.

3.5. Nonlinear Principal Component (PC) and Cluster Analysis

The nonlinear principal component analysis was performed to identify the most discriminative variables among the pigeonpea genotypes. A total of 98% of the variation explained by the qualitative and quantitative traits were explained by the first three principal components (Table 12). In general, traits such as GYD, DTF, DTM, leaf hairiness (LH), leaf shape (LS), and NRP contributed much to the phenotypic variation in the PCs. However, GYD, LH, NPP, HSWT, and NSP were the highest contributors (with contributions of 0.86, 0.63, 0.63, 0.51, and 0.45, respectively) on PC1. The second principal component accounted for 73% of the total variation, with NRP and pod form (PF) being the highest (0.74 and 0.62) positive contributors. Conversely, traits including LS, flowering pattern (FP), and flower main color (FMC) negatively correlated with PC2 exhibiting negative (−0.63, −0.57, and −0.44, respectively) PC scores. DTF and DTM were the positive contributors to the observed phenotypic variation on PC3 with PC loadings of 0.83 and 0.79, respectively.

Table 12. Principal components showing variation and contribution by 24 phenotypic traits among 81 pigeonpea genotypes assessed in six environments in Malawi.

Trait	Dimension		
	1	2	3
FMC	−0.026	−0.435	−0.269
FP	0.016	−0.568	0.050
FSC	−0.101	−0.172	0.291
FSP	0.357	−0.095	−0.080
GH	−0.077	0.227	0.151
LH	0.629	−0.435	−0.112
LS	−0.386	−0.626	0.143
PC	0.010	0.175	−0.310
PF	0.203	0.616	0.043
SCP	−0.050	0.327	−0.235
SEC	−0.038	0.236	−0.345
SMC	0.020	−0.144	−0.060
SSH	0.082	−0.134	0.186
STC	−0.042	0.000	0.023
DTF	0.186	−0.069	0.827
DTM	0.236	−0.190	0.793
PH	0.294	−0.357	0.118
NPB	0.037	0.208	0.239
NSB	0.160	0.398	0.204
NPP	0.626	0.086	−0.073
PL	0.252	0.353	0.104
NRP	0.001	0.735	−0.117
NSP	0.476	0.076	−0.075
HSWT	0.508	0.219	−0.084
GYD	0.863	0.146	−0.109
Eigen value	3.404	2.967	2.163
Variance %	39	34	25
Cumulative	39	73	98

DTF = days to 50% flowering, DTM = days to 75% maturity, PH = plant height, NPB = number of primary branches, NSB = number of secondary branches per plant, NRP = number of racemes per plant, NPP = number of pods per plant, GYD = grain yield, HSWT = 100 seed.

Figure 2 shows the variable correlation plot showing positive association between PC1 and DTF traits such as DTM, GYD, HSWT, NSP, plant height (PH), and pod length (PL). Conversely, seed eye color (SEC), pod color (PC), number of secondary branches per plant (NSB), and FMC exhibited negative associations with PC1. Quantitative traits such as plant habit (PH), NPP, HSWT, DTF, DTM, NRP, and GYD were positively correlated to each other as exhibited by their vectors, which were in the same direction and separated by acute angles between them. Similarly, the qualitative traits such as SEC, PC, and FMC were positively correlated to each other. However, the quantitative traits were positively correlated with GYD, while the qualitative traits were negatively correlated with GYD.

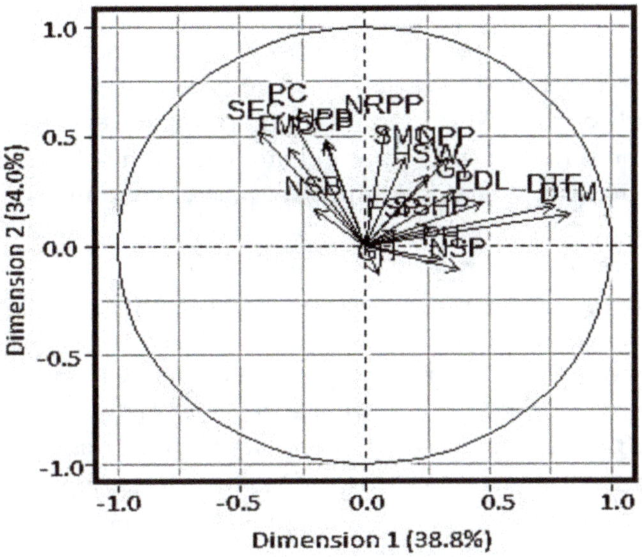

Figure 2. Trait biplot showing the relationship among quantitative and qualitative traits in 81 pigeonpea genotypes evaluated in six environments in Malawi. Dim 1 = dimension 1, Dim2 = dimension 2, PCA = principal component analysis. For trait code description, refer to Table 3.

Assessment of the phenotypic diversity using morphological attributes delineated the genotypes into three distinct clusters (Figure 3). The first cluster had the highest number (51) of genotypes. The second cluster had 27 genotypes, and the third cluster had three genotypes. However, the composition of genotypes in all the three clusters consisted of mixtures of landraces, breeding lines, and cultivars.

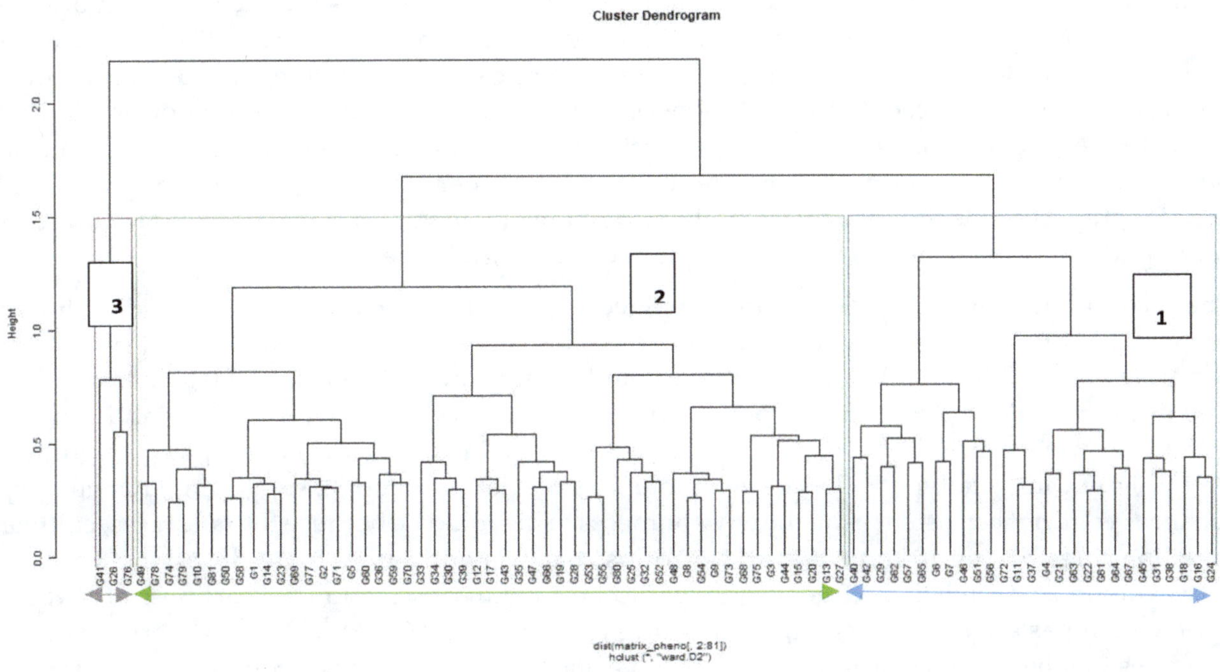

Figure 3. Hierarchical cluster dendogram showing genetic similarity matrix of 81 genotypes evaluated in six environments in Malawi based on phenotypic traits. Cluster 1, in pink color, cluster 2 in green color and cluster 3 in blue color. See Table 1 for the genotypes codes.

4. Discussion

The current study evaluated 81 pigeonpea genotypes across six environments to assess the genetic diversity and yield stability, and to select complementary and unique genotypes for breeding. The genotypes exhibited wide and significant variation in qualitative traits (Table 4), which indicated that the tested germplasm could harbor important genetic variation that underpins the morphological variation. Similarly, [16] reported significant variation in qualitative traits among pigeonpea accessions sourced from ICRISAT's international genebank. The variation in qualitative traits such as growth habit and seed color is important for breeding cultivars that meet farmer expectations. For instance, the variation present in growth habit is important to identify genotypes with compact growth habit for intercropping to maximize space utilization and productivity in moisture-limited environments. Farmers often intercrop pigeonpea with cereal crops such as maize and sorghum, and legumes such as groundnuts. Hence, pigeonpea genotypes with a spreading growth habit may not be suitable for mixed cropping systems [17]. The diversity in pigeonpea seed color helps to identify genotypes that are preferred by local farmers. For instance, farmers in Malawi prefer pigeonpea varieties with a cream seed color, which they associate with good cooking quality. Similar findings were reported by [18], who reported a predominance of cream and light gray pigeonpea varieties in Benin, reflecting the farmers' color preferences. Knowledge of variability in qualitative traits among the accessions and understanding farmer preferences are important as a basis for the development of direct breeding objectives and appropriate breeding strategies.

The significant genetic variation exhibited in the quantitative traits (Table 5) highlights the genetic diversity available for exploitation during cultivar development. The genotype performances were also affected by significant genotype × environment interactions, suggesting that genotype performances were not consistent in all the environments. Genotypic variation is underpinned by differences in genetic constitution among the genotypes, which is important for crop improvement [19]. The environment influences phenotypic expression through variation in factors such as temperature, humidity, and soil fertility. The significant impact of the environment on phenotypic expression is known to reduce genotype–phenotype correlation [20], which complicates the identification of stable and superior genotypes. However, significant genotype × environment interaction on yield and yield components of legumes such as common bean, cowpea, and pigeonpea has been previously reported [21–23]. In the present study, the genotypes that matured early were shorter with low numbers of branches and pods per plant and low grain yields compared to the medium to late maturing genotypes that grew taller, produced more branches and pods per plant, and had higher grain yields. Similarly, [24] reported that cultivars with higher numbers of primary branches, secondary branches, number of pods per plant, and taller plant height had higher grain yields.

There was limited genetic variation among pigeonpea landraces, cultivars, and breeding lines in this study (Table 6). This could be attributed to gene flow arising from the exchange of germplasm between Malawi and Tanzania. In addition, there could be high level of genotype relatedness since the breeding lines and cultivars were developed from the landraces collected from Malawi and Tanzania by ICRISAT. However, the genotype performance in terms of days to maturity and plant height were affected by significant genotype × environment interactions (Tables 7 and 8), suggesting that genotype performances were not consistent in all the environments. This could be because the landraces, cultivars, and breeding lines belong to three maturity groups: early, medium, and late duration. The early maturity exhibited by the ICRISAT genotypes could be a result of selection for earliness at ICRISAT in Kenya, which has advanced pigeonpea breeding programs and has developed a number of elite breeding lines that have been distributed in several East and Southern African countries for evaluation. The TARI and DARS genotypes are comprised of landraces and cultivars that are medium to late maturing. Similarly, [21] also reported that traditionally grown pigeonpea cultivars and landraces are represented by varieties from medium to long maturity groups (150 to 280 days), which are high yielding but very sensitive to photoperiod.

The positive and moderate correlation between GYD and HSWT ($r = 0.50$, $p < 0.01$) (Table 11) indicated that HSWT could be used for the direct selection for GYD. The moderate positive correlation between DTF, DTM, and PH revealed that selection for earliness can be based on the plant height. Although pigeonpea is relatively drought-tolerant, there is a need to develop early flowering and maturing cultivars to fit in the cropping cycles of sub-Saharan Africa, which are becoming progressively shorter due to climate change. The positive correlations exhibited by most secondary traits show that multiple trait selection would be possible. However, the weak correlations among the traits would result in an inefficient selection or low genetic gains. A strong correlation ($r = 0.858$) between grain yield and the number of pods per plant was reported by [25]. In addition, [26] reported moderate to weak correlations between grain yield and days to 50% flowering ($r = 0.58$), days to maturity ($r = 0.59$), and plant height ($r = 0.42$). Conversely, [27] and [28] reported a negative association between 100 seed weight and grain yield. The significant relationship between DTF, DTM, HSWT, PH, NPP, and GYD is useful when selecting for high grain yield [16]. Direct selection for these traits would result in yield improvement in pigeonpea.

The nonlinear principal component analysis enabled the identification of important traits with high variability among the genotypes. In this study, GDY, DTF, DTM, NPP, NRP LH, HSWT, and NSP were identified as the most important traits due to their high contribution on PC1 and PC2 (Table 12). This suggests that these traits are useful for selection. Accessions that exhibit high and desirable mean performances based on the target traits would be selected for improvement. Other reports indicated that trait contribution to different PCs varies with genetic diversity within the tested germplasm and the number of traits evaluated [16,25]. The results further revealed that DTF, NPP, NPB, NSB, PH, PL, and HSWT are important secondary traits for the indirect selection for GYD due to their positive association with GYD and their high contribution on the PCs.

The cluster analysis delineated the accessions into three groups (Figure 3), suggesting the presence of considerable genetic variation among the genotypes. However, a mixture of breeding lines, landraces, and cultivars in each group could be attributed to the geographical proximity between Malawi and Tanzania (where some of the landraces were collected). The level of natural outcrossing in pigeonpea is very high and varies from 5 to 70% depending on the prevailing weather conditions and insect activities for pollination [29]. In addition, the breeding lines from ICRISAT were developed using some parents selected from the landraces from Tanzania and Malawi. East Africa is known as a center of diversity for pigeonpea—hence the close genetic relatedness. The present finding is in agreement with [30], who reported little variation among the cultivated pigeonpea collected in Africa based on diversity array technology (DArT) markers.

5. Conclusions

The study examined 81 pigeonpea genotypes for their diversity and yield stability. The genotypes exhibited a wide genetic variation in qualitative traits such as growth habit, flower main color, flower streak pattern, pod color, and seed traits. The combined analysis revealed significant genotype × environment interaction effects for most traits, suggesting the need for selection for specific adaptation. A lack of significant variation in quantitative traits among landraces, cultivars, and breeding lines indicate that there is potentially high gene flow among the different categories of germplasm, which could present genetic bottlenecks during breeding. Traits such as GDY, DTF, DTM, NPP, NRP, HSWT, LH, and NSP with high scores on PC1 and PC2 are useful selection indices for pigeonpea improvement. Accessions that exhibited high and desirable mean performances in the target traits such as early maturing (MWPLR 14, ICEAP 01170, ICEAP 871091, ICEAP 01285) and high yielding (Kachangu, MWPLR 16, TZA 5582, No. 40, and MWPLR 14) would be recommended as parental lines for the breeding program. The genetic diversity analysis using morphological traits has enabled the identification of promising parents and heterotic clusters for breeding.

Author Contributions: Each author has participated sufficiently in the completion of this work. H.S., M.L., I.M., and A.S. contributed to the experimental design, data analysis, and review of this manuscript. E.N.Y. implemented the experiments and wrote up the manuscript. All authors have read and agreed to the published version of the manuscript.

Acknowledgments: The authors acknowledge the station managers and technical staff of the Chitedze, Bvumbwe, and Makoka Research Stations in Malawi for technical assistance and for the overall research support. The Department of Agricultural Research Services (DARS)/Malawi, Tanzania Agricultural Research Institute, International Crops Research Institute for Semi-Arid Tropics, and Malawi Genebank are sincerely thanked for providing the germplasm used in the study.

References

1. Ade-Omowaye, B.; Tucker, G.; Smetanska, I. Nutritional potential of nine underexploited legumes in Southwest Nigeria. *Int. Food. Res. J.* **2015**, *22*, 798.
2. Subbarao, G.; Chauhan, Y.; Johansen, C. Patterns of osmotic adjustment in pigeonpea—Its importance as a mechanism of drought resistance. *Eur. J. Agron.* **2000**, *12*, 239–249. [CrossRef]
3. Giller, K.E. *Nitrogen Fixation in Tropical Cropping Systems*; Cabi: Wallingford, UK, 2001.
4. Kwena, K.; Karuku, G.; Ayuke, F.; Esilaba, A. Nitrogen Deficiency in Semi-Arid Kenya: Can Pigeonpea fix it? *East Afr. Agric For. J* **2019**, *83*, 322–340. [CrossRef]
5. Mula, M.; Saxena, K. *Lifting the Level of Awareness on Pigeonpea—A Global Perspective*; International Crops Research Institute for the Semi-Arid Tropics: Patancheru, India, 2010.
6. Food and Agriculture Organisation of the United Nations. FAOSTAT. Crop Statistics. 2017. Available online: www.fao.org/faostat/en/#data/QC (accessed on 20 January 2019).
7. Kananji, G.; Yohane, E.; Mviha, P.J.Z.; Siambi, M.; Silim, S. *A Guide to Pigeonpea Production in Malawi*; Department of Agricultural Research Services: Lilongwe, Malawi, 2016.
8. Bohra, A.; Dubey, A.; Saxena, R.K.; Penmetsa, R.V.; Poornima, K.; Kumar, N.; Farmer, A.D.; Srivani, G.; Upadhyaya, H.D.; Gothalwal, R. Analysis of BAC-end sequences (BESs) and development of BES-SSR markers for genetic mapping and hybrid purity assessment in pigeonpea (Cajanus spp.). *Biomed. Cent. BMC Plant Bio.* **2011**, *11*, 56. [CrossRef] [PubMed]
9. Saxena, R.K.; Von Wettberg, E.; Upadhyaya, H.D.; Sanchez, V.; Songok, S.; Saxena, K.; Kimurto, P.; Varshney, R.K. Genetic diversity and demographic history of Cajanus spp. illustrated from genome-wide SNPs. *PLoS ONE* **2014**, *9*, e88568. [CrossRef]
10. Upadhyaya, H.D.; Reddy, K.N.; Sharma, S.; Dwivedi, S.L.; Ramachandran, S. Enhancing the value of genetic resources for use in pigeonpea improvement. *Legume Persp.* **2016**, *11*, 13–16.
11. IBPGR; ICRISAT. *Descriptors for Pigeonpea [Cajanus cajan (L.) Millsp.]*; International Board of Plant Genetic Resources: Rome, Italy; International Crops Research Institute for Semi-Arid Tropics: Patancheru, India, 1993.
12. George, D.; Mallery, P. *IBM SPSS Statistics 26 Step by Step: A Simple Guide and Reference*; Routledge: Abbingdon, UK, 2019.
13. Payne, R.; Murray, D.; Harding, S.; Baired, D.; Soultar, D. *Genstat Windows*, 17th ed.; VSN International: Hemel Hempstead, UK, 2017.
14. Shimelis, H.; Shiringani, R. Variance components and heritabilities of yield and agronomic traits among cowpea genotypes. *Euphytica* **2010**, *176*, 383–389. [CrossRef]
15. Meulman, J.J.; Van der Kooij, A.J.; Heiser, W.J. Principal components analysis with nonlinear optimal scaling transformations for ordinal and nominal data. *Sage Handb. Quant. Methodol. Soc. Sci.* **2004**, *3*, 49–72.
16. Upadhyaya, H.; Reddy, K.; Gowda, C.; Singh, S. Phenotypic diversity in the pigeonpea (Cajanus cajan) core collection. *Genet. Res. Crop Evol* **2007**, *54*, 1167–1184. [CrossRef]
17. Manyasa, E.; Silim, S.; Christiansen, J. Variability patterns in Ugandan pigeonpea landraces. *J. Sat. Agric. Res.* **2009**, *7*, 1–9.
18. Ayenan, M.A.T.; Ofori, K.; Ahoton, L.E.; Danquah, A. Pigeonpea [(Cajanus cajan (L.) Millsp.)]

production system, farmers' preferred traits and implications for variety development and introduction in Benin. *Agric. Food Sec.* **2017**, *6*, 48. [CrossRef]
19. Grausgruber, H.; Sailer, C.; Ghambashidze, G.; Bolyos, L.; Ruckenbauer, P. Genetic variation for plant breeding. In Proceedings of the 17th EUCARPIA General Congress, Tulln, Austria, 8–11 September 2004.
20. Bustos-Korts, D.; Romagosa, I.; Borràs-Gelonch, G.; Casas, A.; Slafer, G.; Van Eeuwijk, F. *Genotype by Environment Interaction and Adaptation; In Encyclopedia of Sustainability Science and Technology*; Springer: New York, NY, 2018; pp. 29–71.
21. Vales, M.; Srivastava, R.; Sultana, R.; Singh, S.; Singh, I.; Singh, G.; Patil, S.; Saxena, K. Breeding for earliness in pigeonpea: Development of new determinate and nondeterminate lines. *Crop Sci.* **2012**, *52*, 2507–2516. [CrossRef]
22. Kimaro, D. Genetic Improvement of Pigeonpea (Cajanus cajan (L.) Millsp.) for Fusarium wilt resistance in Tanzania. Ph.D. Thesis, University of Kwazulu-Natal, Pitermaritzburg, South Africa, December 2016.
23. Gerrano, A.S.; Van Rensburg, W.S.J.; Mathew, I.; Shayanowako, A.I.T.; Bairu, M.W.; Venter, S.L.; Swart, W.; Mofokeng, A.; Mellem, J.J.; Labuschagne, M. Genotype and genotype 3 environment interaction effects on the grain yield performance of cowpea genotypes in dryland farming system in South Africa. *Euphytica* **2020**, *216*, 80. [CrossRef]
24. Rekha, R.; Prasanthi, L.; Sekhar, M.; Priya, M. Studies on selection indices in pigeonpea [Cajanus cajan (L.) Millsp]. *Int. J. Appl. Biol. Pharm.* **2013**, *4*, 291–294.
25. Saroj, S.; Singh, M.; Kumar, R.; Singh, T.; Singh, M. Genetic variability, correlation and path analysis for yield attributes in pigeonpea. *Bioscan* **2013**, *8*, 941–944.
26. Sreelakshmi, C.; Shivani, D.; Kumar, C. Genetic divergence and stability analysis in Pigeonpea (Cajanus cajan L.). *Electr. J. Plant Breed* **2010**, *1*, 530–535.
27. Hemavathy, A.T.; Bapu, J.; Priyadharshini, C. Principal component analysis in pigeonpea (Cajanus cajan (L.) millsp.). *Electr. J. Plant Breed* **2017**, *8*, 1133–1139. [CrossRef]
28. Narayanan, S.L.; Manivannan, N.; Mahalingam, A. Correlation and Path Analyses of Yield and its Component Traits in Pigeonpea [Cajanus cajan(L.) Millsp.]. *Int. J. Curr. Microbiol. App. Sci.* **2018**, *7*, 614–618. [CrossRef]
29. Choudhary, A.K.; Sultana, R.; Ontagodi, T.; Singh, I.; Bhatt, B. Recent advances in breeding pigeonpea [Cajanus cajan (L.) Millsp.]. In Proceedings of the National Conference on Global Research Initiatives for Sustainable Agriculture & Allied Sciences (GRISASS-2015), Gwalior, MP, India, 12–13 December 2015.
30. Yang, S.; Pang, W.; Ash, G.; Harper, J.; Carling, J.; Wenzl, P.; Huttner, E.; Zong, X.; Kilian, A. Low level of genetic diversity in cultivated pigeonpea compared to its wild relatives is revealed by diversity arrays technology. *Appl. Gen.* **2006**, *113*, 585–595. [CrossRef] [PubMed]

3

Genome-Wide Characterization and Analysis of Metallothionein Family Genes that Function in Metal Stress Tolerance in *Brassica napus* L.

Yu Pan [1,2,†], Meichen Zhu [2,3,†], Shuxian Wang [2,3], Guoqiang Ma [2,3], Xiaohu Huang [2,3], Cailin Qiao [2,3], Rui Wang [2,3], Xinfu Xu [2,3], Ying Liang [2,3], Kun Lu [2,3], Jiana Li [2,3,*] and Cunmin Qu [2,3,*]

1 Key Laboratory of Horticulture Science for Southern Mountainous Regions, Ministry of Education, Southwest University, No. 2 Tiansheng Road, Beibei, Chongqing 400715, China; panyu1020@swu.edu.cn
2 Academy of Agricultural Sciences, Southwest University, Chongqing 400715, China; zmc0809@email.swu.edu.cn (M.Z.); wsxsummer@email.swu.edu.cn (S.W.); mgqdw123456@email.swu.edu.cn (G.M.); hxh9305@email.swu.edu.cn (X.H.); qcl123@email.swu.edu.cn (C.Q.); ruiwang71@163.com (R.W.); xinfuxu@swu.edu.cn (X.X.); yliang@swu.edu.cn (Y.L.); drlukun@swu.edu.cn (K.L.)
3 Chongqing Rapeseed Engineering Research Center, College of Agronomy and Biotechnology, Southwest University, No. 2 Tiansheng Road, Beibei, Chongqing 400715, China
* Correspondence: ljn1950@swu.edu.cn (J.L.); drqucunmin@swu.edu.cn (C.Q.)

† These authors contributed equally to this paper.

Abstract: *Brassica* plants exhibit both high biomass productivity and high rates of heavy metal absorption. Metallothionein (MT) proteins are low molecular weight, cysteine-rich, metal-binding proteins that play crucial roles in protecting plants from heavy metal toxicity. However, to date, MT proteins have not been systematically characterized in *Brassica*. In this study, we identified 60 MTs from *Arabidopsis thaliana* and five *Brassica* species. All the MT family genes from Brassica are closely related to *Arabidopsis* MTs, encoding putative proteins that share similar functions within the same clades. Genome mapping analysis revealed high levels of synteny throughout the genome due to whole genome duplication and segmental duplication events. We analyzed the expression levels of 16 *Brassica napus* MTs (*BnaMTs*) by RNA-sequencing and real-time RT-PCR (RT-qPCR) analysis in plants under As^{3+} stress. These genes exhibited different expression patterns in various tissues. Our results suggest that *BnaMT3C* plays a key role in the response to As^{3+} stress in *B. napus*. This study provides insight into the phylogeny, origin, and evolution of MT family members in *Brassica*, laying the foundation for further studies of the roles of MT proteins in these important crops.

Keywords: metallothionein; *Brassica*; *Brassica napus*; As^{3+} stress

---

1. Introduction

Heavy metals are essential micronutrients for various physiological processes in plants. However, excess amounts of essential (copper, zinc) and non-essential metals (cadmium) are toxic to plants, as they inhibit plant growth, impair root development, and decrease chlorophyll contents, resulting in chlorosis [1,2]. Therefore, plants have evolved a suite of mechanisms involving the chelation and sequestration of heavy metals by various amino acids, organic acids, phytochelatins (PCs) and metallothioneins (MTs) [3,4]. These compounds play crucial roles in protecting plants from heavy metal toxicity [5–9].

MTs, low-molecular-weight proteins (7–10 kDa) with a high percentage of cysteine (Cys) residues [10,11], have been widely characterized in various prokaryotic and eukaryotic organisms. Plant MTs are classified into four types according to the arrangement of their Cys residues [12], including the *MT1*, *MT2*, *MT3*, and *MT4* subfamilies [10,13]. MTs play crucial roles in ion homeostasis and tolerance in plants. Seven functional MT genes have been isolated from the model plant *Arabidopsis thaliana*. Of these, *AtMT1a*, *AtMT2a*, *AtMT2b*, and *AtMT3* enhance plant tolerance of Cu ions, especially in leaves [14,15], as well as Cd tolerance in transformed yeast and fava bean (*Vicia faba*) guard cells [16,17]. *AtMT4* modulates Zn homeostasis in seeds and is highly expressed during the late stages of development [18]. Additionally, various *MT* genes show significantly different expression patterns in plants under heavy metal stress. For example, *MT2a* and *MT2b* genes are more highly expressed in the roots of the heavy metal hyperaccumulator *Noccaea caerulescens* than in those of *A. thaliana*, while *MT3* is more highly expressed in shoots of *N. caerulescens* than in the non-hyperaccumulator *Thlaspi arvense* [19,20]. *MT4* mRNAs are primarily expressed in ripening fruits and developing seeds [14]. Therefore, plant MTs are likely involved in many physiological processes, such as seed development and germination [18,21,22], fruit ripening [14], and root development [21,23].

Brassica plants are considered to be highly tolerant to heavy metals (e.g., Cd, Cu, Ni, Zn, Pb, and Se), making them ideal plants for studying metal accumulation in phytoremediation studies [24–27]. Indian mustard (*Brassica juncea*) is a high-biomass-producing crop with the potential to take up and accumulate heavy metals [1,23,27,28]. However, this plant accumulates Cd less effectively than other crops such as maize (*Zea mays*), rice (*Oryza sativa*), and sugar beet (*B. vulgaris*) when it is present at low concentrations in the soil [29]. Rapeseed (*B. napus*) has many advantages for this type of analysis due to its rapid growth, high biomass productivity and efficient heavy metal absorption, and it is therefore also widely used to investigate heavy metal tolerance [26,30–32]. Indeed, while heavy metal tolerance has been well studied in various *Brassica* species, the mechanisms that contribute to the tolerance of these plants to heavy metals remain unclear.

In the present study, we screened the draft genome sequences of *A. thaliana* and various *Brassica* species (*Brassica rapa*, *Brassica oleracea*, *B. napus*, *Brassica juncea* and *Brassica nigra*) for *MT* genes that participate in heavy metal detoxification. We identified 60 *MT* genes and performed a detailed analysis of their duplication patterns, classifications, and chromosomal distribution and motifs, as well as a phylogenetic analysis. Finally, we verified the differential expression profiles of selected rapeseed *MT* genes in different *B. napus* tissues at various developmental stages. We also investigated the expression patterns of *MT* genes in *B. napus* seedlings exposed to heavy metals. Our results provide important information about the origin and evolution of the *MT* gene family in *Brassica* and provide a basis for further studies of the functions of MT family proteins in rapeseed.

2. Results

2.1. Identification and Multiple Sequence Alignment of MT Family Genes

Using the protein sequences of the *MT* family genes from the TAIR10 database (Table 1) as queries, we identified 60 *MT* genes in *A. thaliana* and various *Brassica* plants (*B. rapa*, *B. oleracea*, *B. napus*, *B. juncea* and *B. nigra*). These genes were classified into four subgroups (Figure 1, Table 1). Of these, seven were *MT1* subfamily members, five of which were identified from the corresponding genome databases, except *BolMT1* and *BjuMT1*. *BraMT1* has been reported in *B. rapa* with 45 amino-acid proteins [33]; the difference between these sequences requires further study. In addition, 37 were *MT2* subfamily members, encoding deduced proteins ranging from 56 to 103 amino acids in length; nine were *MT3* subfamily members, encoding proteins from 65 to 69 amino acids in length; and seven were *MT4* subfamily members, encoding proteins from 85 to 120 amino acids in length (Table 1). Of the *MT1* subfamily members, three homologs were identified in *A. thaliana*, while *BnaMT1* and *BraMT1* were identified in *B. napus* and *B. rapa*, respectively. No *MT1* subfamily members were found in *B. oleracea*, *B. juncea*, or *B. nigra*, whereas *BolMT1* and *BjuMT1* have been reported previously [13], indicating that genome gaps may still emerge in *B. oleracea*, and *B. juncea*. Six Cys-X-Cys motifs were equally distributed on the *N*- and *C*-termini of MT1 family proteins, except in the case of BraMT1 and AtMT1B (Figure 2A).

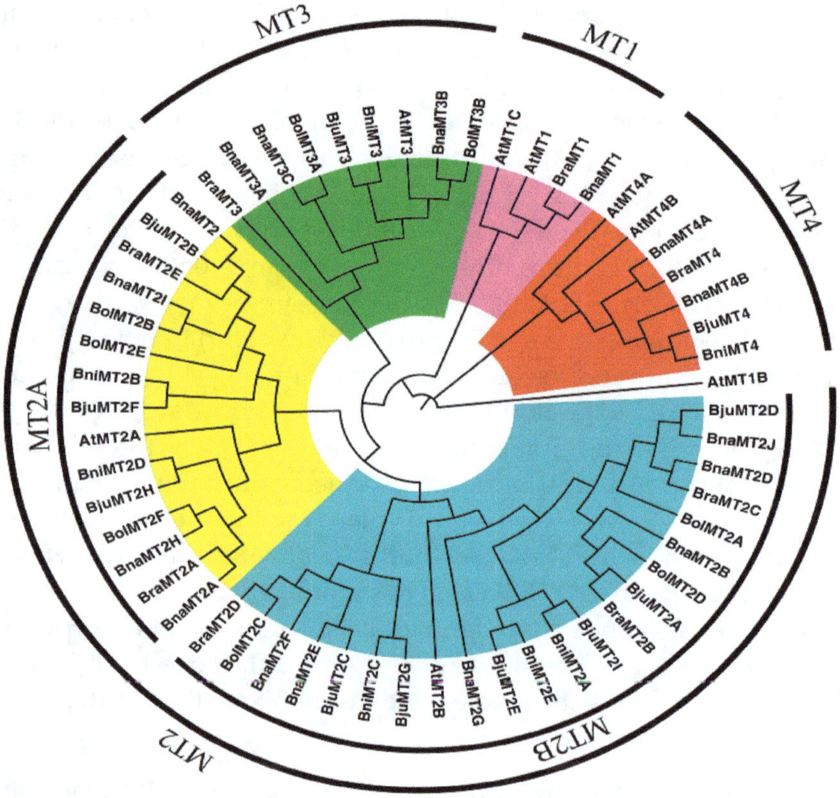

Figure 1. Neighbor-Joining (NJ) phylogenetic tree showed the relationships of *Metallothionein* (*MT*) family genes from *A. thaliana* and various *Brassica* species. The rooted neighbor-joining phylogenetic tree was constructed using MEGA6 and visualized using Figure Tree v1.4.2. The MTs were divided into four subfamilies (MT1–MT4), which are indicated by different colors. Organism name and gene accession numbers are shown in Table 1.

We identified 37 *MT2* subfamily genes in *A. thaliana* (*AtMT2A* and *AtMT2B*) and *Brassica* (five in *B. rapa*, six in *B. oleracea*, ten in *B. napus*, nine in *B. juncea*, and five in *B. nigra*), which were divided into the *MT2A* and *MT2B* subgroups (Figure 2B, Table 1), pointing to the extensive triplication and expansion of these genomes during their evolution in *Brassica* plants. Furthermore, one Cys–Cys and two Cys–X–Cys motifs were almost always present in the N-terminal regions of these proteins, and three Cys–X–Cys were almost always present in their C-terminal regions (Figure 2B). In addition, MT2 subfamily genes encode a deduced protein with the MSCCGGN/S sequence in their N-termini, which is consistent with previous findings [13,34]. Three variant regions were found in the MT2A subgroup and three in the MT2B subgroup (Figure 2B), which might be associated with their roles in metal tolerance.

We identified *MT3* subfamily genes, including one each in *A. thaliana*, *B. rapa*, *B. juncea*, and *B. nigra*, two in *B. oleracea* and three in *B. napus* (Figure 1, Table 1). The alignment of the MT3 amino acid sequences showed a completely conserved sequence, CXXCDCX$_5$C, located in the N-terminus of each protein, and a highly conserved consensus sequence with eight Cys residues at the C-terminus (Figure 2C). In addition, 30–40 amino acids were detected in the Cys-poor linker region between the *N*- and *C*-terminal regions, in accordance with the MT2 subfamily (Figure 2B,C), pointing to a possible evolutionary relationship between the *MT2* and *MT3* family genes.

Table 1. List of *Metallothionein* (*MT*) genes identified in the *A. thaliana* and *Brassica* genomes.

Groups	Name	Gene ID	Chr.	Start (bp)	End (bp)	Length (bp)	Length (aa)	MW (KDa)	pIs	Exon	Intron
	AtMT1	AT1G07600	AtChr1	2338904	2339321	138	45	4.580	4.23	2	1
	AtMT1C	AT1G07610	AtChr1	2341542	2342123	138	45	4.495	4.54	2	1
	AtMT1B	AT5G56795	AtChr5	22972042	22972449	156	51	5.428	10.25	2	1
MT1	*BraMT1*	Bra015594	BraA10	766670	772706	450	149	16.77	9.32	4	3
	BnaMT1	BnaA10g04950D	BnaA10	2673266	2673770	138	45	4.480	3.92	2	1
	BolMT1	DK501359	UN	UN	UN	138	45	4.412	3.92	UN	UN
	BjuMT1	EF471214	UN	UN	UN	138	45	4.439	3.92	UN	UN
	AtMT2A	AT3G09390	AtChr3	2889486	2890229	246	81	8.163	4.35	2	1
	AtMT2B	AT5G02380	AtChr5	506498	507244	234	77	7.766	4.54	2	1
	BraMT2A	Bra001309	BraA03	15803598	15803933	246	81	8.197	4.17	2	1
	BraMT2B	Bra005720	BraA03	275453	275766	243	80	8.033	4.29	2	1
	BraMT2C	Bra009595	BraA10	16182058	16182453	243	80	8.031	4.29	2	1
	BraMT2D	Bra028875	BraA02	269835	270,238	246	81	8.386	4.20	2	1
	BraMT2E	Bra029765	BraA05	23082833	23083178	243	80	8.024	4.35	2	1
	BolMT2A	Bol000591	Scaffold000521	37950	38342	243	80	8.031	4.29	2	1
	BolMT2B	Bol011307	Scaffold000212	445183	445551	243	80	8.054	4.58	2	1
	BolMT2C	Bol012825	BolC02	305134	305536	246	81	8.386	4.20	2	1
	BolMT2D	Bol015273	BolC03	101213	101535	243	80	8.137	4.29	2	1
	BolMT2E	Bol023080	BolC01	37533909	37534079	171	56	5.920	4.15	1	0
	BolMT2F	Bol033925	Scaffold000040	316614	316949	246	81	8.147	4.15	2	1
	BnaMT2A	BnaA03g30680D	BnaA03	14857530	14858170	240	79	7.966	3.81	2	1
	BnaMT2B	BnaA03g54880D	A03_random	44751	45330	243	80	8.077	4.29	2	1
	BnaMT2C	BnaA05g29010D	BnaA05	20416060	20416808	243	80	8.024	4.35	2	1
	BnaMT2D	BnaA10g27170D	BnaA10	17170773	17171571	243	80	8.031	4.29	2	1
	BnaMT2E	BnaAnng00330D	Ann_random	321245	322046	246	81	8.370	4.20	2	1
MT2	*BnaMT2F*	BnaC02g03550D	BnaC02	1685618	1686330	246	81	8.386	4.20	2	1
	BnaMT2G	BnaC03g00710D	BnaC03	346945	347282	216	71	7.284	4.08	2	1
	BnaMT2H	BnaC03g35960D	BnaC03	21778025	21778740	240	79	7.950	3.79	2	1
	BnaMT2I	BnaC05g43490D	BnaC05	40327486	40328304	243	80	8.028	4.58	2	1
	BnaMT2J	BnaCnng40400D	Cnn_random	38972401	38973067	243	80	8.031	4.29	2	1
	BjuMT2A	BjuA008858	BjuA03	321697	322020	243	80	8.033	4.29	2	1
	BjuMT2B	BjuA020647	BjuA05	29317680	29319366	312	103	10.661	4.24	3	2
	BjuMT2C	BjuA040818	BjuA02	430697	431097	246	81	8.370	4.20	2	1
	BjuMT2D	BjuA044587	BjuA10	19619066	19619458	243	80	8.031	4.29	2	1
	BjuMT2E	BjuB001621	BjuB08	169755	170152	243	80	8.059	4.29	2	1

Table 1. *Cont.*

Groups	Name	Gene ID	Chr.	Start (bp)	End (bp)	Length (bp)	Length (aa)	MW (KDa)	pIs	Exon	Intron
	BjuMT2F	BjuB005939	BjuB01	43381550	43383641	297	98	10.057	3.93	3	2
	BjuMT2G	BjuB012072	BjuB05	20387272	20387699	246	81	8.353	4.11	2	1
	BjuMT2H	BjuB031838	BjuB03	12766757	12767103	222	73	7.596	4.35	2	1
	BjuMT2I	BjuB044439	BjuB02	52757313	52757655	243	80	8.061	4.29	2	1
	BniMT2A	BniB001954-PA	BniB08	30524166	30524483	243	80	8.075	4.29	2	1
	BniMT2B	BniB007929-PA	BniB05	2189985	2190357	243	80	7.978	4.35	2	1
	BniMT2C	BniB023579-PA	BniB02	28213025	28213453	246	81	8.353	4.11	2	1
	BniMT2D	BniB039464-PA	BniB07	32240383	32240733	222	73	7.596	4.35	2	1
	BniMT2E	BniB045064-PA	BniB03	44318189	44318591	225	74	7.569	4.08	2	1
MT3	*AtMT3*	AT3G15353	AtChr3	5180642	5181586	210	69	7.373	4.35	3	2
	BraMT3	Bra027254	BraA05	20683494	20683920	204	67	7.183	4.17	3	2
	BolMT3A	Bol011145	BolC05	28120530	28120940	204	67	7.158	4.15	3	2
	BolMT3B	Bol025753	BolC03	17061982	17062438	198	65	7.016	4.40	3	2
	BnaMT3A	BnaA05g24200D	BnaA05	18177871	18178669	204	67	7.127	4.15	3	2
	BnaMT3B	BnaC03g39060D	BnaC03	24091592	24092154	198	65	7.016	4.40	3	2
	BnaMT3C	BnaC05g38240D	BnaC05	37017131	37017881	204	67	7.158	4.15	3	2
	BjuMT3	BjuB025665	BjuB01	40434253	40434691	204	67	7.187	4.15	3	2
	BniMT3	BniB008959-PA	BniB05	4957466	4957899	204	67	7.187	4.15	3	2
MT4	*AtMT4B*	AT2G23240	AtChr2	9895855	9896325	261	86	8.437	5.58	2	1
	AtMT4A	AT2G42000	AtChr2	17529243	17530443	366	121	12.229	7.62	4	3
	BraMT4	Bra000590	BraA03	11951235	11951571	261	86	8.480	7.37	2	1
	BnaMT4A	BnaA03g23200D	BnaA03	11067719	11068213	261	86	8.480	7.37	2	1
	BnaMT4B	BnaC03g27400D	BnaC03	15895032	15895647	261	86	8.468	6.97	2	1
	BjuMT4	BjuO006263	Contig407_1_341981	122112	122451	261	86	8.500	6.97	2	1
	BniMT4	BniB049568-PA	BniB03	29749524	29749863	261	86	8.472	6.97	2	1

Note, At, *A. thalinana*; Bra, *B. rapa*; Bol, *B. oleracea*; Bni, *B. nigra*; Bna, *B. napus*; and Bju, *B. juncea*; Chr., Chromosome; UN, unknown.

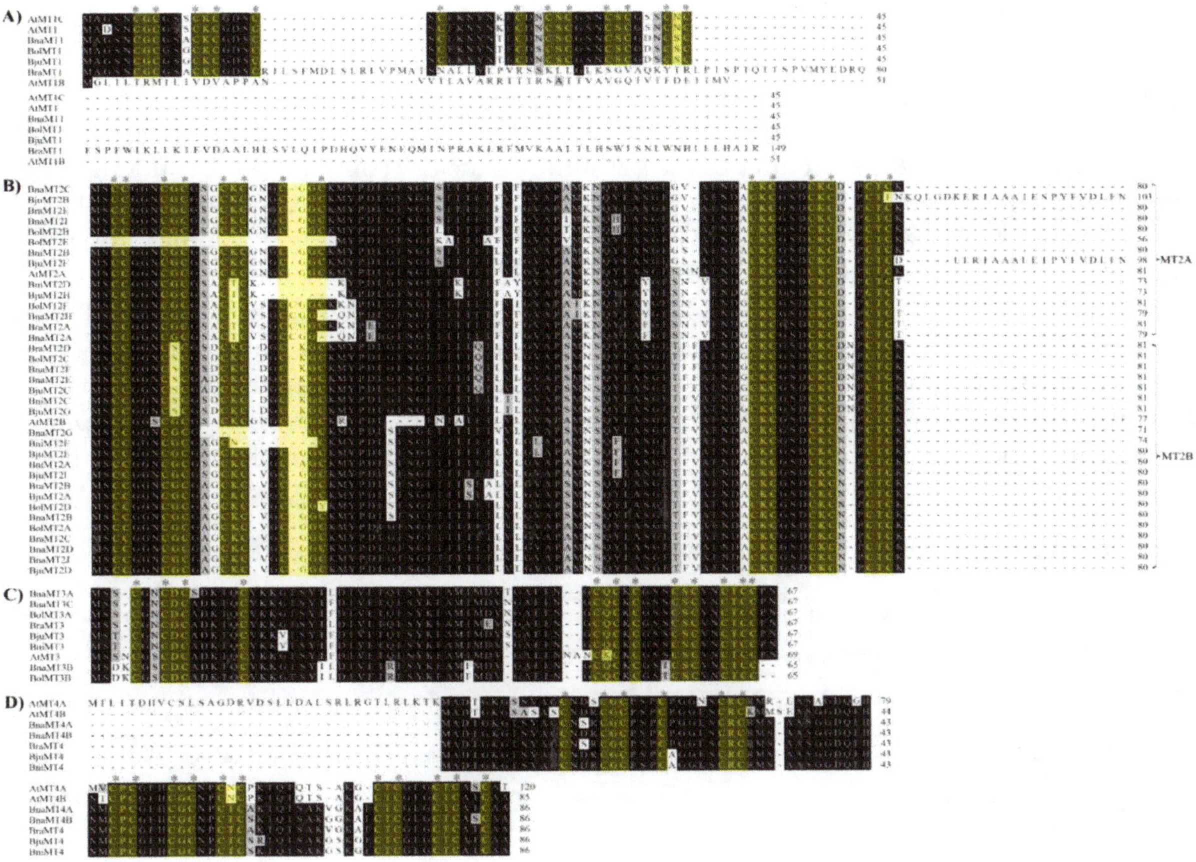

Figure 2. Alignment of MT protein sequences from *A. thaliana* and various *Brassica* species. Black and light gray shading indicate identical and conserved amino acid residues, respectively. (**A**) The MT1 protein sequences; (**B**) the MT2 protein sequences; (**C**) the MT3 protein sequences; (D) the MT4 protein sequences. The conserved cysteines regions are highlighted by asterisks and light yellow. The MTs were preliminarily classified by Cobbett and Goldsbrough reported [10]; detailed information is provided in Table 1.

MT4 subfamily genes, which are homologous to plant EC metallothionein-like genes, are different from *MT1*–*MT3* subfamily members. Two *MT4* subfamily members were found in *A. thaliana* and *B. napus*, one each in *B. rapa*, *B. juncea*, and *B. nigra*, and none in *B. oleracea* (Figure 1, Table 1). These proteins contain three Cys-poor linkers comprising 12–15 amino acids, as well as two Cys-rich regions with a highly conserved consensus sequence among them (Figure 2D).

2.2. Phylogenetic Analysis of MT Family Genes

Based on the multiple sequence alignment of the deduced MT1–MT4 proteins, it was found that Cys-rich regions are widely distributed among MT family proteins. These regions are characterized by conserved consensus sequences, with motifs such as Cys–G–Cys, Cys–K–Cys, and Cys–S–Cys (Figure 2). To investigate the evolutionary relationships among MT family genes from *A. thaliana* and various *Brassica* species, we constructed a NJ phylogenetic tree based on the alignment of MT domains. Based on the phylogenetic tree, the 58 MT domains were classified into four subfamilies (*MT1*, *MT2*, *MT3*, and *MT4*), except for *BolMT1* and *BjuMT1*, which were not annotated in the genome databases, and most genes were grouped with the *AtMTs* (Figure 2). *AtMT1B* represents the outgroup in the phylogenetic tree. In addition, the *MT2* subfamily was classified into two sister groups (Figure 1), *MT2A* and *MT2B*, which is highly consistent with the results of multiple sequence alignments of whole proteins (Figure 2). For example, two sister groups were also identified and found to contain eight and six *MT* family genes, respectively (Figure 2), which also contain highly conserved consensus sequences (Figure 1). These results will be helpful in identifying the functions of *MT* family genes via orthology analysis.

2.3. Genomic Structure and Conserved Motif Analysis of the MT Gene Family

We characterized the gene structures of the *MT* family genes by comparing the full-length CDS and the corresponding genomic DNA sequences using GSDS 2.0 (http://gsds.cbi.pku.edu.cn/index.php). Of the 58 *MT* genes, 44 contain a single intron with a highly conserved structure in each group, i.e., the *MT1*, *MT2* and *MT4* subfamilies. Additionally, *MT3* subfamily members contain two introns, which were also found in *BjuMT2B* and *BjuMT2F* (Figure 3). *BolMT2E* lacks an intron and belongs to the *MT2A* gene family, while *AtMT4A* and *BraMT1* contain three introns with distinct sizes (Figure 3). Most genes in the same subfamily exhibit similar exon–intron structures, but the genomic structures of *BjuMT2B* and *BjuMT2F* are similar to those of the *MT3* subfamily, providing further support for the evolutionary relationship and classification of the *MT* gene family members identified in this study.

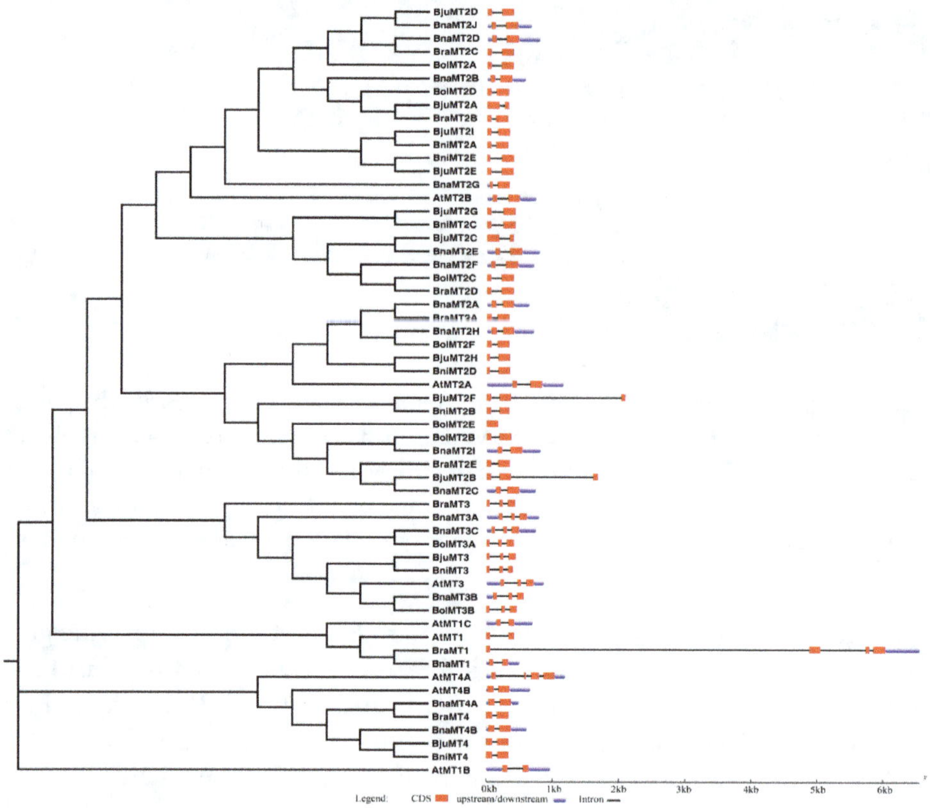

Figure 3. Phylogenetic relationships and genomic structures of the *MT* genes from *A. thaliana* and various *Brassica* species. The red boxes represent exons, solid lines represent introns (connecting two exons), and blue boxes represent untranslated regions (UTRs). The lengths of the *MT* genes are indicated by horizontal lines (kb).

Using MEME v4.12.0 (http://meme-suite.org/tools/meme), six, eight, three, and four conserved motifs were detected in the *MT1*, *MT2*, *MT3*, and *MT4* subfamilies, respectively, in *B. rapa*, *B. oleracea*, *B. napus*, *B. juncea*, and *B. nigra* (Figure 4A–D); the detailed structures of the motifs are shown in Figure S1A–D. All members of the *MT1* subfamily except for *AtMT1B* contain motif 1 (Figure 4A and Figure S1A). All members of the *MT2* subfamily contain motif 1, whereas all members of the *MT2A* subfamily contain motifs 2, 4, and 6, but motif 3 is found only in the *MT2B* subfamily members. The motifs in the *MT2B* subfamily members are more variable than those of the other *MTs* (Figure 4B and Figure S1B), pointing to the triplication and expansion of *Brassica* genomes. All nine genes in the *MT3* subfamily contain motif 1, while the *MT3A* subfamily genes contain motif 2 and the *MT3B* subfamily genes contain motif 3 (Figure 4C and Figure S1C). The *MT4* subfamily genes contain motifs 1, 2, and 3, indicating that these motifs are conserved among these genes (Figure 4D and Figure S1D). In summary, the same conserved motifs are widely found in paralogous/orthologous genes, suggesting that they might have similar functions at the protein level.

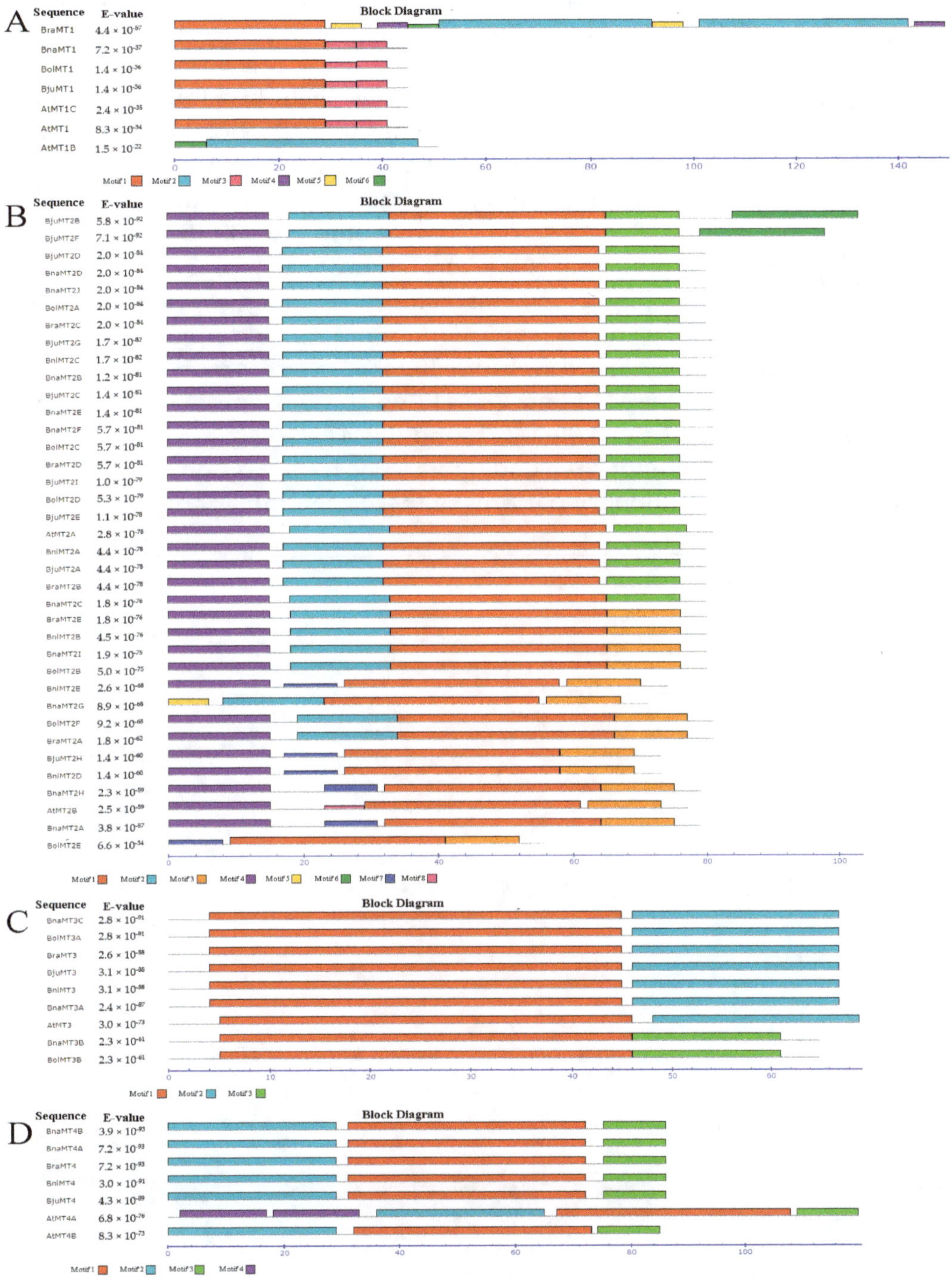

Figure 4. Putative conserved motifs in MT family proteins in various *Brassica* species identified using the MEME search tool. (**A**) the conserved motifs in MT1 family; (**B**) the conserved motifs in MT2 family; (**C**) the conserved motifs in MT3 family; (**D**) the conserved motifs in MT4 family. Different motifs are represented by different colors, and protein names and combined p values are shown on left side of this figure. The best possible matched motifs, their functional annotation, and motif width are shown in Figure S1.

2.4. Chromosome Locations and Duplication of *MT* Genes in Brassica

Brassica includes three diploid species, *B. rapa* (AA, 2*n* = 20), *B. oleracea* (CC, 2*n* = 18), and *B. nigra* (BB, 2*n* = 20) and three allotetraploid species, *B. napus* (AACC, 2*n* = 38), *B. juncea* (AABB, 2*n* = 36), and *B. carinata* (BBCC, 2*n* = 34), and the evolution and relationships between the members of *Brassica* can be well understood according to the U-triangle theory [35]. Five of these species have been completely sequenced, and their sequences are available in the *Brassica* database (BRAD) database. To identify the physical positions of the *MT* genes, we mapped them to the chromosomes in the corresponding *Brassica* species. The 43 *MT* genes are located on 27 chromosomes in the five *Brassica* species with available whole-genome sequences, including four chromosomes (BraA02, BraA03, BraA05, and BraA10) in *B. rapa*, four chromosomes (BolC01, BolC02, BolC03 and BolC05) in *B. oleracea* and five chromosomes in *B. nigra* (BniB02, BniB03, BniB05, BniB07 and BniB08) (Figure 5). Further, we detected high levels of synteny among *MT* family genes in these species. For example *BraMT2D* on chromosome BraA02, *BjuMT2C* on chromosome BjuA02, *BnaMT2F* on chromosome BnaC02, and *BolMT2C* on chromosome BolC02 are located near the top of the chromosomes and are classified into the same subgroups (Figure 2), suggesting that these genes might have undergone whole-genome duplication events during the evolutionary process and might have similar functions. However, some of these genes, e.g., *BnaMT4A* and *BraMT4* on chromosome A03, *BnaMT3A* and *BraMT3* on chromosome A05, and *BnaMT1* and *BraMT1* on chromosome A10 might have undergone segmental duplications (Figure 5). Finally, *BjuMT2H* and *BniMT2E* on chromosome B03 and *BjuMT2E* and *BniMT2A* on chromosome B08 might have undergone gene transposition (Figure 5). Taken together, these results shed light on the evolutionary patterns of these subfamilies among adjacent species.

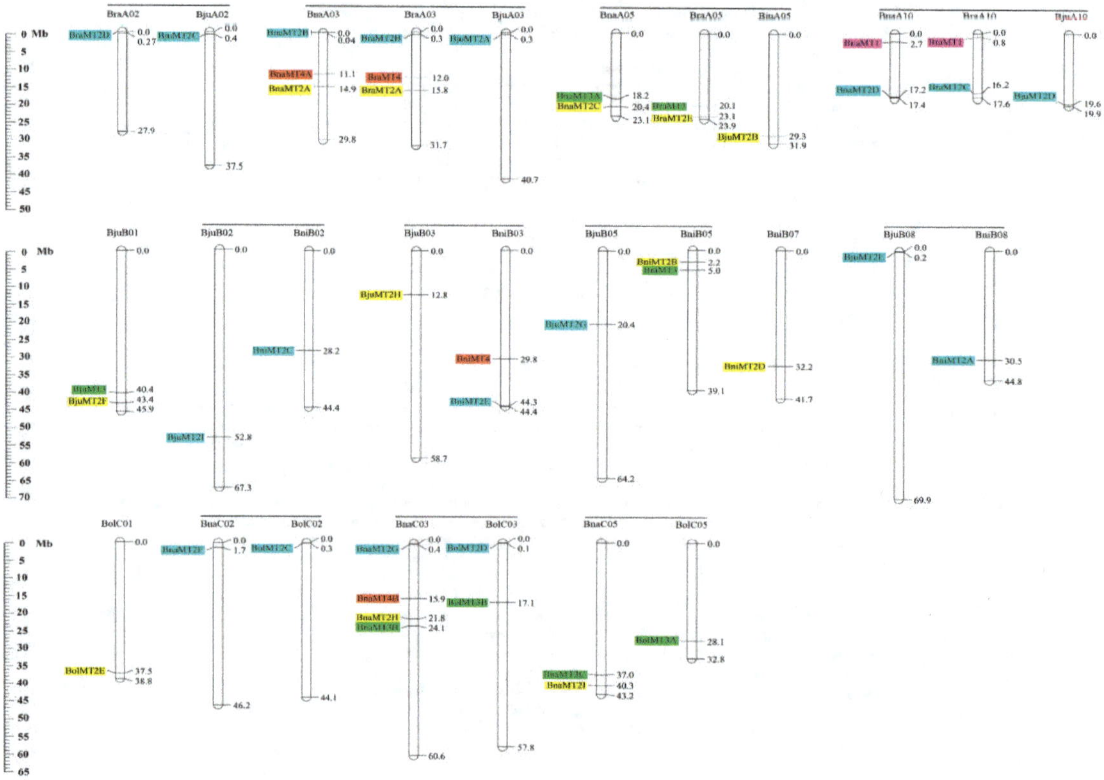

Figure 5. Chromosomal distribution and analysis of duplication events in *MT* family genes among *Brassica* species. Genes from the same subgroups are indicated by the same color, which is consistent with the corresponding family in the phylogenetic tree (Figure 1). The labels on the corresponding chromosomes indicate the names of the source organism and the subgenome. The scales indicate the sizes of various *Brassica* plant genomes (Mb). Bra, *B. rapa*; Bol, *B. oleracea*; Bni, *B. nigra*; Bna, *B. napus*; and Bju, *B. juncea*. The genes located on the scaffold are not shown in the Figure 5.

2.5. Expression Profiles of BnaMT Family Genes in B. napus

Based on the transcriptome sequencing datasets from *B. napus* ZS11 (BioProject ID PRJNA358784), we characterized the expression profiles of the *BnaMT* genes in eight different tissues, covering all stages of rapeseed development (Figure 6, Tables S1 and S2). Among the 16 *BnaMT* genes, *BnaMT1* was more highly expressed in the stems, leaves, and siliques 30 days after pollination than in other tissues (Figure 6). Among *MT2* genes, *BnaMT2A* and *BnaMT2H* were specifically expressed in buds; *BnaMT2C* and *BnaMT2I* were expressed at higher levels in roots, hypocotyls, cotyledons, and buds than in others tissues; *BnaMT2B*, *BnaMT2D*, and *BnaMT2J* were highly expressed throughout plant development, whereas *BnaMT2G* was expressed at low levels; and *BnaMT2E* and *BnaMT2F* were more highly expressed in stems and leaves than in other tissues (Figure 6). *BnaMT3A*, *BnaMT3B*, and *BnaMT3C* were more highly expressed in stems, leaves, and siliques before day 30 than in other tissues (Figure 6). Finally, *BnaMT4A* and *BnaMT4B* were mainly expressed in ripening seeds (Figure 6). The expression patterns of *MT* family genes correspond with the results of the phylogenetic analysis (Figure 2). For example, the expression patterns were similar for *BnaMT2A* and *BnaMT2H*, *BnaMT2C* and *BnaMT2I*, and *BnaMT2B*, *BnaMT2D*, and *BnaMT2J*, which were classified into the same sister groups.

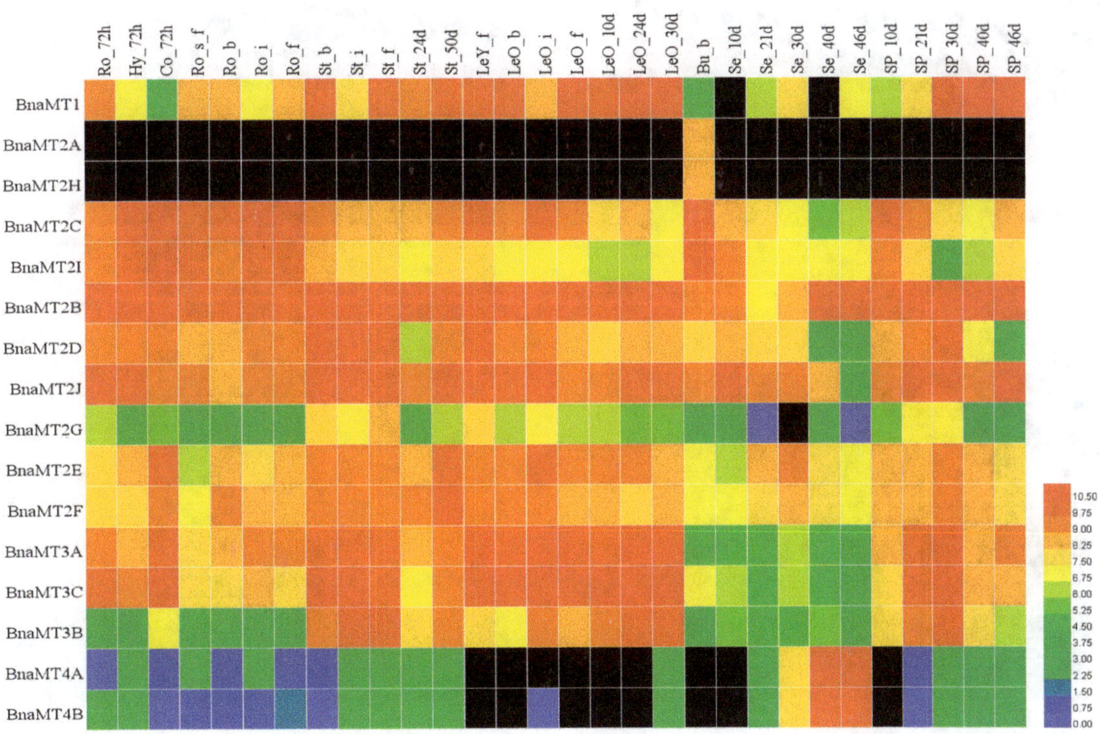

Figure 6. Heatmap of the expression profiles of *BnaMT* family genes in different tissues and organs. The abbreviations above the heatmap indicate the different tissues and organs/developmental stages of *B. napus* ZS11 (listed in Table S1). The expression data was gained from the RNA-seq data and shown as log2, as calculated by fragments per kilo base of exon model permillion (FPKM) values. Black boxes indicate that no expression was detected by RNA-seq analysis.

2.6. Expression Analysis of BnaMT Genes in Response to Metal Treatment

MTs are the best-characterized heavy-metal-binding ligands in plants. To analyze the roles of *BnaMTs* in metal tolerance, we compared the expression profiles of *BnaMTs* in the roots, hypocotyls, and cotyledons of *B. napus* plants under As^{3+} stress versus normal conditions via real-time RT-PCR

(RT-qPCR). Under normal conditions, the expression patterns of the *BnaMTs* were similar to the patterns identified by RNA-seq, with different expression profiles detected among different rapeseed varieties (Figure 7, Table S3). For example, *BnaMT2B*, *BnaMT2C*, *BnaMT2D*, and *BnaMT2J* were highly expressed in all tissues; *BnaMT1* and *BnaMT4B* were expressed at lower levels in roots, hypocotyls, and cotyledons; *BnaMT2A*, *BnaMT2F*, *BnaMT2G*, *BnaMT2H*, *BnaMT2I*, *BnaMT3A*, and *BnaMT3B* were expressed at lower levels in roots and hypocotyls than in cotyledons (Figure 7); and *BnaMT2A* and *BnaMT2H* did not exhibit tissue-specific expression in *B. napus* (Figures 6 and 7). After As^{3+} treatment, all *BnaMT* genes were expressed at higher levels in roots than in hypocotyls but were expressed at the highest levels in cotyledons (Figure 7). For example, *BnaMT1* was upregulated by As^{3+} treatment, and *BnaMT2A*, *BnaMT2B*, *BnaMT2F*, *BnaMT2J*, and *BnaMT3B* were more significantly upregulated in cotyledons than in roots and hypocotyls (Figure 7). Importantly, *BnaMT3C* was more highly expressed in varieties B33 and B34 than in B93 and B113 (Figure 7).

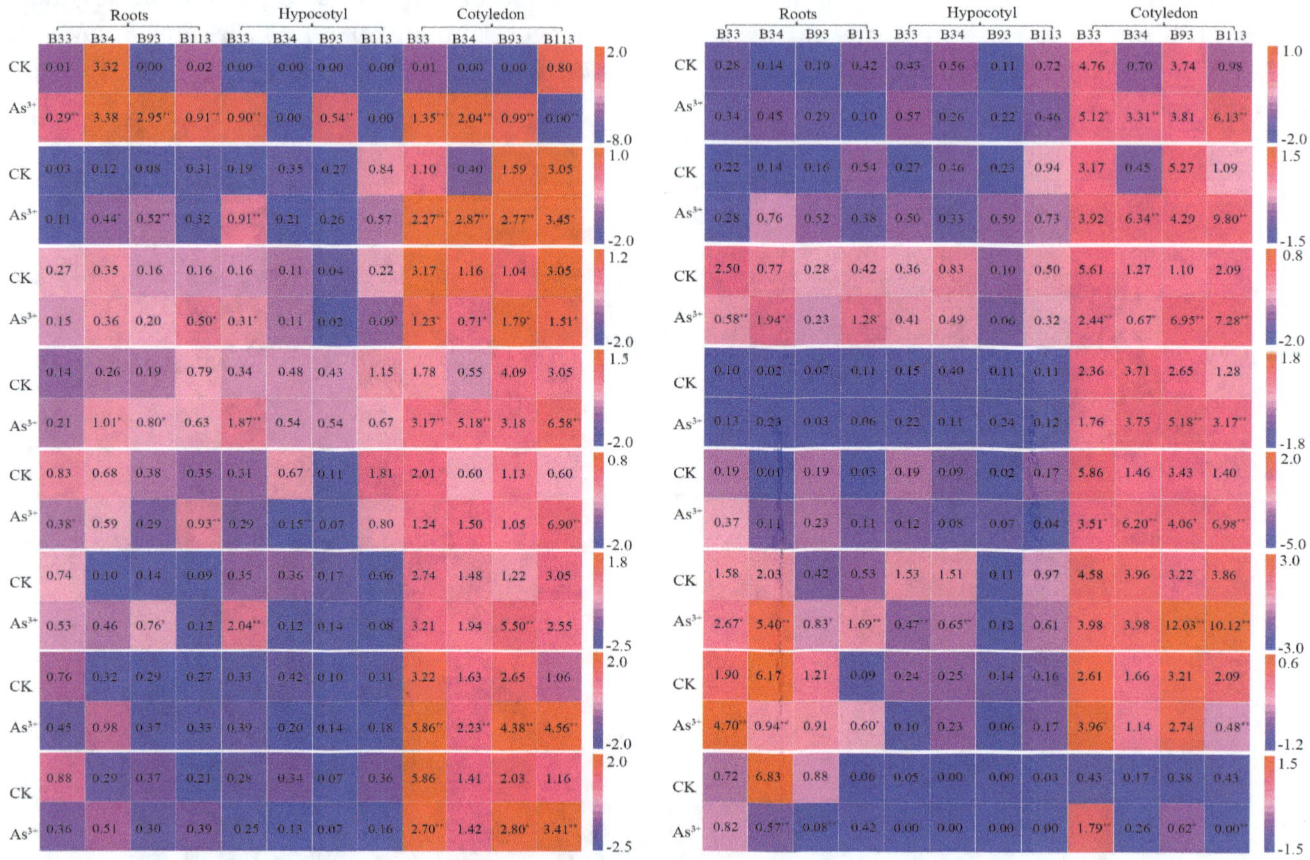

Figure 7. Expression analysis of *BnaMT* family genes in different tissues under control and As^{3+} treatment via real-time RT-PCR (RT-qPCR). Three biological replicates per sample were used for analysis, and three technical replicates were analyzed per biological replicate. Values represent the average of three biological replicates with three technical replicates of each tissue (Table S3). The expression data was gained from the real-time RT-PCR (RT-qPCR) analysis data and shown as log2 as calculated by average values normalized to that of the reference gene *BnACTIN7* (EV116054). *, ** indicates a significance level at 0.05 and 0.01, respectively.

3. Discussion

The high-affinity heavy metal chelators, PCs and MTs, play crucial roles in maintaining metal homeostasis during plant development [36–41]. Moreover, the *Brassica* plants had high biomass productivity and high levels of heavy metal absorption, as analyzed in *B. juncea* [1,23], *B. rapa* [33], and *B. napus* [42]. In addition, seven putative *MT* genes have been identified in *Arabidopsis* [15,16], but no comprehensive study of these genes has been reported. *Brassica* species, which were derived from a common ancestor, are ideal model systems for analyzing polyploid evolution and genome duplication [43]. Many analyses have focused on the model plant *A. thaliana* and various *Brassica* species (*B. rapa, B. nigra, B. oleracea, B. napus, B. juncea* and *B. carinata*). The whole genome sequences of all species except *B. carinata* are available in BRAD (the *Brassica* Database, http://brassicadb.org/brad/downloadOverview.php). In the present study, we identified 52 *MT* genes from various *Brassica* species based on *A. thaliana MT* gene sequences (Table 1). Phylogenetic analysis revealed that all *MT* family genes are closely associated with *AtMTs* (Figure 2), suggesting that they share similar functions or have undergone gene fusion [44]. Of these, *MT1* subfamily genes from *B. nigra, B. oleracea*, and *B. juncea* have not been identified, but *BolMT1* and *BjuMT1* have been identified [13]. The number of *MT2* (10) and *MT3* (3) genes in *B. napus* is nearly equal to the sum of these genes in *B. rapa* (5 *MT2* and *1 MT3*) and *B. oleracea* (6 *MT2* and 2 *MT3*), and most genes showed high levels of synteny throughout the genome, reflecting the fact that whole genome duplications and segmental duplications were a major contributor to the expansion of *MTs* during evolution. However, the deduced protein sequence of *BraMT1* in *B. rapa* is longer than the previously published sequence [33], indicating the need for further study to confirm *BraMT1*. In addition, homologs of *BnaMT4A* and *BraMT4, BnaMT3A* and *BraMT3*, and *BnaMT1* and *BraMT1* were not detected in the corresponding genomes (Figure 5, Table 1), in accordance with the finding that gene loss typically occurs after polyploidization in eukaryotes [45–47]. Furthermore, Cys-rich regions were almost completely conserved among MTs, and the distinct spacer sequences in the Cys-poor linkers were also well-conserved, comprising 7 amino acids in *Brassica* MT1s, 40–42 amino acids in *Brassica* MT2s, 32–34 amino acids in *Brassica* MT3s, and 14–15 amino acids in *Brassica* MT4s (Figure 1). These results are in close agreement with previous predictions that Cys-rich regions will show highly conserved *MT* family genes [13,33], and we infer that the variations in the MTs might be associated with their different functions in plants [48].

To date, plant MTs have been widely characterized, exhibiting different tissue-specific expression patterns [8,14,49–52]. For example, *AtMT1A* and *AtMT2B* were predominantly expressed in roots and leaves, while *AtMT2A* and *AtMT3* were highly expressed in roots and young leaves [14]. Likewise, notable differences in expression patterns were also found among *MT1, MT2A*, and *MT2B* subfamily genes in *B. napus* (Figure 6). *BnaMT1* was expressed at the highest level in siliques of 30D, except for roots and leaves, whereas *B. napus MT2A* genes (with four members; *BnaMT2A, BnaMT2C, BnaMT2H* and *BnaMT2I*) and *MT2B* genes (with six members; *BnaMT2B, BnaMT2D, BnaMT2E, BnaMT2F, BnaMT2G* and *BnaMT2J*; Figure 2) showed variable expression patterns. For example, *BnaMT2A* and *BnaMT2H* were preferentially expressed in buds, and *BnaMT2G* was expressed at low levels in all organs (Figure 6). These differences may be attributed to concentrations and species differences in future works. However, *B. napus MT3* and *MT4* subfamily genes shared similar expression patterns with *AtMT3* and *AtMT4* [14]. *B. napus MT3s* were mainly expressed in stems, roots, and leaves, and *B. napus MT4s* were primarily expressed in developing seeds (Figure 6). The expression patterns of these *BnaMTs* revealed by RT-qPCR corresponded well with the patterns obtained by transcriptome analysis under normal conditions, although there were differences among *B. napus* varieties (Figure 7). Although no comprehensive heavy metal tolerance mechanisms have been uncovered in *Brassica*, distinctive expression patterns were identified among the *B. napus MT* family members in this work, laying the foundation for investigating the biochemical and physiological functions of MTs in plants.

Heavy metal (Cu, Cd, and As) pollution in agricultural soils has become a critical problem affecting crop production and quality. The absorption of these heavy metals by plants plays an important role in the entry of these metals into the food chain [42]. Further, *MTs* have been shown to play an important role in metal homeostasis and tolerance in plants [4,6,9,10,21,26,33,34]. Strikingly, *Brassica* plants exhibit efficient heavy metal uptake and translocation, as well as a high tolerance to heavy metals [31,42], and several *MT* genes in *Brassica* have been reported, especially in Indian mustard (*Brassica juncea* L.) [23,27,33,53]. Recently, three *B. rapa* metallothionein genes (*BrMT1–3*) displayed differential expression levels under various exogenous stress factors [54], and MT-like, protein-encoding gene transcription was obviously induced in roots and leaves of *B. napus* under As treatment [55]. Here, we investigated the expression profiles of *BnaMT* family genes in *B. napus* under normal conditions and As^{3+} stress. The *BnaMTs* were obviously induced by the As^{3+} treatments, which is in accordance with findings that MTs are involved in the chelation and sequestration of heavy metals [3,4,55]. Like other plant *MT* genes [4,41,54,55], however, these genes also had different expression profiles in different *B. napus* varieties and in different tissues, with high expression levels in cotyledons and low expression levels in hypocotyls, such as *BnaMT1*, *BnaMT2A*, *BnaMT2B*, *BnaMT2C*, *BnaMT2F*, *BnaMT2J*, *BnaMT3B*, and *BnaMT3C* (Figure 7). These results suggest that hypocotyls might merely be involved in the transport of heavy metal ions, but that these ions accumulate in roots and cotyledons. In addition, *BnaMT3C* was obviously increased in the roots and hypocotyls of B33 and B34, but the higher expression levels of B93 and B113 in cotyledons (Figure 7), which comply with the finding that B33 and B34 exhibited better growth than B93 and B113 under As^{3+} treatment (Figure S2), indicate that they play crucial roles in the response to As^{3+} stress in *B. napus*. Our results provide important information for further functional studies of MT family genes in *B. napus*.

4. Materials and Methods

4.1. Identification of MT Family Genes in Brassica

The amino acid sequences of MTs from the Arabidopsis Information Resource (TAIR10) database (ftp://ftp.arabidopsis.org) were used as queries for the BLASTp analysis against the whole genome sequences in the *Brassica* database [56]. The candidate sequences with *E*-values $\leq 1 \times 10^{-20}$ were identified and confirmed using the Hidden Markov model (HMM) searches program (HMMER v3.0, http://hmmer.janelia.org/), and the BLAST analysis of the MTs was performed against a *Brassica* protein database constructed using Geneious v4.8.5 software (http://www.geneious.com/, Biomatters, Auckland, New Zealand). The coding sequences (CDS) of the MTs were identified by BLASTn searches against the *Brassica* genome database. The candidate proteins were named using the species abbreviation of the source organism (italicized), the gene family name, and the positions in the subtribe, e.g., *AtMT1B* and *BnaMT1A*. Physicochemical properties, including the molecular weight (kDa), isoelectric point (pI), and the grand average of hydropathy (GRAVY) value of each deduced protein were determined using the online ExPASy-ProtParam tool (http://web.expasy.org/protparam/).

4.2. Multiple Sequence Alignment and Phylogenetic Analysis of MTs in Brassica

The deduced amino acid sequences of MT proteins from *A. thaliana* and various *Brassica* species, including *B. rapa*, *B. oleracea*, *B. napus*, *B. juncea*, and *B. nigra*, were subjected to multiple protein sequence alignment using the ClustalW software with default settings [57]. To illustrate the evolutionary relationships of MTs in *Brassica*, a neighbor-joining (NJ) phylogenetic tree was generated with the MEGA v6.0 program (Tokyo Metropolitan University, Tokyo, Japan) using the JTT+I+G substitution model and a bootstrap test with 1000 replicates [58]. The phylogenetic trees were visualized using FigTree v1.4.2 (http://tree.bio.ed.ac.uk/software/figtree/).

4.3. Conserved Motif Recognition and Gene Structure Analysis

The CDS of the MTs from the *Brassica* species were retrieved based on their protein sequences, and the corresponding genomic sequences were extracted from the *Brassica* genome sequences. The exon–intron structures of the MTs were analyzed online using the Gene Structure Display Server (GSDS v2.0, http://gsds.cbi.pku.edu.cn/index.php). Conserved motifs were identified using Multipel Expectation Maximization for Motif Elucidation (MEME v4.12.0, http://meme-suite.org/tools/meme) with the following parameters: number of repetitions, any; maximum number of motifs, 15; and optimum width of each motif, between 6 and 300 residues [59]. Each motif with an *E*-value $< 1 \times 10^{-10}$ was retained for motif detection.

4.4. Chromosomal Locations of *MT* Family Genes in *B. napus*

The *MT* family genes were mapped to the rapeseed chromosomes according to their physical distances in the GFF genome files, which were downloaded from the *B. napus* genome database (http://www.genoscope.cns.fr/brassicanapus/) [43]. A map of the chromosomal locations of the MTs was constructed using MapChart v2.0 (https://www.wur.nl/en/show/Mapchart.htm) [60].

4.5. Plant Materials and Metal Stress Treatments

B. napus seeds were collected from the Rapeseed Engineering Research Center of Southwest University in Chongqing, China (CERCR). Fifty healthy seeds were selected and soaked in a dish (the diameter was 90 mm) containing deionized water for 24 h. Then morphologically uniform seedlings were selected and plugged into a hydroponic system with a float tray (60 cm × 40 cm × 10 cm) for 7 days. Here, the seedlings were exposed to distilled water and 35 μM As^{3+} solutions, respectively. Meanwhile, they were cultivated under long-day conditions (16 h light/8 h dark, 5000 Lux) at 25 °C. After 7 days, the whole roots, hypocotyls, and cotyledons were sampled to analyze the *MT* gene expression patterns; the tissues were snap frozen in liquid nitrogen and stored at −80 °C prior to total RNA extraction. All experiments were repeated three times.

4.6. Total RNA Extraction and RT-qPCR Analysis

As the *B. napus* cultivars B33 and B34 grow better than B93 and B113 under heavy metal treatment (Figure S2), they were therefore used for expression analysis. Total RNA was isolated from the samples using a DNAaway RNA Mini-Prep Kit (Sangon Biotech, Shanghai, China). For the tissue-specific expression analysis, RNA was extracted from the roots, hypocotyls, and cotyledons and pretreated with gDNA Eraser (Takara, Dalian, China). Subsequently, 1 μg of the total RNA was used to synthesize first-strand cDNA with an RNA PCR Kit (AMV) Ver. 3.0 (Takara, Dalian, China). The cDNA was subjected to RT-qPCR analysis using SYBR Premix Ex Taq II (Takara, Dalian, China) on a Bio-Rad CFX96 Real Time System (Bio-Rad Laboratories, Hercules, CA, USA) as previously described [61]. *BnACTIN7* (EV116054) was employed as a reference gene to normalize MT gene expression levels via the $2^{-\Delta\Delta Ct}$ method [62]. All experiments were performed with three technical replicates, and the values represent the average ± standard error (SE). The specific primer sequences used in this study were obtained from the qPCR Primer Database [63] and are listed in Supplementary Table S4.

4.7. Statistical Analysis

All experiments were repeated three times (three biological replicates). All data were statistically analyzed using the Student's *t*-test with the statistical analysis software package SPSS v15.0 (IBM Corp, Armonk, NJ, USA).

5. Conclusions

In this study, we identified 60 *MTs* from *A. thaliana* and five *Brassica* species. The phylogenetic analysis showed that all *MT* family genes are closely associated with the *AtMTs*. Genome-mapping analysis revealed high levels of synteny throughout the genome due to whole genome duplication and segmental duplication events. In addition, all 16 *BnaMTs* were induced by heavy metal stress, especially in cotyledons versus roots and hypocotyls. Finally, *BnaMT3C* might improve the response to As^{3+} stress in *B. napus*. Our results provide a basis for the further functional analysis of the molecular functions of *MT* family genes in *B. napus*.

Author Contributions: C.Q. (Cunmin Qu) and J.L. conceived and designed the experiments; Y.P. and M.Z. conducted the experiments; S.W., G.M. and X.H. collected and analyzed the data; Y.L., C.Q. (Cailin Qiao), R.W. and X.X. carried out the experiments and performed the software analyses; K.L. and C.Q. (Cunmin Qu) wrote the manuscript; and C.Q. (Cunmin Qu) and J.L. reviewed the manuscript.

Acknowledgments: We would also like to thank Kathy Farquharson for critical reading of this manuscript.

References

1. Ebbs, S.; Uchil, S. Cadmium and zinc induced chlorosis in Indian mustard [*Brassica juncea* (L.) czern] involves preferential loss of chlorophyll b. *Photosynthetica* **2008**, *46*, 49–55. [CrossRef]
2. Lingua, G.; Franchin, C.; Todeschini, V.; Castiglione, S.; Biondi, S.; Burlando, B.; Parravicini, V.; Torrigiani, P.; Berta, G. *Arbuscular mycorrhizal* fungi differentially affect the response to high zinc concentrations of two registered poplar clones. *Environ. Pollut.* **2008**, *153*, 137–147. [CrossRef] [PubMed]
3. Hall, J.L. Cellular mechanisms for heavy metal detoxification and tolerance. *J. Exp. Bot.* **2002**, *53*, 1–11. [CrossRef] [PubMed]
4. Lv, Y.; Deng, X.; Quan, L.; Xia, Y.; Shen, Z. Metallothioneins *BcMT1* and *BcMT2* from *Brassica campestris* enhance tolerance to cadmium and copper and decrease production of reactive oxygen species in *Arabidopsis thaliana*. *Plant Soil* **2013**, *367*, 507–519. [CrossRef]
5. Huang, G.Y.; Wang, Y.S. Expression and characterization analysis of type 2 metallothionein from grey mangrove species (*Avicennia marina*) in response to metal stress. *Aquat. Toxicol.* **2010**, *99*, 86–92. [CrossRef] [PubMed]
6. Mir, G.; Domenech, J.; Huguet, G.; Guo, W.J.; Goldsbrough, P.; Atrian, S.; Molinas, M. A plant type 2 metallothionein (MT) from cork tissue responds to oxidative stress. *J. Exp. Bot.* **2004**, *55*, 2483–2493. [CrossRef] [PubMed]
7. Gall, J.E.; Boyd, R.S.; Rajakaruna, N. Transfer of heavy metals through terrestrial food webs: A review. *Environ. Monit. Assess.* **2015**, *187*, 201. [CrossRef] [PubMed]
8. Gu, C.S.; Liu, L.Q.; Deng, Y.M.; Zhu, X.D.; Huang, S.Z.; Lu, X.Q. The heterologous expression of the *Iris lactea* var. chinensis type 2 metallothionein IlMT2b gene enhances copper tolerance in *Arabidopsis thaliana*. *Bull. Environ. Contam. Toxicol.* **2015**, *94*, 247–253. [CrossRef] [PubMed]
9. Hassinen, V.H.; Tuomainen, M.; Peraniemi, S.; Schat, H.; Karenlampi, S.O.; Tervahauta, A.I. Metallothioneins 2 and 3 contribute to the metal-adapted phenotype but are not directly linked to Zn accumulation in the metal hyperaccumulator, *Thlaspi caerulescens*. *J. Exp. Bot.* **2009**, *60*, 187–196. [CrossRef] [PubMed]
10. Cobbett, C.; Goldsbrough, P. Phytochelatins and metallothioneins: Roles in heavy metal detoxification and homeostasis. *Annu. Rev. Plant Biol.* **2002**, *53*, 159–182. [CrossRef] [PubMed]
11. Bourdineaud, J.P.; Baudrimont, M.; Gonzalez, P.; Moreau, J.L. Challenging the model for induction of metallothionein gene expression. *Biochimie* **2006**, *88*, 1787–1792. [CrossRef] [PubMed]
12. Robinson, N.J.; Tommey, A.M.; Kuske, C.; Jackson, P.J. Plant metallothioneins. *Biochem. J.* **1993**, *295*, 1–10. [CrossRef] [PubMed]
13. Leszczyszyn, O.I.; Imam, H.T.; Blindauer, C.A. Diversity and distribution of plant metallothioneins: A review of structure, properties and functions. *Metallomics* **2013**, *5*, 1146–1169. [CrossRef] [PubMed]
14. Guo, W.J.; Bundithya, W.; Goldsbrough, P.B. Characterization of the *Arabidopsis* metallothionein gene family: Tissue-specific expression and induction during senescence and in response to copper. *New Phytol.* **2003**, *159*, 369–381. [CrossRef]
15. Benatti, M.R.; Yookongkaew, N.; Meetam, M.; Guo, W.J.; Punyasuk, N.; Abuqamar, S.; Goldsbrough, P. Metallothionein deficiency impacts copper accumulation and redistribution in leaves and seeds of *Arabidopsis*. *New Phytol.* **2014**, *202*, 940–951. [CrossRef] [PubMed]
16. Guo, W.J.; Meetam, M.; Goldsbrough, P.B. Examining the specific contributions of individual *Arabidopsis* metallothioneins to copper distribution and metal tolerance. *Plant Physiol.* **2008**, *146*, 1697–1706. [CrossRef] [PubMed]
17. Lee, J.; Shim, D.; Song, W.Y.; Hwang, I.; Lee, Y. Arabidopsis metallothioneins 2a and 3 enhance resistance to cadmium when expressed in *Vicia faba* guard cells. *Plant Mol. Biol.* **2004**, *54*, 805–815. [CrossRef] [PubMed]
18. Ren, Y.; Liu, Y.; Chen, H.; Li, G.; Zhang, X.; Zhao, J. Type 4 metallothionein genes are involved in regulating

Zn ion accumulation in late embryo and in controlling early seedling growth in *Arabidopsis*. *Plant Cell Environ.* **2012**, *35*, 770–789. [CrossRef] [PubMed]
19. Roosens, N.; Leplae, R.; Bernard, C.; Verbruggen, N. Variations in plant metallothioneins: The heavy metal hyperaccumulator *Thlaspi caerulescens* as a study case. *Planta* **2005**, *222*, 716–729. [CrossRef] [PubMed]
20. Hammond, J.P.; Bowen, H.C.; White, P.J.; Mills, V.; Pyke, K.A.; Baker, A.J.; Whiting, S.N.; May, S.T.; Broadley, M.R. A comparison of the *Thlaspi caerulescens* and *Thlaspi arvense* shoot transcriptomes. *New Phytol.* **2006**, *170*, 239–260. [CrossRef] [PubMed]
21. Yuan, J.; Chen, D.; Ren, Y.; Zhang, X.; Zhao, J. Characteristic and Expression Analysis of a Metallothionein Gene, *OsMT2b*, Down-Regulated by Cytokinin Suggests Functions in Root Development and Seed Embryo Germination of Rice. *Plant Physiol.* **2008**, *146*, 1637–1650. [CrossRef] [PubMed]
22. Zhou, Y.; Chu, P.; Chen, H.; Li, Y.; Liu, J.; Ding, Y.; Tsang, E.W.; Jiang, L.; Wu, K.; Huang, S. Overexpression of *Nelumbo nucifera* metallothioneins 2a and 3 enhances seed germination vigor in *Arabidopsis*. *Planta* **2012**, *235*, 523–537. [CrossRef] [PubMed]
23. An, Z.; Li, C.; Zu, Y.; Du, Y.; Andreas, W.; Gromes, R.; Rausch, T. Expression of *BjMT2*, a metallothionein 2 from *Brassica juncea*, increases copper and cadmium tolerance in *Escherichia coli* and *Arabidopsis thaliana*, but inhibits root elongation in *Arabidopsis thaliana* seedlings. *J. Exp. Bot.* **2006**, *57*, 3575–3582.
24. Kumar, P.B.; Dushenkov, V.; Motto, H.; Raskin, I. Phytoextraction: The use of plants to remove heavy metals from soils. *Environ. Sci. Technol.* **1995**, *29*, 1232–1238. [CrossRef] [PubMed]
25. Salt, D.E.; Blaylock, M.; Kumar, N.P.; Dushenkov, V.; Ensley, B.D.; Chet, I.; Raskin, I. Phytoremediation: A novel strategy for the removal of toxic metals from the environment using plants. *Biotechnology* **1995**, *13*, 468–474. [CrossRef] [PubMed]
26. Cojocaru, P.; Gusiatin, Z.M.; Cretescu, I. Phytoextraction of Cd and Zn as single or mixed pollutants from soil by rape (*Brassica napus*). *Environ. Sci. Pollut. Res.* **2016**, *23*, 10693–10701. [CrossRef] [PubMed]
27. Gasic, K.; Korban, S.S. Expression of *Arabidopsis* phytochelatin synthase in Indian mustard (*Brassica juncea*) plants enhances tolerance for Cd and Zn. *Planta* **2007**, *225*, 1277–1285. [CrossRef] [PubMed]
28. Sridhar, B.B.M.; Diehl, S.V.; Han, F.X.; Monts, D.L.; Su, Y. Anatomical changes due to uptake and accumulation of Zn and Cd in Indian mustard (*Brassica juncea*). *Environ. Exp. Bot.* **2005**, *54*, 131–141. [CrossRef]
29. Ishikawa, S.; Noriharu, A.; Murakami, M.; Wagatsuma, T. Is *Brassica juncea* a suitable plant for phytoremediation of cadmium in soils with moderately low cadmium contamination?—Possibility of using other plant species for Cd-phytoextraction. *Soil Sci. Plant Nutr.* **2006**, *52*, 32–42. [CrossRef]
30. Marchiol, L.; Assolari, S.; Sacco, P.; Zerbi, G. Phytoextraction of heavy metals by canola (*Brassica napus*) and radish (*Raphanus sativus*) grown on multicontaminated soil. *Environ. Pollut.* **2004**, *132*, 21–27. [CrossRef] [PubMed]
31. Solhi, M.; Shareatmadari, H.; Hajabbasi, M.A. Lead and Zinc Extraction Potential of Two Common Crop Plants, *Helianthus Annuus* and *Brassica napus*. *Water Air Soil Pollut.* **2005**, *167*, 59–71. [CrossRef]
32. Touiserkani, T.; Haddad, R. Cadmium-induced stress and antioxidative responses in different *Brassica napus* cultivars. *J. Agric. Sci. Technol.* **2012**, *14*, 929–937.
33. Kim, S.H.; Lee, H.S.; Song, W.Y.; Choi, K.S.; Hur, Y. Chloroplast-targeted BrMT1 (*Brassica rapa* type-1 metallothionein) enhances resistance to cadmium and ROS in transgenicatabidopsis plants. *J. Plant Biol.* **2007**, *50*, 1–7. [CrossRef]
34. Zhang, M.; Takano, T.; Liu, S.; Zhang, X. Abiotic stress response in yeast and metal-binding ability of a type 2 metallothionein-like protein (PutMT2) from *Puccinellia tenuiflora*. *Mol. Biol. Rep.* **2014**, *41*, 5839–5849. [CrossRef] [PubMed]
35. Nagaharu, U. Genome analysis in Brassica with special reference to the experimental formation of *B. napus* and peculiar mode of fertilisation. *J. Jpn. Bot.* **1935**, *7*, 389–452.
36. Von Ruecker, A.A.; Wild, M.; Rao, G.S.; Bidlingmaier, F. Atrial natriuretic peptide protects hepatocytes against damage induced by hypoxia and reactive oxygen. Possible role of intracellular free ionized calcium. *J. Clin. Chem. Clin. Biochem.* **1989**, *27*, 531–537. [PubMed]
37. Palmiter, R.D. The elusive function of metallothioneins. *Proc. Natl. Acad. Sci. USA* **1998**, *95*, 8428–8430. [CrossRef] [PubMed]
38. Tottey, S.; Rondet, S.A.; Borrelly, G.P.; Robinson, P.J.; Rich, P.R.; Robinson, N.J. A copper metallochaperone for photosynthesis and respiration reveals metal-specific targets, interaction with an importer, and alternative sites for copper acquisition. *J. Biol. Chem.* **2002** *277*, 5490–5497. [CrossRef] [PubMed]

39. Coyle, P.; Philcox, J.; Carey, L.; Rofe, A.M. Metallothionein: The multipurpose protein. *Cell. Mol. Life Sci.* **2002**, *59*, 627–647. [CrossRef] [PubMed]
40. Guo, S.Y.; Xu, D.; Yang, P.; Yu, W.T.; Lv, M.K.; Yuan, D.R.; Yang, Z.H.; Zhang, G.H.; Sun, S.Y.; Wang, X.Q.; et al. A Novel Nonlinear Optical Complex Crystal with an Organic Ligand Coordinated Through an O Atom: Tetrathiocyanatocadmiummercury-Dimethyl Sulfoxide. *Cryst. Res. Technol.* **2001**, *36*, 609–614. [CrossRef]
41. Ansarypour, Z.; Shahpiri, A. Heterologous expression of a rice metallothionein isoform (OsMTI-1b) in *Saccharomyces cerevisiae* enhances cadmium, hydrogen peroxide and ethanol tolerance. *Braz. J. Microbiol.* **2017**, *48*, 537–543. [CrossRef] [PubMed]
42. Benáková, M.; Ahmadi, H.; Dučaiová, Z.; Tylová, E.; Clemens, S.; Tůma, J. Effects of Cd and Zn on physiological and anatomical properties of hydroponically grown *Brassica napus* plants. *Environ. Sci. Pollut. Res.* **2017**, *24*, 20705–20716. [CrossRef] [PubMed]
43. Chalhoub, B.; Denoeud, F.; Liu, S.; Parkin, I.A.; Tang, H.; Wang, X.; Chiquet, J.; Belcram, H.; Tong, C.; Samans, B.; et al. Early allopolyploid evolution in the post-neolithic *Brassica napus* oilseed genome. *Science* **2014**, *345*, 950–953. [CrossRef] [PubMed]
44. Sun, H.; Fan, H.-J.; Ling, H.-Q. Genome-wide identification and characterization of the bHLH gene family in tomato. *BMC Genom.* **2015**, *16*, 9. [CrossRef] [PubMed]
45. Sankoff, D.; Zheng, C.; Zhu, Q. The collapse of gene complement following whole genome duplication. *BMC Genom.* **2010**, *11*, 313. [CrossRef] [PubMed]
46. Wang, X.; Wang, H.; Wang, J.; Sun, R.; Wu, J.; Liu, S.; Bai, Y.; Mun, J.-H.; Bancroft, I.; Cheng, F.; et al. The genome of the mesopolyploid crop species *Brassica Rapa*. *Nat. Genet.* **2011**, *43*, 1035–1039. [CrossRef] [PubMed]
47. Qu, C.; Zhao, H.; Fu, F.; Wang, Z.; Zhang, K.; Zhou, Y.; Wang, X.; Wang, R.; Xu, X.; Tang, Z.; et al. Genome-wide survey of flavonoid biosynthesis genes and gene expression analysis between black- and yellow-seeded *Brassica napus*. *Front. Plant Sci.* **2016**, *7*. [CrossRef] [PubMed]
48. Zimeri, A.M.; Dhankher, O.P.; Mccaig, B.; Meagher, R.B. The Plant MT1 Metallothioneins are Stabilized by Binding Cadmiums and are Required for Cadmium Tolerance and Accumulation. *Plant Mol. Biol.* **2005**, *58*, 839–855. [CrossRef] [PubMed]
49. Framond, A.J.D. A metallothionein-like gene from maize (*Zea mays*) Cloning and characterization. *FEBS Lett.* **1991**, *290*, 103–106. [CrossRef]
50. Zhou, J.; Goldsbrough, P.B. Structure, organization and expression of the metallothionein gene family in *Arabidopsis*. *Mol. Gen. Genet.* **1995**, *248*, 318–328. [CrossRef] [PubMed]
51. Foley, R.C.; Singh, K.B. Isolation of a Vicia faba metallothionein-like gene, expression in foliar trichomes. *Plant Mol. Biol.* **1994**, *26*, 435–444. [CrossRef] [PubMed]
52. Shawber, C.; Nofziger, D.; Hsieh, J.J.; Lindsell, C.; Bögler, O.; Hayward, D.; Weinmaster, G. Notch signaling inhibits muscle cell differentiation through a CBF1-independent pathway. *Development* **1996**, *122*, 3765–3773. [PubMed]
53. Liu, J.X.; Zu, Y.G.; Shi, X.G.; Ai, Y.Z.; Du, Y.J.; Fu, Y.J.; An, Z.G. BjMT2, a metallothionein type-2 from *Brassica juncea*, may effectively remove excess lead from erythrocytes and kidneys of rats. *Environ. Toxicol. Pharmacol.* **2007**, *23*, 168–173. [CrossRef] [PubMed]
54. Ahn, Y.O.; Kim, S.H.; Lee, J.; Kim, H.; Lee, H.S.; Kwak, S.S. Three *Brassica rapa* metallothionein genes are differentially regulated under various stress conditions. *Mol. Biol. Rep.* **2012**, *39*, 2059–2067. [CrossRef] [PubMed]
55. Farooq, M.A.; Gill, R.A.; Ali, B.; Wang, J.; Islam, F.; Ali, S.; Zhou, W. Subcellular distribution, modulation of antioxidant and stress-related genes response to arsenic in *Brassica napus* L. *Ecotoxicology* **2016**, *25*, 350–366. [CrossRef] [PubMed]
56. Altschul, S.F.; Madden, T.L.; Schäffer, A.A.; Zhang, J.; Zhang, Z.; Miller, W.; Lipman, D.J. Gapped BLAST and PSI-BLAST: A new generation of protein database search programs. *Nucleic Acids Res.* **1997**, *25*, 3389–3402. [CrossRef] [PubMed]
57. Larkin, M.A.; Blackshields, G.; Brown, N.; Chenna, R.; McGettigan, P.A.; McWilliam, H.; Valentin, F.; Wallace, I.M.; Wilm, A.; Lopez, R. Clustal W and Clustal X version 2.0. *Bioinformatics* **2007**, *23*, 2947–2948. [CrossRef] [PubMed]
58. Tamura, K.; Stecher, G.; Peterson, D.; Filipski, A.; Kumar, S. MEGA6: Molecular evolutionary genetics analysis version 6.0. *Mol. Biol. Evol.* **2013**, *30*, 2725–2729. [CrossRef] [PubMed]

59. Bailey, T.L.; Boden, M.; Buske, F.A.; Frith, M.; Grant, C.E.; Clementi, L.; Ren, J.; Li, W.W.; Noble, W.S. MEME SUITE: Tools for motif discovery and searching. *Nucleic Acids Res.* **2009**, *37*, W202–W208. [CrossRef] [PubMed]
60. Voorrips, R. MapChart: Software for the graphical presentation of linkage maps and QTLs. *Heredity* **2002**, *93*, 77–78. [CrossRef]
61. Qu, C.; Fu, F.; Lu, K.; Zhang, K.; Wang, R.; Xu, X.; Wang, M.; Lu, J.; Wan, H.; Zhanglin, T.; et al. Differential accumulation of phenolic compounds and expression of related genes in black-and yellow-seeded *Brassica napus*. *J. Exp. Bot.* **2013**, *64*, 2885–2898. [CrossRef] [PubMed]
62. Wu, G.; Li, Z.; Yuhua, W.; Yinglong, C.; Changming, L. Comparison of Five Endogenous Reference Genes for Specific PCR Detection and Quantification of *Brassica napus*. *J. Agric. Food Chem.* **2010**, *58*, 2812–2817. [CrossRef] [PubMed]
63. Lu, K.; Li, T.; He, J.; Chang, W.; Zhang, R.; Liu, M.; Yu, M.; Fan, Y.; Ma, J.; Sun, W. qPrimerDB: A thermodynamics-based gene-specific qPCR primer database for 147 organisms. *Nucleic Acids Res.* **2018**, *46*, D1229–D1236. [CrossRef] [PubMed]

4

iTRAQ-Based Proteomic Analysis of Ogura-CMS Cabbage and its Maintainer Line

Fengqing Han [1], Xiaoli Zhang [2], Limei Yang [1], Mu Zhuang [1], Yangyong Zhang [1], Zhansheng Li [1], Zhiyuan Fang [1] and Honghao Lv [1,*]

[1] Institute of Vegetables and Flowers, Chinese Academy of Agricultural Sciences, Key Laboratory of Biology and Genetic Improvement of Horticultural Crops, Ministry of Agriculture, Beijing 100081, China; feng857142@163.com (F.H.); yanglimei@163.com (L.Y.); zhuangmu@163.com (M.Z.); zhangyangyong@163.com (Y.Z.); lizhansheng@163.com (Z.L.); fangzhiyuan@163.com (Z.F.)

[2] Tianjin Kernel Vegetable Research Institute, The National Key Laboratory of Vegetable Germplasm Innovation, The Enterprise key Laboratory of Tianjin Vegetable Genetics and Breeding, Jinjing Road, Xiqing District, Tianjin 300384, China; zxl19871009@163.com

* Correspondence: lvhonghao@caas.cn

Abstract: Ogura cytoplasmic male sterility (CMS) contributes considerably to hybrid seed production in *Brassica* crops. To detect the key protein species and pathways involved in Ogura-CMS, we analysed the proteome of the cabbage Ogura-CMS line CMS01-20 and its corresponding maintainer line F01-20 using the isobaric tags for the relative and absolute quantitation (iTRAQ) approach. In total, 162 differential abundance protein species (DAPs) were identified between the two lines, of which 92 were down-accumulated and 70 were up-accumulated in CMS01-20. For energy metabolism in the mitochondrion, eight DAPs involved in oxidative phosphorylation were down-accumulated in CMS01-20, whereas in the tricarboxylic acid (TCA) cycle, five DAPs were up-accumulated, which may compensate for the decreased respiration capacity and may be associated with the elevated O_2 consumption rate in Ogura-CMS plants. Other key protein species and pathways involved in pollen wall assembly and programmed cell death (PCD) were also identified as being male-sterility related. Transcriptome profiling revealed 3247 differentially expressed genes between the CMS line and the fertile line. In a conjoint analysis of the proteome and transcriptome data, 30 and 9 protein species/genes showed the same and opposite accumulation patterns, respectively. Nine noteworthy genes involved in sporopollenin synthesis, callose wall degeneration, and oxidative phosphorylation were presumably associated with the processes leading to male sterility, and their expression levels were validated by qRT-PCR analysis. This study will improve our understanding of the protein species involved in pollen development and the molecular mechanisms underlying Ogura-CMS.

Keywords: *Brassica oleracea*; Ogura-CMS; iTRAQ; transcriptome; pollen development

---

1. Introduction

Cabbage (*Brassica oleracea* L. var. *capitata*) is an important leafy vegetable cultivated worldwide, and it provides substantial amounts of fibre, vitamins, mineral elements and health-promoting nutrients. The Food and Agriculture Organization of the United Nations reported that the global harvested area of vegetables in 2014 was 20,119,000 ha, with cabbage and other cole crops accounting for approximately 12% of the total (2,470,000 ha; see http://faostat3.fao.org/).

Commercially available cabbage mainly consists of hybrid cultivars because of their significant levels of heterosis. Cross-pollination in hybrid seed production is mainly accomplished using male sterility and self-incompatibility. Self-incompatibility has several limitations, such as poor seed purity and high costs of parental reproduction, whereas male sterility is generally more reliable and economically effective [1,2]. Male sterility encompasses genic male sterility (GMS), caused by nuclear genes, and cytoplasmic male sterility (CMS), caused by interactions between mitochondrial and nuclear genes [3]. Currently, CMS represents the most widely used breeding tool in cabbage hybrid seed production [4,5]. Ogura cytoplasmic male sterility (Ogura-CMS) was discovered in radish [6] and has been transferred to several *Brassica* species [5,7–9]. The original Ogura-CMS *B. oleracea* line exhibits poor agronomic traits, which have been improved by protoplast fusion [10]. Ogura-CMS is stable and easy to transfer between species; thus, it has become one of the most important types of CMS in *B. oleracea* [11].

In addition to the crucial breeding role of CMS in harnessing heterosis, it provides important materials for studying gametophyte development and mitochondrial–nuclear interactions, etc. [3]. CMS has been observed in approximately 200 species and is inherited maternally [12]. At least 17 CMS-related genes have been studied at the genetic and molecular levels [13]. These loci share similar characteristics, with CMS primarily caused by either novel mitochondrial open reading frames (ORFs), which generally result from rearrangements or recombination events in the mitochondrial genome [14]. Fertility restoration is mediated by nuclear-encoded fertility restorer (Rf) genes, most of which encode pentatricopeptide repeat (PPR) proteins that counteract the influence of CMS-associated genes [15,16].

Two hypothetical CMS pathways have been proposed: (I) CMS inhibits energy production by disrupting the mitochondrial electron transport chain complex; and (II) CMS impairs the normal growth of cells, at least in *E. coli* and/or yeast, because its products are cytotoxic [13,17]. In *Brassica* and *Raphanus* species, Ogura-CMS is caused by a mitochondrial gene named *orf138*, which encodes a subunit of a large mitochondrial membrane complex ORF138 protein [18–20]. Both plant membrane fractionation and analyses of *E. coli* have suggested that the ORF138 protein forms oligomers in the inner mitochondrial membrane of male-sterile plants, which is similar to another CMS protein, T-URF1321. Although the ORF138 protein severely inhibits bacterial growth, it does not affect respiration [21]. In a later study, Duroc et al. reported that the complex formed by the ORF138 protein in the inner mitochondrial membrane exerted an uncoupling effect because the mitochondria isolated from sterile plants consumed more oxygen than those of fertile plants, and this uncoupling effect was compensated at the cell and tissue levels, especially in vegetative tissues/organs, although the compensatory effects were apparently not efficient in male reproductive organs [17]. Despite these genetic and molecular studies, the mechanisms underlying the interference exerted by CMS genes on male gametophyte development are largely unknown.

High-throughput next-generation sequencing (NGS) has facilitated transcriptome analyses of Ogura-CMS materials (including Chinese cabbage, cabbage, broccoli) and may help provide a comprehensive understanding of the mechanisms underlying Ogura-CMS [22–25]. Xing et al. performed transcriptome and proteome analyses (focused on the transcriptome) and identified gibberellin, and sporopollenin synthesis as important pathways in Ogura-CMS cabbage [25]. Other studies performing proteomic analyses using two-dimensional gel electrophoresis (2-DE) have identified distinct differences in the proteomes of Ogura-CMS and fertile plants [26,27], including the down-accumulated protein species associated with processes that include carbohydrate and energy metabolism, cell wall remodelling, flavonoid synthesis and up-accumulated protein species like protease inhibitors.

Our group developed several elite Ogura-CMS cabbage lines with excellent agronomic performance [5], and they have been successfully used for the hybrid seed production of many elite varieties. F01-20 is an elite cabbage line originally introduced from Canada [28], and its Ogura-CMS line CMS01-20 was bred through crosses with a different Ogura-CMS line using F01-20 as the male parent and subsequent recurrent backcrossing with F01-20 for more than 20 generations. The very similar genetic backgrounds of this CMS line and its maintainer line make them ideal materials for cabbage breeding as well as for studying the molecular mechanisms of Ogura-CMS.

Herein, we describe the isobaric tags for the relative and absolute quantitation (iTRAQ)-based proteome analysis of Ogura-CMS using the cabbage lines CMS01-20 and F01-20. Our goals were to identify essential differential abundance protein species (DAPs) and pathways between male-sterile and male-fertile lines and investigate their potential mechanistic roles in Ogura-CMS.

2. Results

2.1. Morphological and Microscopic Examination

CMS01-20 showed degenerated anthers and no visible pollen compared with its maintainer line F01-20 (Figure 1A,I). We further observed male gametophytes of the two lines at different developmental stages using light microscopy (Figure 1B–H,J–P). No obvious phenotypic differences were observed before the tetrad stage. At the tetrad stage, certain tetrads exhibited irregular shapes. At the late tetrad stage or shortly after the release of microspores, CMS01-20 tapetal cells were swollen and vacuolated and the separation of microspores was delayed, indicating defects in the dissolution of the callose and tetrad walls. At the uninucleate to bicellular microspore stage, the tapetum layers of CMS01-20 showed earlier degradation, and we also observed the aggregation of abnormal and vacuolated immature pollen microspores as described in previous studies [23,29], although these microspores may be held together by the residue of degraded tapetum cells rather than by the tapetal layers. At the mature microspore stage, the aborted microspores were completely degenerated, no pollen was viable in mature locules, and the anthers did not dehisce.

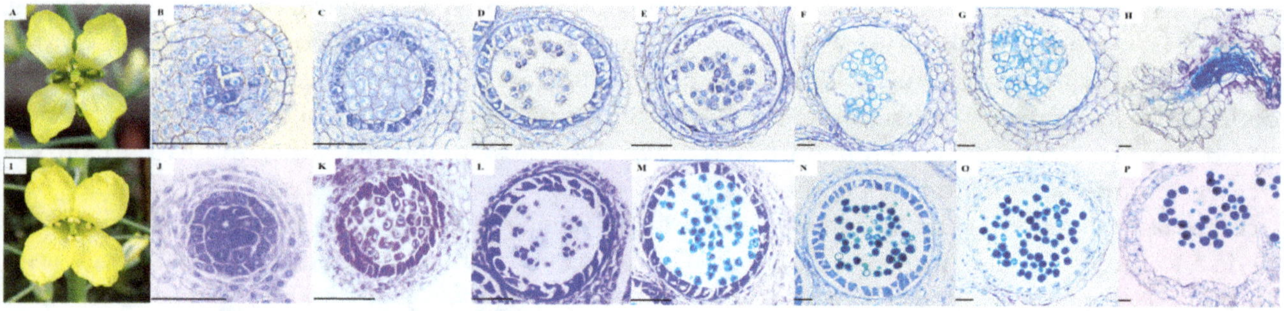

Figure 1. Phenotypes of Ogura cytoplasmic male sterility (Ogura-CMS) line CMS01-20 and its corresponding maintainer line F01-20. (**A–H**) CMS01-20; I-P: F01-20. (**A**) CMS01-20 shows degenerated anthers and no visible pollen; (**I**) F01-20 is fertility; (**B,J**) microsporocyte stage; (**C,K**) meiotic stage; (**D,L**) tetrad stage; (**E,M**) uninucleate stage; (**F,N**) bicellular microspore stage to trinucleate microspores stage. (**G,H,O,P**) mature pollen satge. Bar = 50 μm.

2.2. Overview of the Protein Species Identified Using iTRAQ Data

iTRAQ-based proteomic analysis was employed to assess protein changes between the buds of F01-20 and CMS01-20. A total of 197,216 spectra were generated. After filtering the data with Mascot, 53,777 spectra were matched to known sequences, over half of which (28,865) were unique (Figure 2A). By searching against the cabbage A2 reference genome database, 12,062 unique peptides were identified within 4188 protein species (Figure 2A), which represented 11.8% of all predicted protein-coding loci in the genome. Most of the identified peptides had lengths between 8 and 16 amino acids. More than 60% of the protein species had masses between 30 and 70 kDa (Figure 2B), and approximately 70% of the protein species contained at least two mapped peptides.

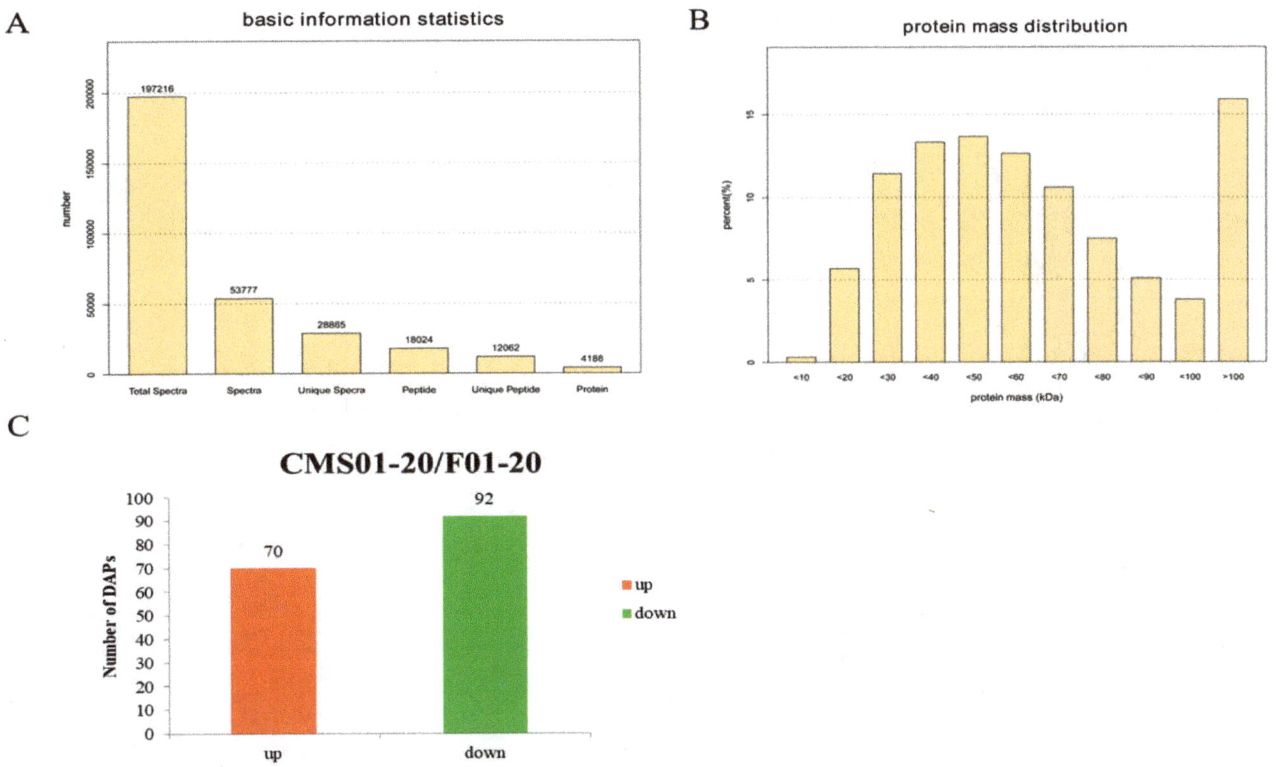

Figure 2. Protein species identification by the isobaric tags for the relative and absolute quantitation (iTRAQ) approach. (**A**) Number of spectra, peptide and protein; (**B**) percentage of protein mass distribution; (**C**) differential abundance protein species (DAPs) between CMS01-20 and F01-20.

To understand the functions of these 4188 protein species in cabbage buds, analyses were conducted with the Gene Ontology (GO), the Kyoto Encyclopaedia of Genes and Genomes (KEGG) and the Clusters of Orthologous Groups of proteins (COG) databases (Figure S1). The GO analysis showed that the biological process terms "metabolic process" (79.0%), "cellular process" (76.3%), "response to stimulus" (79.0%) and "single-organism process" (42.5%) were the most highly overrepresented functional groups; 10.8% of the protein species were involved in "reproductive process/reproduction"; the most overrepresented among the cellular component terms were "cell & cell part" (87.0%), "organelle" (72.6%), "organelle part" (38.6%) and "membrane" (34.9%); and the main for molecular function terms were "binding" (61.7%) and "catalytic activity" (57.4%).

2.3. *Overview of the DAPs between CMS01-20 and F01-20*

Following the criteria of a fold difference $\geq$1.2 and p value $\leq$ 0.05, we identified 162 DAPs between CMS01-20 and F01-20, of which 92 were down-accumulated and 70 were up-accumulated in the CMS line (Figure 2C, Table S1).

GO annotations were performed based on the TAIR GO Slim method provided by blast2GO. The GO annotations for 153 (94.4%) DAPs were divided into 35 functional groups, of which 16 were biological process GO terms (the largest category was "metabolic process"); cellular components accounted for 12 GO terms (the largest category was "cell"); and molecular functions accounted for 7 GO terms (the largest category was "catalytic activity") (Figure 3A). Of the DAPs, 112 (70.4%) were assigned to 62 KEGG pathways, and enriched in 14 KEGG pathways (p-value $<$ 0.05) including peroxisome (7.89%), cutin, suberine and wax biosynthesis (3.51%), sulphur metabolism (4.4%), ribosome (16.7%), tryptophan metabolism (5.26%), linoleic acid metabolism (1.75%), alpha-linolenic acid metabolism (3.51%), tropane, piperidine and pyridine alkaloid biosynthesis (3.51%), photosynthesis-antenna proteins (3.51%), carbon metabolism (14.9%), glyoxylate and dicarboxylate metabolism (6.1%), glutathione metabolism (5.3%), photosynthesis (5.3%), citrate cycle (the tricarboxylic acid) (4.4%) as listed in Table S2.

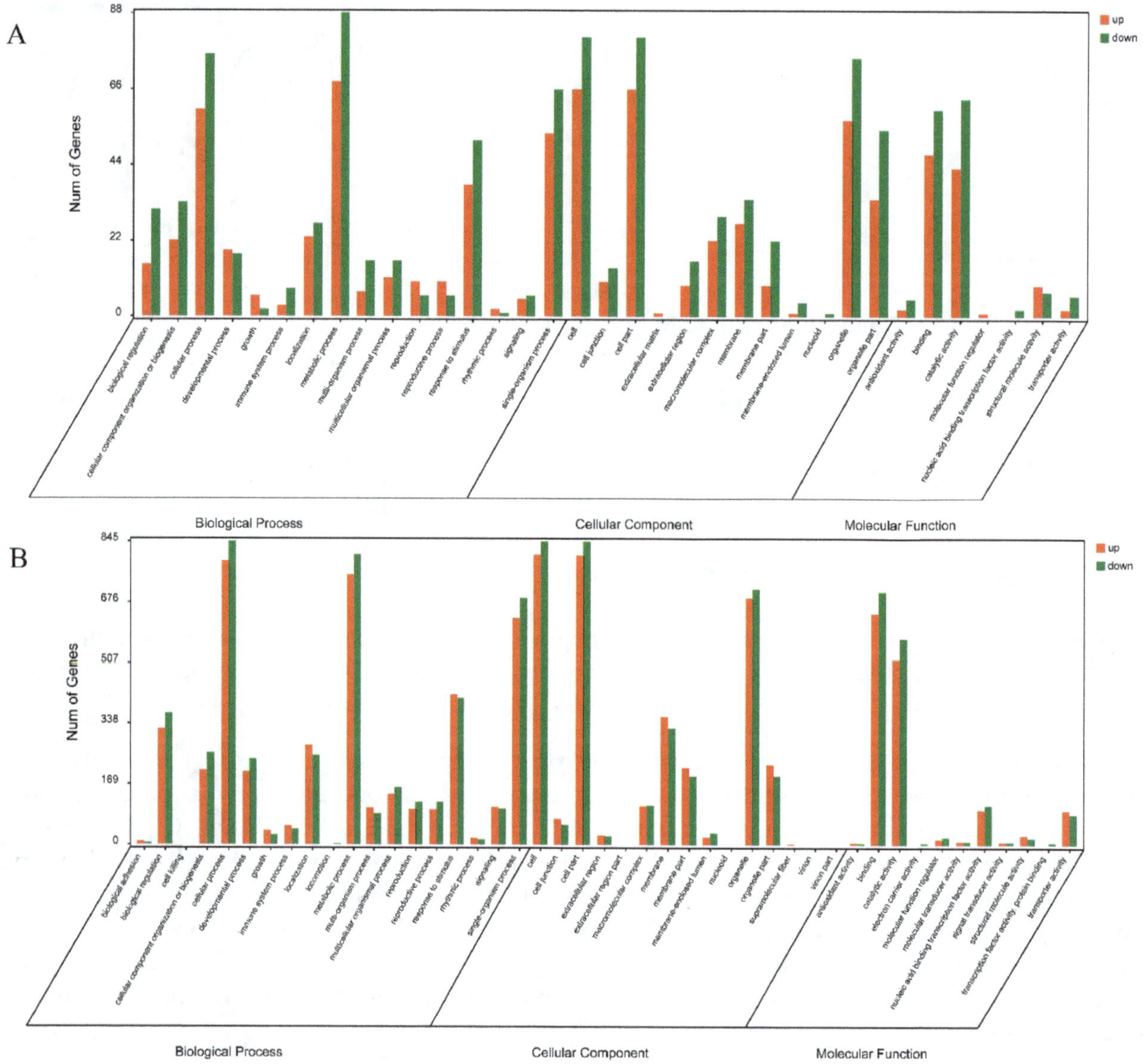

Figure 3. (**A**) Gene ontology categories for differential abundance protein species in the proteome data; (**B**) gene ontology categories for differentially expressed genes in the transcriptome data.

2.4. DAPs Involved in Oxidative Phosphorylation and TCA Cycle

Most CMS are associated with disturbances in the energy metabolism. The ORF138 protein formed a mitochondrial membrane complex that exhibited uncoupling effect, and affected oxygen consumption [17,21]. Thus oxidative phosphorylation and the TCA cycle are important pathways possibly affected by ORF138 protein. We found that all of the eight DAPs involved in oxidative phosphorylation were down-accumulated in CMS01-20 (Table 1), including one ETC complex I protein (Bol015119, NADH-ubiquinone oxidoreductase B18 subunit), one cytochrome *c* protein (Bol012326, cytochrome *c*), one complex IV protein (Bol010838, cytochrome *c* oxidase subunit Vc), and five complex V proteins (Bol009135 and Bol025034, ATP synthase subunit d; Bol017288, Bol025922 and Bol015469, ATP synthase 6 KD subunit). These results suggest CMS02-10 may have a decreased energy-generation capacity.

The citrate cycle (TCA cycle) provides NADH, $FADH_2$ and H^+ for oxidative phosphorylation. Five DAPs were annotated to be involved in TCA cycle: pyruvate dehydrogenase (Bol022522 and Bol008536) catalyses the synthesis of acetyl-CoA; citrate synthase 4 (Bol02950) catalyses the condensation of acetyl-CoA and oxaloacetate yielding citrate and CoA; aconitate hydratase (Bol029048) catalyses the conversion of citrate to isocitrate; 2-oxoglutarate dehydrogenase (Bol008657, 2-oxoglutarate dehydrogenase E1 component) catalyses the conversion of 2-oxoglutarate to succinyl-CoA. Interestingly, all the five DAPs were up-accumulated in CMS01-20, which may represent a compensatory mechanism triggered by the uncoupling effect of ORF138 protein [17], and this compensatory mechanism may be the reason for the increased oxygen consumption of male sterile plants [17]. The oxidative phosphorylation and TCA cycle network are shown in Figure 4.

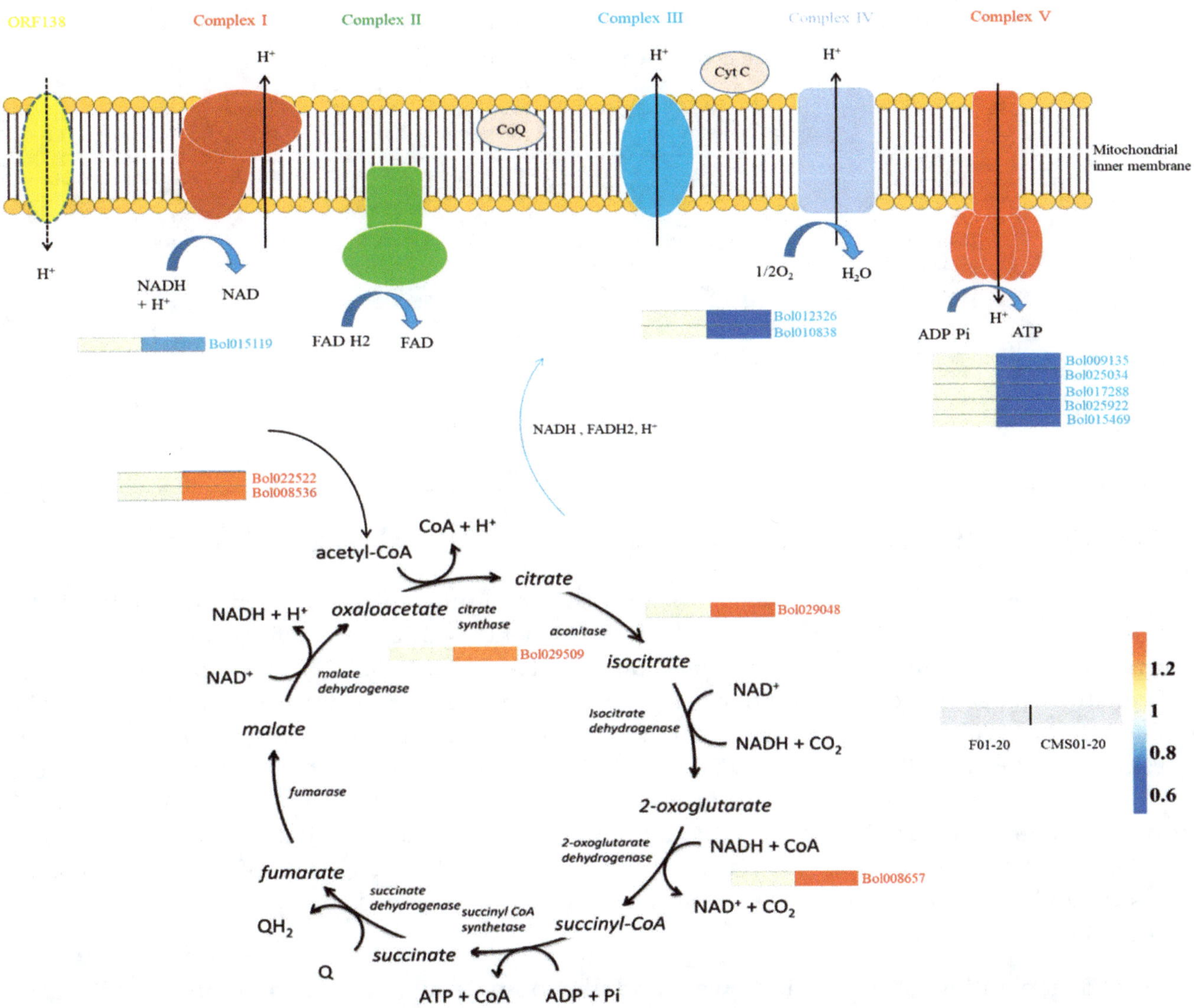

Figure 4. Differential abundance protein species involved in the oxidative phosphorylation and the tricarboxylic acid cycle. The possible uncoupling role of the ORF138 protein was also indicated on mitochondrial inner membrane. The fold changes of differential abundance protein species are indicated by the colour filled in the squares.

Table 1. Differential abundance protein species involved in oxidative phosphorylation, the tricarboxylic acid cycle, pollen wall, tetrad wall and programmed cell death.

	ID	Description	Up/Down in CMS Line
	Bol015119	NADH-ubiquinone oxidoreductase B18 subunit	down
	Bol009135	ATP synthase subunit d, mitochondrial-like	down
	Bol025034	ATP synthase subunit d, mitochondrial-like	down
Oxidative phosphorylation	Bol017288	mitochondrial ATP synthase 6 KD subunit	down
	Bol025922	mitochondrial ATP synthase 6 KD subunit	down
	Bol015469	ATP synthase 6 kDa subunit	down
	Bol012326	cytochrome c	down
	Bol010838	cytochrome c oxidase subunit Vc	down
	Bol022522	pyruvate dehydrogenase E1 component subunit beta-2	up
	Bol008536	pyruvate dehydrogenase E1 component subunit beta-2	up
TCA cycle	Bol008657	2-oxoglutarate dehydrogenase	up
	Bol029048	aconitate hydratase 1	up
	Bol029509	citrate synthase 4	up
	Bol013698	LAP5; Chalcone and stilbene synthase family protein	down
	Bol025267	LAP6; Chalcone and stilbene synthase family protein	down
pollen wall	Bol007277	MS2; fatty acyl-CoA reductase	down
	Bol034656	LAP5; Chalcone and stilbene synthase family protein	down
	Bol040704	cytochrome P450 703A2	down
	Bol010336	MS2; fatty acyl-CoA reductase	down
	Bol009974	probable glucan endo-1,3-beta-glucosidase A6	down
tetrad wall	Bol037314	*O*-Glycosyl hydrolases family 17 protein;	down
	Bol033052	beta-D-xylosidase 1	up
	Bol030909	beta-glucosidase 43 isoform X2	down
	Bol006999	catalase-3	down
	Bol026973	catalase-3	down
	Bol035942	allene oxide synthase	down
PCD	Bol037061	peroxisomal	down
	Bol005496	stromal ascorbate peroxidase	down
	Bol004624	glutathione S-transferase F9	down
	Bol033376	glutathione S-transferase F9	down

2.5. Other Ogura-CMS Related DAPs and Pathways

Notably, terms related to cell wall assembly were significantly enriched among the DAPs. Fifty-five protein species were involved in cellular component organization or biogenesis, among which 33 were down-accumulated and 22 were up-accumulated in CMS01-20. We focused on protein species involved in the assembly (or degeneration) of pollen exine and the tetrad wall. We identified six DAPs involved in the synthesis of pollen exine. In the flavonoid biosynthesis pathway, LAP5 (Bol013698, Bol034656), LAP6 (Bol025267) and CYP703A/CYP703A2 (Bol040704) were down-accumulated in CMS01-20. In unsaturated fatty acid and fatty acid elongation pathways, MS2 (Bol010336, Bol007277) was down-accumulation in CMS01-20 (Table 1, Figure 5). These genes are vital for the development of viable pollen; therefore, they may be partially responsible for the phenotype of Ogura-CMS plants, such as the underdeveloped exine and the aberrant/aborted microspores observed here and in previous studies [23,29].

The degeneration of the callose wall and the outer wall (or pollen mother cell wall) is required for the release of microspores. Pectinase, endo-β-1,3-glucanases, exo-β-1,3-glucanase, and endo-β-1,4-glucanase enzymes are likely involved in this degeneration process [30]. We identified four DAPs that may be involved in the degradation of callose walls (Table 1, Figure 5). Bol009974 (predicted probable glucan endo-1,3-beta-glucosidase A6), Bol037314 (*O*-Glycosyl hydrolases family 17 protein) and Bol030909 (predicted beta-glucosidase) were down-accumulated in CMS01-20. Bol033052 (predicted beta-D-xylosidase, glycosyl hydrolase family 3) was up-accumulated in CMS01-20. They may be responsible for the pollen separation defects of CMS01-20 (shown in Figure 1). However, we did not identify any DAPs associated with pectin degradation.

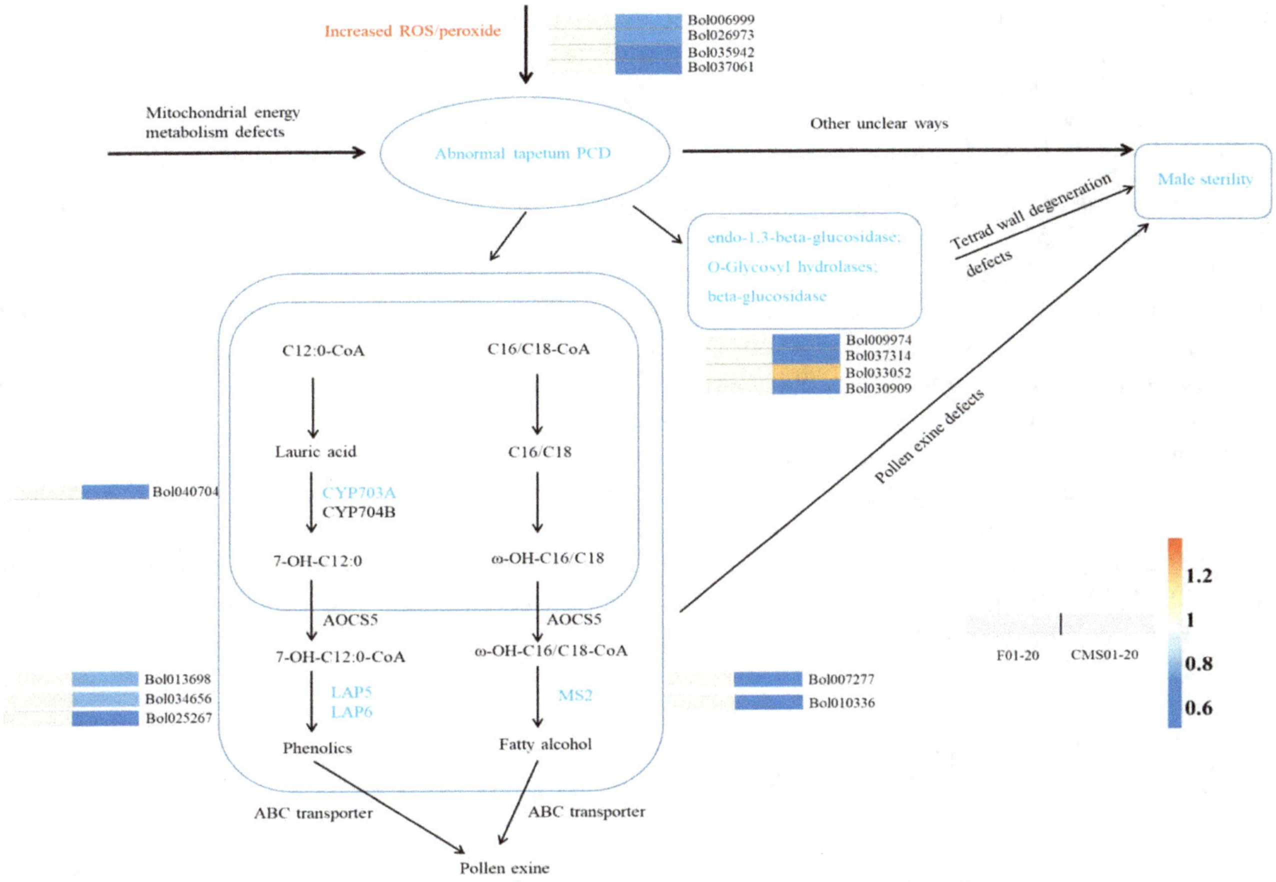

Figure 5. A possible network according to which abnormal tapetal programmed cell death is triggered by reactive oxygen species (ROS), resulting in male sterility. The differential abundance protein species involved in pollen exine formation and tetrad wall degeneration are also shown. The fold changes of differential abundance protein species are indicated by the colour filled in the squares.

Several previous studies have proposed a hypothetical mechanism in which CMS proteins trigger abnormal programmed cell death (PCD), which is usually associated with an increase in reactive oxygen species (ROS) and the release of cytochrome *c* in male organs, such as the tapetum [31–33]. Additionally, tapetal PCD often depends on the generation of ROS that can be detoxified by antioxidative enzymes including catalases [34,35]. In this study, we identified nine DAPs associated with PCD (Table 1, Figure 5). In the glyoxylate and dicarboxylate metabolism pathway, two catalase-3 proteins (Bol006999 and Bol026973) and one glycolate oxidase (Bol037061) were down-accumulated in CMS01-20. In the alpha-linolenic acid metabolism/linolenic acid metabolism pathway, four DAPs were down-accumulated in CMS01-20, among which allene oxide synthase (Bol035942) was a key enzyme catalysing the dehydration of the hydroperoxide to an unstable allene oxide. In the glutathione metabolism pathway, five down-accumulated DAPs and one up-accumulated DAP were identified. Among these five down-accumulated DAPs, Bol005496, Bol004624 and Bol033376 showed peroxidase activity. ROS may burst due to the down-accumulation of these enzymes and may trigger the abnormal PCD of the tapetum. Additionally, the disruption of allene oxide synthase DDE2 (the homologue of Bol035942) in *Arabidopsis* resulted in male sterility. Two oxygen-evolving enhancer protein species (Bol023353, Bol041074) were also associated with PCD and down-accumulated in CMS01-20, but they were predicted to be involved in the photosynthesis pathway and thus may not be CMS-related protein species.

We also identified DAPs and pathways that were similarly found in previous proteomic and/or transcriptomic analysis cases for male-sterile plants [25,36,37]. For example, the ribosome pathway (12 up-accumulated and 7 down-accumulated) and protein processing in the endoplasmic reticulum pathway (4 up-accumulated and 0 down-accumulated) was also identified in male-sterile cabbage and soybean [25,36,37], but their roles are largely unknown. For all the remaining DAPs and pathways, we did not find clues as to their possible roles in Ogura-CMS.

2.6. Joint Proteome–Transcriptome Analysis

To better understand the mechanisms underlying Ogura-CMS in cabbage, we analysed the transcriptomes of the CMS line CMS01-20 and its maintainer F01-20 via NGS. The RNA-seq libraries for F01-20 and CMS01-20 produced 91,952,648 and 98,470,304 clean reads, respectively. In total, 32,687 transcripts were identified for F01-20, and 32,680 transcripts for CMS01-20. A total of 3247 differentially expressed genes (DEGs) ($p < 0.05$) were identified, including 1525 up-accumulated and 1722 down-accumulated genes (Table S3). The GO analysis classified these DEGs into 45 GO categories, which showed a similar pattern to that of the DAPs (Figure 3B).

Integrative analyses comparing proteome and transcriptome data were performed between CMS01-20 and F01-20. The genes fell into nine groups based on the calculated $\log_2$ accumulation ratios of their protein species and transcripts (Figure 6A,B). These results showed poor correlation between the mRNA and protein species accumulation patterns as has been described in many previous studies [36,38,39]; this may be related to a combination of translational regulation, protein localization, protein modification, degradation, and other factors.

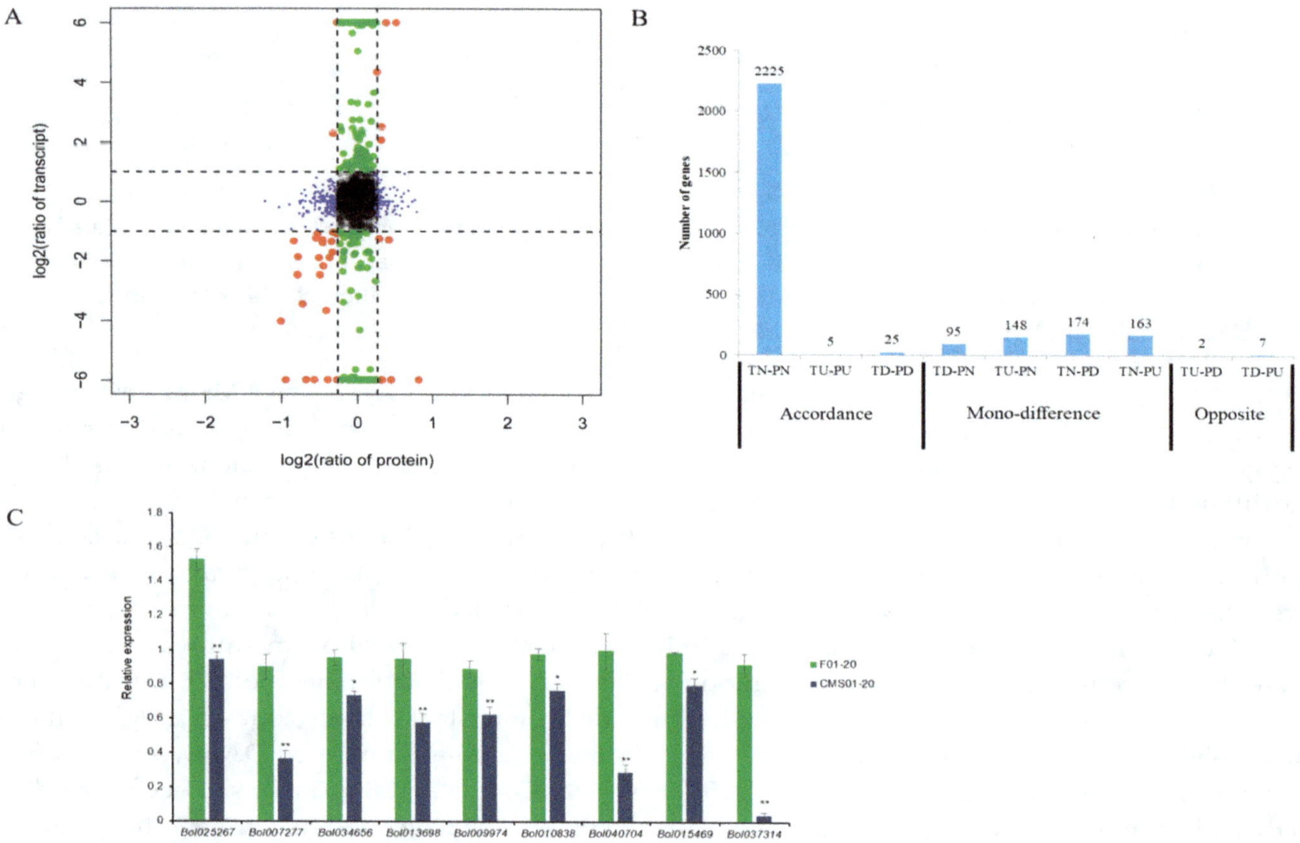

Figure 6. Integrative analyses comparing proteome and transcriptome data. (**A**) Genes were divided into nine groups according to $\log_2$ ratios of the protein species (y-axis) and transcripts (x-axis); (**B**) Number of genes among the nine groups in (**A**). T: transcript; P: protein species; N, no difference; U: up-accumulation; D, down-accumulation. (**C**) Expression validation for nine key genes by qRT-PCR. * $p < 0.05$, ** $p < 0.01$.

Although this conjoint analysis showed little overlap between the gene accumulation at the transcript and protein levels, certain noteworthy genes involved in sporopollenin synthesis (Bol013698,

Bol025267, Bol034656, Bol040704, Bol007277), callose wall degeneration (Bol009974, Bol037314), and oxidative phosphorylation (Bol015469, Bol010838) showed accordance patterns. The expression levels of these nine genes were validated by qRT-PCR analysis, which revealed that all of the genes showed expression patterns consistent with the RNA-seq data (Figure 6C).

3. Materials and Methods

3.1. Plant Materials and Sample Preparation

Cabbage CMS line 01-20 (CMS01-20) and its maintainer line F01-20 were used in this study. These lines were sown on 20 August, transferred to a cold frame for vernalisation on 20 November and finally transplanted to a greenhouse on 3 March 2016 for bolting and flowering. All plant materials were obtained from the Institute of Flowers and Vegetables of the Chinese Academy of Agriculture Sciences (IVFCAAS, Beijing, China). During the flowering stage, flower buds with different lengths were sampled to observe a range of microspore developmental stages and identify differences between CMS01-20 and F01-20 using an Olympus CX31 optical microscope (Olympus Japan Co., Tokyo, Japan). Based on the microscopic examination results, flower buds before the bicellular microspore stage ($\leq$3.5 mm) were collected for the transcriptome and proteome analyses. All collected buds were immediately frozen in liquid nitrogen and stored at -80 °C. Three biological replicate were performed for all experiments.

3.2. Microscopy

Flower buds with different lengths were fixed in formalin-aceto-alcohol (FAA), dehydrated in an ethanol series, embedded in paraffin, sectioned into 3–5 μm transverse slices using a microtome and stained with 1% toluidine blue as described by Lou et al. [40]. Then the anther transverse sections were observed with an Olympus CX31 optical microscope (Olympus Japan Co., Tokyo, Japan) and photographed with a Nikon 550D camera (Canon, Tokyo, Japan).

3.3. iTRAQ Analysis and Protein Species Annotation

The total protein species was extracted and subjected to iTRAQ labelling, strong cation exchange (SCX) separation and LC-electrospray ionization tandem mass spectrometry (LC-MS/MS) analysis using the same method as described by Chu et al. [41].

After converting them into MGF files, the raw iTRAQ data files were used for protein species identification and quantification. Database searches were performed using Mascot version 2.3.02 (Matrix Science, Boston, MA, USA) against a cabbage database, including 35,400 sequences from the *B. oleracea* genome A2 [42]. The search parameters were set as previously described [38].

Protein species with a fold change $\geq$1.2 (CMS01-20 vs. F01-20) and a false discovery rate (FDR) < 0.05 in at least two replicates were defined as differential abundance protein species (DAPs). All protein species identified were functionally annotated and classified based on Gene Ontology (GO) annotations (http://www.geneontology.org/), the Clusters of Orthologous Groups of proteins (COG) database (http://www.ncbi.nlm.nih.gov/COG/) and the Kyoto Encyclopaedia of Genes and Genomes (KEGG) database (http://www.genome.jp/kegg/pathway.html). DAPs were further analysed using the GO and KEGG databases to identify significantly enriched functional subcategories and metabolic pathways.

3.4. RNA-Seq Analysis and Conjoint Analysis with Proteome Data

Total RNA was extracted using an RNAprep pure Plant Kit (TIANGEN, Beijing, China) following the manufacturer's instructions. High-quality RNA from each sample was used for cDNA library construction and RNA sequencing on an Illumina HiSeq 2500TM platform (Gene Denovo Biotechnology Co., Guangzhou, China). To obtain clean high-quality reads, adapter sequences, low-quality reads (>50% bases with Q-value $\leq$ 20) and unknown bases (>10% N bases) were removed from the raw reads. The short read alignment tool Bowtie2 [43] was used to map reads to a ribosomal RNA (rRNA) database. After removing the rRNA mapped reads, each sample read was then mapped to the reference genome (ftp://brassicadb.org/Brassica_oleracea/) with TopHat2 (version 2.0.3.12) [44]. Gene expression levels were normalized using the FPKM (fragments per kilobase of transcript per million mapped reads) method [45]. The edgeR package (http://www.rproject.org/) was used to identify differentially expressed genes (DEGs) between two samples. We defined genes with a fold change ≥ 2 and an FDR < 0.05 as significant DEGs. DEGs were then analysed for the enrichment of GO functions and KEGG pathways. GO terms or pathways with FDR ≤ 0.05 were defined as significantly enriched in DEGs.

For the conjoint analysis of DAPs and DEGs, the transcriptome and proteome data were combined using the same *B. oleracea* genome A2 database. Thresholds of "FDR ≤ 0.05, $|\log 2FC| \geq 1$" and "$p \leq 0.05$, $|FC| \geq 1.5$" were set to select DEGs and DAPs, respectively. The correlation between the expression levels of the DAPs and their corresponding mRNAs were analysed by Pearson correlation tests.

3.5. Quantitative RT-PCR Analysis

Quantitative real-time RT-PCR (qRT-PCR) analyses were performed to validate the results from the DEGs. Total RNA was extracted from the buds of CMS01-20 and F01-20 plants using an RNAprep pure Plant Kit (TIANGEN, Beijing, China) according to the manufacturer's instructions. RNA was treated with RNase-free DNase I (Fermentas, Harrington, QC, Canada) to remove genomic DNA. First-strand cDNA was synthesized using a PrimeScript 1st Strand cDNA Synthesis Kit (Takara, Kyoto, Japan). qRT-PCR reactions were conducted using SYBR Premix Ex Taq II (Tli RNase H Plus; Takara, Dalian, China) with a CFX96 Touch Real-Time PCR Detection System (Bio-Rad, Hercules, CA, USA). Three biological replicates (with three technical replicates for each biological replicate) were analysed for each gene. The relative expression level of each gene was estimated by the $2^{-\Delta\Delta Ct}$ method [46]. The *B. oleracea actin* gene (GenBank accession number AF044573.1) [23] was used as an internal control.

4. Discussion

As an important type of CMS in *Brassica* and *Raphanus* species, Ogura-CMS contributes significantly to hybrid seed production. Researchers have long been interested in Ogura-CMS, although the mechanisms underlying its ability to interfere with pollen development remain unclear. Previous transcriptome and proteome analyses of Ogura-CMS provided basic knowledge of the DAPs between Ogura-CMS plants and their maintainer lines [22–27]. iTRAQ is also an efficient and reliable approach for the relative and absolute quantification of protein species, and it has been applied in proteome analyses of CMS and GMS in several plants, including cabbage [25,37] soybean [36], cotton [38], and cybrid pummelo [47]. Herein, we reported the iTRAQ analysis of an elite cabbage Ogura-CMS line and its corresponding maintainer line. Many more protein species were identified using this method compared to the traditional 2-DE technique [26,27]. In total, 162 DAPs were identified, with 92 protein species down-accumulated and 70 up-accumulated in CMS01-20. These DAPs are mainly involved in carbon metabolism, energy metabolism, and cell wall assembly, etc.

Many CMS-related proteins are associated with deficiencies in the ETC and oxidative phosphorylation [3]. In sunflower, a chimeric mitochondrial ORF522 protein has been described that likely decreases ATP hydrolysis via mitochondrial ATP synthase [48]. In a CMS tobacco, the ATP/ADP ratio is significantly decreased in the floral buds of male-sterile plants [49]. In wild beet, mitochondrial gene *G* alters the molecular weight of a respiratory chain complex subunit, and the male-sterile *G* cytoplasm plants exhibit severely reduced cytochrome *c* oxidase activity [50]. In the HL CMS line of rice, ORFH79 disrupts the F_0F_1-ATPase, and reduced protein quantity and enzyme activity are observed in sterile plants [51]. In a more recent study, ORFH79 was confirmed to decrease the enzymatic activity of the mitochondrial ETC complex III by interacting with P61, a subunit of the ETC complex III, which resulted in a deficiency in ATP production and an increase in reactive oxygen species

(ROS) content [52]. In the present study, we identified eight predicted ETC components differentially accumulated between the floral buds of CMS01-20 and F01-20. Five of these genes are predicted to be components of the mitochondrial ATP synthase complex and were down-accumulated in the CMS line, similar to a previous report in rapeseed [27]. Additionally, two cytochrome *c* (oxidase) genes and one NADH-ubiquinone oxidoreductase were significantly down-accumulated in sterile plants. These results suggest CMS02-10 may have a decreased respiration capacity, although Duroc et al. reported that ORF138 protein does not impair the capacities of electron transport chain complexes I, II, IV, or ATP synthase [17]. Interestingly, all DAPs involved in the TCA cycle were up-accumulated in CMS01-20. The TCA cycle provides precursors for many biochemical pathways and produces energy in the form of ATP largely via oxidative phosphorylation. The up-accumulation of TCA genes may compensate for the decreased respiration capacity and may be associated with the elevated O_2 consumption rate observed in mitochondria from Ogura-CMS plants [17].

Pollen grains are covered by an exine wall that consists of sporopollenin, which provides essential protection from the environment and is involved in interactions with female stigma cells. Although the mechanisms of exine formation are not well understood, dozens of genes regulating sporopollenin biosynthesis have been characterized [53]. Six genes involved in sporopollenin synthesis were identified in the present study. Except for *Bol010336*, these genes were all down-accumulated at both the transcript and protein levels in CMS01-20, which was confirmed by the qRT-PCR analysis, whereas *Bol010336* was down-accumulated at only the protein level. *LAP5* (homologue to *Bol013698*, *Bol034656*) encodes an anther-specific chalcone and stilbene synthase (CHS) family protein, but does not present CHS activity in vitro, and it may act as a multifunctional enzyme or could be involved in a novel pathway for sporopollenin synthesis [54,55]. The mutation of this gene results in abnormal exine patterning. *LAP6* (homologue to *Bol025267*) is similar to *LAP5*, and double mutants of *LAP5* and *LAP6* exhibit strong male sterility because of a lack of exine on the surface of the pollen grains. *CYP703A/CYP703A2* (homologue to *Bol040704*) is specifically expressed in the anthers of land plants, and is involved in catalysing medium-chain saturated fatty acids and thus is essential for sporopollenin synthesis; moreover, *cyp703a* mutants produces pollen grains without exine, and they display a partial male-sterile phenotype [56]. *Male sterility 2* (homologue to *Bol010336*) is the first gene identified through a genetic approach using mutants with exine defects [57], and it encodes a fatty acid reductase that is responsible for the accumulation of C16 and C18 fatty alcohols, which are essential for pollen exine wall biosynthesis [58]. The homologue of this gene in moss shows a conserved function, suggesting that *MS2* is a core component of the sporopollenin biosynthetic pathway [59]. However, in Ogura-CMS plants, the down-accumulation of these protein species/genes may not explain the vacuolated and early degenerated tapetum phenotypes. Conversely, these genes may be down-accumulated because of the abnormal tapetum development (possibly caused by abnormal programmed cell death) because the tapetum supplies necessary metabolites, nutrients, and sporopollenin precursors for the normal development of the male gamete [60]. Indeed, Ogura-CMS tapetal cells showed reduced secretory activity [29]. Thus, other genes are likely involved in the impaired function of Ogura-CMS tapetal cells.

At the late tetrad stage, some key enzymes are secreted from tapetal cells to dissolve the callose wall and the outer wall. Three *quartet* (*QRT*) genes have been identified in *A. thaliana*, and they are involved in the pectin degradation of the pollen mother wall [61,62]. Mutants of these three *QRT* genes produce microspore tetrads that fail to separate, although these adhered microspores are viable [61,62]. Tratt reported other enzymes that are likely involved in dissolving the pollen tetrad walls, including endo-β-1,3-glucanases, exo-β-1,3-glucanase, and endo-β-1,4-glucanase [30]. In the present study, a light microscopy examination suggested a delayed separation of CMS01-20 microspores, which may be caused by these tetrad wall degradation-related genes. We also identified four down-accumulated genes that may be involved in the degradation of tetrad walls. In Ogura-CMS *Brassica napus*, Sheoran et al. also reported the down-accumulation of protein species associated with cell wall remodelling, including β-1,3-glucanase and pectinesterase using a 2-DE approach [27]. We also considered the down-regulation of these protein species/genes as a result of abnormal tapetal cells.

Premature PCD in the tapetum was observed in PET1-CMS cytoplasm sunflowers as indicated by cell condensation, oligonucleosomal cleavage of nuclear DNA, chromatin separation into delineated masses, and initial mitochondrial persistence [31]. This early PCD may be caused by the release of cytochrome *c* from the mitochondria into the cytosol of tapetal cells [31]. HL CMS rice showed a PCD phenotype in microspores accompanied by inner mitochondrial membrane disruption [32] that was triggered by chronic oxidative stress caused by increased ROS levels and reduced superoxide dismutase (SOD), ascorbate peroxidase (APX) and catalase activity in mitochondria. In wild abortive CMS (CMS-WA) rice, Luo et al. reported that WA352 interacts with nuclear cytochrome *c* oxidase 11 (COX11) to inhibit its function in peroxide metabolism, and this interaction was demonstrated to be responsible for premature PCD in the tapetum and male sterility [33].

Although PCD in the tapetum is a feature of normal development, premature or delayed tapetum PCD usually results in male sterility due to tapetum's crucial role in pollen development [13,33]. Similar to several previous studies of Ogura-CMS plants, we observed a premature PCD phenotype in the tapetum [23,29], which may be responsible for the down-accumulation of many pollen wall assembly and pollen development-related genes, ultimately leading to male sterility. Therefore, identifying the key genes involved in tapetum PCD is crucial. In this study, we identified nine DAPs associated with PCD, especially, four protein species that are highly related to PCD: the two predicted catalase-3 proteins (Bol006999, and Bol026973, which catalyses the breakdown of hydrogen peroxide into water and oxygen), one allene oxide synthase (Bol035942, which catalyses dehydration of the hydroperoxide to an unstable allene oxide in the JA biosynthetic pathway), and one glycolate oxidase (Bol037061, which encodes a glycolate oxidase that modulates reactive oxygen species-mediated signal transduction).

Decreased accumulation of these protein species, especially catalase and allene oxide synthase, may trigger PCD in the tapetum mediated by hydrogen peroxide or other ROS [13,63,64]. In addition, the homologue of Bol035942 in *Arabidopsis* is DDE2 (AT5G42650), which is an enzyme involved in jasmonic acid biosynthesis, and *dde2*-2 mutants show male sterility and exhibit filament elongation and defects in anther dehiscence [65]. Thus, *Bol037061* may be associated with the small indehiscent anther phenotype of Ogura-CMS.

5. Conclusions

The present study provided an iTRAQ-based proteome analysis of Ogura-CMS using the cabbage Ogura-CMS line CMS01-20 and its isogenic maintainer line F01-20. A total of 4188 proteins were identified, and 162 were designated as DAPs. Key pathways and DAPs involved in processes including energy metabolism in mitochondrion, assembly/degeneration of pollen exine and the tetrad wall, and programmed cell death were found to be closely related to male sterility. Transcriptome profiling revealed 3247 differentially expressed genes between the CMS line and the fertile line. Additionally, the integrative analyses of the transcriptome and proteome data revealed nine Ogura-CMS-related genes showing accordance accumulation patterns at the transcript and protein levels, and the expression levels of these nine genes were validated by qRT-PCR. This study improves our understanding of the genes associated with pollen development and the molecular mechanisms of Ogura-CMS.

Author Contributions: Data curation, F.H.; Formal analysis, F.H.; Funding acquisition, Z.F.; Project administration, Z.F.; Resources, L.Y.; Supervision, Z.F. and H.L.; Validation, L.Y., M.Z., Y.Z., Z.L. and H.L.; Writing—original draft, F.H. and X.Z.; Writing—review & editing, F.H.

Acknowledgments: The work reported here was performed in the Key Laboratory of Biology and Genetic Improvement of Horticultural Crops, Ministry of Agriculture, Beijing 100081, China.

References

1. Fang, Z.Y.; Liu, Y.M.; Yang, L.M.; Wang, X.W.; Zhuang, M.; Zhang, Y.Y.; Sun, P.T. A survey of research in genetic breedings of cabbage in China. *Acta Hort. Sin.* **2002**, *29*, S657–S663.
2. Prakash, C.; Verma, T. Heterosis in cytoplasmic male sterile lines of cabbage. *Crucif. Newslett.* **2004**, *25*, 49–50.
3. Chen, L.; Liu, Y.G. Male sterility and fertility restoration in crops. *Annu. Rev. Plant Biol.* **2014**, *65*, 579–606. [CrossRef] [PubMed]
4. Fang, Z.Y.; Sun, P.T.; Liu, Y.M.; Yang, L.M.; Wang, X.W.; Hou, A.F.; Bian, C.S. A male sterile line with dominant gene (*MS*) in cabbage (*Brassica oleracea* var. *capitata*) and its utilization for hybrid seed production. *Euphytica* **1997**, *97*, 265–268.
5. Fang, Z.Y.; Sun, P.T.; Liu, Y.M. Investigation of different types of male sterility and application of dominant male sterility in cabbage. *China Veg.* **2001**, *1*, 6–10.
6. Ogura, H. Studies of a new male-sterility in Japanese radish, with special reference to the utilization of this sterility towards the practical raising of hybrid seeds. *Mem. Fac. Agric. Kagoshima Univ.* **1968**, *6*, 39–78.
7. Bannerrot, H.; Boulidard, L.; Couderon, Y.; Temple, J. Transfer of cytoplasmic male sterility from *Raphanus sativus* to *Brassica oleracea*. *Proc. Eucarpia Meet. Crucif.* **1974**, *25*, 52–54.
8. Sigareva, M.; Earle, E. Direct transfer of a cold-tolerant Ogura male-sterile cytoplasm into cabbage (*Brassica oleracea* ssp. *capitata*) via protoplast fusion. *Theor. Appl. Genet.* **1997**, *94*, 213–220. [CrossRef]
9. Walters, W.T.; Mutschler, A.M.; Earle, D.E. Protoplast fusion-derived Ogura male-sterile cauliflower with cold tolerance. *Plant Cell Rep.* **1992**, *10*, 624–628. [CrossRef] [PubMed]
10. Pelletier, G.; Primard, C.; Vedel, F.; Chetrit, P.; Remy, R.; Renard, M. Intergeneric cytoplasmic hybridization in Cruciferae by protoplast fusion. *Mol. Genet. Genomics* **1983**, *191*, 244–250. [CrossRef]
11. Wang, Q.B.; Zhang, Y.Y.; Fang, Z.Y.; Liu, Y.M.; Yang, L.M.; Zhuang, M. Chloroplast and mitochondrial SSR help to distinguish allo-cytoplasmic male sterile types in cabbage (*Brassica oleracea* L. var. *capitata). Mol. Breed.* **2012**, *30*, 709–716. [CrossRef]
12. Chase, C.D. Cytoplasmic male sterility: A window to the world of plant mitochondrial-nuclear interactions. *Trends Genet.* **2007**, *23*, 81–90. [CrossRef] [PubMed]
13. Hu, J.; Huang, W.; Huang, Q.; Qin, X.; Yu, C.; Wang, L.; Li, S.; Zhu, R.; Zhu, Y. Mitochondria and cytoplasmic male sterility in plants. *Mitochondrion* **2014**, *19*, 282–288. [CrossRef] [PubMed]
14. Tuteja, R.; Saxena, R.K.; Davila, J.; Shah, T.; Chen, W.; Xiao, Y.; Fan, G.; Saxena, K.B.; Alverson, A.J.; Spillane, C.; et al. Cytoplasmic male sterility-associated chimeric open reading frames identified by mitochondrial genome sequencing of four *Cajanus* genotypes. *DNA Res.* **2013**, *20*, 485–495. [CrossRef] [PubMed]
15. Carlsson, J.; Glimelius, K. Cytoplasmic male-sterility and nuclear encoded fertility restoration. *Plant Mitochond.* **2011**, *1*, 469–491.
16. Woodson, J.D.; Chory, J. Coordination of gene expression between organellar and nuclear genomes. *Nat. Rev. Genet.* **2008**, *9*, 383–395. [CrossRef] [PubMed]
17. Duroc, Y.; Hiard, S.; Vrielynck, N.; Ragu, S.; Budar, F. The Ogura sterility-inducing protein forms a large complex without interfering with the oxidative phosphorylation components in rapeseed mitochondria. *Plant Mol. Biol.* **2009**, *70*, 123–137. [CrossRef] [PubMed]
18. Bonhomme, S.; Budar, F.; Lancelin, D.; Small, I.; Defrance, M.C.; Pelletier, G. Sequence and transcript analysis of the *Nco2.5* Ogura specific fragment correlated with cytoplasmic male sterility in *Brassica* cybrids. *Mol. Gen. Genet.* **1992**, *235*, 340–348. [CrossRef] [PubMed]
19. Grelon, M.; Budar, F.; Bonhomme, S.; Pelletier, G. Ogura cytoplasmic male-sterility (CMS)-associated *orf138* is translated into a mitochondrial membrane polypeptide in male-sterile *Brassica* cybrids. *Mol. Gen. Genet.* **1994**, *243*, 540–547. [CrossRef] [PubMed]
20. Krishnasamy, S.; Makaroff, C.A. Characterization of the radish mitochondrial *orfB* locus: Possible relationship with male sterility in Ogura radish. *Curr. Genet.* **1993**, *24*, 156–163. [CrossRef] [PubMed]
21. Duroc, Y.; Gaillard, C.; Hiard, S.; Defrance, M.C.; Pelletier, G.; Budar, F. Biochemical and functional characterization of ORF138, a mitochondrial protein responsible for Ogura cytoplasmic male sterility in Brassiceae. *Biochimie* **2005**, *87*, 1089–1100. [CrossRef] [PubMed]
22. Dong, X.; Kim, W.K.; Lim, Y.P.; Kim, Y.K.; Hur, Y. Ogura-CMS in Chinese cabbage (*Brassica rapa* ssp. *pekinensis*) causes delayed expression of many nuclear genes. *Plant Sci.* **2013**, *199*, 7–17. [PubMed]

23. Wang, S.; Wang, C.; Zhang, X.X.; Chen, X.; Liu, J.J.; Jia, X.F.; Jia, S.Q. Transcriptome de novo assembly and analysis of differentially expressed genes related to cytoplasmic male sterility in cabbage. *Plant Physiol. Biochem.* **2016**, *105*, 224–232. [CrossRef] [PubMed]
24. Shu, J.; Zhang, L.; Liu, Y.; Li, Z.; Fang, Z.; Yang, L.; Zhuang, M.; Zhang, Y.; Lv, H. Normal and abortive buds transcriptomic profiling of broccoli ogu cytoplasmic male sterile line and its maintainer. *Int. J. Mol. Sci.* **2018**, *19*, 2501. [CrossRef] [PubMed]
25. Xing, M.; Sun, C.; Li, H.; Hu, S.; Lei, L.; Kang, J. Integrated analysis of transcriptome and proteome changes related to the Ogura cytoplasmic male sterility in cabbage. *PLoS ONE* **2018**, *13*, e0193462. [CrossRef] [PubMed]
26. Mihr, C.; Baumgärtner, M.; Dieterich, J.H.; Schmitz, U.K.; Braun, H.P. Proteomic approach for investigation of cytoplasmic male sterility (CMS) in *Brassica*. *J. Plant Physiol.* **2001**, *158*, 787–794. [CrossRef]
27. Sheoran, I.S.; Sawhney, V.K. Proteome analysis of the normal and Ogura (ogu) CMS anthers of *Brassica napus* to identify proteins associated with male sterility. *Botany* **2010**, *88*, 217–230. [CrossRef]
28. Liu, X.; Han, F.; Kong, C.; Fang, Z.; Yang, L.; Zhang, Y.; Zhuang, M.; Liu, Y.; Li, Z.; Lv, H. Rapid introgression of the Fusarium wilt resistance gene into an elite cabbage line through the combined application of a microspore culture, genome background analysis, and disease resistance-specific marker assisted foreground selection. *Front. Plant Sci.* **2017**, *8*, 354. [CrossRef] [PubMed]
29. González-Melendi, P.; Uyttewaal, M.; Morcillo, C.N.; Mora, J.R.H.; Fajardo, S.; Budar, F.; Lucas, M.M. A light and electron microscopy analysis of the events leading to male sterility in Ogu-INRA CMS of rapeseed (*Brassica napus*). *J. Exp. Bot.* **2008**, *59*, 827–838. [CrossRef] [PubMed]
30. Tratt, J. Identifying the Wall-Degrading Enzymes Responsible for Microspore Release from the Pollen Tetrad. Ph.D. Thesis, University of Bath, Bath, UK, 2016.
31. Balk, J.; Leaver, C.J. The PET1-CMS mitochondrial mutation in sunflower is associated with premature programmed cell death and cytochrome *c* release. *Plant Cell* **2001**, *13*, 1803–1818. [CrossRef] [PubMed]
32. Li, S.; Wan, C.; Kong, J.; Zhang, Z.; Li, Y.; Zhu, Y. Programmed cell death during microgenesis in a Honglian CMS line of rice is correlated with oxidative stress in mitochondria. *Funct. Plant Biol.* **2004**, *31*, 369–376. [CrossRef]
33. Luo, D.P.; Xu, H.; Liu, Z.L.; Guo, J.X.; Li, H.Y.; Chen, L.T.; Fang, C.; Zhang, Q.Y.; Bai, M.; Yao, N.; et al. A detrimental mitochondrial-nuclear interaction causes cytoplasmic male sterility in rice. *Nat. Genet.* **2013**, *45*, 573–577. [CrossRef] [PubMed]
34. Yi, J.; Moon, S.; Lee, Y.S.; Zhu, L.; Liang, W.; Zhang, D.; Jung, K.H.; An, G. Defective Tapetum Cell Death 1 (DTC1) regulates ROS levels by binding to metallothionein during tapetum degeneration. *Plant Physiol.* **2016**, *170*, 1611–1623. [CrossRef] [PubMed]
35. Min, L.; Zhu, L.; Tu, L.; Deng, F.; Yuan, D.; Zhang, X. Cotton *GhCKI* disrupts normal male reproduction by delaying tapetum programmed cell death via inactivating starch synthase. *Plant J.* **2013**, *75*, 823–835. [CrossRef] [PubMed]
36. Li, J.; Ding, X.; Han, S.; He, T.; Zhang, H.; Yang, L.; Yang, S.; Gai, J. Differential proteomics analysis to identify proteins and pathways associated with male sterility of soybean using iTRAQ-based strategy. *J. Proteomics* **2016**, *138*, 72–82. [CrossRef] [PubMed]
37. Ji, J.; Yang, L.; Fang, Z.; Zhuang, M.; Zhang, Y.; Lv, H.; Liu, Y.; Li, Z. Complementary transcriptome and proteome profiling in cabbage buds of a recessive male sterile mutant provides new insights into male reproductive development. *J. Proteomics* **2018**, *179*, 80–91. [CrossRef] [PubMed]
38. Liu, J.; Pang, C.; Wei, H.; Song, M.; Meng, Y.; Ma, J.; Fan, S.; Yu, S. iTRAQ-facilitated proteomic profiling of anthers from a photosensitive male sterile mutant and wild-type cotton (*Gossypium hirsutum* L.). *J. Proteomics* **2015**, *126*, 68–81. [CrossRef] [PubMed]
39. Pan, Z.; Zeng, Y.; An, J.; Ye, J.; Xu, Q.; Deng, X. An integrative analysis of transcriptome and proteome provides new insights into carotenoid biosynthesis and regulation in sweet orange fruits. *J. Proteomics* **2012**, *75*, 2670–2684. [CrossRef] [PubMed]
40. Lou, P.; Kang, J.; Zhang, G.; Bonnema, G.; Fang, Z.; Wang, X. Transcript profiling of a dominant male sterile mutant (*Ms-cd1*) in cabbage during flower bud development. *Plant Sci.* **2007**, *172*, 111–119. [CrossRef]
41. Chu, P.; Yan, G.; Yang, Q.; Zhai, L.; Zhang, C.; Zhang, F.; Guan, R. iTRAQ-based quantitative proteomics analysis of *Brassica napus*, leaves reveals pathways associated with chlorophyll deficiency. *J. Proteomics* **2015**, *113*, 244–259. [CrossRef] [PubMed]

42. Liu, S.; Liu, Y.; Yang, X.; Tong, C.; Edwards, D.; Parkin, I.A.; Zhao, M.; Ma, J.; Yu, J.; Huang, S.; et al. The *Brassica oleracea* genome reveals the asymmetrical evolution of polyploid genomes. *Nat. Commun.* **2014**, *5*, 3930. [CrossRef] [PubMed]
43. Langmead, B.; Salzberg, S.L. Fast gapped-read alignment with Bowtie 2. *Nat. Methods* **2012**, *9*, 357–359. [CrossRef] [PubMed]
44. Kim, D.; Pertea, G.; Trapnell, C.; Pimentel, H.; Kelley, R.; Salzberg, S. TopHat2: Accurate alignment of transcriptomes in the presence of insertions, deletions and gene fusions. *Genome Biol.* **2013**, *14*, R36. [CrossRef] [PubMed]
45. Trapnell, C.; Williams, B.A.; Pertea, G.; Mortazavi, A.; Kwan, G.; van Baren, M.J.; Salzberg, S.L.; Wold, B.J.; Pachter, L. Transcript assembly and quantification by RNA-Seq reveals unannotated transcripts and isoform switching during cell differentiation. *Nat. Biotechnol.* **2010**, *28*, 511–515. [CrossRef] [PubMed]
46. Livak, K.J.; Schmittgen, T.D. Analysis of relative gene expression data using real-time quantitative PCR and the $2^{-\Delta\Delta Ct}$ method. *Methods* **2001**, *25*, 402–408. [CrossRef] [PubMed]
47. Zheng, B.B.; Fang, Y.N.; Pan, Z.Y.; Sun, L.; Deng, X.X.; Grosser, J.W.; Guo, W.W. iTRAQ-based quantitative proteomics analysis revealed alterations of carbohydrate metabolism pathways and mitochondrial proteins in a male sterile *Cybrid pummelo*. *J. Proteome Res.* **2014**, *13*, 2998–3015. [CrossRef] [PubMed]
48. Sabar, M.; Gagliardi, D.; Balk, J.; Leaver, C.J. ORFB is a subunit of F1F(O)-ATP synthase: Insight into the basis of cytoplasmic male sterility in sunflower. *EMBO Rep.* **2003**, *4*, 381–386. [CrossRef] [PubMed]
49. Bergman, P.; Edqvist, J.; Farbos, I.; Glimelius, K. Male-sterile tobacco displays abnormal mitochondrial atp1 transcript accumulation and reduced floral ATP/ADP ratio. *Plant Mol. Biol.* **2000**, *42*, 531–544. [CrossRef] [PubMed]
50. Ducos, E.; Touzet, P.; Boutry, M. The male sterile *G*, cytoplasm of wild beet displays modified mitochondrial respiratory complexes. *Plant J.* **2001**, *26*, 171–180. [CrossRef] [PubMed]
51. Zhang, H.; Li, S.; Yi, P.; Wan, C.; Chen, Z.; Zhu, Y. A Honglian CMS line of rice displays aberrant F_0 of F_0 F_1-ATPase. *Plant Cell Rep.* **2007**, *26*, 1065–1071. [CrossRef] [PubMed]
52. Wang, K.; Gao, F.; Ji, Y.; Liu, Y.; Dan, Z.; Yang, P.; Zhu, Y.; Li, S. ORFH79 impairs mitochondrial function via interaction with a subunit of electron transport chain complex III in Honglian cytoplasmic male sterile rice. *New Phytol.* **2013**, *198*, 408–418. [CrossRef] [PubMed]
53. Ariizumi, T.; Toriyama, K. Genetic regulation of sporopollenin synthesis and pollen exine development. *Annu. Rev. Plant Biol.* **2011**, *62*, 437–460. [CrossRef] [PubMed]
54. Shirley, B.W.; Kubasek, W.L.; Storz, G.; Bruggemann, E.; Koornneef, M.; Ausubel, F.M.; Goodman, H.M. Analysis of *Arabidopsis* mutants deficient in flavonoid biosynthesis. *Plant J.* **1995**, *8*, 659–671. [CrossRef] [PubMed]
55. Dobritsa, A.A.; Lei, Z.; Nishikawa, S.I.; Urbanczyk-Wochniak, E.; Huhman, D.V.; Preuss, D.; Sumner, L.W. *LAP5* and *LAP6* encode anther-specific proteins with similarity to chalcone synthase essential for pollen exine development in *Arabidopsis thaliana*. *Plant Physiol.* **2010**, *110*, 157446. [CrossRef]
56. Morant, M.; Jørgensen, K.; Schaller, H.; Pinot, F.; Møller, B.L.; Werck-Reichhart, D.; Bak, S. CYP703 is an ancient cytochrome P450 in land plants catalyzing in-chain hydroxylation of lauric acid to provide building blocks for sporopollenin synthesis in pollen. *Plant Cell* **2007**, *19*, 1473–1487. [CrossRef] [PubMed]
57. Aarts, M.G.M.; Hodge, R.; Kalantidis, K.; Florack, D.; Wilson, Z.A.; Mulligan, B.J.; Stiekema, W.J.; Scott, R.; Pereira, A. The *Arabidopsis MALE STERILITY 2* protein shares similarity with reductases in elongation/condensation complexes. *Plant J.* **1997**, *12*, 615–623. [CrossRef] [PubMed]
58. Chen, W.; Yu, X.H.; Zhang, K.; Shi, J.; De Oliveira, S.; Schreiber, L.; Shanklin, J.; Zhang, D. *Male Sterile2* encodes a plastid-localized fatty acyl carrier protein reductase required for pollen exine development in *Arabidopsis*. *Plant Physiol.* **2011**, *157*, 842–853. [CrossRef] [PubMed]
59. Wallace, S.; Chater, C.C.; Kamisugi, Y.; Cuming, A.C.; Wellman, C.H.; Beerling, D.J.; Fleming, A.J. Conservation of *Male Sterility* 2 function during spore and pollen wall development supports an evolutionarily early recruitment of a core component in the sporopollenin biosynthetic pathway. *New Phytol.* **2015**, *205*, 390–401. [CrossRef] [PubMed]
60. Piffanelli, P.; Ross, J.H.E.; Murphy, D.J. Biogenesis and function of the lipidic structures of pollen grains. *Sexual Plant Reprod.* **1998**, *11*, 65–80. [CrossRef]
61. Preuss, D.; Rhee, S.Y.; Davis, R.W. Tetrad analysis possible in *Arabidopsis* with mutation of the *QUARTET* (*QRT*) genes. *Science* **1994**, *264*, 1458–1460. [CrossRef] [PubMed]

62. Rhee, S.Y.; Osborne, E.; Poindexter, P.D.; Somerville, C.R. Microspore separation in the *quartet 3* mutants of *Arabidopsis* is impaired by a defect in a developmentally regulated polygalacturonase required for pollen mother cell wall degradation. *Plant Physiol.* **2003**, *133*, 1170–1180. [CrossRef] [PubMed]
63. Jiang, P.; Zhang, X.; Zhu, Y.; Zhu, W.; Xie, H.; Wang, X. Metabolism of reactive oxygen species in cotton cytoplasmic male sterility and its restoration. *Plant Cell Rep.* **2007**, *26*, 1627–1634. [CrossRef] [PubMed]
64. Gechev, T.S.; Hille, J. Hydrogen peroxide as a signal controlling plant programmed cell death. *J. Cell Biol.* **2005**, *168*, 17–20. [CrossRef] [PubMed]
65. Von Malek, B.; van der Graaff, E.; Schneitz, K.; Keller, B. The *Arabidopsis* male-sterile mutant *dde2-2* is defective in the *ALLENE OXIDE SYNTHASE* gene encoding one of the key enzymes of the jasmonic acid biosynthesis pathway. *Planta* **2002**, *216*, 187–192. [CrossRef] [PubMed]

5

Interactions between WUSCHEL- and CYC2-like Transcription Factors in Regulating the Development of Reproductive Organs in *Chrysanthemum morifolium*

Yi Yang [1], Ming Sun [1], Cunquan Yuan [1], Yu Han [1], Tangchun Zheng [1], Tangren Cheng [1], Jia Wang [1] and Qixiang Zhang [1,2,*]

1 Beijing Key Laboratory of Ornamental Plants Germplasm Innovation & Molecular Breeding, National Engineering Research Center for Floriculture, Beijing Laboratory of Urban and Rural Ecological Environment, Engineering Research Center of Landscape Environment of Ministry of Education, Key Laboratory of Genetics and Breeding in Forest Trees and Ornamental Plants of Ministry of Education, School of Landscape Architecture, Beijing Forestry University, Beijing 100083, China; yiyang921124@126.com (Y.Y.); 13683295193@163.com (M.S.); yuancunquan@163.com (C.Y.); hanyu19880514@126.com (Y.H.); zhengtangchun@126.com (T.Z.); chengtangren@163.com (T.C.); wangjia8248@163.com (J.W.)

2 Beijing Advanced Innovation Center for Tree Breeding by Molecular Design, Beijing Forestry University, Beijing 100083, China

* Correspondence: zqxbjfu@126.com

Abstract: *Chrysanthemum morifolium* is a gynomonoecious plant that bears both female zygomorphic ray florets and bisexual actinomorphic disc florets in the inflorescence. This sexual system is quite prevalent in Asteraceae, but poorly understood. CYCLOIDEA (CYC) 2 subclade transcription factors, key regulators of flower symmetry and floret identity in Asteraceae, have also been speculated to function in reproductive organs and could be an entry point for studying gynomonoecy. However, the molecular mechanism is still unclear. On the other hand, the *Arabidopsis* WUSCHEL (WUS) transcription factor has been proven to play a vital role in the development of reproductive organs. Here, a *WUS* homologue (*CmWUS*) in *C. morifolium* was isolated and characterized. Overexpression of *CmWUS* in *A. thaliana* led to shorter siliques and fewer stamens, which was similar to *CYC2*-like genes reported before. In addition, both *CmWUS* and *CmCYC2* were highly expressed in flower buds during floral organ differentiation and in the reproductive organs at later development stages, indicating their involvement in the development of reproductive organs. Moreover, CmWUS could directly interact with CmCYC2d. Thus, our data suggest a collaboration between CmWUS and CmCYC2 in the regulation of reproductive organ development in chrysanthemum and will contribute to a further understanding of the gynomonoecious sexual system in Asteraceae.

Keywords: *Chrysanthemum morifolium*; WUS; CYC2; gynomonoecy; reproductive organ; flower symmetry

---

1. Introduction

The inflorescence of *C. morifolium* (Asteraceae) is always comprised of two kinds of florets: the bilaterally symmetric female ray florets and radially symmetric bisexual disc florets [1]. Different sex expression and flower symmetry in ray and disc florets are significant features of chrysanthemum inflorescence. This gynomonoecious sexual system is quite prevailing in Asteraceae and has been considered to play a pivotal role in reducing herbivore damage and pollen-pistil interference, as well as in attracting pollinators [2,3]. However, the genetic mechanism of gynomonoecy is poorly understood.

Interestingly, the connection between shifts in flower symmetry and the development of reproductive organs has been discovered and recorded in many species, including Asteraceae members [4–6], and the flower symmetry genes have been speculated to be involved in breeding system [6–8].

CYCLOIDEA (CYC) 2 subclade transcription factors, which belong to ECE-CYC/TB1 clade of plant-specific TCP family [9,10], have been proven to be essential for the regulation of flower symmetry in angiosperms [11] and inflorescence architecture in Asteraceae [6,12]. *CYC* of *Antirrhinum* was the first gene isolated in this subclade and is expressed in the dorsal domain of floral meristem from initiation and maintained throughout the differentiation of petals and stamens [13]. *CYC* promotes the growth of dorsal petals and arrests the development of dorsal stamen to form a staminode [13,14]. Gaudin et al. [15] speculated that *CYC* could directly or indirectly suppress the expression of *cyclinD3b* and other cell cycle genes in the staminode. Studies in *Opithandra* further indicated the negative effects of *OpdCYC* on *OpdcyclinD3* genes and the correlation between the expressions of *OpdCYC* and the abortion of both dorsal and ventral stamen [16]. On the other hand, Preston et al. [17] found that expression patterns of *CYC2*-like genes were not corelated with patterns of stamen arrest in *Veronica montana* and *Gratiola officinalis*. In contrast, in *Papaveracea*, *CYC2*-like genes promote stamen initiation and growth [18].

Previous studies have shown that *CYC2*-like genes in gerbera (*Gerbera hybrida*) are functionally redundant in regulating ray floret identity by promoting ligule growth and suppressing stamen development [19–21]. In addition to stamens, *CYC2*-like genes have also been speculated to have late functions in the development of ovaries and carpels in Asteraceae [6,12]. Both in gerbera and sunflower (*Helianthus annuus*), *CYC2*-like genes are highly expressed in ovary, stigma and style tissues [22]. Expression levels of *AcCYC2a* and *AcCYC2d* are also increased in the developing ovules of *Anacyclus clavatus* [23]. Moreover, constitutive expression of all the gerbera *CYC2*-like genes, except *GhCYC2*, in *A. thaliana* leads to shorter siliques with fewer seeds. In addition, stamen development is also severely disrupted in the transgenic lines ectopically expressing *GhCYC4* and *GhCYC7* [20]. Also, different from the empty achenes in the zygomorphic ray florets of wild type plants, actinomorphic ray florets can produce filled achenes through hand pollination in the *turf* mutant [24], which is caused by insertion of TEs in the TCP domain of *HaCYC2c* in sunflower [25–27]. Still, the molecular relationship between *CYC2*-like genes and the development of reproductive organs awaits more research to elaborate.

In addition to *CYC2*-like genes, our previous comparative transcriptome analysis between ray and disc florets in *C. morifolium* [28] has predicted other candidate transcription factor genes during inflorescence development and organ determination for further studies. Among them, a *WUSCHEL*-like gene, which was highly expressed in the central disc florets, has attracted our attention. *WUSCHEL* (*WUS*) is a member of the WUSCHEL-RELATED HOMEOBOX (WOX) family [29] and takes part in several regulatory networks in shoot and floral meristems [30–32]. In *A. thaliana*, *WUS* is expressed in anther stomium cells during early stages and is required for anther development [33]. In ovules, *WUS* is confined to be expressed in the nucellus and is essential for the initiation of integument [34,35]. Reduced stamens and disappeared carpel in *wus* mutants of *Arabidopsis* also suggest crucial functions of *WUS* in the development of reproductive organs [36]. In *Cucumis sativus* (cucumber), CsWUS directly interacts with CsSPL, a vital factor in male and female fertility, and takes part in the regulatory network that controls the development of reproductive organs [37].

We have identified six *CmCYC2* genes in *C. morifolium* before, and they were also found to be strongly expressed in ray florets [38]. In contrast to *CmCYC2*, *CmWUS* was highly expressed in disc florets [28]. These two distinct expression patterns between ray and disc florets suggested their involvements in the development of inflorescence. In particular, whether *CmWUS* and *CmCYC2* are connected in reproductive organ development of chrysanthemum is an interesting problem worthy of study. Here, we isolated *CmWUS* and expressed it ectopically in *A. thaliana* for functional analysis. Additionally, expression patterns of *CmWUS* and *CmCYC2* during inflorescence development were compared. Furthermore, we performed yeast two-hybrid (Y2H) and bimolecular fluorescence complementation (BiFC) assays to determine protein-protein interactions between CmWUS and CmCYC2. These results show a direct interaction between CmWUS and CmCYC2 and may help to understand the genetic and molecular mechanisms of reproductive organ development in Asteraceae.

2. Results

2.1. Identification and Phylogenetic Analysis of CmWUS

To identify the function of *CmWUS* in reproductive organ development, we isolated the ORF of *CmWUS* (912 bp) from inflorescences of *C. morifolium* 'Fen Ditan'. The encoded CmWUS protein (303 amino acids) was aligned with WUS-like sequences from other species. As shown in Supplementary Figure S1A, the WOX domain [29] was highly conserved. The signature motifs WUS-box and EAR-like motif [39] were also identified in CmWUS at the carboxyl terminus. A neighbor joining phylogenetic tree (Supplementary Figure S1B) was constructed based on the full length of amino acid sequences of 15 WOX family members from *A. thaliana* and WUS from other species. As described by Graaff, Laux and Rensing [29], these WOX members could be divided into three clades: the ancient clade, the intermediate clade and the WUS clade. The phylogenetic analysis confirmed that CmWUS belongs to WUS clade of WOX family and is closely related to WUS-like from other species of Asterceae: *H. annuus*, *L. sativa* and *C. cardunculus*.

2.2. Overexpression of CmWUS in A. thaliana Inhibits the Development of Reproductive Organs and Affects Flower Symmetry

The *CmWUS* ORF was overexpressed in *A. thaliana* (Columbia) for functional analysis during floral development. The transgenic lines in which *CmWUS* was highly expressed were confirmed by qPCR assay. Three *35S::CmWUS* lines (line 6, 8,13) with higher and consistent expression levels were selected for detailed analysis. The wild type *Arabidopsis* flowers are polysymmetric with four sepals, four petals, four medial and two lateral stamens and two fused carpels (Figure 1A). Meanwhile, in our transgenic lines, the flowers were changed into monosymmetric with one symmetry plane. The petals on both sides of the lateral stamens were arranged close to each other and the development of the lateral stamens was also inhibited (Figure 1B–E). As listed in Figure 1K, the number of stamens in three *35S::CmWUS* lines were reduced to 4 to 5. In addition, they produced shorter siliques than the wild type. In addition to these three transgenic lines, line 1 showed a stronger phenotype, with flower meristems that were ectopically initiated on the surface of inflorescence stems (Figure 1H,I), which was consistent with the phenotype of *Arabidopsis* overexpressing *AtWUS* [40,41]. Furthermore, petals were slightly curled at the edges (Figure 1F) and siliques were much shorter than wild type (Figure 1G) in line 1.

J

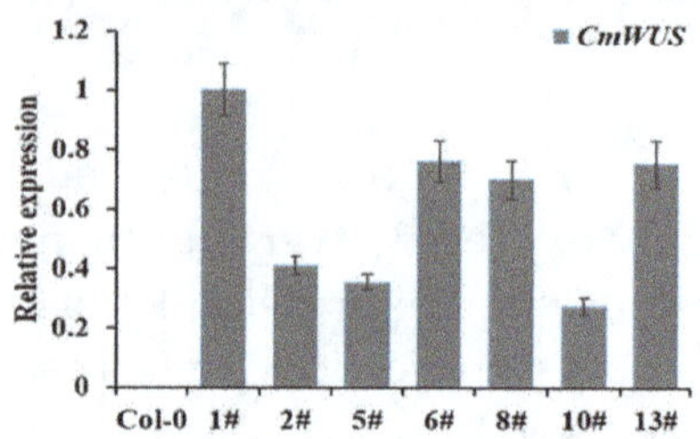

K

Lines	Length of siliques (cm)	Number of stamens
Col-0	1.45±0.08 a	6.0±0.0 a
1#	0.44±0.03 b	6.0±0.0 a
6#	1.16±0.11 c	4.2±0.42 b
8#	1.05±0.16 d	4.6±0.51 c
13#	0.97±0.08 d	4.5±0.52 bc

Figure 1. Ectopic expression of *CmWUS* in *A. thaliana* (Columbia). (**A**) Flower of wild type *A. thaliana* (Columbia). (**B–E**) Flower phenotypes in transgenic line 6, 8 and 13. Petals on both sides of the two lateral stamens were arranged close to each other and the development of the lateral stamens was also inhibited. The number of stamens were reduced to 4 (**B**,**C**) and 5 (**D**,**E**). Stamens are marked with white arrows. (**F**) Slightly curled petals at the edges of the flowers in transgenic line 1. (**G**) Siliques of transgenic line 1 (left) were much shorter than wild type (right). (**H**,**I**) Ectopic initiated flower buds on the surface of inflorescence stems in transgenic line 1. (**J**) qPCR detection of *CmWUS* transcripts in wild type (WT) and transgenic lines of *A. thaliana*. The endogenous *Arabidopsis ACTIN* was chosen as a housekeeper gene. (**K**) Statistics of silique length and stamen number in wild type and transgenic lines of *Arabidopsis*. Statistically significant differences are indicated with lowercase letters (Fisher's LSD, $p < 0.05$). Bars = 1 mm.

2.3. High Expression of CmWUS and CmCYC2 in the Reproductive Organs of C. morifolium

Three developmental phases of flower buds—initiation of floral primordia (I), differentiation of floral organs (II) and growth of floral organs (III) [38]—were selected (Figure 2A–C) to analyze the expression of *CmWUS* at early stages of inflorescence development in *C. morifolium*. As shown in Figure 2D, the expression level of *CmWUS* increased from stage I to stage II and then decreased to the lowest at stage III, which was similar to *CmCYC2* genes reported previously [38] and indicates their involvement in floral organ differentiation.

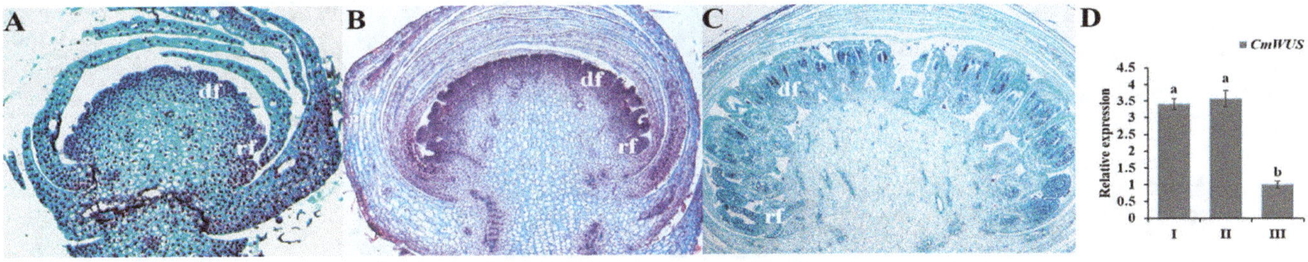

Figure 2. Expression patterns of *CmWUS* in flower buds of *C. morifolium* 'Fen Ditan' at early stages of inflorescence development. (**A–C**) Morphological characteristics of flower buds at three stages: I, initiation of floral primordia (**A**); II, differentiation of floral organs (**B**); and III, growth of floral organs (**C**) were analyzed at a histological level. Abbreviations: rf = ray florets, df = disc florets. (**D**) Expression levels of *CmWUS* in flower buds at stage I, II and III of inflorescence development. The expression levels are relative to the flower buds at stage III. Expression levels of *PP2Acs* are utilized for normalization. Error bars show the standard deviation of three biological replicates. Statistically significant differences are indicated with different lowercase letters (Fisher's LSD, $p < 0.05$).

qPCR assays were also performed to compare the expression patterns of *CmWUS* and *CmCYC2* at later stages of inflorescence development (Figure 3) between ray and disc florets. As shown in Figure 4A, *CmWUS* was expressed extremely highly in disc florets, especially at stage 1. The expression of *CmWUS* in ray florets was also detected, but was pretty weak compared to disc florets. Unlike *CmWUS*, *CmCYC2* genes, especially *CmCYC2c* and *CmCYC2d*, were expressed at relatively higher levels in ray florets than disc florets. To further explore the possible roles of *CmWUS* and *CmCYC2* genes, their expression levels in different tissues of *C. morifolium* 'Fen Ditan' at late development stages were studied. As shown in Figure 4B, *CmCYC2* and *CmWUS* were primarily expressed in floral organs and were strongly expressed in pistils (including ovary, style and stigma). *CmWUS* was also expressed in stamens, but the expression level was not as high as in pistils like *CmCYC2d*. *CmCYC2* genes were also expressed at high levels in petals, especially in ray petals, while *CmWUS* was not, which may explain the differences in expression levels of *CmWUS* and *CmCYC2* between ray and disc florets. Thus, we speculate that *CmWUS* and *CmCYC2* genes are all involved in the regulation of reproductive organ (especially the pistils) development.

Figure 3. Inflorescence morphology of *C. morifolium* 'Fen Ditan' and five later stages of inflorescence development.

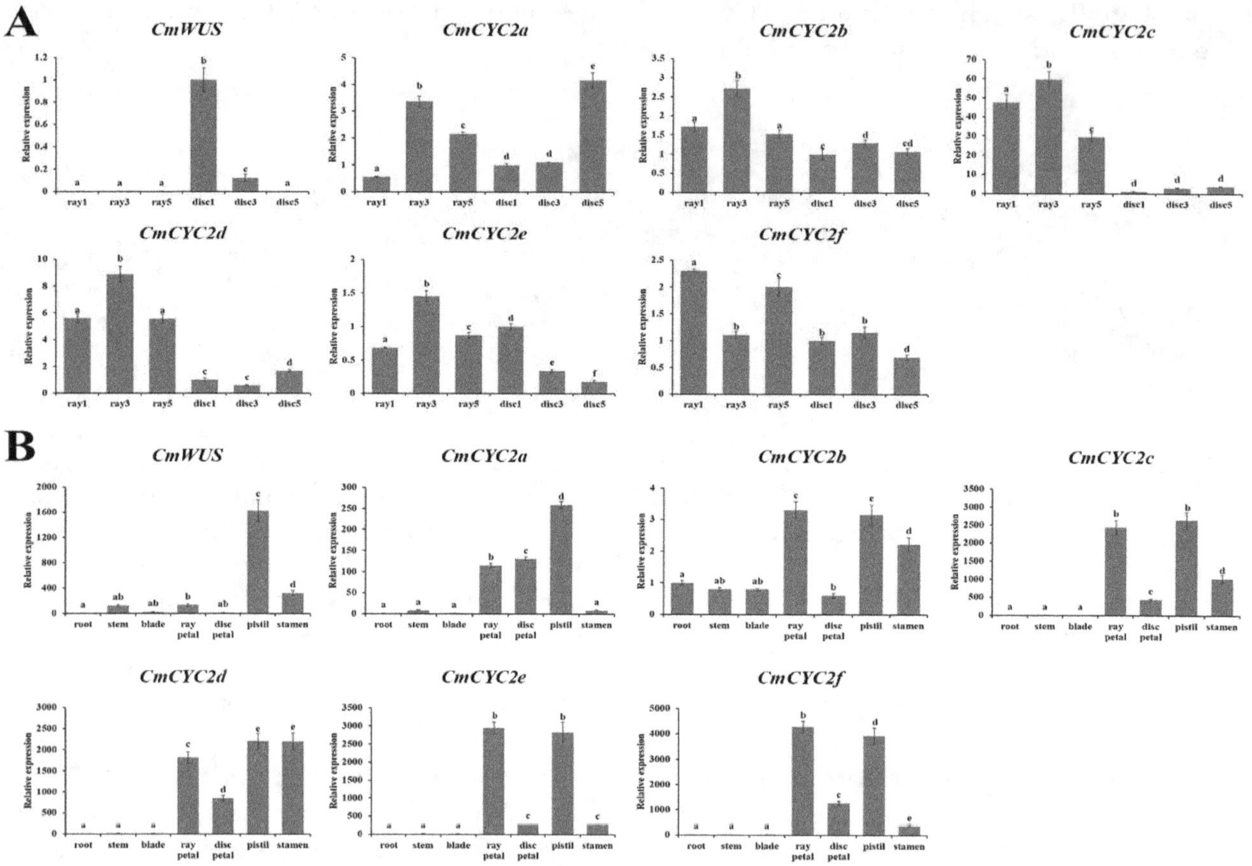

Figure 4. Comparative expression analysis of *CmWUS* and *CmCYC2* genes in *C. morifolium* 'Fen Ditan' at later stages of inflorescence development. (**A**) Gene expression patterns between ray and disc florets at later stages (stage 1, 3 and 5) of inflorescence development. The expression levels are relative to the disc florets at stage 1. (**B**) Relative expression levels of *CmWUS* and *CmCYC2* genes in different tissues of *C. morifolium* 'Fen Ditan'. Tissues analyzed including: root, stem, blade, ray petal, disc petal, pistil (including stigma, style and ovary) and stamen. The expression levels are relative to the root sample. Expression levels of *PP2Acs* are utilized for normalization. Error bars show the standard deviation of three biological replicates. Statistically significant differences are indicated with different lowercase letters (Fisher's LSD, $p < 0.05$).

2.4. Protein-Protein Interactions between CmWUS and CmCYC2

Since *CmWUS* and *CmCYC2* were both highly expressed in the reproductive organs, we further examined the interactions between CmWUS and CmCYC2 to reveal their relationship. The GFP and DAPI fluorescence indicated that CmWUS and CmCYC2 were mainly localized to the cell nucleus (Figure 5). In yeast two-hybrid (Y2H) assays, CmWUS had no autoactivation activity and was used as a bait. The results are shown in Figure 6. CmWUS could not form a homodimer, which was the opposite to the results in *Arabidopsis*, and this may be caused by the differences in the homodimerization interacting amino acids at the central part of the CmWUS sequence (Supplementary Figure S1A) [42]. Furthermore, CmWUS could dimerize with CmCYC2b and CmCYC2d, and the interactions with CmCYC2c, CmCYC2e and CmCYC2f were quite weak. Bimolecular fluorescence complementation (BiFC) assays were performed to provide further evidence for the interactions. There was no interaction in YFP^N/YFP^C, CmCYC2-YFP^N/YFP^C, CmWUS-YFP^N/YFP^C or YFP^N/CmWUS-YFP^C combinations. As shown in Figure 7, only in the combination of CmCYC2d-YFP^N/CmWUS-YFP^C, YFP fluorescence was detected. Taken together, CmWUS could directly interact with CmCYC2d, and the CmWUS-CmCYC2d complex is localized to the cell nucleus.

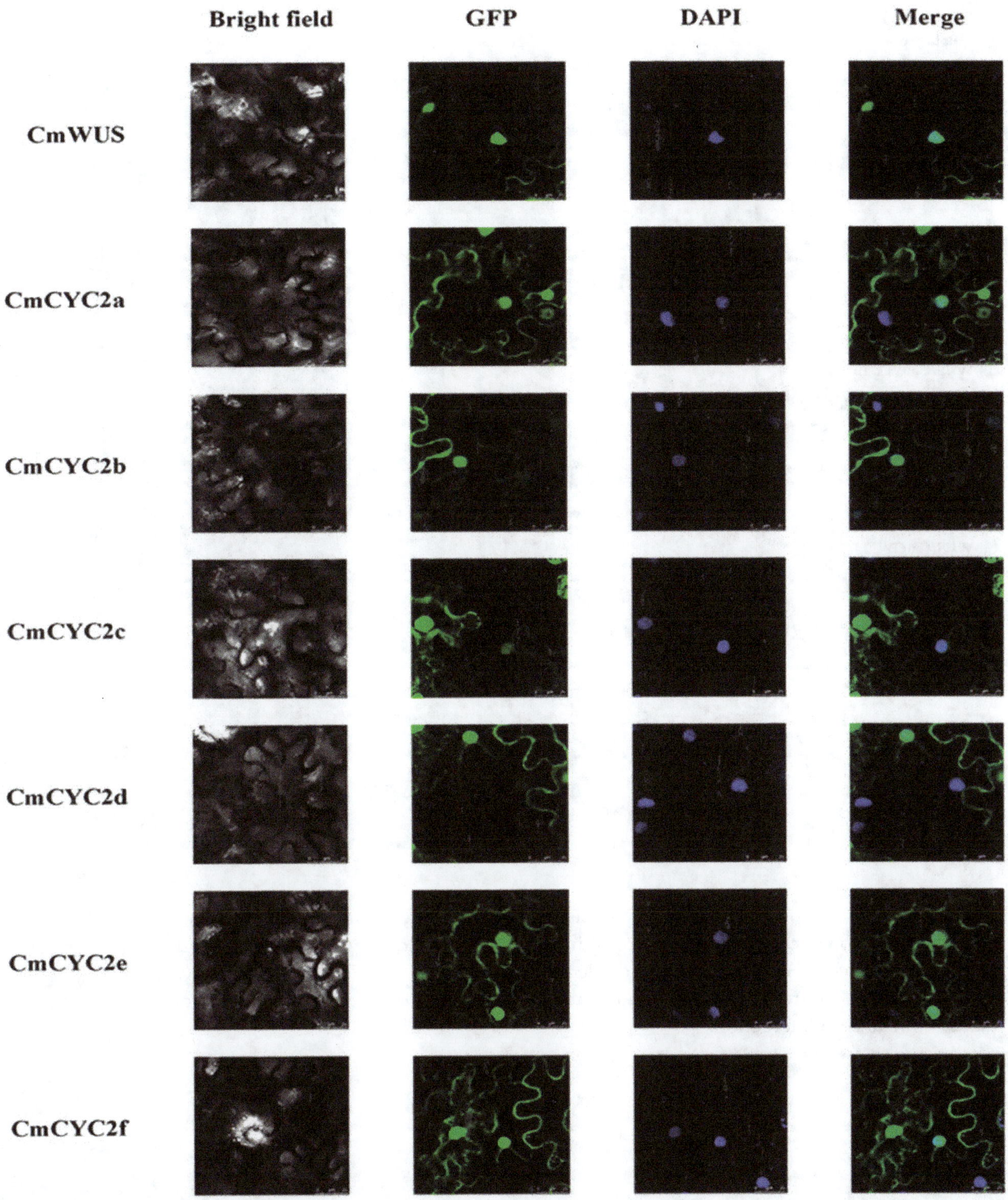

Figure 5. Subcellular localization of CmWUS and CmCYC2. pSuper1300-CmWUS and pSuper1300-CmCYC2 constructs were transiently transformed into the leaves of *Nicotiana benthamiana*. The fusion proteins (CmWUS-GFP and CmCYC2-GFP) were observed under the confocal laser scanning microscope. The merge pictures were made up of the GFP and DAPI pictures. The green and blue fluorescence show the position of proteins and nuclei, respectively. Bars = 25 μm.

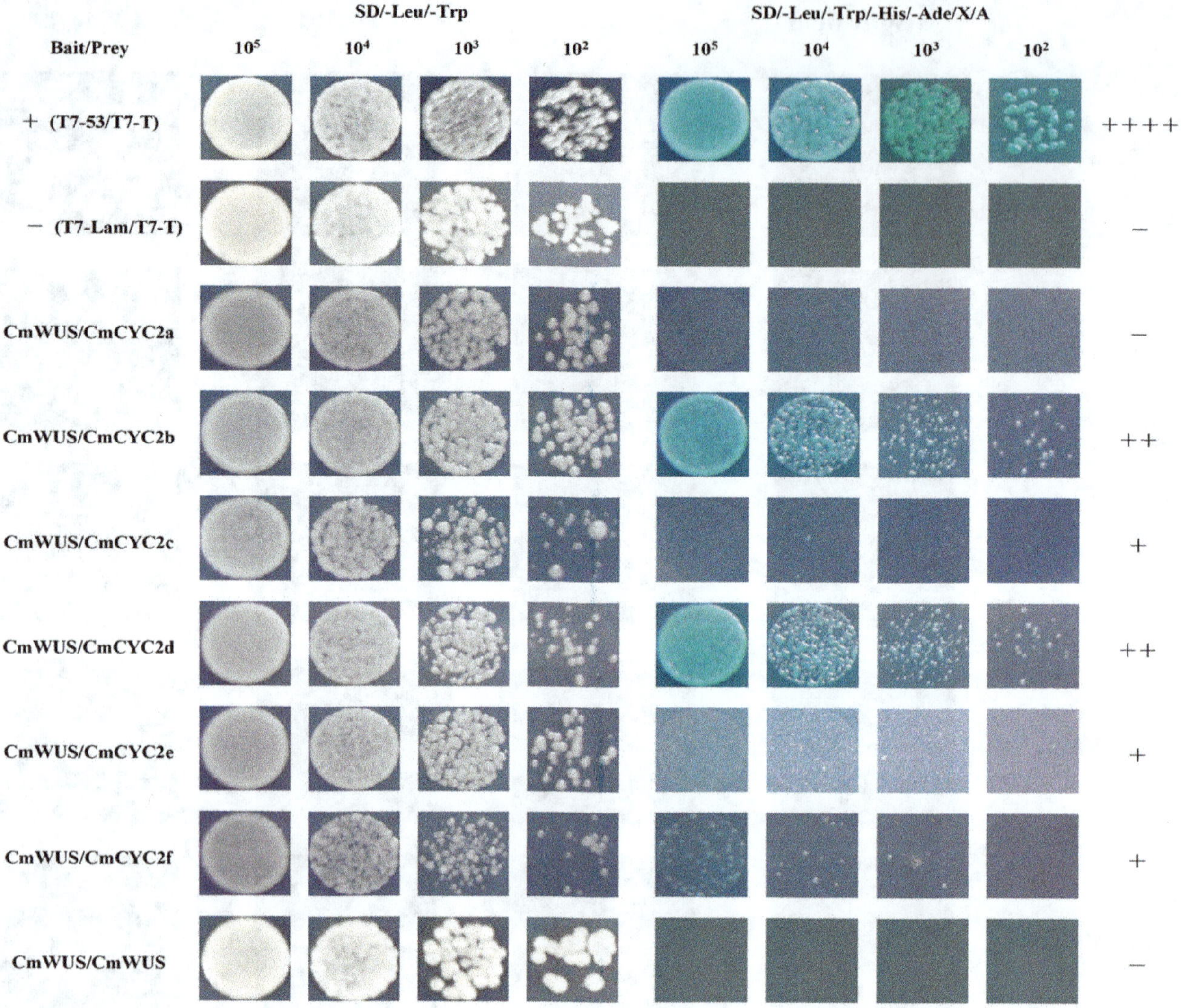

Figure 6. Yeast two-hybrid (Y2H) analysis of protein-protein interactions between CmWUS and CmCYC2. Clones containing each combination of bait and prey vectors were cultured on both nonselective media (SD/-Trp/-Leu) and selective media (SD/-Leu/-Trp/-His/-Ade/X/A). T7-53/T7-T and T7-Lam/T7-T are the positive and negative control. "+" represents the intensity of the interaction and "-" means no interaction.

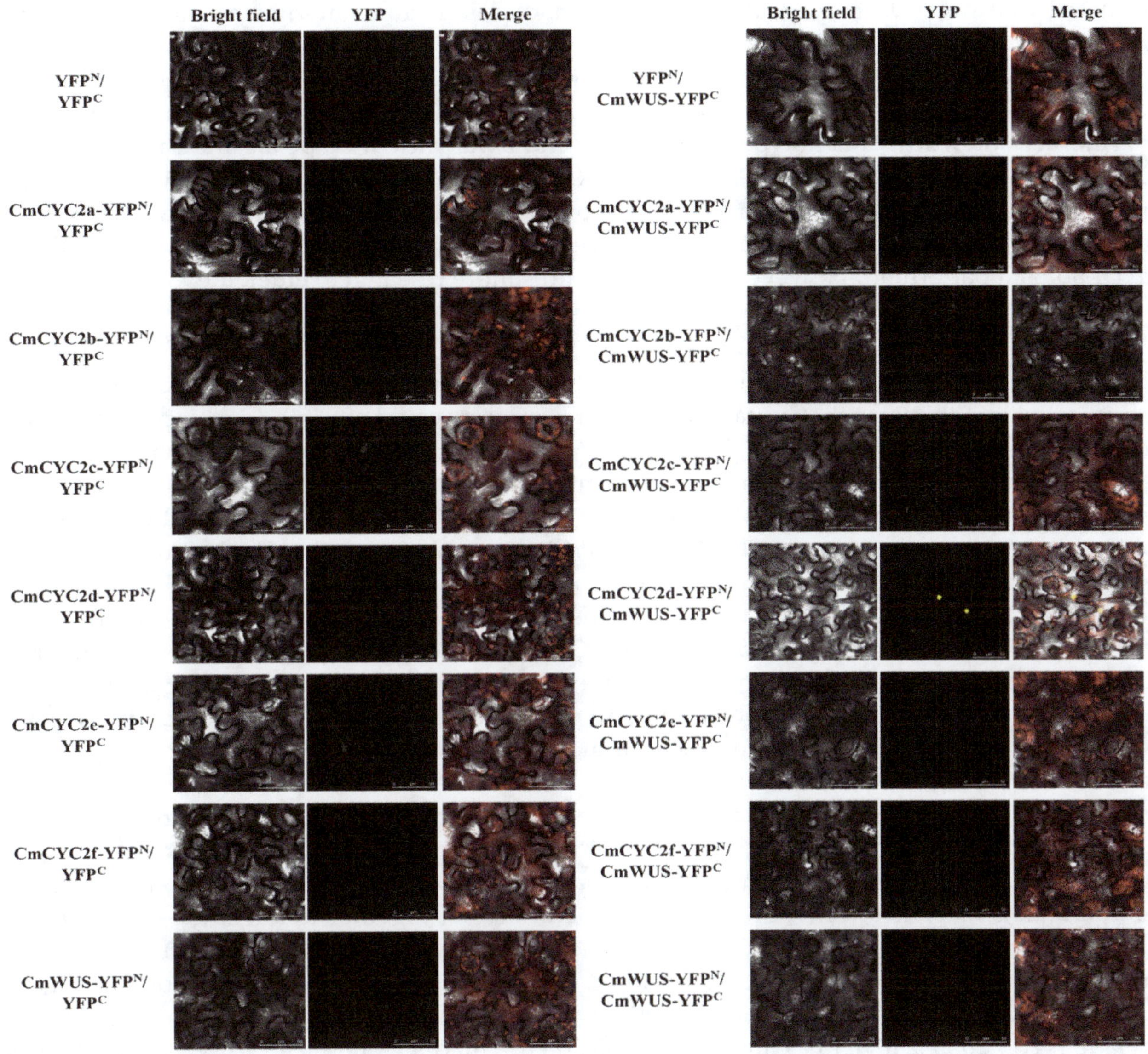

Figure 7. Bimolecular fluorescence complementation (BiFC) analysis of the interactions between CmWUS and CmCYC2 proteins in the epidermal cells of *N. benthamiana* leaves. CmCYC2 and CmWUS were fused to the N-terminal and C-terminal fragment of pCambia1300-YFP respectively and then co-transformed into *N. benthamiana* leaf cells. The confocal laser scanning microscope was used for visualizing. The yellow fluorescence shows the position of protein. Bars = 50 μm.

3. Discussion

3.1. Ectopic Expression of CmWUS in A. thaliana Indicates Possible Conserved Functions in Floral Meristems

Bifunctional transcription factor WUS plays a vital role in the stem cell maintenance of shoot and floral meristems and has been proven to be sufficient for the meristem reestablishment in the inflorescence stem [39–41]. To elucidate the functions of *CmWUS*, we first analyzed the sequence in detail. The WUS-box motif, which was elementary for WUS function in both shoot and floral meristems [39], was highly conserved. Also, the transcriptional repression related EAR motif [43] was identified at the carboxyl terminus. We further explored the function of *CmWUS* during flower development through overexpression in *A. thaliana*. In our transgenic line 1, clustered flower buds were ectopically initiated on the inflorescence stems. This phenotype was consistent with *sef*, a gain-of-function mutant caused by the overexpression of endogenous *WUS* [41]. Therefore, we speculated that *CmWUS* may retain conserved functions in floral meristems. In *sef* mutant, the floral identity gene *LFAFY (LFY)* was also activated [41] and it could cooperate with *WUS* to activate *AGAMOUS (AG)*, a MADS-box gene which specifies the identity of carpel and stamen [30,39,44]. This *WUS/LFY-AG* regulatory loop could be a possible explanation of the ectopic floral buds [40,41].

Another noteworthy phenotype in transgenic line 1 was the curled petals, indicating more active cell proliferation in abaxial side. *WOX1* and *WOX3*, which belong to WUS clade of WOX family [29], have been reported to regulate leaf and floral organ development and affect the abaxial-adaxial balance [45,46]. Thus, *CmWUS* may also be involved in petal morphogenesis through the regulation of abaxial-adaxial patterning. However, this still requires more research to elucidate.

3.2. Proposed Interaction between CmWUS and CmCYC2 in Regulating Reproductive Organ Development

Changes in the number of stamens always come after the shifts in flower symmetry, and it has been reported in Asteraceae that mutations of floret symmetry could affect the development of stamens and carpels [6,47]. *CYC2*-like genes, key factors of flower symmetry, are vital in determining floret identity and regulating floral organ development in Asteraceae [12]. In the transgenic *Arabidopsis* lines with constitutive expression of gerbera *CYC2*-like genes, the siliques were shorter than wild type. Moreover, overexpression of *GhCYC4* and *GhCYC7* could disrupt the development of petals and stamens and carpels were unable to produce normal siliques [20]. In this study, *35S::CmWUS* lines also produced shorter siliques and fewer stamens with variations in flower symmetry. In addition, the transcriptional level of *CmWUS* and *CmCYC2* genes during inflorescence development were compared in chrysanthemum. All the genes were highly expressed at the early stages of flower bud differentiation [38] and may be involved in floral organ development. At later stages, tissue-specific expression analysis revealed that they were all highly expressed in reproductive organs. In general, based on the transgenic *Arabidopsis* phenotypes and gene expression patterns, we conclude that *CmWUS* and *CmCYC2* genes may play an important role in the development of reproductive organs in chrysanthemum. Furthermore, Y2H and BiFC analyses indicated that CmWUS directly interact with CmCYC2d, an ortholog of GhCYC3 that has been proven to suppress stamen development in gerbera [20,38]. Hence, CmWUS and CmCYC2d may act together to affect the development of reproductive organs. This may further explain the mechanism of *CYC2*-like genes in the regulation of reproductive organ development. In addition, previous studies of CYC2-like proteins in gerbera and sunflower have shown redundant functions and higher capacity to form dimers within CYC2 subclade [12,19,20,22,26]. Thus, CmCYC2d could be the mediator between CmWUS and CmCYC2 and a complex regulatory network involving CmWUS and CmCYC2 subclade may exist in regulating reproductive organ development in chrysanthemum.

3.3. WUS Can Be a Bridge to Connect MADS-box and ECE (CYC/TB1)

It has been speculated that the flower morphology-related ECE and MADS-box genes may be closely linked [6,9,12,48]. In *Antirrhinum*, B-class MADS-box gene *DEF* and C-class gene *PLENA* are suggested to be required in the maintenance of *CYC* in whorl 2 and whorl 3, respectively [49]. *CYC2*-like genes are also indicated to be involved in regulating sepal identity by suppressing B-class genes in *Cysticapnos* [18]. In the *mtaga mtagb* double mutant of *Medicago truncatula*, the abnormal petals are related to the upregulation of *CYC2*-like genes [50]. Also, *GhSOC1* is thought to function upstream of CYC2 subclade genes in *Gerbera* [12,51]. However, the regulatory connections between MADS-box and ECE genes still remain to be illustrated. On the other hand, WUS acts as an activator in regulating the expression of C-class MADS-box gene *AG* in floral patterning and *AG* represses *WUS* directly or indirectly through activation of *KNUCKLES* at later stages of floral development in turn [39,44,52]. Furthermore, an A-class gene, *APETALA2 (AP2)*, antagonizes *AG* through promoting the expression of *WUS* in the floral meristem [53]. In this study, *CmWUS* and *CmCYC2* were found to be highly expressed in the reproductive organs of chrysanthemum and CmWUS could directly interact with CmCYC2d. A connection between WUS and ECE was established. Taken together, WUS, ECE and

MADS-box may be linked together during floral development and WUS acts as the adaptor to connect MADS-box and ECE.

In conclusion, this study characterized a *WUS*-like gene, *CmWUS*, in *C. morifolium* and revealed a remarkable link between CmWUS and CmCYC2 subclade. Since the significant function of CmWUS in reproductive organ development, our findings will help fill in the missing link of CmCYC2 in regulating the development of reproductive organs, especially in pistils, and contribute to a further understanding of the molecular mechanisms of gynomonoecy in Asteraceae.

4. Materials and Methods

4.1. Plant Materials and Growth Condition

C. morifolium 'Fen ditan' (Figure 3) and *A. thaliana* were cultivated in a greenhouse of Beijing Forestry University, China. They were grown under photoperiods of 8 h light (24 °C)/16 h dark (20 °C) and 16 h light (22 °C)/8 h dark (19 °C), respectively.

4.2. Gene Cloning

Total RNA was extracted from the inflorescences of *C. morifolium* 'Fen Ditan' with Plant RNA Kit (Omega, Norcross, GA, USA), and then used as template to synthesize first strand of cDNA with TransScript One-Step gDNA Removal and cDNA Synthesis SuperMix (Transgen, Beijing, China). Partial sequence of *CmWUS* in chrysanthemum was retrieved from our previously published RNAseq data [28]. SMARTerTM RACE 5′/3′ Kit (Clontech, Mountain View, CA, USA) was used for 5′ and 3′ RACE. 5′-GSP and 3′-GSP (Supplementary Table S1), gene-specific primers for RACE, were designed according to the instructions. Based on the 5′- and 3′-ends, *CmWUS*-F1 and *CmWUS*-R1 (Supplementary Table S1) were designed to amplify the open reading frame (ORF) sequence of *CmWUS*. Six *CmCYC2* genes (GenBank ID: *CmCYC2a*, KU595430.1; *CmCYC2b*, KU595431.1; *CmCYC2c*, KU595428.1; *CmCYC2d*, KU595426.1; *CmCYC2e*, KU595427.1; *CmCYC2f*, KU595429.1) were amplified with primers reported before [38]. All the PCR products were cloned into pCloneEZ-Blunt TOPO vectors (Taihe, Beijing, China), transformed into *Escherichia coli* DH5α cells (Tiangen, Beijing, China) and sequenced by Taihe (Beijing, China). The coding sequence of *CmWUS* (GenBank accession number: MK124768) has been uploaded to the NCBI database.

4.3. Bioinformatics Analysis

ClustalX software was used to perform alignment of multiple sequences, including CmWUS and WUS-like sequences from other species. GeneDoc software was used to edit the alignment. A phylogenetic tree was constructed by MEGA 7 based on the neighbor-joining method with 1000 bootstrap replicates, using the full length of the amino acid sequences of WUS homologs from various species and 15 WOX family members from *A. thaliana*. The accession numbers of sequences used here were as follows: AtWUS, *A. thaliana*, NM_127349.4; AtWOX1, AY251394.1; AtWOX2, NM_125325.3; AtWOX3, NM_128422.3; AtWOX4, FJ440850.1; AtWOX5, AY251398.1; AtWOX6, AY251399.2; AtWOX7, NM_120659.2; AtWOX8, AY251400.1; AtWOX9, AY251401.1; AtWOX10, NM_101923.1; AtWOX11, AY251402.1; AtWOX12, AY251403.1; AtWOX13, AY251404.1; AtWOX14, NM_101922.3; AmWUS, *Antirrhinum majus*, AAO23113.1; BnWUS, *Brassica napus*, XM_013803833.2; CcWUS, *Cynara cardunculus* var. *scolymus*, XM_025106474.1; CsWUS, *Citrus sinensis*, NM_001288918.1; GmWUS, *Glycine max*, XP_003517180.2; HaWUS, *Helianthus annuus*, HE616565.1; LsWUS, *Lactuca sativa*, XM_023909093.1; MtWUS, *Medicago truncatula*, XP_003612158.1; NtWUS, *Nicotiana tabacum*, XM_016619508.1; SlWUS, *Solanum lycopersicum*, ADZ13564.1; StWUS, *Solanum tuberosum*, XP_006340731.1; VvWUS, *Vitis vinifera*, XM_002266287.3.

4.4. Overexpression of *CmWUS* in *A. thaliana*

CmWUS was amplified using primers *CmWUS*-F2 and *CmWUS*-R2 (Supplementary Table S1) and subcloned into *Nco*I/*Bst*EII-cleaved pCambia1304 vector under the CaMV35S promoter using In-Fusion® HD Cloning Kit System (Clontech, Mountain View, CA, USA). The resulting pCambia1304-*CmWUS* vector was transformed into *A. thaliana* (Columbia) via *Agrobacteriaum tumefaciens* GV3101with the floral dip method [54]. The seeds were selected on MS medium containing hygromycin B (50 mg/L; Roche, Basel, Switzerland). qRT-PCR was performed using young leaves to confirm positive lines with primers *CmWUS*-F3/R3 and *AtACTIN*-F/R (Supplementary Table S1). Three independent homozygous T_3 lines with higher and consistent expression levels were selected for floral phenotype analysis. Forty flowers were analyzed and the significant differences were determined according to Fisher's LSD ($p < 0.05$) with SPSS 20.0.

4.5. Microscope Observations

The floral buds of *C. morifolium* 'Fen Ditan' at different stages were fixed in FAA (50% ethanol: acetic acid: formaldehyde = 90:5:5, v/v), dehydrated with a graded ethanol series (50%–100%) and then transferred into xylene (100%). All the samples were embedded in paraffin and cut into 8 μm sections using a microtome (Leica, Wetzlar, Germany). After that, paraffin was removed from the sections with xylene, and then safranin (1%) and fast green (0.5%) were used for histological staining. All the sections were examined and photographed under a light microscope (Zeiss, Jena, Germany) after sealed with neutral gum.

4.6. Gene Expression Analysis in *C. morifolium*

Floral buds of *C. morifolium* 'Fen Ditan' at different stages were collected for analysis of gene expression patterns. To compare expression patterns of *CmWUS* and *CmCYC2* genes at later stages of inflorescence development between ray and disc florets, samples were pooled from the flower heads of *C. morifolium* 'Fen Ditan' at different stages (Figure 4). To analyze tissue-specific expression of *CmWUS* and *CmCYC2* genes, vegetative and reproductive tissues were collected from the inflorescences of *C. morifolium* 'Fen Ditan' at stage 4 and 5 of inflorescence development (Figure 4). Particularly, pistil samples were dissected from both ray and disc florets, while stamen samples were pooled from disc florets only. Total RNA was extracted as described above and PrimeScriptTM RT reagent Kit (Perfect Real Time; TaKaRa, Shiga, Japan) was used to synthesize the first strand of cDNA. Quantitative real-time PCR experiments were performed using the PikoReal real-time PCR system (Thermo Fisher Scientific, Waltham, MA, USA) with a 10 μL mix of SYBR Premix ExTaq II (5 μL; Takara, Shiga, Japan), forward and reverse primers (10 μM, 0.5 μL each), cDNA (2 μL) and sterile distilled water (2 μL). The qPCR primers of *CmCYC2* genes and the reference gene *PP2Acs* were reported before [28,38,55]. *CmWUS*-F3/R3 (Supplementary Table S1) was used as qPCR primer of *CmWUS*. Three biological replicates were conducted with three technical replicates each. $2^{-\Delta\Delta Ct}$ method [56] was used to calculate the relative expression levels.

4.7. Subcellular Localization

CmWUS and *CmCYC2* genes were amplified and subcloned into *SalI*/*SpeI*- cleaved pSuper1300-*GFP* vectors to generate the transformation plasmids *35S::CmWUS::GFP* and *35S::CmCYC2::GFP*. The plasmids were transformed into *A. tumefaciens* and injected into the leaves of *Nicotiana benthamiana* following the procedure reported before [57]. TCS SP8 (Leica, Wetzlar, Germany) confocal laser scanning microscope was used to assess subcellular localization at 488 and 408 nm for GFP and DAPI fluorescence, respectively. Primers used for subcellular localization are listed in Supplementary Table S1.

4.8. Y2H Assay

Matchmaker Gold Yeast Two-Hybrid System (Clontech, Mountain View, CA, USA) was used to carry out Y2H assays. *CmCYC2* and *CmWUS* were amplified and subcloned into the pGADT7 (prey) and pGBKT7 (bait) vectors. The reconstructed pGADT7 and pGBKT7 vectors were transformed into Y187 and Y2H gold yeast strains and cultured on SD/-Leu and SD/-Trp plates, respectively. If the colonies containing bait vector are significantly smaller than colonies containing the empty pGBKT7 vector on SD/-Trp plates, then the bait is toxic to the yeast cells. To test the bait for autoactivation, Y2H gold yeast cells containing pGBKT7-*CmWUS* vector were cultured on SD/-Trp, SD/-Trp/X-α-Gal SD/-Trp/X-α-Gal/Aureobasidin A (AbA) and SD–Trp/–His/–Ade plates. If the colonies grow on both SD/-Trp and SD/-Trp/X-α-Gal plates, but not on SD/-Trp/X-α-Gal/AbA and SD–Trp/–His/–Ade plates, then the bait cannot autoactivate the AbA^r and His3/Ade2 reporter. After the testing of toxicity and autoactivation, diploid mating was conducted as described previously [58], and the transformed colonies were cultured on SD/-Trp/-Leu and SD/-Leu/-Trp/-His/-Ade/X-α-Gal/AbA (SD/-Leu/-Trp/-His/-Ade/X/A) plates to test for possible interactions. Y2H screenings were performed in triplicate. Primers used for Y2H assays are listed in Supplementary Table S1.

4.9. BiFC Assay

CmCYC2 and *CmWUS* genes were amplified and subcloned into the pCambia1300-YFP^N and pCambia1300-YFP^C vectors. Co-expression was conducted in the leaves of tobacco (*N. benthamiana*) as described in Subcellular Localization. TCS SP8 (Leica, Wetzlar, Germany) confocal laser scanning microscope was used to detect YFP fluorescence at 514 nm. Primers used for BiFC assays are listed in Supplemental Table S1.

Author Contributions: Conceptualization, Y.Y. and Q.Z.; Formal analysis, Y.Y.; Investigation, Y.Y.; Methodology, Y.Y., Y.H. and T.Z.; Project administration, Q.Z.; Resources, M.S., C.Y., T.C. and J.W.; Supervision, C.Y. and Q.Z.; Validation, Y.Y.; Writing—original draft, Y.Y.; Writing—review & editing, C.Y., Y.H. and T.Z.

Acknowledgments: Conceptualization, Y.Y. and Q.Z.; Data curation, Y.Y.; Formal analysis, Y.Y.; Investigation, Y.Y.; Methodology, Y.Y. and Y.H.; Resources, M.S., C.Y., T.Z., T.C., J.W. and Q.Z.; Software, T.C. and J.W.; Supervision, M.S., C.Y. and Q.Z; Writing—original draft, Y.Y.; Writing—review & editing, C.Y., Y.H. and T.Z.

References

1. Gillies, A.C.M.; Cubas, P.; Coen, E.S.; Abbott, R.J. Making rays in the Asteraceae: Genetics and evolution of radiate versus discoid flower heads. In *Developmental Genetics and Plant Evolution*; Quentin, C.B., Cronk, R.M.B., Julie, A.H., Eds.; Taylor & Francis: London, UK, 2002; Volume 65, pp. 233–246.
2. Bertin, R.I.; Kerwin, M.A. Floral sex ratios and gynomonoecy in Aster (Asteraceae). *Am. J. Bot.* **1998**, *85*, 235–244. [CrossRef] [PubMed]
3. Bertin, R.I.; Connors, D.B.; Kleinman, H.M. Differential herbivory on disk and ray flowers of gynomonoecious asters and goldenrods (Asteraceae). *Biol. J. Linn. Soc.* **2010**, *101*, 544–552. [CrossRef]
4. Sun, M.; Ganders, F.R. Outcrossing rates and allozyme variation in rayed and rayless morphs of *Bidens pilosa*. *Heredity* **1990**, *64*, 139–143. [CrossRef]
5. Andersson, S. Pollinator and nonpollinator selection on ray morphology in *Leucanthemum vulgare* (oxeye daisy, Asteraceae). *Am. J. Bot.* **2008**, *95*, 1072–1078. [CrossRef]
6. Fambrini, M.; Pugliesi, C. CYCLOIDEA 2 clade genes: Key players in the control of floral symmetry, inflorescence architecture, and reproductive organ development. *Plant Mol. Biol. Rep.* **2016**, *35*, 20–36. [CrossRef]

7. Kalisz, S.; Ree, R.H.; Sargent, R.D. Linking floral symmetry genes to breeding system evolution. *Trends Plant Sci.* **2006**, *11*, 568–573. [CrossRef]
8. Hileman, L.C.; Cubas, P. An expanded evolutionary role for flower symmetry genes. *J. Biol.* **2009**, *8*, 90. [CrossRef] [PubMed]
9. Howarth, D.G.; Donoghue, M.J. Phylogenetic analysis of the "ECE" (CYC/TB1) clade reveals duplications predating the core eudicots. *Proc. Natl. Acad. Sci. USA* **2006**, *103*, 9101–9106. [CrossRef]
10. Martin-Trillo, M.; Cubas, P. TCP genes: A family snapshot ten years later. *Trends Plant Sci.* **2010**, *15*, 31–39. [CrossRef]
11. Hileman, L.C. Bilateral flower symmetry—How, when and why? *Curr. Opin. Plant Biol.* **2014**, *17*, 146–152. [CrossRef]
12. Broholm, S.K.; Teeri, T.H.; Elomaa, P. Molecular control of inflorescence development in Asteraceae. In *Advances in Botanical Research*; Fornara, F., Ed.; Academic Press: Oxford, UK, 2014; Volume 72, pp. 297–333.
13. Luo, D.; Carpenter, R.; Vincent, C.; Copsey, L.; Coen, E. Origin of floral asymmetry in *Antirrhinum*. *Nature* **1996**, *383*, 794–799. [CrossRef]
14. Luo, D.; Carpenter, R.; Copsey, L.; Vincent, C.; Clark, J.; Coen, E. Control of organ asymmetry in flowers of *Antirrhinum*. *Cell* **1999**, *99*, 367–376. [CrossRef]
15. Gaudin, V.; Lunness, P.A.; Fobert, P.R.; Towers, M.; Rioukhamlichi, C.; Murray, J.A.H.; Coen, E.; Doonan, J.H. The Expression of *D-Cyclin* Genes Defines Distinct Developmental Zones in Snapdragon Apical Meristems and Is Locally Regulated by the *Cycloidea* Gene. *Plant Physiol.* **2000**, *122*, 1137–1148. [CrossRef]
16. Song, C.F.; Lin, Q.B.; Liang, R.H.; Wang, Y.Z. Expressions of ECE-CYC2 clade genes relating to abortion of both dorsal and ventral stamens in *Opithandra* (Gesneriaceae). *BMC Evol. Biol.* **2009**, *9*, 244. [CrossRef]
17. Preston, J.C.; Kost, M.A.; Hileman, L.C. Conservation and diversification of the symmetry developmental program among close relatives of snapdragon with divergent floral morphologies. *New Phytol.* **2009**, *182*, 751–762. [CrossRef]
18. Zhao, Y.; Pfannebecker, K.; Dommes, A.B.; Hidalgo, O.; Becker, A.; Elomaa, P. Evolutionary diversification of CYC/TB1-like TCP homologs and their recruitment for the control of branching and floral morphology in *Papaveraceae* (basal eudicots). *New Phytol.* **2018**, *220*, 317–331. [CrossRef]
19. Broholm, S.K.; Tahtiharju, S.; Laitinen, R.A.; Albert, V.A.; Teeri, T.H.; Elomaa, P. A TCP domain transcription factor controls flower type specification along the radial axis of the Gerbera (Asteraceae) inflorescence. *Proc. Natl. Acad. Sci. USA* **2008**, *105*, 9117–9122. [CrossRef]
20. Juntheikki-Palovaara, I.; Tahtiharju, S.; Lan, T.; Broholm, S.K.; Rijpkema, A.S.; Ruonala, R.; Kale, L.; Albert, V.A.; Teeri, T.H.; Elomaa, P. Functional diversification of duplicated CYC2 clade genes in regulation of inflorescence development in *Gerbera hybrida* (Asteraceae). *Plant J.* **2014**, *79*, 783–796. [CrossRef]
21. Elomaa, P.; Zhao, Y.; Zhang, T. Flower heads in Asteraceae—Recruitment of conserved developmental regulators to control the flower-like inflorescence architecture. *Hortic. Res.* **2018**, *5*, 36. [CrossRef]
22. Tahtiharju, S.; Rijpkema, A.S.; Vetterli, A.; Albert, V.A.; Teeri, T.H.; Elomaa, P. Evolution and diversification of the CYC/TB1 gene family in Asteraceae—A comparative study in Gerbera (Mutisieae) and sunflower (Heliantheae). *Mol. Biol. Evol.* **2012**, *29*, 1155–1166. [CrossRef]
23. Bello, M.A.; Cubas, P.; Alvarez, I.; Sanjuanbenito, G.; Fuertes-Aguilar, J. Evolution and expression patterns of CYC/TB1 genes in *Anacyclus*: Phylogenetic insights for floral symmetry genes in Asteraceae. *Front. Plant Sci.* **2017**, *8*, 589. [CrossRef]
24. Mizzotti, C.; Fambrini, M.; Caporali, E.; Masiero, S.; Pugliesi, C. A *CYCLOIDEA*-like gene mutation in sunflower determines an unusual floret type able to produce filled achenes at the periphery of the pseudanthium. *Botany* **2015**, *93*, 171–181. [CrossRef]
25. Fambrini, M.; Salvini, M.; Pugliesi, C. A transposon-mediate inactivation of a *CYCLOIDEA*-like gene originates polysymmetric and androgynous ray flowers in *Helianthus Annuus*. *Genetics* **2011**, *139*, 1521–1529. [CrossRef]
26. Chapman, M.A.; Tang, S.; Draeger, D.; Nambeesan, S.; Shaffer, H.; Barb, J.G.; Knapp, S.J.; Burke, J.M. Genetic analysis of floral symmetry in Van Gogh's sunflowers reveals independent recruitment of *CYCLOIDEA* genes in the Asteraceae. *PLoS Genet.* **2012**, *8*, e1002628. [CrossRef]
27. Fambrini, M.; Basile, A.; Salvini, M.; Pugliesi, C. Excisions of a defective transposable CACTA element (*Tetu1*) generate new alleles of a *CYCLOIDEA*-like gene of *Helianthus annuus*. *Gene* **2014**, *549*, 198–207. [CrossRef]

28. Liu, H.; Sun, M.; Du, D.; Pan, H.; Cheng, T.; Wang, J.; Zhang, Q.; Gao, Y. Whole-transcriptome analysis of differentially expressed genes in the ray florets and disc florets of *Chrysanthemum morifolium*. *BMC Genom.* **2016**, *17*, 398. [CrossRef]
29. Graaff, E.V.D.; Laux, T.; Rensing, S.A. The WUS homeobox-containing (WOX) protein family. *Genom. Biol.* **2009**, *10*, 248. [CrossRef]
30. Lohmann, J.U.; Hong, R.L.; Hobe, M.; Busch, M.A.; Parcy, F.; Simon, R.; Weigel, D. A Molecular link between stem cell regulation and floral patterning in *Arabidopsis*. *Cell* **2001**, *105*, 793–803. [CrossRef]
31. Yadav, R.K.; Perales, M.; Gruel, J.; Girke, T.; Jönsson, H.; Reddy, G.V. WUSCHEL protein movement mediates stem cell homeostasis in the *Arabidopsis* shoot apex. *Genes Dev.* **2012**, *25*, 2025–2030. [CrossRef]
32. Zhou, Y.; Yan, A.; Han, H.; Li, T.; Geng, Y.; Liu, X.; Meyerowitz, E.M. HAIRY MERISTEM with WUSCHEL confines CLAVATA3 expression to the outer apical meristem layers. *Science* **2018**, *361*, 502–506. [CrossRef]
33. Deyhle, F.; Sarkar, A.K.; Tucker, E.J.; Laux, T. *WUSCHEL* regulates cell differentiation during anther development. *Dev. Biol.* **2007**, *302*, 154–159. [CrossRef]
34. Doerks, T.; Copley, R.R.; Schultz, J.; Ponting, C.P.; Bork, P. Systematic identification of novel protein domain families associated with nuclear functions. *Genom. Res.* **2002**, *12*, 47–56. [CrossRef]
35. Yamada, T.; Sasaki, Y.; Hashimoto, K.; Nakajima, K.; Gasser, C.S. *CORONA*, *PHABULOSA* and *PHAVOLUTA* collaborate with BELL1 to confine WUSCHEL expression to the nucellus in *Arabidopsis* ovules. *Development* **2016**, *143*, 422–426. [CrossRef]
36. Laux, T.; Mayer, K.F.; Berger, J.; Jürgens, G. The *WUSCHEL* gene is required for shoot and floral meristem integrity in *Arabidopsis*. *Development* **1996**, *122*, 87–96.
37. Liu, X.; Ning, K.; Che, G.; Yan, S.; Han, L.; Gu, R.; Li, Z.; Weng, Y.; Zhang, X. CsSPL functions as an adaptor between HD-ZIP III and CsWUS transcription factors regulating anther and ovule development in *Cucumis sativus* (cucumber). *Plant J.* **2018**, *94*, 535–547. [CrossRef]
38. Huang, D.; Li, X.; Sun, M.; Zhang, T.; Pan, H.; Cheng, T.; Wang, J.; Zhang, Q. Identification and Characterization of *CYC*-Like Genes in Regulation of Ray Floret Development in *Chrysanthemum morifolium*. *Front. Plant Sci.* **2016**, *7*, 1633. [CrossRef]
39. Ikeda, M.; Mitsuda, N.; Ohme-Takagi, M. *Arabidopsis* WUSCHEL is a bifunctional transcription factor that acts as a repressor in stem cell regulation and as an activator in floral patterning. *Plant Cell* **2009**, *21*, 3493–3505. [CrossRef]
40. Gallois, J.L.; Nora, F.R.; Mizukami, Y.; Sablowski, R. *WUSCHEL* induces shoot stem cell activity and developmental plasticity in the root meristem. *Genes Dev.* **2004**, *18*, 375–380. [CrossRef]
41. Xu, Y.Y.; Wang, X.M.; Li, J.; Li, J.H.; Wu, J.S.; Walker, J.C.; Xu, Z.H.; Chong, K. Activation of the *WUS* gene induces ectopic initiation of floral meristems on mature stem surface in *Arabidopsis thaliana*. *Plant Mol. Biol.* **2005**, *57*, 773–784. [CrossRef]
42. Rodriguez, K.; Perales, M.; Snipes, S.; Yadav, R.K.; Diaz-Mendoza, M.; Reddy, G.V. DNA-dependent homodimerization, sub-cellular partitioning, and protein destabilization control *WUSCHEL* levels and spatial patterning. *Proc. Natl. Acad. Sci. USA* **2016**, *113*, 6307–6315. [CrossRef]
43. Paponov, I.A.; Teale, W.; Lang, D.; Paponov, M.; Reski, R.; Rensing, S.A.; Palme, K. The evolution of nuclear auxin signalling. *BMC Evol. Biol.* **2009**, *9*, 126. [CrossRef]
44. Lenhard, M.; Bohnert, A.; Jurgens, G.; Laux, T. Termination of stem cell maintenance in *Arabidopsis* floral meristems by interactions between *WUSCHEL* and *AGAMOUS*. *Cell* **2001**, *105*, 805–814. [CrossRef]
45. Vandenbussche, M.; Horstman, A.; Zethof, J.; Koes, R.; Rijpkema, A.S.; Gerats, T. Differential recruitment of WOX transcription factors for lateral development and organ fusion in Petunia and *Arabidopsis*. *Plant Cell* **2009**, *21*, 2269–2283. [CrossRef]
46. Honda, E.; Yew, C.L.; Yoshikawa, T.; Sato, Y.; Hibara, K.I.; Itoh, J.I. *LEAF LATERAL SYMMETRY1*, a member of the *WUSCHEL-RELATED HOMEOBOX3* gene family, regulates lateral organ development differentially from other paralogs, *NARROW LEAF2* and *NARROW LEAF3* in Rice. *Plant Cell Physiol.* **2018**, *59*, 376–391. [CrossRef]
47. Berti, F.; Fambrini, M.; Turi, M.; Bertini, D.; Pugliesi, C. Mutations of corolla symmetry affect carpel and stamen development in *Helianthus annuus*. *Can. J. Bot.* **2005**, *83*, 1065–1072. [CrossRef]
48. Preston, J.C.; Hileman, L.C. Parallel evolution of TCP and B-class genes in Commelinaceae flower bilateral symmetry. *Evodevo* **2012**, *3*, 6. [CrossRef]

49. Clark, J.I.; Coen, E.S. The *cycloidea* gene can respond to a common dorsoventral prepattern in *Antirrhinum*. *Plant J.* **2002**, *30*, 639–648. [CrossRef]
50. Zhu, B.; Li, H.; Wen, J.; Mysore, K.S.; Wang, X.; Pei, Y.; Niu, L.; Lin, H. Functional specialization of duplicated *AGAMOUS* homologs in regulating floral organ development of *Medicago truncatula*. *Front. Plant Sci.* **2018**, *9*, 854. [CrossRef]
51. Ruokolainen, S.; Ng, Y.P.; Albert, V.A.; Elomaa, P.; Teeri, T.H. Over-expression of the *Gerbera hybrida At-SOC1-like1* gene *Gh-SOC1* leads to floral organ identity deterioration. *Ann. Bot.* **2011**, *107*, 1491–1499. [CrossRef]
52. Liu, X.; Kim, Y.J.; Muller, R.; Yumul, R.E.; Liu, C.; Pan, Y.; Cao, X.; Goodrich, J.; Chen, X. *AGAMOUS* terminates floral stem cell maintenance in *Arabidopsis* by directly repressing *WUSCHEL* through recruitment of Polycomb Group proteins. *Plant Cell* **2011**, *23*, 3654–3670. [CrossRef]
53. Huang, Z.; Shi, T.; Zheng, B.; Yumul, R.E.; Liu, X.; You, C.; Gao, Z.; Xiao, L.; Chen, X. *APETALA2* antagonizes the transcriptional activity of *AGAMOUS* in regulating floral stem cells in *Arabidopsis thaliana*. *New Phytol.* **2017**, *215*, 1197–1209. [CrossRef]
54. Clough, S.J.; Bent, A.F. Floral dip: A simplified method for *Agrobacterium*-mediated transformation of *Arabidopsis thaliana*. *Plant J.* **1998**, *16*, 735–743. [CrossRef]
55. Gu, C.; Chen, S.; Liu, Z.; Shan, H.; Luo, H.; Guan, Z.; Chen, F. Reference gene selection for quantitative real-time PCR in Chrysanthemum subjected to biotic and abiotic stress. *Mol. Biotechnol.* **2011**, *49*, 192–197. [CrossRef]
56. Schmittgen, T.D.; Livak, K.J. Analyzing real-time PCR data by the comparative C(T) method. *Nat. Protoc.* **2008**, *3*, 1101–1108. [CrossRef]
57. Zhao, K.; Zhou, Y.; Ahmad, S.; Xu, Z.; Li, Y.; Yang, W.; Cheng, T.; Wang, J.; Zhang, Q. Comprehensive cloning of *Prunus mume* dormancy associated MADS-Box genes and their response in flower bud development and dormancy. *Front. Plant Sci.* **2018**, *9*, 17. [CrossRef]
58. Zhou, Y.; Xu, Z.; Yong, X.; Ahmad, S.; Yang, W.; Cheng, T.; Wang, J.; Zhang, Q. SEP-class genes in *Prunus mume* and their likely role in floral organ development. *BMC Plant Biol.* **2017**, *17*, 10. [CrossRef]

6

Identification of Quantitative Trait Loci Associated with Nutrient use Efficiency Traits, using SNP Markers in an Early Backcross Population of Rice (*Oryza sativa* L.)

Zilhas Ahmed Jewel [1,†], Jauhar Ali [1,*,†], Anumalla Mahender [1,†], Jose Hernandez [2], Yunlong Pang [1,3] and Zhikang Li [4]

1 Rice Breeding Platform, International Rice Research Institute (IRRI), Los Baños, Laguna 4031, Philippines; jeweluplb@gmail.com (Z.A.J.); m.anumalla@irri.org (A.M.); y.pang@sdau.edu.cn (Y.P.)

2 College of Agriculture, University of the Philippines Los Baños, Laguna 4031, Philippines; joehernandez56@gmail.com

3 College of Agronomy, Shandong Agricultural University, Taian 271018, China

4 Institute of Crop Sciences, Chinese Academy of Agricultural Sciences, Beijing 100081, China; zhkli1953@126.com

* Correspondence: J.Ali@irri.org

† These authors contributed equally to this work.

Abstract: The development of rice cultivars with nutrient use efficiency (NuUE) is highly crucial for sustaining global rice production in Asia and Africa. However, this requires a better understanding of the genetics of NuUE-related traits and their relationship to grain yield. In this study, simultaneous efforts were made to develop nutrient use efficient rice cultivars and to map quantitative trait loci (QTLs) governing NuUE-related traits in rice. A total of 230 BC_1F_5 introgression lines (ILs) were developed from a single early backcross population involving Weed Tolerant Rice 1, as the recipient parent, and Hao-an-nong, as the donor parent. The ILs were cultivated in field conditions with a different combination of fertilizer schedule under six nutrient conditions: minus nitrogen (–N), minus phosphorus (–P), (–NP), minus nitrogen phosphorus and potassium (–NPK), 75% of recommended nitrogen (75N), and NPK. Analysis of variance revealed that significant differences ($p < 0.01$) were noted among ILs and treatments for all traits. A high-density linkage map was constructed by using 704 high-quality single nucleotide polymorphism (SNP) markers. A total of 49 main-effect QTLs were identified on all chromosomes, except on chromosome 7, 11 and 12, which are showing 20.25% to 34.68% of phenotypic variation. With further analysis of these QTLs, we refined them to four top hotspot QTLs (QTL harbor-I to IV) located on chromosomes 3, 5, 9, and 11. However, we identified four novel putative QTLs for agronomic efficiency (AE) and 22 QTLs for partial factor productivity (PFP) under –P and 75N conditions. These interval regions of QTLs, several transporters and genes are located that were involved in nutrient uptake from soil to plant organs and tolerance to biotic and abiotic stresses. Further, the validation of these potential QTLs, genes may provide remarkable value for marker-aided selection and pyramiding of multiple QTLs, which would provide supporting evidence for the enhancement of grain yield and cloning of NuUE tolerance-responsive genes in rice.

Keywords: nutrient use efficiency; quantitative trait loci (QTLs), molecular markers; agronomic efficiency; partial factor productivity

1. Introduction

Rice (*Oryza sativa* L.) is one of the most prominent staple food crops in the world and it has significantly contributed to global food security [1]. By 2050, rice production has to be increased by more than 60% to meet the rapid increase in food demand [2,3]. Approximately 90% of rice cultivation is carried out in irrigated and lowland rainfed ecosystems in Asia, where we still face problems such as decreasing arable land and freshwater availability, and increased labor costs and biotic and abiotic stress factors that impede production and productivity. Further, the drastic rise in fertilizer costs drives the search for suitable rice cultivars that are efficient in grain yield production [4–8]. Nutrient fertilizers are one of the most predominant factors that influence the genetic enhancement of grain yield productivity in most agricultural regions [6].

In the last four decades, the amount of nitrogen (N) applied to crops has risen 12 to 104 teragrams per year (Tg year^{-1}), and 30%–50% of the N has been harvested in the grain and the remaining 70%–50% of the fertilizer has been lost through a combination of leaching, denitrification, volatilization, surface runoff, and microbial consumption [9]. Therefore, the excessive amount and long-term use of fertilizers cause severe environmental pollution, and this is one of the significant costly inputs for poor farmers [10–14]. Hence, the identification of crops with higher NuUE that are less dependent on NPK fertilizers (nitrogen, phosphorous, and potassium) is crucial for the sustainability of agriculture. Kant et al. [15] estimated that an increase of 1% in NuUE could save about US$ 1.1 billion annually. The Green Revolution period witnessed a modest increase of 13% in harvested rice area in Asia, whereas grain yield production more than doubled, from 240 to 513 million tons, with consumption of fertilizers high in NPK increasing from 6.7 to 69.0 million tons. According to the FAO electronic database, particularly in Asian countries, China, Sri Lanka, Indonesia, and India had the highest average fertilizer consumption of 565.2 kg ha^{-1}, 251.7 kg ha^{-1}, 185.141 kg ha^{-1}, and 157.48 kg ha^{-1}, respectively, from 2005 to 2014 (http://ricestat.irri.org:8080/wrsv3/entrypoint.htm#). Therefore, the identification of superior cultivars with improved nutrient use efficiency (NuUE) is an essential target research area for plant breeders, for attaining higher grain yield and also reducing production costs [16].

In the improvement of the tolerance of rice cultivars of single biotic and abiotic stresses, tremendous progress has been achieved [8,17–21]. However, these tolerant varieties have not been able to succeed to attain higher grain yields under both the irrigated and rainfed ecosystems. To overcome these conditions, breeding aspects need to focus on developing multiple-stress-tolerant cultivars (MSTC) and improving nutrient use efficiency, which are vital for the sustainability of grain yield productivity, which could have a more significant impact on higher yield, especially under low-fertilizer-input conditions, besides being more beneficial to poor farmers. Several critical components of morphological, physiological, and agronomic traits are reducing grain yield and decreasing biomass content under nutrient-deficient conditions. These traits become altered due to changes in their molecular mechanism and physiological pathways, leading to their susceptibility [22]. The exploitation of rice genetic resources, using advanced genomic technologies, is essential for the identification of rice cultivars with NuUE for increasing crop grain yield under low-input conditions.

Developing rice varieties with stress tolerance and NuUE through conventional breeding approaches is extremely slow because of several factors that influence the molecular genetics and physiological mechanisms underlying low-input tolerance, the complexity of NuUE traits, and the absence of efficient breeding selection criteria [23–25]. The combination of advanced molecular marker-assisted breeding and conventional breeding platforms could speed up the breeding procedure for varietal development and identifying trait-associated QTLs. Over the past two decades, several types of traditional molecular markers, such as restriction fragment length polymorphism (RFLP), randomly amplified polymorphic DNA (RAPD), amplified fragment length polymorphism (AFLP) and simple sequence repeats (SSRs), have been used for QTL identification for NuUE-related traits under low-input conditions of nitrogen [26–30], phosphorus [7,31–38], and potassium [39,40]. However, these molecular markers have some disadvantages vis-à-vis SNP markers, such as partial chromosome coverage, labor-intensiveness, low resolution, and higher cost. Among these various molecular markers, SNP markers have a comprehensive range of applications in the construction of genetic maps, uniform distribution, and cloning of QTLs. Their clear advantages, such as high density and accurate and reliable approaches, have been used to perform high-throughput genotyping based on SNPs across the genome [41–44]. In a genomics era, high-throughput genotyping with next-generation sequencing (NGS) and various array-based SNP detection platforms have become excellent tools for the dissecting of complex traits and identification of trait-associated genes and alleles in rice [45–48]. As compared to NGS, a SNP array can be used for many samples within a short period; it has a low cost and analysis of data interpretation is relatively easy [47,49]. The recently developed SNP array has been successfully used in diversity studies and genome-wide association studies (GWAS), and in the identification of numerous QTLs and genes in rice [43,49–51].

Over the years, the identification of QTLs for varied agro-morphological traits related to NuUE, by using different populations, hardly resulted in any significant impact regarding rice crop improvement and the development of varieties. The QTLs were identified related to NPK use efficiency using recombinant inbred lines (RILs) [34,48,52–54], backcross inbred lines (BILs) [55], doubled haploids (DHs) [39,56–59], BC_2F_3 [33], introgression lines (ILs) [60,61], and chromosome segment substitution lines (CSSLs) [35] populations. These QTLs have a smaller effect, and many of them exhibit significant epistatic and QTL x Environment interactions, making them less amenable to breeding programs.

Several QTLs may correspond to known genes in the N or P metabolic pathway, for example, Qyd-2b for N deficiency tolerance was located in the vicinity of the gene encoding cytosolic glutamine synthetase (GS1) [62], and Qyd-3b and Qpn-3 were nearby the genes for glutamate dehydrogenase (GDH2) [26]. Interestingly, Qyd-12 was detected only in low-P conditions and was co-localized with a major QTL, Pup1, on chromosome 12, which was reported to be involved in P absorption [63]. These results indicate that the QTLs specially detected under single N or P deficiency conditions may be involved in different pathways of N and P metabolism. The tightly linked markers have breeding potential in pyramiding elite genes and QTLs for N and P use efficiency.

Although N and P use efficient QTLs and linked traits were reported earlier in different genetic backgrounds of mapping populations, to date, there are no reports of different combinations of NPK fertilizers and their response to the foremost NuUE traits, such as grain yield response, agronomic efficiency, and partial factor productivity in rice. In the present study, taking advantage of the rapid development of SNP technologies, we used a 6K SNP array to genotype a BC_1F_5 introgression line (IL) derived from a cross between Weed Tolerant Rice 1 (WTR-1), as the recipient parent, and Hao-an-nong (HAN), as the donor parent. In this study, we identified the NuUE-related QTLs by subjecting the population to varying nutrient rates and understanding the performance of yield and its components.

2. Results

2.1. Phenotypic Variation of NuUE Traits and Their Correlation among Traits

A total of 230 BC_1F_5 ILs were evaluated in six different NPK combinations (–N, –P, –NP, –NPK, 75N and NPK) during the dry season of 2014 in the experimental fields of International Rice Research Institute (IRRI). The key traits of NuUE were analyzed in the total of ILs and were compared with those of the parents and checks (Figure 1). The descriptive statistics for five traits as grain yield (GY), 1000-grain weight (1000-Gwt), percentage of spikelet fertility (PSPF), biomass yield (BY), filled grains per plant (FGN) mean values and testing of significance are presented in Tables 1 and 2. The highest GY range of 19.29–50.26 g/plant was observed in NPK, followed by 20.82–48.83 g/plant in 75N, and 12.81–42.95 g/plant in –NPK, and the lowest GY range of 13.89–34.96 g/plant was detected in –N conditions. Among these six nutrient conditions, 18 ILs in NPK, 23 ILs in 75N, 4 ILs in –P, and 1 IL in both –NPK and –NP had more than 40 g/plant. In –N conditions, the lowest GY was observed (34.96 g/plant). The best performing ILs under each of the six nutrient conditions with the highest grain yields were *GSR IR2-1-RF6-NU7-NU2-NU76-NU96* -WTR 1-RF6 in NPK that gave 50.26 g/plant;

and 42.95 g/plant (*GSR IR2-1-RF6-NU7-NU3-NU82-NU97*-WTR 1-RF6) in –NPK; 48.83 g/plant (*GSR IR2-1-Y17-NU2-NU5-NU6-NU9*-WTR 1-Y17) in 75N; 34.96 g/plant (*GSR IR2-1-L3-NU1-NU1-NU1-NU1*-WTR 1-LI3) in –N; 40.94 g/plant (*GSR IR2-1-RF6-NU4-NU9-NU14-NU66*-WTR 1-RF6) in –P; and 40.43 g/plant (*GSR IR2-1-RF6-NU7-NU2-NU77-NU94*-WTR 1-RF6) in –NP conditions, respectively. Across all six nutrient conditions, the lowest GYs were observed in WTR 1-RF13 (*GSR IR2-1-RF13-NU2-NU8-NU4-NU12*), WTR 1-RF14 (*GSR IR2-1-RF14-NU2-NU4-NU6-NU3*) with 11.85 g/plant; 12.43 g/plant detected in –NP; and followed by WTR 1-LI3 (*GSR IR2-1-L3-NU1-NU1-NU3-NU5*) and WTR 1-LI12 (*GSR IR2-1-L12-NU1-NU4-NU7-NU5*) with 12.86 g/plant and 12.81 g/plant in –NPK conditions, respectively. Under –NP and –NPK conditions, the recipient parent (Weed Tolerant Rice 1) gave 19.87 g/plant and 20.94 g/plant, respectively, while the donor parent (Hao-an-nong) gave 15.73 g/plant and 16.02 g/plant, respectively.

After further analysis of ILs with higher grain yield in each NuUE condition, we selected the first ten highest GY lines, and GY ranged from 28.86 to 50.26 g/plant. Interestingly, we identified promising lines, WTR 1-RF6 (*GSR IR2-1-RF6-NU7-NU3-NU82-NU97*), that showed higher GY in four different nutrient conditions: –N (29.11 g/plant), –P (38.40 g/plant), –NP (31.90 g/plant), and –NPK (42.95 g/plant). Similarly, introgression line WTR 1-RF6 (*GSR IR2-1-RF6-NU7-NU2-NU77-NU94*) had higher GY in three different NuUE conditions, –NP (40.43 g/plant), –NPK (31.43 g/plant), and NPK (42.76 g/plant); and WTR 1-RF6 (*GSR IR2-1-RF6-NU7-NU2-NU76-NU96*) in –P (40.09 g/plant), –NP (33.09 g/plant), and NPK (50.26 g/plant) conditions. Likewise, WTR 1-Y17 (*GSR IR2-1-Y17-NU2-NU5-NU6-NU8* and *GSR IR2-1-Y17-NU2-NU5-NU6-NU9*) had higher GY in 75N (42.00 g/plant and 48.83 g/plant, respectively), –NP (35.73 g/plant and 33.41 g/plant, respectively), and –NPK (33.29 g/plant and 34.75 g/plant, respectively) conditions, The results for NuUE key traits, such as GY (–NP, –NPK), FGN (75N, –NP, –NPK), and BY (NPK, 75N, –N, –P, –NP, –NPK), had a high level of phenotypic variation with a coefficient of variation (CV) of above 20%. The correlations of five traits in six different nutrient conditions were identified to show significant relationships between the respective traits. GY was found to be significantly and positively correlated with BY; FGN in NPK; FGN in 75N; GWT, BY, and FGN in –P; PSPF, BY, and FGN in –N; and BY and FGN in –NP ($p < 0.01$) conditions. However, GY showed itself to be negatively correlated with 1000-Gwt in NPK and with BY in 75N conditions. Among the six nutrient conditions, GY was negatively correlated with BY under 75N. However, the remaining GWT, PSPF and FGN traits followed with significant positive correlation between the traits.

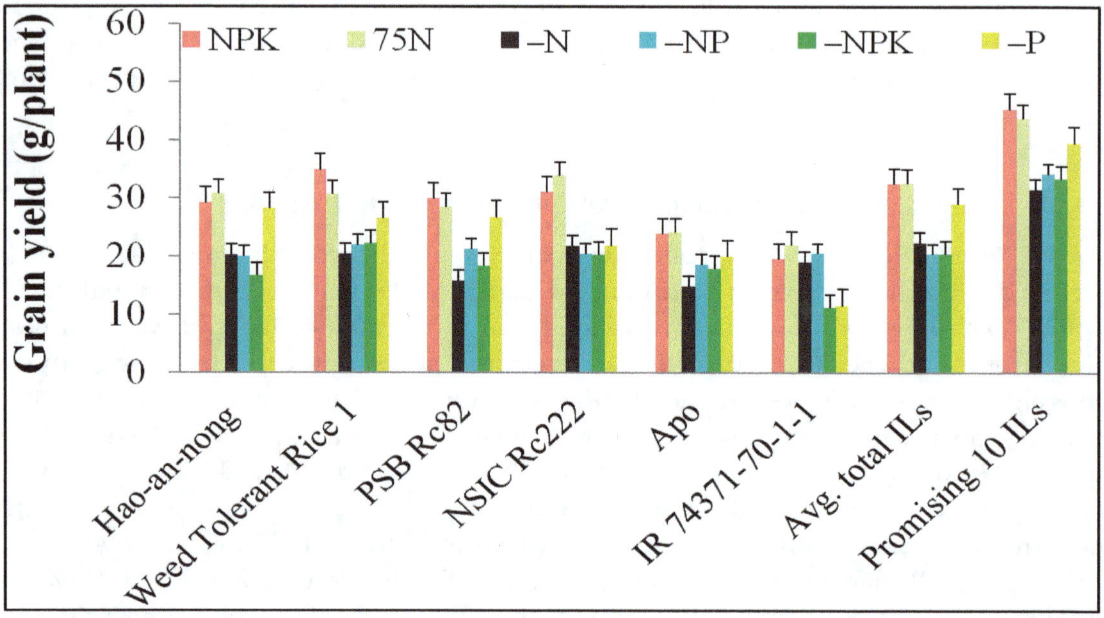

Figure 1. Grain yield performances of introgression lines, parents and checks in all six nutrient conditions.

Table 1. Statistical description of nutrient use efficiency-introgression lines for critical traits under six NPK combinations.

Traits	NuUE Condition	Mean ± Std. Error of Mean	Range (Min–Max)	SD	Variance (S^2)	CV%
GY	NPK	32.48 ± 0.34	19.29–50.26	5.21	27.19	16.04
	75N	32.58 ± 0.35	20.82–48.83	5.44	29.69	16.70
	–N	21.61 ± 0.24	13.89–34.96	3.79	14.37	17.54
	–P	26.74 ± 0.31	17.57–42.19	4.78	22.86	17.88
	–NP	20.41 ± 0.29	11.86–40.43	4.49	20.20	22.00
	–NPK	20.49 ± 0.29	12.81–42.95	4.54	20.66	22.16
1000 Gwt	NPK	27.35 ± 0.13	19.20–31.75	2.05	4.21	7.50
	75N	27.35 ± 0.14	15.18–39.50	2.24	5.03	8.19
	–N	26.41 ± 0.11	20.65–30.60	1.82	3.31	6.89
	–P	27.13 ± 0.16	14.80–33.75	2.52	6.36	9.29
	–NP	26.88 ± 0.14	15.60–31.10	2.14	4.58	7.96
	–NPK	27.16 ± 0.12	21.50–33.35	1.83	3.37	6.76
PSPF	NPK	87.02 ± 0.28	68.05–95.24	4.34	18.84	4.99
	75N	87.14 ± 0.24	72.39–94.35	3.76	14.15	4.32
	–N	89.24 ± 0.29	68.50–96.59	4.44	19.78	4.98
	–P	87.60 ± 0.27	72.95–95.95	4.21	17.75	4.81
	–NP	89.93 ± 0.28	70.52–97.37	4.30	18.54	4.78
	–NPK	89.88 ± 0.26	77.27–97.75	4.09	16.76	4.55
FGN	NPK	1518.90 ± 17.95	956.83–2621.83	273.41	74,756.36	18.00
	75N	1402.42 ± 19.23	783.00–3177.50	293.04	85,876.01	20.90
	–N	1012.30 ± 11.88	686.50–1660.83	180.97	32,750.56	17.88
	–P	1318.12 ± 16.54	786.50–2251.17	252.01	63,512.83	19.12
	–NP	947.57 ± 13.26	445.33–1704.50	202.11	40,848.47	21.33
	–NPK	1040.31 ± 14.50	565.83–1675.83	220.94	48,814.58	21.24
BY	NPK	79.72 ± 1.14	35.92–137.23	17.43	303.83	21.86
	75N	210.24 ± 1.29	41.52–30,066.06	1968.81	3,876,231.42	936.46
	–N	50.47 ± 1.48	23.46–336.63	22.54	508.41	44.66
	–P	69.26 ± 1.18	38.31–144.27	18.07	326.83	26.09
	–NP	45.01 ± 0.84	18.16–89.43	12.90	166.49	28.66
	–NPK	47.29 ± 0.91	18.09–116.50	13.95	194.736	29.50

GY—grain yield; 1000-Gwt—1000-grain weight; PSPF—percentage of spikelet fertility; FGN—filled grains per plant; BY—biomass yield (BY); CV—coefficient of variance; SD—standard deviation.

2.2. ANOVA and Interaction with the Environment

The fixed effect model was used for the analysis of variance with Satterthwaite denominator, and the summary of ANOVA is given in Table 2. Among the six nutrient conditions, significant genotypic effects were observed under –N (F = 1.44, p = 0.0005), –NP (F = 1.76, p < 0.0001), and –NPK (F = 2.47, p = 0.0007) conditions at the 1% level. In contrast, under 75N (F = 1.2, p = 0.053), –P (F = 1.38, p = 0.0658), and NPK (F = 1.07, p = 0.4293) conditions, the genotypes were not significantly different from each other. It was noted that the treatments showing significant genotypic effects are those that lack the nitrogen fertilizer component, indicating that some of these materials are nutrient use efficient. In the significant effect of fertilizers with genotype, the study used a −2 log-likelihood ratio test and found a significant environmental effect at the 1% level (χ^2 = 15.23, p = 0.0001). However, genotype and environment showed non-significant interactions (χ^2 = 0.00, p = 0.9992) (Table 3). This suggests that the different fertilizer applications significantly affect yield performance of the ILs as a whole. On the other hand, the non-significant genotypes by environment interactions indicate the consistent performance of genotypes across environments.

Table 2. ANOVA for the testing of significance of genotype effect per fertilizer condition.

S. No.	Environment	Degrees of Freedom	Sum of Squares	Mean Squares	F-Value	Satterthwaite Denominator	Pr (>F)
1	NPK	230	12,493.16	54.56	1.07	34.38	0.4293
2	75N	230	12,736.71	55.62	1.20	451.78	0.0530
3	–N	230	6540.20	28.56	1.44	451.61	0.0005 ***
4	–P	230	9273.85	40.50	1.38	62.36	0.0658
5	–NP	230	9288.00	40.56	1.76	229.00	0.0000 ***
6	–NPK	230	9503.51	41.50	2.47	37.76	0.0007 ***

Significant codes: 0 '***', 0.001 '**', 0.01 '*' 0.0.

Table 3. Testing for significance of fertilizer and its combined effect with genotype using −2 log-likelihood ratio test.

Effect	Model	AIC	BIC	Log-Likelihood	Chi Square	Degrees of Freedom	Pr (>Chisq)
Environment	1	17,071.23	18,457.21	−8301.62			
	2	17,058.00	18,449.91	−8294.00	15.2284	1	0.0001 ***
Genotype X Environment	3	17,056.00	18,441.98	−8294.00			
	4	17,058.00	18,449.91	−8294.00	0	1	0.9992

AIC—Akaike information criterion; BIC—Bayes information criterion; YLD—Yield; Deg—Degrees of freedom; Env—Environment; Rep—Replication; Blck—Block; Model 1: YLD~1 + Deg + (1 | Env) + (1 | Rep:Env) + (1 | Rep:Blck:Env) + (1 | Deg:Env). Model 2: YLD~1 + Deg + (1 | Rep:Env) + (1 | Rep:Blck:Env) + (1 | Deg:Env). Model 3: YLD~1 + Deg + (1 | Env) + (1 | Rep:Env) + (1 | Rep:Blck:Env) + (1 | Deg:Env). Model 4: YLD~1 + Deg + (1 | Env) + (1 | Rep:Env) + (1 | Rep:Blck:Env).

2.3. Analysis of Agronomic Efficiency (AE) and Partial Factor Productivity (PFP)

In the NuUE study, agronomic use efficiency and partial factor productivity were calculated using the fertilizer application rate of NPK (160–50–50 kg ha^{-1}). The computation using the equations (i to iii) revealed 15.06–34.11 kg of grain for each kg of N derived from an NPK fertilizer given in 3.2:1:1 ratio, respectively. However, using this equation, we found 117 ILs with >15 kg of grain per kg^{-1} N (NPK), along with 33 ILs (–P) and 161 ILs (75N). These were also compared with parents and we identified 42 ILs, 23 ILs, and 79 ILs that had a higher AE with respect to each equation (Table 4). Similarly, analysis of partial factor productivity helped to identify 16 ILs (NPK), 4 ILs (–P), and 151 ILs (75N) that had a >50 kg of grain per kg^{-1} N (NPK) condition. In comparison with parents, 25 ILs (NPK), 61 ILs (–P), and 117 ILs (75N) showed more than 50 kg of grain per kg^{-1} N in each condition.

Table 4. Determination of NPK fertilizer efficiency in ILs under experiment on NuUE.

Agronomic Efficiency (AE)			
AE applied nitrogen	AE formula	>15 kg grain kg^{-1} nitrogen applied	>Parents
AE(N) = grain yield (N fertilized–0NPK unfertilized) in kg ha^{-1}/Fertilizer N in kg ha^{-1}	$AE_{(N)} = (Y_{NPK} - Y_{0NPK}) \div F_N$	117	74
	$AE(N) = (Y_{NK} - Y_{0NPK}) \div F_N$	33	28
	$AE(N) = (Y_{75N} - Y_{0NPK}) \div F_{75N}$	161	86
Partial Factor Productivity (PFP)			
PFP Applied Nitrogen	PFP formula	>50 kg grain kg^{-1} nitrogen applied	>Parents
PFP(N) = grain yield N fertilized in kg ha^{-1}/ Fertilizer N in kg ha^{-1}	$PFP_{(N)} = Y_{(+NPK)} \div F_N$	16	25
	$PFP_{(N)} = Y_{(-P)} \div F_N$	4	61
	$PFP_{(N)} = Y(_{75N}) \div F_N$	151	117

2.4. *Construction of Linkage Map and Segregation of SNP Markers*

A total of 704 high-quality polymorphic SNP markers, with an average for each chromosome of 58 SNP markers, were used to genotype the 230 ILs, and a high-density genetic linkage map was developed covering 1526.8 cM, with an average of 127.2 cM per SNP marker (Table 5). The logarithm of the odds (LOD) thresholds and explained phenotypic variance for six nutrient conditions ranged from 2.52% to 17.76% and 5.87% to 34.68%, respectively. Using single marker analysis (SMA), a total of 261 QTLs were mapped on all 12 chromosomes, with an average of 21 QTLs for yield attributed to seven key component traits of NuUE in rice. Among the 12 chromosomes, the highest number of QTLs was located on chromosome 6 (26 QTLs), whereas the lowest number of QTLs was observed on chromosome 12 (6 QTLs), and all QTLs are listed in Table S1. The hotspot QTL regions are shown in Figures 2 and 3. The majority of the QTLs were associated with 1000-Gwt, and PSPF traits had a negative additive effect of 61.9% and 61.5%, indicating that alleles from the recipient parent WTR-1 contributed to increasing phenotype.

Table 5. Distribution of 704 polymorphic single nucleotide polymorphism (SNP) markers distributed in across the 12 chromosomes, with their average distance, genome size, coverage percentage, genetic distance, and physical distance per cM.

S. No.	Chr	Marker No.	Average Distance (Kb)	Genome Size (Kb)	Genome Size (Gramene)	Coverage Percentage	Genetic Distance (cM)	Physical Distance per (Kb)
1	Chr01	76	564.0	42,492.4	43,270.92	98.20	181.8	238.01
2	Chr02	45	797.4	35,401.9	35,937.25	98.51	157.9	227.59
3	Chr03	72	497.7	35,824.4	36,413.81	98.38	166.4	218.83
4	Chr04	84	405.0	33,864.4	35,502.69	95.39	129.6	273.94
5	Chr05	50	479.8	29,100.3	29,958.43	97.14	122.3	244.96
6	Chr06	73	422.7	30,809.5	31,248.78	98.59	124.4	251.20
7	Chr07	74	393.9	28,942.5	29,697.62	97.46	118.6	250.40
8	Chr08	43	654.5	27,809.9	28,443.02	97.77	121.1	234.87
9	Chr09	41	521.7	21,348.9	23,012.72	92.77	93.5	246.13
10	Chr10	43	464.0	19,635.6	23,207.28	84.61	83.8	276.94
11	Chr11	58	492.2	28,312.7	29,021.10	97.56	117.9	246.15
12	Chr12	45	603.7	27,023.4	27,531.85	98.15	109.5	251.43

Chr—chromosome; Kb—kilo base pairs; cM—CentiMorgan.

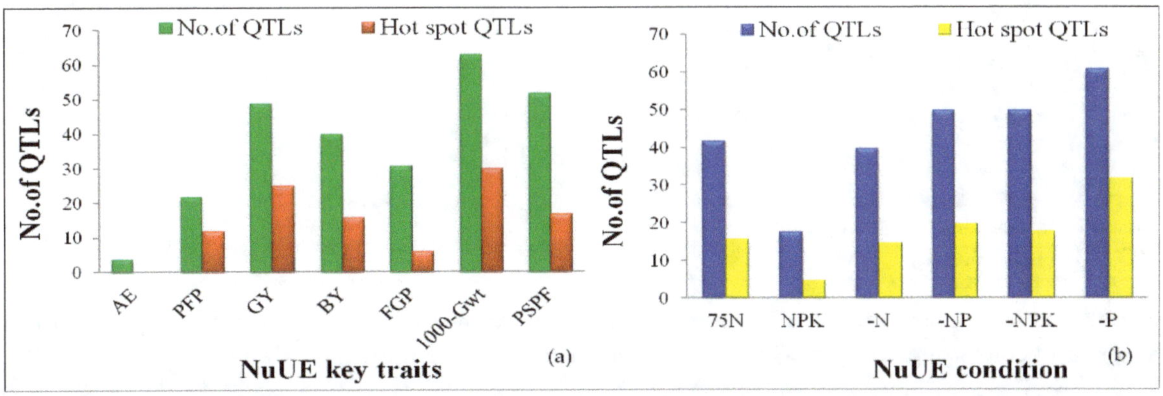

Figure 2. Distribution of trait-wise (**a**); and NPK combinations (**b**), associated with a total number of quantitative trait loci (QTLs) and hotspot QTLs in different nutrient conditions in rice.

Out of the 261 QTLs identified, 62.5% (163 QTLs) had a positive additive effect contributed by WTR-1, and the remaining 37.5% followed with the negative allele from donor parent Hao-an-nong. QTLs with >20% phenotypic variation explained (PVE) were determined as major QTLs with an LOD score of >9.50, while others were considered as minor QTLs. In summary, a total of 261 QTLs for seven traits under six different combinations of NPK nutrients were identified, out of which 49 were major QTLs and 212 were minor QTLs (Table S1). Out of these NuUE conditions, a large number of QTLs were observed in –P (61 QTLs), –NP, and –NPK (50 QTLs), and followed with 75N (42 QTLs) and –N (40 QTLs), and the fewest were detected in NPK conditions with 18 QTLs. The PVE of significant major QTLs ranged from 20.25% to 34.68% at 9.55 to 17.76 LOD value, whereas, for minor QTLs, the explained PV ranged from 5.87% to 19.98% at 2.52 to 9.49 LOD value. The distribution of all major and minor QTLs, represented on all 12 chromosomes, is shown in Figure 3, and the hotspot QTLs (>5 QTLs), identified over nine chromosomes (chr01, chr02, chr03, chr04, chr05, chr07, chr08, chr09, and chr11), were linked to six essential nutrient traits (PFP, GY, BY, FGP, 1000-Gwt, and PSPF) in rice.

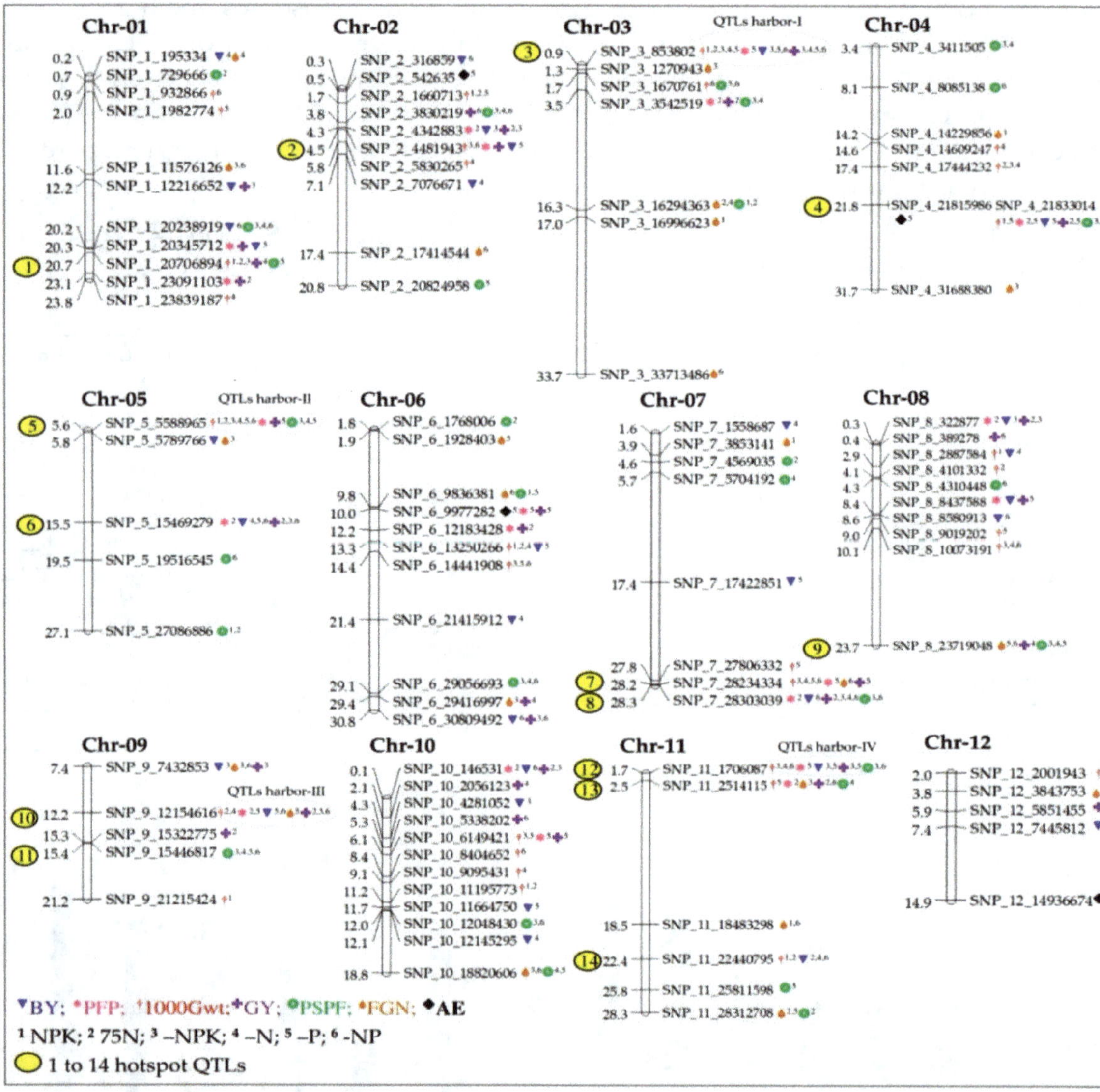

Figure 3. Linkage map of 261 QTLs distributed on 12 chromosomes, with respective polymorphic markers and colors depicting the QTLs governing crucial different nutrient traits.

Out of the 704 polymorphic markers, SNP_2_4481943 had the highest LOD value recorded (17.76), which explained PV of 34.68%, followed by two other markers, SNP_10_6149421 (LOD 16.28) and SNP_8_10073191 (LOD 15.31), which explained PV of 32.32% and 30.73%, respectively. Interestingly, these QTLs were significantly associated with 1000-Gwt under —NPK conditions, and their negative value of the allele was exhibited from donor parent Hao-an-nong. In contrast, the lowest LOD (2.52) was recorded in SNP_6_9836381, which explained PV of 5.87%, followed by three other SNP markers, SNP_12_14936674 (2.55), SNP_12_7445812 (2.55), and SNP_12_5851455 (2.53), which explained PV of 5.92%, 5.92%, and 5.89%, respectively. These four QTLs were linked with different key traits such as FGN, AE, BY, and GY under –NP, 75N, –N, and –NPK conditions. The lowest LOD value of two markers (SNP_6_9836381 and SNP_12_14936674) was contributed by the negative allele from donor parent Hao-an-nong, and two other markers (SNP_12_7445812 and SNP_12_5851455) had a positive allele from recipient parent WTR-1.

2.5. QTLs for NuUE Traits

2.5.1. Agronomic Efficiency (AE)

Agronomic efficiency (AE) was determined in the experiment on NuUE, and we found four QTLs (*qAE_2.1*, *qAE_4.1*, *qAE_6.1* and *qAE_12.1*) that were located on chromosomes 2, 4, 6, and 12, respectively (Table 6). For the agronomic efficiency of applied nitrogen in terms of –P conditions (applied nitrogen and potassium fertilizer), three QTLs were detected with an LOD score and phenotypic variation of 2.77, 4.01, and 4.52 and 6.43%, 9.17%, and 10.27%, respectively, with a positive additive effect of 3.16, 2.13, and 2.28, which indicates that the progeny carried the trait from the recipient parent WTR-1. QTL *qAE_6.1* consisted of a peak marker (SNP_6_9977282) that was also associated with GY and PFP under –P conditions, with explained phenotypic variation of 17.6%. In 75N conditions, one QTL (*qAE_12.1*) was detected with a peak marker (SNP_12_14936674) located on 14936674 bp of chromosome 12, whereas LOD score and phenotypic variation were 2.55% and 5.92%, respectively, since the additive value of –2.8 shows that the progeny came from donor parent Hao-an-nong.

Table 6. Putative QTLs identified for AE and PFP for six different nutrient conditions using 230 nutrient use efficient ILs.

S. No.	NuUE Condition [a]	Trait [b]	QTLs [c]	Chr	Position (bp) [d]	Peak Marker [e]	LOD Value	PVE% [f]	Additive Effect
1	–P	AE	*qAE_2.1*	2	542,635	SNP_2_542635	2.77	6.43	3.16
2	–P	AE	*qAE_4.1*	4	21,815,986	SNP_4_21815986	4.01	9.17	2.13
3	–P	AE	*qAE_6.1*	6	9,977,282	SNP_6_9977282	4.52	10.27	2.28
4	75N	AE	*qAE_12.1*	12	14,936,674	SNP_12_14936674	2.55	5.92	−2.80
5	–P	PFP	*qPFP_1.1*	1	20,345,712	SNP_1_20345712	8.64	18.71	2.87
6	–P	PFP	*qPFP_2.2*	2	4,481,943	SNP_2_4481943	11.68	24.44	−3.11
7	–P	PFP	*qPFP_3.1*	3	853,802	SNP_3_853802	8.93	19.28	3.05
8	–P	PFP	*qPFP_4.1*	4	21,833,014	SNP_4_21833014	10.59	22.44	3.03
9	–P	PFP	*qPFP_5.1*	5	5,588,965	SNP_5_5588965	3.39	7.81	2.10
10	–P	PFP	*qPFP_6.1*	6	9,977,282	SNP_6_9977282	8.07	17.60	2.71
11	–P	PFP	*qPFP_7.1*	7	28,234,334	SNP_7_28234334	7.35	16.16	2.82
12	–P	PFP	*qPFP_8.2*	8	8,437,588	SNP_8_8437588	9.90	21.14	−2.89
13	–P	PFP	*qPFP_9.2*	9	12,154,616	SNP_9_12154616	8.73	18.89	3.15
14	–P	PFP	*qPFP_10.2*	10	6,149,421	SNP_10_6149421	12.15	25.28	−3.16
15	–P	PFP	*qPFP_11.1*	11	1,706,087	SNP_11_1706087	5.93	13.25	2.29
16	75N	PFP	*qPFP_1.2*	1	23,091,103	SNP_1_23091103	5.89	13.17	3.45
17	75N	PFP	*qPFP_2.1*	2	4,342,883	SNP_2_4342883	9.44	20.25	−3.99
18	75N	PFP	*qPFP_3.2*	3	3,542,519	SNP_3_3542519	7.32	16.09	4.16
19	75N	PFP	*qPFP_4.1*	4	21,833,014	SNP_4_21833014	7.60	16.66	3.68
20	75N	PFP	*qPFP_5.2*	5	15,469,279	SNP_5_15469279	9.78	20.91	−4.05
21	75N	PFP	*qPFP_6.2*	6	12,183,428	SNP_6_12183428	4.46	10.14	2.92
22	75N	PFP	*qPFP_7.2*	7	28,303,039	SNP_7_28303039	7.21	15.89	4.04
23	75N	PFP	*qPFP_8.1*	8	322,877	SNP_8_322877	7.09	15.64	−3.50
24	75N	PFP	*qPFP_9.1*	9	12,154,616	SNP_9_12154616	7.87	17.19	4.23
25	75N	PFP	*qPFP_10.1*	10	146,531	SNP_10_146531	9.13	19.68	−3.92
26	75N	PFP	qPFP_11.2	11	2,514,115	SNP_11_2514115	3.66	8.41	2.57

[a] NPK condition trait: –P (negative phosphorus) and 75N (75% of nitrogen). [b] Trait name: AE (agronomic efficiency), PFP (partial factor productivity). [c] Name of identified QTL. [d] Nucleotide position (bp) of the SNP detected on each chromosome. [e] Peak marker of identified QTL. [f] Explanation of phenotypic variation.

2.5.2. Partial Factor Productivity (PFP)

A total of 22 QTLs were identified across all chromosomes, except chromosome 12, under two different nutrient conditions (–P and 75N). Eleven QTLs were associated with PFP with LOD scores ranging from 3.39 (SNP_5_5588965) to 12.15 (SNP_10_6149421), which explained phenotypic variation from 7.81% to 25.28% in –P conditions. Among these 11 QTLs, the highest LOD value of three peak markers, SNP_2_4481943 (LOD 11.68), SNP_8_8437588 (LOD 9.9), and SNP_10_6149421 (LOD 12.15), was contributed by the Hao-an-nong allele on chromosomes 2, 8, and 10, and eight QTLs remained on chromosomes 1, 3, 4, 5, 6, 7, 9, and 11 that had a positive additive value, indicating the contribution of recipient parent WTR-1. Under –P conditions, one of the peak markers (SNP_1_20345712) was located at the 20345712 bp position on chromosome 1, which is associated with BY and GY, with an LOD score of 4.01 and 8.63, explaining a PV of 9.17% and 18.71%, respectively (Table 6). Similarly, we discovered 11 significant QTLs for PFP under 75N conditions. Of the 11 QTLs detected, four (*qPFP_2.1*, *qPFP_5.2*, *qPFP_8.1*, and *qPFP_10.1*) were contributed by the donor parent allele of Hao-an-nong, and the remaining seven QTLs by the WTR-1 allele. An average LOD value of 7.22 and PV of 15.82% were explained by 11 QTLs. The two highest peak markers were SNP_5_15469279 and SNP_2_4342883, located on chromosomes 2 and 5, respectively, which explained PV of 20.25% and 20.91%.

2.5.3. Grain Yield (GY)

Forty-nine QTLs were identified for GY. Each QTL explained 5.89%~25.28% of the phenotypic variation, with respective LOD scores of 2.53~12.15. Among the total QTLs of GY, nine QTLs showed more than 20% of the phenotypic variance and were located on chromosome 2, 4, 5, 8, and 10. Four GY QTLs (*qGY_2.4, qGY_4.2, qGY_8.4*, and *qGY_10.5*) were detected in –P, two QTLs (*qGY_5.4, qGY_10.4*) in –NP and one QTL (*qGY_5.3*) significantly expressed under –NPK condition. Similarly, two GY QTLs (*qGY_2.2* and *qGY_5.2*) were identified in 75N nutrient condition. The peak marker of SNP_5_15469279 was located on chromosome 5, and it was associated with GY under three different nutrient conditions, such as 75N, –NPK, and –NP. Whereas LOD and phenotypic variations of this marker was recorded 9.78, 9.97, and 9.92 and 20.91%, 21.26%, and 21.18%, respectively, since additive value shows negative allele contributing from a donor parent Hao-an-nong.

2.5.4. Biomass Yield (BY)

A total of 40 QTLs were discovered in six nutrient conditions, except for NPK. Under deficiency of –N, –P, –NP, and –NPK conditions, 39 QTLs were linked with BY on all 12 chromosomes. Chromosomes 1, 2, 5, 8, 10, and 11 had more than three QTLs controlling BY. The contributions of these QTLs showed phenotypic variation, ranging from 5.99% to 17.63%, and their LOD values ranged from 2.57 to 8.08, respectively. These QTLs had the largest effects, with negative additive values showing them to be contributed by donor parent Hao-an-nong to the progeny. Under 75N conditions, one QTL (*qBY_11.3*), located on chromosome 11, at the position of 22440795 bp, had a PVE of 7.37% and it was contributed by the Hao-an-nong allele.

2.5.5. Percentage of Spikelet Fertility (PSPF)

In total, 52 QTLs were identified for six nutrient conditions and they were distributed on all chromosomes except for chromosome 12. These QTLs explained phenotypic variation ranging from 5.98% to 29.76%. Of these 52 QTLs, 6 QTLs for 75N, 3 QTLs for NPK, 10 QTLs for –P, and 11 QTLs for each nutrient condition (–N, –NP, and –NPK) were significantly expressed, with an LOD score ranging from 2.57 to 14.73. Under –N deficiency conditions, four QTLs (*qPSPF_3.3, qPSPF_6.4, qPSPF_8.2*, and *qPSPF_9.1*) and another three QTLs (*qPSPF_3.4, qPSPF_6.5*, and *qPSPF_9.2*) for —NPK conditions were recorded with PVE of >20%. The peak markers were located on chromosomes 3, 6, 8, and 9 with the favorable allele contribution from WTR-1.

2.5.6. 1000-Gwt

Out of the 261 QTLs, 63 were identified in 1000-Gwt for six nutrient (–N, –P, –NP, –NPK, 75N, and NPK) conditions. These QTLs together explained an average phenotypic variation ranging from 5.95% to 34.68% and an LOD score of 2.56 to 17.76. Among these QTLs, 24 (38%) were from donor parent Hao-an-nong, which was lower than WTR-1's contribution with 39 QTLs (62%). The QTLs *q1000Gwt_2.4, q1000Gwt_8.5*, and *q1000Gwt_10.1* had an LOD score greater than the threshold of 15 and explained PV of 17.76%, 15.31%, and 16.26%, respectively, under –NPK conditions. The three highest LOD values (17.76, 15.31, and 16.28) were observed on chromosomes 2, 8, and 10 and were contributed by the WTR-1 allele

2.5.7. Filled Grain Number (FGN)

A total of 31 QTLs had an LOD score of 2.52 to 7.59 (higher than the threshold level) and were located across all 12 chromosomes identified under six different nutrient conditions. The QTLs had phenotypic variation ranging from 5.87% to 15.84%. The QTLs *qFGN_4.2*, *qFGN_5.1*, and *qFGN_7.1* on chromosomes 4, 5, and 7 had a large additive effect (−118.1, −127.97, and −139.93), accounting for 7.16%, 10.17%, and 7.25% of the phenotypic variation, respectively. The alleles for increasing FGN came from both Hao-an-nong (10 QTLs) and WTR-1 (21 QTLs). Only two QTLs (*qFGN_1.1* and *qFGN_3.4*) controlling FGN on chromosomes 1 and 3 were detected, and explained 10.23% and 8.82% of the phenotypic variation under –N conditions, respectively. Out of 31 QTLs, 10 were contributed with an allele from Hao-an-nong, and the remaining 21 QTLs carried an allele from recipient parent WTR-1.

2.5.8. Hotspot QTLs for Multiple Traits

Among the total of 261 QTLs, 106 were identified as promising QTLs located over 14 hotspot regions of nine chromosomes (1, 2, 3, 4, 5, 7, 8, 9, and 11), and each chromosome had more than five QTLs contributing to single SNP peak markers (Figure 2a,b). The QTLs identified under each of the six nutrient conditions showed 32 QTLs identified in –P conditions, and these were significantly linked with six traits (PFP, GY, BY, FGP, 1000-Gwt, and PSPF) that explained PV ranging from 5.99% to 27.9% with LOD values of 2.57 to 13.64. Subsequently, several QTLs were identified under five nutrient conditions, –N (15 QTLs), –NP (20 QTLs), –NPK (18 QTLs), 75N (16 QTLs), and NPK (5 QTLs), and these were discovered across nine chromosomes. On each chromosome, the hotspot regions were identified that contributed to different traits under six different nutrient conditions. The peak marker SNP_1_20706894 on chromosome 1 was found to be linked to five QTLs associated with three traits, 1000-Gwt (75N, NPK, and –NPK), GY (–N), and PSPF (–P) conditions. Similarly, five QTLs located on chromosomes 2, 9, and 11, with peak markers SNP_2_4481943, SNP_9_15446817, and SNP_11_2514115, respectively, were associated with five traits: 1000-Gwt, PFP, BY, GY, and PSPF. Further analysis of the remaining hotspot regions, chr04 (eight QTLs), chr05 (seven QTLs), chr07 (eight QTLs-SNP_7_28303039 and six QTLs-SNP_7_28234334), chr08 (six QTLs) and chr11 (six QTLs), showed association with six prominent critical traits: 1000-Gwt, PFP, BY, GY, PSPF, and FGN. One to 14 hotspot QTLs per chromosome were classified (Table S1 and Figure 3) for the discovery of the top hotspot QTLs (a QTL that contributes to ≥10 QTLs). Out of these 14 hotspot QTLs, four QTLs were located on chromosome 3 (harbor-I QTLs), 11 QTLs on chromosome 5 (harbor-II QTLs), and 10 QTLs on chromosome 9 (harbor-III QTLs) and chromosome 11 (harbor-IV QTLs) at the nucleotide positions of 853,802; 5,588,965; 12,154,616; and 1,706,087 bp, respectively. The top hotspot regions of each chromosome were designated as QTLs harbor-I to -IV. While analyzing these top QTLs, harbor-I to -IV, we identified a positive allele contributed from recipient parent WTR-1 (Figure 3). Of these harbor-I to -IV QTLs, 16 QTLs were expressed on chr03, chr05, chr09, and chr11 for –P; nine QTLs on chr03, chr05, and chr11 for –NPK; seven QTLs on chr03, chr05, chr09, and chr11 for –NP; six QTLs on chr03, chr09, and chr11 for –N; five QTLs on chr03, chr05, and chr09 for 75N, and two QTLs on chr03 and chr05 for NPK, respectively. As a result, the top four hotspot QTLs explained phenotypic variation ranging from 6.09% to 26.97% and LOD scores ranging from 2.62 to 13.11.

2.5.9. Fine-Tuning of QTLs Harbor-I to -IV

The top four hotspot QTLs of chromosomes 3, 5, 9, and 11 were refined through QTARO databases. The first QTL (harbor-I) was identified on chromosome 3, between peak markers SNP_3_853802 and SNP_3_16294363, covering a span of 15.44 Mbp regions. In this region, nine genes (*pez1*, *OsIRO3*, *Mit*, *OsApx1*, *RPN10*, *OsFRDL1*, *OsMTP8.1*, *OsGS1; 2*, and *OsPT2*) and four QTLs (*qRFWw3*, *n-p3*, *qDLR3*, and *qZNT-3*), documented based on earlier reports, were associated with soil stress tolerance mechanism-related traits, for uptake and translocation from roots to shoots, for various nutrients, such as iron, cadmium, phosphate, and manganese (Table S2). Two transcription factors (*IDEF2* and *OsHsfA4a*) and one gene (*OsZIP5*), identified in the region spanning 21.5 Mbp on chromosome 5 (QTL harbor-II); two QTLs (*qLBI-9* and *qALSRL-9*) and one gene (*OsSTR1*), located on chromosome 9 covering a 13.78 Mbp region (QTL harbor-III); and two QTLs (*qDLR11* and *qRRE-11*), identified on chromosome 11 covering a region of 26.62 Mbp (QTL harbor-IV), were identified. Of these eight QTLs, QTL harbors distributed on three chromosomes (3, 9, and 11) accounted for PV ranging from 3.47% (*qDLR3*) to 18.3% (*qALSRL-9*) and LOD value ranging from 2.64 (*qRRE-11*) to 15.23 (*qn-p3*).

3. Discussion

NuUE is a complex trait influenced by several factors, and it is considered a vital trait to improve rice grain yield productivity in marginal and rainfed lowland areas. Rice breeding programs need to incorporate NuUE traits as this helps resource-poor farmers to save on fertilizer, and allows maximization of valuable resources to increase profitability. To date, several morphological and agronomic traits have been identified, and they can be used as indicators of nitrogen (N) [52,64–66], phosphorus (P) [7,38,67–70], and potassium (K) [16,39,40,71–73] deficiency tolerance under both field and hydroponic conditions. Using different genetic backgrounds of mapping populations, such as RILs, NILs, BILs, ILs, CSSLs, DHs, and BC_2F_4, there are several reports on QTL identification with independent studies of low-N, -P, and -K conditions. The previously identified QTLs were mapped on different chromosomes using low-density linkage maps that were constructed by PCR-based molecular markers, such as RFLP, AFLP, SSR, and STS [26,29,31,34,39,48,52,55,58,67,74–76]. However, in this study, we used SNP markers to identify the QTLs.

For the identification of QTLs, single-marker analysis, simple interval mapping, and composite interval mapping are widely used methods [77,78]. The development of nutrient use efficient selective ILs, through an early generation backcross breeding approach, allows both the detection of QTLs for NuUE traits and the simultaneous development of promising breeding materials into varieties. This is the strength of selective introgression breeding first proposed by Tanksley [78], and later demonstrated for rice by Li et al. [79]. However, for QTL detection in selective introgression backcross breeding populations, we could use single marker analysis (SMA) and association mapping with single nucleotide polymorphism (SNP) markers, and use a higher threshold of LOD value of 2.5 to declare a strong QTL.

3.1. Analysis of Critical NuUE Traits

The phenotypic evaluation of 230 BC_1F_5 ILs varied significantly in six different NPK nutrient combinations: N, –P, –NP, –NPK, 75N, and NPK. Grain yield is a quantitative trait and a highly complex character for all crops [80], and is the essence of any breeding program. Various morphological and physiological plant traits contribute to grain yield. Yield components are inter-related with each other, indicating a complex chain of relationships, which is highly influenced by the environment [81]. The breeding strategy in rice mainly depends on the degree of associated characters as well as the magnitude and nature of variation [82,83]. Based on a higher grain yield level, 18 ILs for NPK, 23 ILs for 75N, 4 ILs for –P, and 1 IL for both –NPK and –NP had more than 40 g/plant, and in –N conditions the lowest GY was observed (34.96 g/plant).

An understanding of the correlations between traits is of great importance in breeding programs, especially if the selection of one of them is impaired by low heritability or difficulties of measurement and identification [84]. Information on trait correlations has been helpful as a basis for selection in breeding programs. Ashura [85] showed a positive correlation between the number of filled grains per panicle, number of panicles per plant, and 1000-grain weight for grain yield. A correlation study enables breeders to understand the major traits for which selection can be based on population improvement. GY had a significant positive correlation with BY and FGN in NPK; FGN in 75N; GWT, BY, and FGN in –P; PSPF, BY, and FGN in –N; and BY and FGN in –NP ($p < 0.01$) conditions. A negative correlation was recorded for GY with 1000-Gwt in NPK and for BY in 75N conditions. Under all six nutrient conditions, GY showed a negative relationship with BY and the remaining traits, GWT, PSPF, and FGN, had a significant positive correlation between the traits. Highly correlated traits were indicated to have a common genetic basis, suggesting that these eight relative phenotypic traits could be used for the evaluation of P-deficiency tolerance in BILs at the seedling stage [86].

3.2. Promising Traits of AE-Associated QTLs

The application of AE and PFP traits was essential to plant nutrients for an optimum quantity and right proportion, through the correct method and time of application, and this is the key to increased and sustained crop production [87]. These two traits are important for the measurement of nutrient use efficiency, and this also provides an integrative index that quantifies total economic output relative to the use of all the nutrient resources in the system [88]. In 1996, Cassman et al. [10] defined PFP and AE and showed that they could be increased by raising the amount, uptake, and use of available nutrients, and by increasing the efficiency with which applied nutrients are taken up by the crop and used to produce grain. In this study, we identified 42 ILs with higher AE in NPK, 23 ILs in –P, and 79 ILs in 75N, as compared with their parents, and they were explicit in >15 kg grain kg^{-1} N conditions. Yoshida [89] and Cassman et al. [10], respectively, estimated AE to be 15–25 kg kg^{-1} and 15–20 kg kg^{-1} in the dry season in farmers' fields in the Philippines. In a similar way, Wen-Xia et al. [90] mentioned the AE in two kinds of rice: Jinzao had an AE range of 8.02–20.14 kg grain kg^{-1} N and Shanyou63 had an AE range of 3.40–18.37 kg grain kg^{-1}, differing with N management. In hybrid rice, AE for applied P was 5.2 kg grain kg^{-1} P and for applied K was 11.8 kg grain kg^{-1} K, where the fertilizer application rate for NPK was 200–75–200 and 200–150–200 kg ha^{-1}, respectively; AE for applied P in non-hybrid rice was 2.3 kg grain kg^{-1} P and 4.7 kg grain kg^{-1} P, where the fertilizer rate was the same [81].

Four novel QTLs were identified for AE (*qAE_2.1*, *qAE_4.1*, *qAE_6.1*, and *qAE_12.1*) on chromosomes 2, 4, 6, and 7, which explained phenotypic variation ranging from 5.93% to 10.27%. For these four AE QTLs, each SNP marker position was analyzed at both sides of the 500 kp (up- and down-stream of 1 Mb) regions on the chromosome (Table S3). With the respective positions of the 1 Mb region on chromosome 2, seven genes were identified influencing diverse functional roles as a defense to pathogen resistance and physiological mechanisms. Of the seven genes revealed, *BiP3* for BLB resistance [91]; *OsHPL3* for multiple biotic disease resistance, such as brown planthopper, striped stem borer, and BLB [92]; *BiP1* (*Os06g0622700*) for increased seed storage protein, starch content in the endosperm, and also stress responses [93,94]; *OsHPR1* (*Os02g0104700*) involved in photo-respiratory metabolism [95]; and *OsNOA1* regulating chlorophyll biosynthesis, plastid development, and Rubisco formation in a temperature-dependent manner [95], were identified on chromosome 2. Similarly, on chromosome 6, genes *Pi9* and *Pi2* were noticed [96], which are related to blast disease resistance, along with another three QTLs (*qPHw6-2*, *qSFWd6*, *amy6-1*) related to drought tolerance and amylose content [97,98]. Interestingly, a major QTL for P deficiency tolerance that was previously mapped on chromosome 12 was found to be closely associated with peak marker SNP_12_14936674 [31], and in the same region of markers was linked with drought tolerance QTLs [99].

In PFP analysis, 16 ILs in NPK, 4 ILs in –P, and 151 ILs in 75N conditions yielded >50 kg grain kg^{-1} N and, in comparison with parents, 25 ILs (NPK), 61 ILs (–P), and 117 ILs (75N) yielded more than 50 kg grain kg^{-1} N in each condition. Amanullah et al. [100] showed, with a maize crop, a PFP of applied N of 36.62 kg grain kg^{-1} N and an AE of applied N of 22.49 kg grain kg^{-1} N. After applying DAP and SSP in fields, the AE of two fertilizer applications resulted in 13.01 and 13.71 kg grain kg^{-1} P, and PFP resulted in 63.58 and 61.92 kg grain kg^{-1} P. In this study, 22 QTLs for PFP traits were identified under 75N (11 QTLs) and –P (11 QTLs) conditions, with a PV ranging from 8.41% to 20.91% and 7.81% to 25.28%, respectively. Out of these 22 QTLs, six QTLs at SNP marker positions were analyzed on both sides of 500 kb (up- and down-stream of 1Mb) regions, which were explained by PV of more than 20% on chromosomes 2, 4, 5, 8, and 10. Five genes and six QTLs were identified in the 1 Mb region associated with PFP traits. On chromosome 2, three genes were identified as

OsWRKY71 [101], *Os4CL3* [102], and *OsWRKY45* [103,104], which were functionally related to defense signaling molecules, such as SA, Me, JA and lignin biosynthesis. Two QTLs (*qDSR_8* and *qALRR_8*) on chromosome 8 [105,106] and three QTLs (*qRRE_10, qRFW_10,* and *qRCCL_10*) [107,108] and one gene, *OsAT1/Spl18* [109], located on chromosome 10 revealed having tolerance of aluminum and alkaline stress, and also showed blast disease and ultraviolet-B resistance in rice. On chromosome 5, gene *OsWRKY45* was found to be closely associated with SNP_5_15469279, and it is involved in multiple biotic and abiotic stress tolerance/resistance, such as BLB, blast, sheath blight, drought, salinity, and cold [103,104,110].

3.3. Categorizing NuUE QTLs and Related Traits

Averages of 21.75 QTLs were distributed across all 12 chromosomes. Out of 261 QTLs, the highest numbers were located on chromosome 3 (28 QTLs), chromosome 11 (27 QTLs), chromosome 6 (26 QTLs), chromosome 1 (24 QTLs), and chromosomes 5, 8, and 10 (23 QTLs), and the remaining chromosomes 2, 4, 7, and 9 had a range of 18 to 22 QTLs. All the identified QTLs had PVE ranging from 5.87% to 34.68% and an LOD threshold range of 2.52–17.76. The lowest number of QTLs was observed on chromosome 12, with PVE from 5.89% to 9.49% and LOD value of 2.6 to 11.82. Based on the different nutrient conditions, the largest number of QTLs was detected under P deficiency (–P: 61 QTLs), and the lowest number of QTLs was identified for NPK (18 QTLs) conditions. For other nutrient conditions, the numbers were as follows: –NP (50 QTLs), –NPK (50 QTLs), –N (40 QTLs), and 75N (42 QTLs). A comprehensive literature survey revealed that a majority of the QTLs associated with low phosphorus were reported on chromosomes 1, 2, and 12 (recently reviewed by Mahender et al. [7], Ali et al. [20], and van de Wiel et al. [70], whereas, for low nitrogen conditions, the majority of morpho-physiological trait-linked QTLs in rice were located on chromosomes 3, 5, and 8 [34,52–54,76]. Under six different nutrient conditions, 24.1% of the QTLs were associated with 1000-Gwt, followed by 19.9% for PSPF, 18.7% for GY, 15.3% for BY, 11.8% for FGP, 8.4% for PFP, and 1.5% for AE (Figure 3).

3.4. Consistency and Comparisons of Major QTLs across Different Genetic Backgrounds

In this study, a total of 261 QTLs were identified using a 6K SNP array-based genetic linkage map analysis, in 230 BC_1F_5 introgression lines, that were tested under six different nutrient conditions. Out of 261 QTLs, 49 major QTLs were found with more than 20% PVE and LOD threshold value ranging from 9.44 to 17.76 distributed across all chromosomes, except for chromosomes 7, 11, and 12. Of these 49 QTLs, the highest number (27 QTLs) was associated with 1000-Gwt and the lowest number (6 QTLs) with PFP. For other traits, nine QTLs were associated with GY and seven QTLs with PSPF, which were detected across five nutrient conditions: 75N, –P, –N, –NP, and –NPK. Interestingly, out of these 261 QTLs, 14 QTLs were consistently detected under more than four nutrient treatments, indicating that these were critical traits that were controlled by similar genes under different nutrient conditions. The consistencies of QTLs were identified on nine chromosomes: 1, 2, 3, 5, 7, 8, 9, 10, and 11.

On chromosome 2, SNP_2_4342883 was linked with four QTLs (*qPFP, qBY, qGY,* and *q1000Gwt*), and SNP_5_15469279 was linked with three QTLs (*qPFP, qBY,* and *qGY*) on chromosome 5, which were expressed under four nutrient conditions: –P, –NP, –NPK, and 75N. On chromosome 7, SNP_7_28303039 was linked with four QTLs (*qPFP, qBY, qGY,* and *qPSPF*) and, at another location on the same chromosome, SNP_7_28234334 was associated with four QTLs (*q1000gwt, qPFP, qFGN,* and *qGY*) that were identified in –N, –NP, –NPK, and 75N conditions. In a similar way, on chromosome 8, SNP_8_23719048 was linked with three QTLs (*qFGN, qGY,* and *qPSPF*); on chromosome 10, SNP_10_18820606 was linked with two QTLs (*qFGN* and *qPSPF*); on chromosome 11 with six QTLs (*q1000gwt, qPFP qBY, qGY, qPSPF,* and *qFGN*); and, on chromosome 9, SNP_9_12154616 was linked with four QTLs (*q1000gwt, qPFP, qBY,* and *qFGN*), which were consistently recorded in all nutrient deficiency conditions, such as –N, –P, –NP, and –NPK, respectively. Under –N, –P, –NPK, and 75N conditions, on chromosome 1 (SNP_1_20706894), three QTLs (*q1000gwt, qGY,* and *qPSPF*) were identified. Further, two QTLs (*q1000gwt,* and *qBY*) on chromosome 11 (SNP_11_22440795) and five QTLs (*q1000gwt, qPFP*, *qGY*, *qPSPF*, and *qFGN*) on the same chromosome were consistently expressed in –N, –P, –NP, –NPK, and 75N conditions. However, five QTLs (*q1000Gwt*, *qPFP*, *qBY*, *qFGN*, and *qGY*) located on chromosome 3 (SNP_3_853802) and four QTLs (*q1000Gwt*, *qPFP*, *qGY*, and *qPSPF*) located on chromosome 5 (SNP_5_5588965) were detected consistently under six NuUE conditions: –N, –P, –NP, –NPK, 75N, and NPK. For all 14 of these QTLs, the WTR-1 alleles were significantly associated with key component traits in NuUE, but, in the case of chromosome 10, associated with two QTLs; these were contributed by the Hao-an-nong (HAN) allele in –N, –P, –NP and, –NPK conditions. These results indicated that the promising yield-related traits might share a similar genetic basis under different combinations of NPK treatments.

In previous studies, numerous mapping populations have been used to identify the QTLs associated with NPK deficiency tolerance traits in rice [26,28,29,31,34,48,52,55,58,61,67,74–76,111]. For each chromosome level, we identified seven or more QTLs located on chromosomes 2, 5, 8, and 10 and PVE ranged from 20.25% to 34.68%, and they were significantly associated with 1000-Gwt, PFP, FY, and PSPF under nutrient-deficient conditions of –P, –N, –NP, and –NPK. Several researchers in separate studies mentioned that chromosomes 2, 5, 8, and 10 were markedly associated with several agronomic traits, such as GY, nitrogen use efficiency (NUE), agronomic nitrogen use efficiency (agNUE), PSPF, nitrogen absorption ability (NAA), nitrogen content in shoots (NCS), harvest index (HI), root dry weight (RDW), shoot dry weight (SDW), and number of tillers (NT) QTLs under –N conditions [34,52,54,76]. Similarly, under low-P conditions, 36 QTLs were reported on the same chromosomes 2, 5, 8, and 10, respectively. These chromosomes were linked with phosphorus deficiency tolerance traits, such as phosphorus uptake (PUP), phosphorus use efficiency (PUE), phosphorus use efficiency for grain yield (PUEg), RDW, relative root length (RRL), panicle number per plant (PNPP), phosphors translocation (PT), phosphors translocation efficiency (PTE), spikelet fertility (SPF), 1000-Gwt, BY, relative root volume (RRV), and relative plant height (RPH), mapped in the different genetic backgrounds of mapping populations [48,55,56,59–61,67]. However, the major QTL, *Pup1*, was located on chromosome 12 at 54.5 cM, which explained PV of 78.8%, contributing to enhancing the uptake capacity of P from soils [31,55], and, further, the specified region of the 278-kbp sequence was significantly directly linked with P deficiency tolerance [112]. In this study, six QTLs were located on chromosome 12, with PVE ranging from 5.89% to 9.49%. Among these, one QTL for AE (*qAE_12.1*) was very close to this region (59.7cM), where *Pup1* is located on chromosome 12. Several researchers identified the same region as contributing toward tolerance of several biotic and abiotic stresses, such as drought, cold [97,113–116], and aluminum toxicity [116] in rice.

Interestingly, the identified associated QTLs governing seven traits (1000-Gwt, FGN, PSPF, BY, GY, AE, and PFP) on chromosome 6, under phosphorus-deficient conditions, significantly contributed to PVE ranging from 7.51% to 27.49%. Earlier independent studies of Ni et al. [31] and Wissuwa et al. [55] mapped major QTLs for RTA, RSDW, and RRDW on the same chromosome 6, and a group of P-responsive genes and transcription factors were also located in this region [113], which may confer tolerance of P deficiency [117]. However, in comparison to *Pup1* on chromosome 12, major QTLs and P-responsive genes were mostly located on chromosome 6, which indicates that both have independent genes and regulatory pathways [118].

In the deficiency of –N, –P, –NP, and –NPK conditions, four QTLs for GY were identified in –P (*qGY_2.4*, *qGY_4.2*, *qGY_8.4*, and *qGY_10.5*), two QTLs in –NP (*qGY_5.4* and *qGY_10.4*), and one QTL in –NPK (*qGY_5.3*), which were detected on different chromosomes (2, 4, 5, 8, and 10) from the analysis of 49 major QTLs. These QTLs were explained by their PV ranges from 20.25% to 25.28%. This was in contrast to Yue et al. [52], who reported three QTLs for 1000-Gwt on chromosome 3 and 7 and another three QTLs for GY located on chromosomes 1, 4, and 9, and the phenotypic variation explained by these QTLs ranged from 4.93% to 26.73% and 5.73% to 6.80%, respectively. However, in the present study, on chromosome 4, peak marker SNP_4_21833014 was found to be close to *qGYP-4* and was also flanked by RM273–RM241 [52]. In the present study of the genomic region of SNP markers from SNP_1_195334 to SNP_1_23839187 on chromosome 1, five QTLs were identified (*qBY*, *qGY*, *qFGN*, *qPSPF*, and *q1000Gwt*), which explained PV ranging from 7.18% to 26.83% under N-deficient conditions. However, in the same genomic region of chromosome 1, several QTLs governing relative grain yield, grain weight, 1000-Gwt, and nitrate transporter, such as *NRT 2.1* and *OsNRT2.3,b* are significantly involved in nitrogen deficiency tolerance and improving N uptake and enhancement of grain yield under nitrogen deficiency [119–123]

3.5. QTL Hotspots

Association with several traits within a single genomic region has immense potential value for the development of desired target trait enhancement through breeding and marker-assisted selection (MAS) applications. A single gene may affect more than two traits through pleiotropy or with closely linked genomic loci [124,125]. In the present study, 14 hotspot QTLs were identified to be located on chromosomes 1, 2, 3, 4, 5, 7, 8, 9, and 11, which had PVE ranges from 5.99% to 34.68% (Figure 3). Of these 14, we revealed the top four QTL harbor regions on chromosomes 3, 5, 9, and 11, and were designated as QTLs harbor-I to -IV. Each of these QTL harbor regions contained more than 10 QTLs on the same genomic regions of SNP markers, and significantly carried the positive allele from recipient parent WTR-1. The top four hotspot QTLs (a total of 45 QTLs) located on chromosomes 3, 5, 9, and 11 had PVE ranging from 6.09% to 26.97% and LOD values of 2.62 to 13.11. The four chromosomal (3, 5, 9, and 11) regions were shared by nitrogen (N) [34,52], phosphorus (P) [39,48,55,59–61], and potassium (K) [39] deficiency QTLs reported in different mapping populations in rice.

Significant reports exist on chromosomes 3, 6, and 11 involved in phosphorus deficiency tolerance in rice. Using BILs derived from inter-specific crosses of *Oryza sativa* L. X *O. rufipogon* Griff. [38], F_3 lines from the crosses between P-deficiency-tolerant variety NERICA10 and sensitive variety Hitomebore [126], and 271 introgression lines [60], identified QTL clusters for SDW, BY, and P uptake on chromosomes 3 and 11, which showed high significance for P-deficiency tolerance. Comparison with previous studies showed a majority of the candidate genes for P-deficiency tolerance (*Pup1*) and *PSTOL1* located on chromosome 12 [63,112,118,127] and P-responsive genes (such as *OsPTF1*) on chromosome 6 [117]. Our present study revealed that a major region of chromosomes 3, 5, 9, and 11 might provide novel loci to discover candidate genes related to P-deficiency tolerance, and these genes/alleles might play a co-regulated role for P-deficiency tolerance. These significant results would be more helpful to breeders and biotechnologists to understand the genetic and molecular basis of the physiological mechanisms of P-deficiency tolerance in rice.

In addition, we compared the presently identified QTL harbor regions of chromosomes 3, 5, 9, and 11 with previously reported genes and loci by using Q-TARO (http://qtaro.abr.affrc.go.jp/), and those are significantly associated with soil stress tolerance mechanisms, such as uptake and transportation of various nutrient elements from the soil to roots and shoots [128–135]. In QTL harbor-I on chromosome 3 (32.86 Mb) and QTL harbor-II (21.5 Mb) on chromosome 5 regions, nine genes and four QTLs were identified, along with IDEF2, *OsHsfA4a*, and *OsZIP* [131,136,137], which are involved in Fe homeostasis, phosphate uptake and translocation, distribution of Zn content from roots to shoots, and tolerance of cadmium stress (Table S2). Four QTLs (*qRFWw3*, *n-p3*, *qDLR3*, and *qZNT-3*) associated with multiple traits of low nitrogen and zinc toxicity tolerance showed resistance to alkaline stress [26,98,106,138]. QTL harbor-III (13.78 Mb) and QTL harbor-IV (26.62 Mb) were located on chromosomes 9 and 11. Two QTLs (*qLBI-9* and *qALSRL-9*) and one gene (*OsSTR1*) located on chromosome 9 were associated with mycorrhizal formation and resistance to Fe and Al toxicity [105,133,139], whereas chromosome 11 contained two QTLs (*qDLR11* and *qRRE-11*) responsible for tolerance of Al toxicity and alkaline stress [106,140].

According to the physical positions of one of the QTL hotspot regions, SNP marker SNP_3_853802 was associated with the highest number of 14 QTLs that were located on chromosome 3, and were distributed as five QTLs in –P (*q1000gwt_3*, *qPFP_3*, *qBY_3*, *qFGN_3*, and *qGY_3*), two QTLs in –N (*q1000gwt_3* and *qGY_3*), two QTLs in –NP (*qBY_3* and *qGY_3*), three QTLs in -NPK (*q1000gwt_3*, *qBY_3*, and *qGY_3*), and one QTL in 75N and NPK (*q1000gwt_3*). Hạnh et al. [141] identified three hotspot regions with response to low-nitrogen QTLs, which were flanked by RM265-RM165 on chromosome 1, RM3199-RM514 on chromosome 3, and RM080-RM281 on chromosome 8, which were significantly associated with the traits as total fresh weight of leaf blades (FW), RDW, SDW, nitrogen concentration in sheaths plus stem (NS), nitrogen concentration in leaf blades (NL), plant height (PH), chlorophyll content index (CCI), NUE, physiological nitrogen use efficiency (pNUE), and agNUE in RIL populations of IR64 and Azucena. In another study, Senthilvel et al. [58], using doubled haploid (DH) lines of IR64/Azucena in a pot experiment with three N doses (native, 0 kg/ha^{-1}; normal, 100 kg/ha^{-1}; and high, 200 kg/ha^{-1}), identified seven main-effect QTLs that were associated with NUE traits and plant grain yield on chromosome 3 [58]. In similar studies, with two different N fertilizer rates (0 N and 130–135 kg N ha^{-1}), in field conditions with RILs derived from two *Oryza sativa*–ssp. *indica* rice varieties (Zhenshan97/Minghui63), Wei et al. [76] reported a major QTL for grain yield (*qRGY3*) on chromosome 3, and another two QTLs (*qRGY7* and *qRGY11*) on chromosomes 7 and 11, detected with PV of 10.8% in 2006 and 16.0% in 2007. Moreover, three QTLs (*qRBM9-1*, *qRBM9-2*, and *qRBM10*) for biomass yield on chromosomes 9 and 10 together explained 33.6% of the total phenotypic variation. However, in corroboration with Senthilvel et al. [58] and Wei et al. [76], QTL mapping studies on DH and RIL populations revealed that chromosome 3 holds a promising trait and was associated with nitrogen use efficiency (NUE) in rice. In P deficiency, several QTLs have been reported on chromosome 3 [55,60,61], on chromosome 5 [48,59,61], on chromosome 9 [59–61], and on chromosome 11 [48,60,61], which explained PV ranging from 4.4% to 16.6% in different mapping populations of BILs, ILs, RILs, and BC_2F_3 in rice. Under low K, Wu et al. [39] identified a total 21 QTLs, related to PH, TN, SDW, RDW, relative potassium concentration in plant (RKC), relative potassium use efficiency (RKUE), and relative potassium uptake (RKUP) on chromosomes 2, 3, 7, and 8, that collectively explained PV from about 8% to about 15% using DH populations. On chromosome 3, *qFGN_3.6* significantly associated with RTN and other traits, such as PH, SDW, RDW, and RKUP, were flanked by RG179-RG403 and RZ284-RZ394, contributing PV of 9.7% to 14.4% [39]. Thus, it is suggested that the identified QTL harbor regions support the reported NPK QTLs, and are refined with chromosomal regions of genes and QTLs that were involved in multiple stress tolerance and transportation of nutrient elements from soil to shoots and grain, which could be valuable information for the understanding of their genetic and physiological mechanisms in future molecular breeding programs in rice.

4. Materials and Methods

4.1. Plant Materials

A BC_1F_5 mapping population contained 230 introgression lines derived from a cross between Weed Tolerant Rice 1 (WTR-1), as the recipient parent, and Hao-an-nong (HAN), as the donor parent. The field experiments were carried out at the experimental farm of the International Rice Research Institute (IRRI), Los Baños, Laguna, Philippines (14.11° N, 121.15° E), during the 2014 dry season (DS). Seeds of 230 ILs, parents, and four checks (PSB Rc82, NSIC Rc222, Apo, and IR74371-70-1-1) were sown in a seedling nursery bed, and 21-day-old seedlings were transplanted with a single seedling per hill. The early generation of the backcross population (BC_1F_2) was grown in one generation under low-input, rainfed, and irrigation conditions during the 2011 wet season (WS), and was followed by four consecutive generations over the DS and WS during 2012 and 2013. Phenotypic screening for efficient selection was practiced based on higher grain yield under six nutrient conditions. Subsequently, this led to the development of 230 ILs. The detailed of the breeding strategies are described in Jewel et al [142]. The NuUE phenotyping experiment was laid out in an alpha lattice design with two replications, using a plot size of two rows × 12 plants/row, and a spacing distance of 0.2 × 0.2 m. NPK nutrients in the form of urea, superphosphate, and muriate of potash were applied at 160, 50, and 50 kg ha^{-1} in the DS, and at 90, 30, and 30 kg ha^{-1} in the WS, respectively. The checks and parents were replicated in all six nutrient conditions. The NPK fertilizers were applied five times in splits, as in basal level, and at 20, 36, 54, and 72 days after transplanting (DAT), respectively, against –N, –P, –NP, –NPK, 75N, and NPK conditions (Figure 4). Field management, including pest control, weeding, and irrigation, followed IRRI's standard experimental farm practices to avoid adverse effects on grain quality.

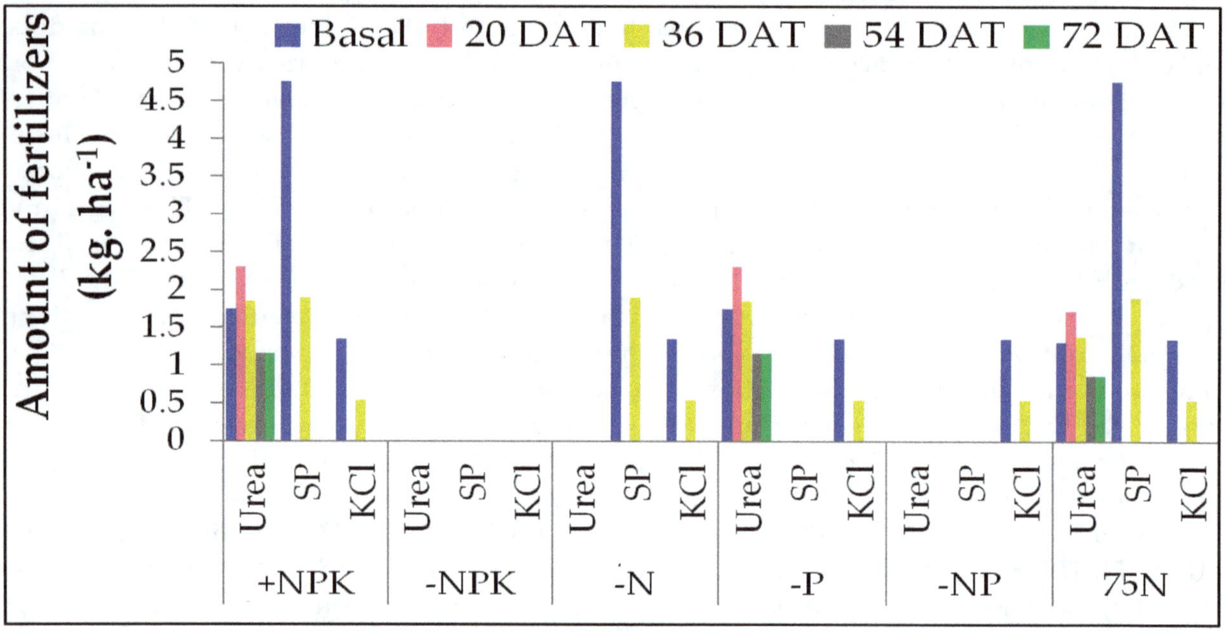

Figure 4. Application of fertilizers in six NuUE conditions with five splits.

4.2. Measurements of Agronomic and Yield-Attributed Traits

Each IL was selected in the middle row of plants, with three replications, and the plants were randomly sampled from each plot for phenotypic evaluation of six agronomic and yield-related traits. The six respective nutrient conditions, –N (PK fertilizer), –P (NK fertilizer), –NP (K fertilizer), –NPK (zero fertilizer), 75N (75% NPK fertilizer), and NPK (all nutrients), were recorded for seven prominent NuUE traits: agronomic efficiency (AE), partial factor productivity (PFP), grain yield (GY), biomass yield (BY), filled grains per plant (FGN), 1000-grain weight (1000-Gwt), and percentage of spikelet fertility (PSPF). Data analysis of the key phenotypic traits was performed with R software [143], and SPSS 17.0 software (IBM, Armonk, NY, USA) was used for the analysis of descriptive means, analysis of variance (ANOVA), Pearson correlation coefficient, and regression analysis, which were calculated among the ILs.

4.3. Calculation of Agronomic Efficiency (AE) and Partial Factor Productivity (PFP)

The increase in grain yield for each kg of fertilizer applied is known as AE. Using the formulas below, we calculated AE and PFP along with nitrogen fertilizer use condition. AE (kg kg^{-1}) = grain yield (N fertilized–N unfertilized) in kg ha^{-1}/Fertilizer N in kg ha^{-1} [144]. AE was computed for all the nutrient use efficient ILs, under six different nutrient conditions. The recommended fertilizer application rate was NPK = 160–50–50 kg ha^{-1}. Among them, three possible AE formulas were depicted:

(i) $AE_{(N)} = (Y_{NPK} - Y_{0NPK})/F_N$

(ii) $AE_{(N)} = (Y_{NK} - Y_{0NPK})/F_N$

(iii) $AE_{(75N)} = (Y_{(75N)} - Y_{0NPK})/F_{75N}$

Similarly, PFP of applied nitrogen is also called nitrogen use efficiency. PFP (N) = YN (crop yield with applied N (kg ha^{-1})/amount of fertilizer N applied (kg ha^{-1}) [144]. Based on applied NPK fertilizer, the performance of the selected and fixed nutrient use efficient ILs, across six different levels of nutrient conditions, was determined by the PFP of applied nitrogen using the following formula:

(iv) $PFP_{(N)} = Y_{(+NPK)}/F_N$
(v) $PFP_{(N)} = Y_{(-P)}/F_N$
(vi) $PFP_{(N)} = Y(_{75N})/F_N$

Y_{+NPK} = crop yield with applied NPK fertilizer (kg ha^{-1}); F_{NPK} = amount of fertilizer NPK applied (kg ha^{-1}); F_N = fertilizer N in kg ha^{-1}; AE = agronomic efficiency applied nitrogen; PFP = partial factor productivity applied nitrogen.

4.4. Genotyping via 6K SNP Array

A total of 230 ILs of plant genomic DNA were extracted from leaf tissue using the cetyl trimethylammonium bromide (CTAB) method [145] and quantified by a NanoDrop 8000 spectrophotometer (ThermoFisher Scientific, Waltham, MA, USA). The concentration of DNA was adjusted to 50 ng μL^{-1}, and approximately 200 ng of DNA from each genotype used in the SNP array. The complete set of the BC_1F_5 mapping population, along with the parents used for the genotyping with 6K SNP array technology, was followed by DNA quantification, incubation, and hybridization of bead chips, staining, and image scanning according to the manufacturer's instructions for the Illumina Infinium assay, and this work was conducted in the Genotyping Services Laboratory of the International Rice Research Institute. The resulting intensity data were processed using the genotyping module V2011.1 of Genome Studio software (Illumina Inc., San Diego, CA, USA) for SNP calling. The generated genotypic data indicated that 704 high-quality SNP markers were identified and further used in the construction of high-density linkage maps for NuUE QTLs, related to agronomic and yield-attributed traits.

4.5. Mapping of QTLs and Hotspot Regions for NuUE

The mapping of QTLs was carried out with IciMapping software (QTL IciMapping version 4.0) [146] using a single marker analysis (SMA) method. The permutation method was used to obtain an empirical threshold for claiming QTLs based on 1000 runs of randomly shuffling the trait values [147], and the logarithm of odds (LOD) value threshold for claiming QTLs was 2.5. The genetic distance (cM) between SNP marker positions was changed to physical distance (kb), with 1 cM equal to 260 kbp [148,149]. For seven promising critical agronomic and yield-related traits under six NuUE conditions, significant markers on the same chromosome, linkage disequilibrium (LD), and phenotypic trait data were calculated by the genetics package in R software [150]. A graphical representation of the linkage map was constructed using MapChart software [151]. Further, for fine-tuning of hotspot QTLs, we used the Q-TARO database (http://qtaro.abr.affrc.go.jp/) for the analysis of previously published co-localization of QTLs and genes related to soil stress tolerance in the rice genome.

5. Conclusions

In many plant breeding programs, the development of rice varieties with NuUE is being considered as a means to reduce the usage of NPK fertilizers by improving their use efficiency, thereby improving grain yield productivity. In the present study, selective introgression lines derived from Weed Tolerant Rice 1, as the recipient parent, and Hao-an-nong, as the donor parent, enabled us to identify a large number of QTLs (261 putative QTLs) linked with NuUE traits, of which 49 QTLs showed high PVE ranging from 20.25% to 34.68%. Among them, 22 QTLs reported as novel QTLs were responsible for PFP and AE traits. They were identified among the promising top four hotspot QTLs (QTLs harbor I-IV), which comprised more than ten QTLs associated with critical NuUE traits, such as 1000-Gwt, PFP, BY, FGN, GY, and PSPF. The hotspot regions of QTLs, expressed across all six NuUE conditions, suggested an underlying uniform basis of genetic mechanisms, contributing to the tolerance of these traits and associated with tightly linked genes or QTLs, or pleiotropic regulations.

Corroboration with several research efforts, of the earlier discovered QTLs for NUE traits, revealed that most of the genomic region was confirmed and their positions precisely confirmed by significant SNP peak markers with high LOD and PVE values, and this could have potential value for introgression of the target using MAS. The list of QTLs and refined hotspot regions will facilitate further validation in systematic breeding for specific adaptability under low-input conditions, and suggest that hotspot genomic regions could be used as targets for a superior understanding of the NuUE mechanism and for improving NuUE traits in rice. In the future, identified promising harbor QTLs would be useful for developing elite introgression breeding lines comprising positive QTLs via marker-assisted selection, and also using them for carrying them forward by a pyramiding approach over the ILs, with desirable QTLs within the same population. Further, this will lead to fine-mapping and molecular cloning of the critical loci that will be useful for enhancing grain yield and quality under low-input fertilizer management conditions.

Author Contributions: J.A. and Z.A.J. conducted a phenotypic experiment; Z.A.J., J.A., J.H., Y.P., and A.M. analyzed genotypic data and mapped the QTLs; Z.A.J., J.A., and A.M. arranged the outline of the content in the manuscript and prepared a draft article; J.A. and Z.L. helped conceive the basic idea, gave suggestions, corrected the entire article, and improved on the prospects in breeding programs.

Acknowledgments: The authors would like to thank the anonymous three reviewers from the International Journal of Molecular Science and two internal reviewers from at the International Rice Research Institute for their valuable suggestions and constructive comments that helped improve this manuscript.

References

1. Godfray, H.C.; Beddington, J.R.; Crute, I.R.; Haddad, L.; Lawrence, D.; Muir, J.F.; Pretty, J.; Robinson, S.; Thomas, S.M.; Toulmin, C. Food security: The challenge of feeding 9 billion people. *Science* **2010**, *327*, 812–818. [CrossRef] [PubMed]
2. FAO. The State of Food Insecurity in the World 2009. Available online: http://www.fao.org/tempref/docrep/fao/012/i0876e/i0876e_flyer.pdf (accessed on 30 October 2009).
3. Tilman, D.; Balzer, C.; Hill, J.; Befort, B.L. Global food demand and the sustainable intensification of agriculture. *Proc. Natl. Acad. Sci. USA* **2011**, *108*, 20260–20264. [CrossRef] [PubMed]
4. Edgerton, M.D. Increasing crop productivity to meet global needs for feed, food, and fuel. *Plant Physiol.* **2009**, *149*, 7–13. [CrossRef] [PubMed]
5. Naresh, R.K.; Gupta, R.K.; Ashok, K.; Singh, B.; Prakash, S.; Kumar, S.; Rathi, R.C. Direct-seeding and reduced-tillage options in the rice-wheat system of the Western Indo-Gangetic Plains. *Int. J. Agric. Sci.* **2011**, *7*, 197–208.
6. Bindraban, P.S.; Dimkpa, C.; Nagarajan, L.; Roy, A.; Rabbinge, R. Revisiting fertilisers and fertilisation strategies for improved nutrient uptake by plants. *Biol. Fertil. Soils* **2015**, *51*, 897–911. [CrossRef]
7. Mahender, A.; Anandan, A.; Pradhan, S.K.; Singh, O.N. Traits-related QTLs and genes and their potential applications in rice improvement under low phosphorus condition. *Arch. Agron. Soil Sci.* **2017**, *64*, 449–464. [CrossRef]
8. Ali, J.; Xu, J.L.; Gao, Y.M.; Ma, X.F.; Meng, L.J.; Wang, Y.; Pang, Y.L.; Guan, Y.S.; Xu, M.R.; Revilleza, J.E.; et al. Harnessing the hidden genetic diversity for improving multiple abiotic stress tolerance in rice (*Oryza sativa* L.). *PLoS ONE* **2017**, *12*, e0172515. [CrossRef]
9. Raun, W.R.; Johnson, G.V. Improving nitrogen use efficiency for cereal production. *Agron. J.* **1999**, *91*, 357–363. [CrossRef]
10. Cassman, K.G.; Gines, G.C.; Dizon, M.A.; Samson, M.I.; Alcanter, J.M. Nitrogen use efficiency in tropical lowland rice systems: Contribution from indigenous and applied nitrogen. *Field Crop. Res.* **1996**, *47*, 1–12. [CrossRef]
11. Giles, J. Nitrogen study fertilizes fears of pollution. *Nature* **2005**, *433*, 791. [CrossRef]
12. Vitousek, P.M.; Naylor, R.; Crews, T.; David, M.B.; Drinkwater, L.E.; Holland, E.; Johnes, P.J.; Katzenberger, J.; Martinelli, L.A.; Matson, P.A.; et al. Agriculture. Nutrient imbalances in agricultural development. *Science* **2009**, *19*, 324, 1519–1520. [CrossRef] [PubMed]

13. Hirel, B.; Tétu, T.; Lea, P.J.; Dubois, F. Improving nitrogen use efficiency in crops for sustainable agriculture. *Sustainability* **2011**, *3*, 1452–1485. [CrossRef]
14. Song, K.; Xue, Y.; Zheng, X.; Weiguang, L.; Qiao, H.; Qin, Q.; Yang, J. Effects of the continuous use of organic manure and chemical fertilizer on soil inorganic phosphorus fractions in calcareous soil. *Sci. Rep.* **2017**, *7*, 1164. [CrossRef] [PubMed]
15. Kant, S.; Bi, Y.; Rothstein, S.J. Understanding plant response to nitrogen limitation for the improvement of crop nitrogen use efficiency. *J. Exp. Bot.* **2011**, *62*, 1499–1509. [CrossRef] [PubMed]
16. Fageria, N.K.; Baligar, V.C.; Li, Y.C. Nutrient uptake and use efficiency by tropical legume cover crops at varying pH of an Oxisol. *J. Plant Nutr.* **2014**, *37*, 294–311. [CrossRef]
17. Das, G.; Rao, G.J.N. Molecular marker assisted gene stacking for biotic and abiotic stress resistance genes in an elite rice cultivar. *Front. Plant Sci.* **2015**, *6*, 1–18. [CrossRef] [PubMed]
18. Anumalla, M.; Roychowdhury, R.; Geda, C.K.; Mazid, M.; Rathoure, A.K. Utilization of plant genetic resources and diversity analysis tools for sustainable crop improvement with special emphasis on rice. *J. Adv. Res.* **2015**, *3*, 1155–1175.
19. Pang, Y.; Chen, K.; Wang, X.; Wang, W.; Xu, J.; Ali, J.; Li, Z. Simultaneous improvement and genetic dissection of salt tolerance of rice (*Oryza sativa* L.) by designed QTL pyramiding. *Front. Plant Sci.* **2017**, *8*, 1275. [CrossRef]
20. Ali, J.; Jewel, Z.A.; Mahender, A.; Anandan, A.; Hernandez, J.; Li, Z. Molecular genetics and breeding for nutrient use efficiency in rice. *Int. J. Mol. Sci.* **2018**, *19*, 1762. [CrossRef]
21. Jewel, Z.; Ali, J.; Pang, Y.; Anumalla, M.; Acero, B.; Hernandez, J.; Xu, J.; Li, Z. Developing Green Super Rice Varieties with High Nutrient Use Efficiency by Phenotypic Selection Under Varied Nutrient Conditions. *Preprints* **2018**. [CrossRef]
22. Novoa, R.; Loomis, R.S. Nitrogen and plant production. *Plant Soil* **1981**, *58*, 177–204. [CrossRef]
23. Baligar, V.C.; Fageria, N.K.; He, Z.L. Nutrient use efficiency in plants. *Commun. Soil Sci. Plant Anal.* **2001**, *32*, 921–950. [CrossRef]
24. McDonald, G.; William, B.; Chunyuan, H.; David, L. Nutrient use efficiency. In *Genomics and Breeding for Climate-Resilient Crops*; Springer: Berlin/Heidelberg, Germany, 2013; Volume 2, pp. 333–393.
25. Stahl, A.; Pfeifer, M.; Frisch, M.; Wittkop, B.; Snowdon, R.J. Recent genetic gains in nitrogen use efficiency in oilseed rape. *Front. Plant Sci.* **2017**, *8*, 963. [CrossRef] [PubMed]
26. Lian, X.; Xing, Y.; Yan, H.; Xu, C.; Li, X.; Zhang, Q. QTLs for low nitrogen tolerance at seedling stage identified using a recombinant inbred line population derived from an elite rice hybrid. *Appl. Genet.* **2005**, *112*, 85–96. [CrossRef] [PubMed]
27. Senthilvel, S.; Govindaraj, P.; Arumugachamy, S.; Latha, R.; Malarvizhi, P.; Gopalan, A.; Maheswaran, M. Mapping genetic loci associated with nitrogen use efficiency in rice (*Oryza sativa*) L. In Proceedings of the 4th International Crop Science Congress, Brisbane, Australia, 26 September–1 October 2004.
28. Srividya, A.; Vemireddy, L.R.; Hariprasad, A.S.; Jayaprada, M.; Sakile, S.; Puram, V.R.R.; Anuradha, G.; Siddiq, E.A. Identification and mapping of landrace derived QTL associated with yield and its components in rice under different nitrogen levels and environments. *Int. J. Plant Breed. Genet.* **2010**, *4*, 210–227. [CrossRef]
29. Tong, H.H.; Chen, L.; Li, W.P.; Mei, H.; Xing, Y.; Yu, X.; Xu, X.; Zhang, S.; Luo, L. Identification and characterization of quantitative trait loci for grain yield and its components under different nitrogen fertilization levels in rice (*Oryza sativa* L.). *Mol. Breed.* **2011**, *28*, 495–509. [CrossRef]
30. Vinod, K.K.; Heuer, S. Approaches towards nitrogen- and phosphorus-efficient rice. *AoB Plants* **2012**, *28*, 1–18. [CrossRef]
31. Ni, J.J.; Wu, P.; Senadhira, D.; Huang, N. Mapping QTLs for phosphorus deficiency tolerance in rice (*Oryza sativa* L.). *Theor. Appl. Genet.* **1998**, *97*, 1361–1369. [CrossRef]
32. Wissuwa, M.; Ae, N. Further characterization of two QTLs that increase phosphorus uptake of rice (*Oryza sativa* L.) under phosphorus deficiency. *Plant Soil* **2001**, *237*, 275–286. [CrossRef]
33. Lang, N.T.; Buu, B.C. Mapping QTLs for phosphorus deficiency tolerance in rice (*Oryza sativa* L.). *Omonrice* **2006**, *14*, 1–9.
34. Cho, Y.I.; Jiang, W.Z.; Chin, J.H.; Piao, Z.; Cho, Y.G.; McCouch, S.; Koh, H.J. Identification of QTLs associated with physiological nitrogen use efficiency in rice. *Mol. Cell* **2007**, *23*, 72–79.
35. Shimizu, A.; Kato, K.; Komatsu, A.; Motomura, K.; Ikehashi, H. Genetic analysis of root elongation induced

by phosphorus deficiency in rice (*Oryza sativa* L.): Fine QTL mapping and multivariate analysis of related traits. *Appl. Genet.* **2008**, *117*, 987–996. [CrossRef] [PubMed]

36. Wang, Y.; Sun, Y.J.; Chen, D.Y.; Yu, S.B. Analysis of quantitative trait loci in response to nitrogen and phosphorus deficiency in rice using chromosomal segment substitution lines. *Acta Agron. Sin.* **2009**, *35*, 580–587. [CrossRef]
37. Chin, J.H.; Gamuyao, R.; Dalid, C.; Bustamam, M.; Prasetiyono, J.; Moeljopawiro, S.; Wissuwa, M.; Heuer, S. Developing rice with high yield under phosphorus deficiency: Pup1 sequence to application. *Plant Physiol.* **2011**, *156*, 1202–1216. [CrossRef] [PubMed]
38. Luo, X.D.; Liu, J.; Dai, L.F.; Zhang, F.T.; Wan, Y.; Xie, J.K. Linkage map construction and QTL identification of P-deficiency tolerance in *Oryza rufipogon* Griff. at early seedling stage. *Euphytica* **2017**, *213*, 96. [CrossRef]
39. Wu, P.; Ni, J.J.; Luo, A.C. QTLs underlying rice tolerance to low-potassium stress in rice seedlings. *Crop Sci.* **1998**, *38*, 1458–1462. [CrossRef]
40. Fageria, N.K.; Dos Santos, A.B.; De Moraes, M.F. Yield, potassium uptake, and use efficiency in upland rice genotypes. *Commun. Soil Sci. Plant Anal.* **2010**, *41*, 2676–2684. [CrossRef]
41. Feltus, F.A.; Wan, J.; Schulze, S.R.; Estill, J.C.; Jiang, N.; Paterson, A.H. An SNP resource for rice genetics and breeding based on subspecies *indica* and *japonica* genome alignments. *Genome Res.* **2004**, *14*, 1812–1819. [CrossRef]
42. Huang, Y.F.; Poland, J.A.; Wight, C.P.; Jackson, E.W.; Tinker, N.A. Using genotyping-by-sequencing (GBS) for genomic discovery in cultivated oat. *PLoS ONE* **2014**, *9*, e102448. [CrossRef]
43. Yu, H.; Xie, W.; Li, J.; Zhou, F.; Zhang, Q. A whole-genome SNP array (RICE6K) for genomic breeding in rice. *Plant Biotechnol. J.* **2014**, *12*, 28–37. [CrossRef]
44. Nadeem, M.A.; Nawaz, M.A.; Shahid, M.Q.; Doğan, Y.; Comertpay, G.; Yıldız, M.; Hatipoğlu, R.; Ahmad, F.; Alsaleh, A.; Labhane, N.; et al. DNA molecular markers in plant breeding: Current status and recent advancements in genomic selection and genome editing. *Biotechnol. Biotechnol. Equip.* **2018**, *32*, 261–285. [CrossRef]
45. Kumar, S.; Banks, T.W.; Cloutier, S. SNP Discovery through next-generation sequencing and its applications. *Int. J. Plant Genom.* **2012**, *2012*, 1–15. [CrossRef] [PubMed]
46. Xu, F.; Wang, W.; Wang, P.; Jun, L.M.; Chung Sham, P.; Wang, J. A fast and accurate SNP detection algorithm for next-generation sequencing data. *Nat. Commun.* **2012**, *3*, 1258. [CrossRef] [PubMed]
47. Thomson, M.J. High-throughput SNP genotyping to accelerate crop improvement. *Plant Breed. Biotechnol.* **2014**, *2*, 195–212. [CrossRef]
48. Wang, K.; Cui, K.; Liu, G.; Xie, W.; Yu, H.; Pan, J.; Huang, J.; Nie, L.; Shah, F. Identification of quantitative trait loci for phosphorus use efficiency traits in rice using a high density SNP map. *BMC Genet.* **2014**, *15*, 155. [CrossRef] [PubMed]
49. Chen, H.; Xie, W.; He, H.; Yu, H.; Chen, W.; Li, J.; Yu, R.; Yao, Y.; Zhang, W.; He, Y.; et al. A high-density SNP genotyping array for rice biology and molecular breeding. *Mol. Plant* **2014**, *7*, 541–553. [CrossRef] [PubMed]
50. Zhao, K.; Tung, C.W.; Eizenga, G.C.; Wright, M.H.; Ali, M.L.; Price, A.H.; Norton, G.J.; Islam, M.R.; Reynolds, A.; Mezey, J.; et al. Genome-wide association mapping reveals a rich genetic architecture of complex traits in *Oryza sativa*. *Nat. Commun.* **2011**, *2*, 467. [CrossRef]
51. Kurokawa, Y.; Noda, T.; Yamagata, Y.; Angeles-Shim, R.; Sunohara, H.; Uehara, K.; Furuta, T.; Nagai, K.; Jena, K.K.; Yasui, H.; et al. Construction of a versatile SNP array for pyramiding useful genes of rice. *Plant Sci.* **2016**, *242*, 131–139. [CrossRef]
52. Feng, Y.; Zhai, R.-R.; Lin, Z.-C.; Cao, L.-Y.; Wei, X.-H.; Cheng, S.-H. Quantitative trait locus analysis for rice yield traits under two nitrogen levels. *Rice Sci.* **2015**, *22*, 108–115.
53. Zhou, Y.; Tao, Y.; Tang, D.; Wang, J.; Zhong, J.; Wang, Y.; Yuan, Q.; Yu, X.; Zhang, Y.; Wang, Y.; et al. Identification of QTL associated with nitrogen uptake and nitrogen use efficiency using high throughput genotyped CSSLs in rice (*Oryza sativa* L.). *Front. Plant Sci.* **2017**, *8*, 1166. [CrossRef]
54. Dai, G.J.; Cheng, S.H.; Hua, Z.T.; Zhang, M.L.; Jiang, H.B.; Feng, Y.; Shen, X.H.; Su, Y.A.; He, N.; Ma, Z.B.; et al. Mapping quantitative trait loci for nitrogen uptake and utilization efficiency in rice (*Oryza sativa* L.) at different nitrogen fertilizer levels. *Genet. Mol. Res.* **2015**, *8*, 10404–10414. [CrossRef] [PubMed]
55. Wissuwa, M.; Yano, M.; Ae, N. Mapping of QTLs for phosphorus-deficiency tolerance in rice (*Oryza sativa* L.). *Appl. Genet.* **1998**, *97*, 777–783. [CrossRef]
56. Feng, M.; Xianwu, Z.; Guohu, M.I.; Ping, H.E.; Lihuang, Z.; Fusuo, Z. Identification of quantitative trait loci affecting tolerance to low phosphorus in rice (*Oryza sativa* L.). *Chin. Sci. Bull.* **2000**, *45*, 520–525.

57. Fang, P.; Wu, P. QTL × N-level interaction for plant height in rice (*Oryza sativa* L.). *Plant Soil* **2001**, *236*, 237–242. [CrossRef]
58. Senthilvel, S.; Vinod, K.K.; Malarvizhi, P.; Maheswaran, M. QTL and QTL environment effects on agronomic and nitrogen acquisition traits in rice. *J. Integr. Plant Biol.* **2008**, *50*, 1108–1117. [CrossRef] [PubMed]
59. Ping, M.U.; Huang, C.; Li, J.X.; Liu, L.F.; Li, Z.C. Yield trait variation and QTL mapping in a DH population of rice under phosphorus deficiency. *Acta Agron. Sin.* **2008**, *34*, 1137–1142.
60. Li, J.; Xie, Y.; Dai, A.; Liu, L.; Li, Z. Root and shoot traits responses to phosphorus deficiency and QTL analysis at seedling stage using introgression lines of rice. *J. Genet. Genom.* **2009**, *36*, 173–183. [CrossRef]
61. Xiang, C.; Ren, J.; Zhao, X.-Q.; Ding, Z.-S.; Zhang, J.; Wang, C.; Zhang, J.-W.; Joseph, C.A.; Zhang, Q.; Pang, Y.-L.; et al. Genetic dissection of low phosphorus tolerance related traits using selected introgression lines in rice. *Rice Sci.* **2015**, *22*, 264–274.
62. Yamaya, T.; Obara, M.; Nakajima, H.; Sasaki, S.; Hayakawa, T.; Sato, T. Genetic manipulation and quantitative trait loci mapping for nitrogen recycling in rice. *J. Exp. Bot.* **2002**, *53*, 917–925. [CrossRef]
63. Wissuwa, M.; Wegner, J.; Ae, N.; Yano, M. Substitution mapping of *Pup1*: A major QTL increasing phosphorus uptake of rice from a phosphorus-deficient soil. *Appl. Genet.* **2002**, *105*, 890–897. [CrossRef] [PubMed]
64. Duan, Y.H.; Zhang, Y.L.; Ye, L.T.; Fan, X.R.; Xu, G.H.; Shen, Q.R. Responses of rice cultivars with different nitrogen use efficiency to partial nitrate nutrition. *Ann. Bot.* **2007**, *99*, 1153–1160. [CrossRef] [PubMed]
65. Wu, L.; Yuan, S.; Huang, L.; Sun, F.; Zhu, G.; Li, G.; Fahad, S.; Peng, S.; Wang, F. Physiological mechanisms underlying the high-grain yield and high-nitrogen use efficiency of elite rice varieties under a low rate of nitrogen application in China. *Front. Plant Sci.* **2016**, *7*, 1024. [CrossRef] [PubMed]
66. Nguyen, H.T.T.; Dang, D.T.; Pham, C.V.; Bertin, P. QTL mapping for nitrogen use efficiency and related physiological and agronomical traits during the vegetative phase in rice under hydroponics. *Euphytica* **2016**, *212*, 473–500. [CrossRef]
67. Shimizu, A.; Yanagihara, S.; Kawasaki, S.; Ikehashi, H. Phosphorus deficiency-induced root elongation and its QTL in rice (*Oryza sativa* L.). *Appl. Genet.* **2004**, *109*, 1361–1368. [CrossRef]
68. Rose, T.J.; Wissuwa, M. Rethinking internal phosphorus utilization efficiency: A new approach is needed to improve PUE in grain crops. *Adv. Agron.* **2012**, *116*, 185–217.
69. Wissuwa, M.; Kretzschmar, T.; Rose, T.J. From promise to application: Root traits for enhanced nutrient capture in rice breeding. *J. Exp. Bot.* **2016**, *67*, 3605–3615. [CrossRef]
70. Wiel van de, C.C.M.; van der Linden, C.G.; Scholten, O.E. Improving phosphorus use efficiency in agriculture: Opportunities for breeding. *Euphytica* **2016**, *207*, 1–22. [CrossRef]
71. Wang, Y.; Wu, W.H. Genetic approaches for improvement of the crop potassium acquisition and utilization efficiency. *Curr. Opin. Plant Biol.* **2015**, *25*, 46–52. [CrossRef]
72. Xue, X.; Lu, J.; Ren, T.; Li, L.; Yousaf, M.; Cong, R.; Li, X. Positional difference in potassium concentration as diagnostic index relating to plant K status and yield level in rice (*Oryza sativa* L.). *Soil Sci. Plant Nutr.* **2016**, *62*, 31–38. [CrossRef]
73. Carmeis Filho, A.C.A.; Crusciol, C.A.C.; Guimarães, T.M.; Calonego, J.C.; Mooney, S.J. Impact of amendments on the physical properties of soil under tropical long-term no till conditions. *PLoS ONE* **2016**, *11*, e0167564. [CrossRef] [PubMed]
74. Tong, H.H.; Mei, H.W.; Yu., X.Q.; Xu., X.Y.; Li, M.S.; Zhang, S.Q.; Luo, L.J. Identification of related QTLs at late developmental stage in rice (*Oryza sativa* L.) under two nitrogen levels. *Acta Genet. Sin.* **2006**, *33*, 458–467. [CrossRef]
75. Piao, Z.; Li, M.; Li, P.; Zhang, C.; Wang, H.; Luo, Z.; Lee, J.; Yang, R. Bayesian dissection for genetic architecture of traits associated with nitrogen utilization efficiency in rice. *Afr. J. Biotechnol.* **2009**, *8*, 6834–6839.
76. Wei, D.; Cui, K.; Ye, G.; Pan, J.; Xiang, J.; Huang, J.; Nie, L. QTL mapping for nitrogen-use efficiency and nitrogen deficiency tolerance traits in rice. *Plant Soil* **2012**, *359*, 281–295. [CrossRef]
77. Tanksley, S.D. Mapping polygenes. *Annu. Rev. Genet.* **1993**, *27*, 205–233. [CrossRef] [PubMed]
78. Tanksley, S.D.; Young, N.D.; Paterson, A.H.; Bonierbale, M.W. RFLP mapping in plant breeding: New tools for an old science. *Nat. Biotechnol.* **1989**, *7*, 257–264. [CrossRef]
79. Li, Z.K.; Fu, B.Y.; Gao, Y.M.; Xu, J.L.; Ali, J.; Lafitte, H.; Jiang, Y.Z.; Rey, J.; Vijayakumar, C.; Maghirang, R.; et al. Genome-wide ILs and their use in genetic and molecular dissection of complex phenotypes in rice (*Oryza sativa* L.). *Plant Mol. Biol.* **2005**, *59*, 33–52. [CrossRef]
80. Ali, J.; Xu, J.L.; Gao, Y.; Fontanilla, M.; Li, Z.K. Breeding for yield potential and enhanced productivity

across different rice ecologies through green super rice (GSR) breeding strategy. In *International Dialogue on Perceptions and Prospects of Designer Rice*; Muralidharan, K., Siddiq, E.A., Eds.; Society for the Advancement of Rice Research, Directorate of Rice Research: Hyderabad, India, 2013; pp. 60–68.
81. Rao, T.N. Improving Nutrient Use Efficiency: The Role of Beneficial Management Practices. In *Better Crops-India*; IPNI–India Program 133: Gurgaon, India, 2007; Volume 1, pp. 6–7.
82. Prasad, B.; Patwary, A.K.; Biswas, P.S. Genetic variability and selection criteria in fine rice (*Oryza sativa* L.). *Pak. J. Biol. Sci.* **2001**, *4*, 1188–1190.
83. Zahid, M.A.; Akhatar, M.; Sabar, N.; Zaheen, M.; Tahir, A. Correlation and path analysis studies of yield and grain traits in Basmati rice (*Oryza sativa* L.). *Asian J. Plant Sci.* **2006**, *5*, 643–645.
84. Cruz, C.D.; Regazzi, A.J. *Biometric Models Applied to Genetic Improvement*, 2nd ed.; Universidade Federal de Viçosa: Viçosa, Brazil, 1997; 390p.
85. Ashura, L. Inter-relationship between yield and some selected agronomic characters in rice. *Afr. Crop Sci. J.* **1998**, *6*, 83–88.
86. Lebreton, C.; LazicJancic, V.; Steed, A.; Pekic, S.Q.S. Identification of QTL for drought responses in maize and their use in testing causal relationships between traits. *J. Exp. Bot.* **1995**, *46*, 853–865. [CrossRef]
87. Cisse, L.; Amar, B. The importance of phosphatic fertilizer for increased crop production in developing countries. In Proceedings of the AFA 6th International Annual Conference, Cairo, Egypt, 31 January–2 February 2000.
88. Yadav, R.L. Assessing on-farm efficiency and economics of fertilizer N, P and K in rice-wheat systems of India. *Field Crop Res.* **2003**, *18*, 39–51. [CrossRef]
89. Yoshida, S. *Fundamentals of Rice Crop Science*; IRRI: Los Baños, Philippines, 1981; 269p.
90. Xie, W.X.; Wang, G.H.; Zhang, Q.C.; Guo, H.C. Effects of nitrogen fertilization strategies on nitrogen use efficiency in physiology, recovery, and agronomy and redistribution of dry matter accumulation and nitrogen accumulation in two typical rice cultivars in Zhejiang. *China J. Zhejiang Univ. Sci. B* **2007**, *8*, 208–216. [CrossRef] [PubMed]
91. Park, C.J.; Bart, R.; Chern, M.; Canlas, P.E.; Bai, W.; Ronald, P.C. Over-expression of the endoplasmic reticulum chaperone BiP3 regulates XA21-mediated innate immunity in rice. *PLoS ONE* **2010**, *5*, e9262. [CrossRef] [PubMed]
92. Ye, N.; Yang, G.; Chen, Y.; Zhang, C.; Zhang, J.; Peng, X. Two hydroxyl pyruvate reductases encoded by OsHPR1 and OsHPR2 are involved in photo respiratory metabolism in rice. *J. Integr. Plant Biol.* **2014**, *56*, 170–180. [CrossRef]
93. Yasuda, H.; Hirose, S.; Kawakatsu, T.; Wakasa, Y.; Takaiwa, F. Over-expression of BiPhas inhibitory effects on the accumulation of seed storage proteins in endosperm cells of rice. *Plant Cell Physiol.* **2009**, *50*, 1532–1543. [CrossRef]
94. Wakasa, Y.; Yasuda, H.; Oono, Y.; Kawakatsu, T.; Hirose, S.; Takahashi, H.; Hayashi, S.; Yang, L.; Takaiwa, F. Expression of ER quality control-related genes in response to changes in BiP1 levels in developing rice endosperm. *Plant J.* **2011**, *65*, 675–689. [CrossRef]
95. Yang, Q.; He, H.; Li, H.; Tian, H.; Zhang, J.; Zhai, L.; Chen, J.; Wu, H.; Yi, G.; He, Z.H.; et al. NOA1 functions in a temperature-dependent manner to regulate chlorophyll biosynthesis and Rubisco formation in rice. *PLoS ONE* **2011**, *6*, e20015. [CrossRef]
96. Qu, S.; Liu, G.; Zhou, B.; Bellizzi, M.; Zeng, L.; Dai, L.; Han, B.; Wang, G.L. The broad-spectrum blast resistance gene *Pi9* encodes a nucleotide-binding site–leucine-rich repeat protein and is a member of a multigene family in rice. *Genetics* **2006**, *172*, 1901–1914. [CrossRef]
97. Amarawathi, Y.; Singh, R.; Singh, A.K.; Singh, V.P.; Mohapatra, T.; Sharma, T.R.; Singh, N.K. Mapping of quantitative trait loci for basmati quality traits in rice (*Oryza sativa* L.). *Mol. Breed.* **2008**, *21*, 49–65. [CrossRef]
98. Cui, K.; Huang, J.; Xing, Y.; Yu, S.; Xu, C.; Peng, S. Mapping QTLs for seedling characteristics under different water supply conditions in rice (*Oryza sativa*). *Physiol. Plant.* **2008**, *132*, 53–68. [CrossRef] [PubMed]
99. Bernier, J.; Kumar, A.; Ramaiah, V.; Spaner, D.; Atlin, G. A large-effect QTL for grain yield under reproductive-stage drought stress in upland rice. *Crop Sci.* **2007**, *47*, 507–517. [CrossRef]
100. Amanullah, M.A.; Almas, L.K.; Amanullaj, J.; Zahir, S.; Rahman, H.; Khalil, S.K. Agronomic efficiency and profitability of P-fertilizers applied at different planting densities of maize in Northwest. *Pak. J. Plant Nutr.* **2012**, *35*, 331–341. [CrossRef]

101. Liu, X.; Bai, X.; Wang, X.; Chu, C. OsWRKY71, a rice transcription factor, is involved in rice defense response. *J. Plant Physiol.* **2007**, *164*, 969–979. [CrossRef] [PubMed]
102. Gui, J.; Shen, J.; Li, L. Functional characterization of evolutionarily divergent 4-coumarate: Coenzyme a ligases in rice. *Plant Physiol.* **2011**, *157*, 574–586. [CrossRef]
103. Tao, Z.; Liu, H.; Qiu, D.; Zhou, Y.; Li, X.; Xu, C.; Wang, S. A pair of allelic WRKY genes play opposite roles in rice-bacteria interactions. *Plant Physiol.* **2009**, *151*, 936–948. [CrossRef] [PubMed]
104. Tao, Z.; Kou, Y.; Liu, H.; Li, X.; Xiao, J.; Wang, S. OsWRKY45 alleles play different roles in abscisic acid signalling and salt stress tolerance but similar roles in drought and cold tolerance in rice. *J. Exp. Bot.* **2011**, *62*, 4863–4874. [CrossRef] [PubMed]
105. Nguyen, V.T.; Nguyen, B.D.; Sarkarung, S.; Martinez, C.; Paterson, A.H.; Nguyen, H.T. Mapping of genes controlling aluminum tolerance in rice: Comparison of different genetic backgrounds. *Mol. Genet. Genom.* **2002**, *267*, 772–780.
106. Qi, D.; Guo, G.; Lee, M.C.; Zhang, J.; Cao, G.; Zhang, S.; Suh, S.C.; Zhou, Q.; Han, L. Identification of quantitative trait loci for the dead leaf rate and the seedling dead rate under alkaline stress in rice. *J. Genet. Genom.* **2008**, *35*, 299–305. [CrossRef]
107. Sato, T.; Ueda, T.; Fukuta, Y.; Kumagai, T.; Yano, M. Mapping of quantitative trait loci associated with ultraviolet-B resistance in rice (*Oryza sativa* L.). *Appl. Genet.* **2003**, *107*, 1003–1008. [CrossRef]
108. Xue, Y.; Wan, J.; Jiang, L.; Wang, C.; Liu, L.; Zhang, Y.M.; Zhai, H. Identification of quantitative trait loci associated with aluminum tolerance in rice (*Oryza sativa* L.). *Euphytica* **2006**, *150*, 37–45. [CrossRef]
109. Mori, M.; Tomita, C.; Sugimoto, K.; Hasegawa, M.; Hayashi, N.; Dubouzet, J.G.; Ochiai, H.; Sekimoto, H.; Hirochika, H.; Kikuchi, S. Isolation and molecular characterization of a Spotted leaf 18 mutant by modified activation-tagging in rice. *Plant Mol. Biol.* **2007**, *63*, 847–860. [CrossRef] [PubMed]
110. Shimono, M.; Koga, H.; Akagi, A.; Hayashi, N.; Goto, S.; Sawada, M.; Kurihara, T.; Matsushita, A.; Sugano, S.; Jiang, C.J.; et al. Rice WRKY45 plays important roles in fungal and bacterial disease resistance. *Mol. Plant Pathol.* **2012**, *13*, 83–94. [CrossRef] [PubMed]
111. Feng, Y.; Cao, L.Y.; Wu, W.M.; Shen, X.H.; Zhan, X.D.; Zhai, R.R.; Wang, R.C.; Chen, D.B.; Cheng, S.H. Mapping QTLs for nitrogen-deficiency tolerance at seedling stage in rice (*Oryza sativa* L.). *Plant Breed.* **2010**, *129*, 652–656. [CrossRef]
112. Heuer, S.; Lu, X.; Chin, J.H.; Tanaka, J.P.; Kanamori, H.; Matsumoto, T.; De Leon, T.; Ulat, V.J.; Ismail, A.M.; Yano, M.; et al. Comparative sequence analyses of the major quantitative trait locus phosphorus uptake 1 (Pup1) reveal a complex genetic structure. *Plant Biotechnol. J.* **2009**, *7*, 456–457. [CrossRef] [PubMed]
113. Babu, R.C.; Nguyen, B.D.; Chamarerk, V.; Shanmugasundaram, P.; Chezhian, P.; Jeyaprakash, P.; Ganesh, S.K.; Palchamy, A.; Sadasivam, S.; Sarkarung, S.; et al. Genetic analysis of drought resistance in rice by molecular markers: Association between secondary traits and field performance. *Crop Sci.* **2003**, *43*, 1457–1469. [CrossRef]
114. Andaya, V.C.; Mackill, D.J. Mapping of QTLs associated with cold tolerance during the vegetative stage in rice. *J. Exp. Bot.* **2003**, *54*, 2579–2585. [CrossRef]
115. Peng, J.W.; Liu, Q.; Rong, X.M.; Zhu, H.M.; Xie, G.X.; Tang, G.R. Effects of different rational ratio of N, P, K fertilizer and amount of N fertilizer on photosynthesis character and yield of rice. *J. Hunan Agric. Univ. Nat. Sci.* **2004**, *30*, 123–127.
116. Wu, P.; Liao, C.Y.; Hu, B.; Yi, K.K.; Jin, W.Z.; Ni, J.J.; He, C. QTLs and epistasis for aluminum tolerance in rice (*Oryza sativa* L.) at different seedling stages. *Appl. Genet.* **2000**, *100*, 1295–1303. [CrossRef]
117. Yi, K.; Wu, Z.; Zhou, J.; Du, L.; Guo, L.; Wu, Y.; Wu, P. OsPTF1, a novel transcription factor involved in tolerance to phosphate starvation in rice. *Plant Physiol.* **2005**, *138*, 2087–2096. [CrossRef]
118. Gamuyao, R.; Chin, J.H.; Pariasca-Tanaka, J.; Pesaresi, P.; Catausan, S.; Dalid, C.; Slamet-Loedin, I.; Tecson-Mendoza, E.M.; Wissuwa, M.; Heuer, S. The protein kinase Pstol1 from traditional rice confers tolerance of phosphorus deficiency. *Nature* **2012**, *488*, 535–539. [CrossRef] [PubMed]
119. Moncada, P.; Martinez, C.P.; Borrero, J.; Chatel, M.; Gauch, H.; Guimaraes, E.; Tohme, J.; McCouch, S.R. Quantitative trait loci for yield and yield components in an *Oryza sativa* × *Oryza rufipogon* BC_2F_2 population evaluated in an upland environment. *Appl. Genet.* **2001**, *102*, 41–52. [CrossRef]
120. Araki, R.; Hasegawa, H. Expression of rice (*Oryza sativa* L.) genes involved in high-affinity nitrate transport during the period of nitrate induction. *Breed. Sci.* **2006**, *56*, 295–302. [CrossRef]
121. Katayama, H.; Mori, M.; Kawamura, Y.; Tanaka, T.; Mori, M.; Hasegawa, H. Production and characterization

of transgenic plants carrying a high-affinity nitrate transporter gene (OsNRT2.1). *Breed. Sci.* **2009**, *59*, 237–243. [CrossRef]

122. Fu, Q.; Zhang, P.; Tan, L.; Zhu, Z.; Ma, D.; Fu, Y.; Zhan, X.; Cai, H.; Sun, C. Analysis of QTLs for yield-related traits in Yuanjiang common wild rice (*Oryza rufipogon* Griff.). *J. Genet. Genom.* **2010**, *37*, 147–157. [CrossRef]
123. Ogawa, S.; Valencia, M.O.; Lorieux, M.; Arbelaez, J.D.; McCouch, M.; Ishitani, M.; Selvaraj, M.G. Identification of QTLs associated with agronomic performance under nitrogen-deficient conditions using chromosome segment substitution lines of a wild rice relative, *Oryza rufipogon*. *Acta Physiol. Plant.* **2016**, *38*, 103. [CrossRef]
124. Hittalmani, S.; Shashidhar, H.E.; Bagali, P.G.; Ning, H.; Sidhu, J.S.; Singh, V.P.; Khush, G.S. Molecular mapping of quantitative trait loci for plant growth, yield and yield related traits across three diverse locations in a doubled haploid rice population. *Euphytica* **2002**, *125*, 207–214. [CrossRef]
125. Pelgas, B.; Bousquet, J.; Meirmans, P.G.; Ritland, K.; Isabel, N. QTL mapping in white spruce: Gene maps and genomic regions underlying adaptive traits across pedigrees, years and environments. *BMC Genom.* **2011**, *12*, 145. [CrossRef]
126. Koide, Y.; Pariasca-Tanaka, J.; Rose, T.; Fukuo, A.; Konisho, K.; Yanagihara, S.; Fukuta, Y.; Wissuwa, M. QTLs for phosphorus-deficiency tolerance detected in upland NERICA varieties. *Plant Breed.* **2013**, *132*, 259–265. [CrossRef]
127. Mukherjee, A.; Sarkar, S.; Chakraborty, A.S.; Yelne, R.; Kavishetty, V.; Biswas, T.; Mandal, N.; Bhattacharyya, S. Phosphate acquisition efficiency and phosphate starvation tolerance locus (PSTOL1) in rice. *J. Genet.* **2014**, *93*, 683–688. [CrossRef]
128. Yokoshom, K.; Yamaji, N.; Ueno, D.; Mitani, N.; Ma, J.F. OsFRDL1 is a citrate transporter required for efficient translocation of iron in rice. *Plant Physiol.* **2009**, *149*, 297–305. [CrossRef] [PubMed]
129. Ai, P.; Sun, S.; Zhao, J.; Fan, X.; Xin, W.; Guo, Q.; Yu, L.; Shen, Q.; Wu, P.; Miller, A.J.; et al. Two rice phosphate transporters, OsPht1;2 and OsPht1;6, have different functions and kinetic properties in uptake and translocation. *Plant J.* **2009**, *57*, 798–809. [CrossRef] [PubMed]
130. Zheng, L.; Ying, Y.; Wang, L.; Wang, F.; Whelan, J.; Shou, H. Identification of a novel iron regulated basic helix-loop-helix protein involved in Fe homeostasis in *Oryza sativa*. *BMC Plant Biol.* **2010**, *11*, 166. [CrossRef] [PubMed]
131. Lee, S.; Jeong, H.J.; Kim, S.A.; Lee, J.; Guerinot, M.L.; An, G. OsZIP5 is a plasma membrane zinc transporter in rice. *Plant Mol. Biol.* **2010**, *73*, 507–517. [CrossRef] [PubMed]
132. Ishimaru, Y.; Kakei, Y.; Shimo, H.; Bashir, K.; Sato, Y.; Nishizawa, N.K. A rice phenolic efflux transporter is essential for solubilizing precipitated apoplasmic iron in the plant stele. *J. Biol. Chem.* **2011**, *286*, 24649–24655. [CrossRef] [PubMed]
133. Gutjahr, C.; Radovanovic, D.; Geoffroy, J.; Zhang, Q.; Siegler, H.; Chiapello, M.; Casieri, L.; An, K.; An, G.; Guiderdoni, E.; et al. The half-size ABC transporters STR1 and STR2 are indispensable for mycorrhizal arbuscule formation in rice. *Plant J.* **2012**, *69*, 906–920. [CrossRef] [PubMed]
134. Chen, Z.; Fujii, Y.; Yamaji, N.; Masuda, S.; Takemoto, Y.; Kamiya, T.; Yusuyin, Y.; Iwasaki, K.; Kato, S.; Maeshima, M.; et al. Mn tolerance in rice is mediated by MTP8.1, a member of the cation diffusion facilitator family. *J. Exp. Bot.* **2013**, *64*, 4375–4387. [CrossRef]
135. Funayama, K.; Kojima, S.; Tabuchi-Kobayashi, M.; Sawa, Y.; Nakayama, Y.; Hayakawa, T.; Yamaya, T. Cytosolic glutamine synthetase1;2 is responsible for the primary assimilation of ammonium in rice roots. *Plant Cell Physiol.* **2013**, *54*, 934–943. [CrossRef]
136. Ogo, Y.; Kobayashi, T.; Nakanishi Itai, R.; Nakanishi, H.; Kakei, Y.; Takahashi, M.; Toki, S.; Mori, S.; Nishizawa, N.K. A novel NAC transcription factor, IDEF2, that recognizes the iron deficiency-responsive element 2 regulates the genes involved in iron homeostasis in plants. *J. Biol. Chem.* **2008**, *19*, 13407–13417. [CrossRef]
137. Shim, D.; Hwang, J.U.; Lee, J.; Lee, S.; Choi, Y.; An, G.; Martinoia, E.; Lee, Y. Orthologs of the class A4 heat shock transcription factor HsfA4a confer cadmium tolerance in wheat and rice. *Plant Cell* **2009**, *21*, 4031–4043. [CrossRef]
138. Dong, Y.J.; Ogawa, T.; Kamiunten, H.; Lin, D.Z.; Cheng, S.H.; Terao, H.; Matsuo, M. Detection of QTLs for zinc toxicity tolerance in rice (*Oryza sativa* L.). *Rice Genet. Newsl.* **2004**, *21*, 33–36.
139. Wan, J.L.; Zhai, H.Q.; Wan, J.M.; Yasui, H.; Yoshimura, A. Detection and analysis of QTLs for some traits associated with tolerance to ferrous iron toxicity in rice (*Oryza sativa* L.), using recombinant inbred lines. *Rice Genet. Newsl.* **2003**, *20*, 55–57.

140. Xue, Y.; Jiang, L.; Su, N.; Wang, J.K.; Deng, P.; Ma, J.F.; Zhai, H.Q.; Wan, J.M. The genetic basis and fine-mapping of a stable quantitative-trait loci for aluminium tolerance in rice. *Planta* **2007**, *227*, 255–262. [CrossRef] [PubMed]
141. Hanh, N.T.T.; Cuòng, P.V.; Pierre, B. Rice nitrogen use efficiency: Genetic dissection. *J. Sci. Dev.* **2013**, *11*, 814–825.
142. Jewel, Z.A.; Ali, J.; Pang, Y.; Mahender, A.; Acero, B.; Hernandez, J.; Xu, J.; Li, Z.K. Developing green super rice varieties with high nutrient use efficiency by phenotypic selection under varied nutrient conditions. *Crop J.* **2019**, in press.
143. R Core Team. *R: A Language and Environment for Statistical Computing*; R Foundation for Statistical Computing: Vienna, Austria, 2013.
144. Dobermann, A.R. *Nitrogen Use Efficiency-State of the Art*; Agronomy-Faculty Publications: Lincoln, NE, USA, 2005; 316p.
145. Murray, M.G.; Thampson, W.F. Rapid isolation of high molecular weight plant DNA. *Nucl. Acids Res.* **1980**, *8*, 4321–4325. [CrossRef]
146. Wang, J.; Li, H.; Zhang, L.; Meng, L. *Users' Manual of QTL IciMapping*; The Quantitative Genetics Group, Institute of Crop Science, Chinese Academy of Agricultural Sciences (CAAS): Beijing, China; Genetic Resources Program, International Maize and Wheat Improvement Center (CIMMYT): Mexico City, Mexico, 2014.
147. Churchill, G.A.; Doerge, R.W. Empirical threshold values for quantitative trait mapping. *Genetics* **1994**, *138*, 963–971.
148. Chen, M.; Presting, G.; Barbazuk, W.B.; Goicoechea, J.L.; Blackmon, B.; Fang, G.; Kim, H.; Frisch, D.; Yu, Y.; Sun, S.; et al. An integrated physical and genetic map of the rice genome. *Plant Cell* **2002**, *14*, 537–545. [CrossRef]
149. Tiwari, S.; Krishnamurthy, S.L.; Kumar, V.; Singh, B.; Rao, A.; Mithra, S.V.A.; Rai, V.; Singh, A.K.; Singh, N.K. Mapping QTLs for salt tolerance in rice (*Oryza sativa* L.) by bulked segregant analysis of recombinant inbred lines using 50K SNP chip. *PLoS ONE* **2016**, *11*, e0153610. [CrossRef]
150. Warnes, G.; Leisch, F. Package genetics: Population Genetics. 2015. Available online: https://cran.r-project.org/web/packages/genetics/genetics.pdf (accessed on 17 May 2016).
151. Voorrips, R.E. MapChart: Software for the Graphical Presentation of Linkage Maps and QTLs. *J Hered.* **2002**, *93*, 77–78. [CrossRef]

7

Normal and Abortive Buds Transcriptomic Profiling of *Broccoli ogu* Cytoplasmic Male Sterile Line and its Maintainer

Jinshuai Shu, Lili Zhang, Yumei Liu *, Zhansheng Li, Zhiyuan Fang, Limei Yang, Mu Zhuang, Yangyong Zhang and Honghao Lv

Institute of Vegetables and Flowers, Chinese Academy of Agricultural Sciences, Key Laboratory of Biology and Genetic Improvement of Horticultural Crops, Ministry of Agriculture, 12 Zhongguancun Nandajie Street, Beijing 100081, China; shujinshuai@caas.cn (J.S.); zhanglilina@126.com (L.Z.); lizhansheng@caas.cn (Z.L.); fangzhiyuan@caas.cn (Z.F.); yanglimei@caas.cn (L.Y.); zhuangmu@caas.cn (M.Z.); zhangyangyong@caas.cn (Y.Z.); lvhonghao@caas.cn (H.L.)

* Correspondence: liuyumei@caas.cn or liuyumeicaas@126.com

Abstract: Bud abortion is the main factor affecting hybrid seeds' yield during broccoli cross breeding when using *ogura* cytoplasmic male sterile (*ogu* CMS) lines. However, the genes associated with bud abortion are poorly understood. We applied RNA sequencing to analyze the transcriptomes of normal and abortive buds of broccoli maintainer and *ogu* CMS lines. Functional analysis showed that among the 54,753 annotated unigenes obtained, 74 and 21 differentially expressed genes in common were upregulated and downregulated in *ogu* CMS abortive buds compared with *ogu* CMS normal buds, maintainer normal, and abortive buds, respectively. Nineteen of the common differentially expressed genes were enriched by GO terms associated with glycosyl hydrolases, reactive oxygen species scavenging, inhibitor, and protein degradation. Ethylene-responsive transcription factor 115 and transcriptional factor basic helix-loop-helix 137 were significantly upregulated; transcription factors DUO1 and PosF21/RF2a/BZIP34 were downregulated in *ogu* CMS abortive buds compared with the other groups. Genes related to polygalacturonase metabolism, glycosyl hydrolases, oxidation reduction process, phenylalanine metabolism, and phenylpropanoid biosynthesis were significantly changed in *ogu* CMS abortive buds. Our results increase our understanding of bud abortion, provide a valuable resource for further functional characterization of *ogu* CMS during bud abortion, and will aid in future cross breeding of *Brassica* crops.

Keywords: broccoli; cytoplasmic male sterile; bud abortion; gene expression; transcriptome; RNA-Seq

1. Introduction

Bud abortion is a very common biological phenomenon in *Brassica* species. During bud abortion, the buds stop growing, convert from green into yellow progressively from the base to the top of the bud, and eventually wither off before flowering [1,2], which is detrimental for hybrid seeds yield during cross breeding. In recent years, bud abortion has attracted considerable attention in *Brassica* species cross breeding [3–5], as it reduces the hybrid seed yield production of many *Brassica* crops and also affects the efficiency of *Brassica* crops cross breeding. In the past decade, to solve the bud abortion problem, many studies have investigated the factors that cause bud abortion [1–8]. However, most of the studies focused on morphology and anatomy, and few studies have reported the factors contributing to *Brassica* bud abortion at the molecular level. Therefore, the fundamental mechanism of bud abortion is poorly understood.

High-throughput sequencing technologies have played important roles in revealing the molecular mechanisms of various organismal biological processes. RNA sequencing (RNA-Seq) technology is an important high throughput sequencing technology that produces functional genomic data for non-model plants that lack genomic sequence data. RNA-Seq technology has been widely applied to assist in determining differentially expressed genes (DEGs) involved in different biological processes in many species and may be a promising method to address the genes associated with bud abortion in *Brassica* species. Surprisingly, no studies on bud abortion using RNA-Seq technology have been documented so far.

Brassica species display obvious heterosis, and using cytoplasmic male sterile (CMS) lines to produce hybrid seeds is an important approach that uses this heterosis. Broccoli (*Brassica oleracea* var. *italica*) is an important vegetable crop and plays a vital role in the vegetable production industry worldwide. More importantly, broccoli is also used to produce health products and develop anti-cancer drugs [9–11]. Thus, broccoli is the most important economical *Brassica* vegetable crop. However, a large problem exists in broccoli cross breeding: the hybrid seed purity cannot reach 100% when using inbred lines or self-incompatible lines to produce hybrid seeds. To improve the purity of broccoli hybrid seeds, many breeders execute wide crosses between broccoli cultivars and other *Brassica* CMS materials to transfer CMS genes. Our group used multiple broccoli inbred lines to with the cabbage CMS material *ogura* $CMSR_3629$ (*ogu* CMS), which was introduced by the Asgrow Seed Co. (USA), which has been used to breed a plurality of broccoli CMS lines with a 100% sterile rate. However, the obtained CMS lines often show serious bud abortion, which leads to delayed flowering time and sharply reduced amounts of flowers, some serious bud abortion CMS lines cannot flower [5]. These problems limit the use of the CMS lines considerably and have reduced their breeding efficiency dramatically. So far, several studies have focused on the causes of bud abortion, e.g., the *RsVPE1* gene encodes a vacuolar processing enzyme that is involved in radish floral bud abortion under heat stress [1]. The expression of stress response, energy metabolism, amino acid synthesizing and processing, signal transduction, disease resistance and senescence, transcription and translation, and transmembrane transport-related genes were different between normal and abortive buds, as detected by cDNA-amplified length polymorphism technology [6–8]. However, as far as we know, no study has investigated the molecular events occurring in bud abortion based on high throughput sequencing during broccoli cross breeding, or any other crop, thus little information is available concerning the genes involved in bud abortion. Thus, it is necessary to determine the elements involved in broccoli bud abortion at the molecular level, which will increase our understanding of the molecular events involved in bud abortion of broccoli and other crops. The objectives of this study were to identify the functional genes involved in bud abortion and to determine the genes expression characteristics related to broccoli bud abortion, based on RNA-Seq technology.

2. Results

2.1. High-Throughput Transcriptome Sequencing and Unigene Assembly

To explore gene expression and gene networks that control bud abortion of *ogu* CMS lines in broccoli, we performed RNA-Seq analysis of normal and abortive buds from *ogu* CMS line CMS93219 and its maintainer line ML93219. The cytoplasmic male sterility original of CMS93219 was cabbage CMS material *ogu* $CMSR_3629$ and the backcross paternal was broccoli ML93219. CMS93219 was stabilized via sixteen generations of backcrossing before this study. There were no visible differences between CMS93219 and ML93219 in terms of the plants and the heading morphologies of the harvested material, except that CMS93219 showed significant bud abortion and ML93219 showed only slight bud abortion before or at the early anthesis stage (Figure 1A–D). The average number of abortive buds per branch of CMS93219 was 35.73 throughout the whole growth phase, accounting for 83.88% of the total bud number, which is extremely significantly higher than the 19.60 per branch (33.33%) in ML93219 (*t*-test, $p < 0.01$) (Figure 1E,F). Neither the abortive nor the normal buds of CMS93219 contained pollen, while the aborted buds of ML93219 contained less pollen than normal buds. Therefore, bud abortion and pollen abortion character were closely connected between ML93219 and CMS93219, and pollen abortion may promote bud abortion. Gene expression changes are usually associated with morphologic changes; therefore, we chose the maximum buds stage (around 3 days before anthesis) for transcriptome analyses (Figure 1B-I,B-II,D-I,D-II).

Figure 1. Morphological characterization of two broccoli lines with different degrees of bud abortion. (**A**,**B**) ML93219, (**C**,**D**) CMS93219. B-I and B-II represent the normal and abortive bud of ML93219, respectively. D-I and D-II represent the normal and abortive bud of CMS93219, respectively. The bars in (**E**,**F**) represent the standard deviation ($n = 15$). Asterisks indicate that the average number of abortive buds and the abortive buds rate per branch are very significantly different between ML93219 and CMS93219 (unpaired t test, $p < 0.01$).

Twelve cDNA libraries from normal buds and abortive buds of the ML93219 and CMS93219, containing three biological replicates for each sample, were subjected to Illumina sequencing. After filtering invalid reads and data cleaning, 663,701,918 clean reads containing 99,555,287,700 nucleotides (99.6 Gb), with a mean length of 150 bp, were obtained. The Q20 and GC percentage were 92.93–94.33% and 46.46–47.12%, respectively (Table 1).

Table 1. Summary of broccoli bud transcriptome sequencing data.

Sample Name	Raw Reads	Clean Reads (Clean/Raw)	Clean Bases (Gb)	Q20 (%)	GC Content (%)	Mapped Reads (Mapped/Clean)
ML_NB1	57,767,246	55,194,896 (95.55%)	8.28	94.30	46.93	41,252,694 (74.74%)
ML_NB2	44,577,570	43,849,450 (98.37%)	6.58	92.93	46.46	32,614,176 (74.38%)
ML_NB3	61,251,852	59,738,162 (97.53%)	8.96	94.03	46.96	44,511,290 (74.51%)
ML_AB1	61,588,740	60,493,582 (98.22%)	9.08	94.32	46.73	45,965,016 (75.98%)
ML_AB2	54,823,836	53,530,342 (97.64%)	8.02	93.90	46.74	40,115,268 (74.94%)
ML_AB3	58,528,852	56,475,844 (96.49%)	8.48	94.33	46.81	42,816,816 (75.81%)
CMS_NB1	54,455,258	53,170,054 (97.64%)	7.98	94.23	47.02	39,776,946 (74.81%)
CMS_NB2	53,265,514	52,284,140 (98.16%)	7.84	93.97	47.12	38,965,364 (74.53%)
CMS_NB3	51,816,268	50,884,086 (98.20%)	7.64	93.79	46.91	37,800,172 (74.29%)
CMS_AB1	64,340,176	63,146,694 (98.15%)	9.48	94.16	47.04	47,846,088 (75.77%)
CMS_AB2	56,566,280	55,530,456 (98.17%)	8.32	94.27	46.78	42,582,890 (76.68%)
CMS_AB3	60,701,770	59,404,212 (97.86%)	8.92	94.09	46.70	44,549,934 (74.99%)

Note: ML_NB and ML_AB represent the normal and abortive bud samples of ML93219, respectively. CMS_NB and CMS_AB represent the normal and abortive bud samples of CMS93219, respectively.

After assembling the clean reads, 97,347 transcripts (91,137,323 nucleotides) and 66,050 unigenes (51,896,834 nucleotides) were obtained. The average length of the transcripts and unigenes were 936 and 786 bp, respectively (Table 2).

Table 2. Summary of de novo transcriptome length distribution.

Length	200–500 bp	500–1 kbp	1 k–2 kbp	>2 kbp	Total	Min (bp)	Mean (bp)	Median (bp)	Max (bp)	N50	Total Nucleotides
No. of transcripts	44,030	21,348	21,467	10,502	97,347	201	936	578	16,361	1510	91,137,323
No. of Unigenes	37,344	12,545	10,691	5470	66,050	201	786	424	16,361	1363	51,896,834

2.2. *Gene Annotation and Functional Classification*

A total of 54,753 unigenes (82.90%) were annotated in at least one database and 6070 unigenes (9.19%) were annotated in seven databases. In the NCBI non-redundant protein sequences (Nr), NCBI non-redundant nucleotide sequences (Nt), KEGG Ortholog (KO), a manually annotated and reviewed protein sequence database (SwissProt), protein family (Pfam), Gene Ontology (GO), eukaryotic ortholog groups (KOG) databases, 44,294 (67.06%), 50,157 (75.93%), 12,403 (18.77%), 29,861 (45.2%), 24,660 (37.33%), 30,285 (45.85%) and 12,492 (18.91%) unigenes were annotated, respectively (Table 3).

Table 3. Annotation of unigene sequences in broccoli buds.

Sequence Database	Number of Annotated Unigenes	Percentage of Annotated Unigene Sequences (%)	Software and Parameters
Annotated in Nr	44,294	67.06	NCBI blast 2.2.28+, *e*-value = 1×10^{-5}
Annotated in Nt	50,157	75.93	NCBI blast 2.2.28+, *e*-value = 1×10^{-5}
Annotated in KO	12,403	18.77	KAAS, KEGG Automatic Annotation Server, *e*-value = 1×10^{-10}
Annotated in SwissProt	29,861	45.2	NCBI blast 2.2.28+, *e*-value = 1×10^{-5}
Annotated in PFAM	24,660	37.33	HMMER 3.0 package, hmmscan, *e*-value = 0.01
Annotated in GO	30,285	45.85	Blast2GO v2.5 [12] and self-write the script, *e*-value = 1×10^{-6}
Annotated in KOG	12,492	18.91	NCBI blast 2.2.28+, *e*-value = 1×10^{-3}
Annotated in all Databases	6070	9.19	-
Annotated in at least one Database	54,753	82.89	-
Total Unigenes	66,050	100	-

We used the KOG functional annotation system for the assembled unigenes to obtain clues to the functions of the unigenes. As a result, 12,492 unigenes have defined, specific protein functions, accounting for 22.82% of the total annotated unigenes and involving 26 KOG functional classes. The five largest categories were "General function prediction only", "Posttranslational modification, protein turnover, chaperones", "Signal transduction mechanisms", "Transcription", "Intracellular trafficking, secretion, and vesicular transport". Two categories, "Unnamed protein" and "Cell motility", accounted for the lowest percentages (Figure 2).

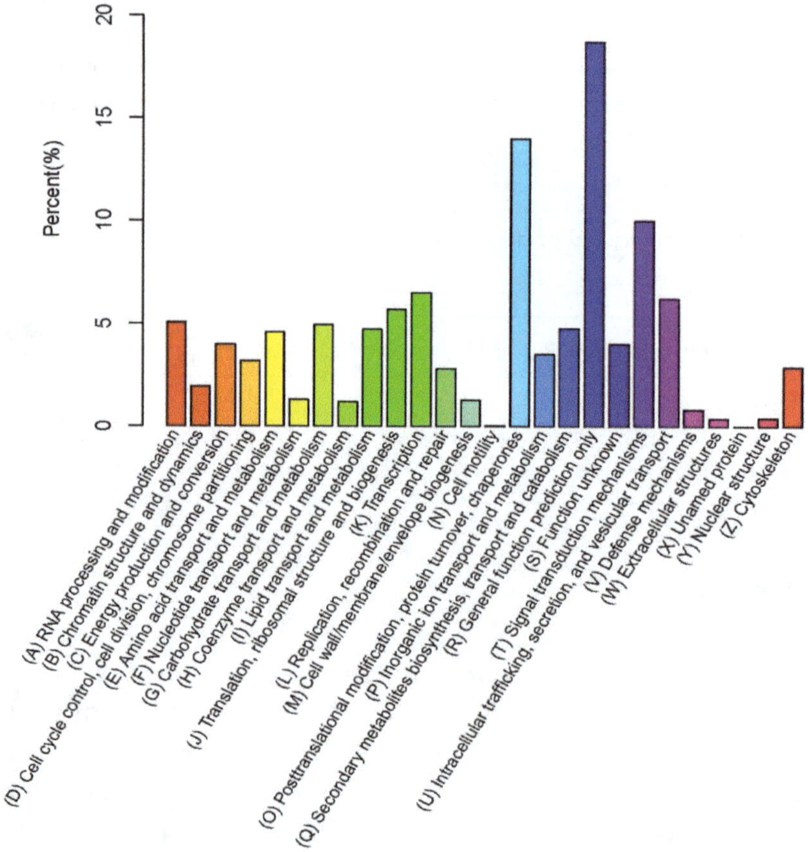

Figure 2. Clusters of eukaryotic orthologous groups (KOG) functional classification of the broccoli bud transcriptome.

To classify the functions of the unigenes, GO assignments were used based on the annotations from the Nr and Pfam databases. The results showed that 30,285 unigenes could be categorized into 57 functional groups, which were separated into three main categories: biological process (25 subcategories), cellular component (18 subcategories), and molecular function (14 subcategories) (Figure 3). The "cellular process," "metabolic process," and "single-organism process" were the major subcategories of the biological processes category; "cell" and "cell part," "binding," and "catalytic activity" were remarkable in the cellular component and molecular function categories, respectively. The classification result indicated that biological processes play a notable role during broccoli bud development, while the terms "biological phase," "cell aggregation," "synapse," "synapse part," "receptor regulator activity," and "metallochaperone activity" were rare.

To predict the activated biochemical pathways in broccoli buds, the unigenes were annotated against the KEGG database (e-value = 1×10^{-10}). A total of 264 KEGG pathways represented by 12,403 unigenes were obtained (Table S1). The main pathways were "Biosynthesis of amino acids [ko01230] (454 unigenes)," "Plant hormone signal transduction [ko04075]" (447 unigenes), "Carbon metabolism [ko01200]" (433 unigenes), and "Starch and sucrose metabolism [ko00500]" (366 unigenes). Moreover, the pathways such as "Endocytosis [ko04144]," "RNA degradation [ko03018]," "ubiquitin mediated proteolysis [ko04120]," "Fatty acid degradation [ko00071]," "Valine, leucine and isoleucine degradation [ko00280]," "Proteasome [ko03050]," "Lysine degradation [ko00310]," and "Apoptosis [ko04210]," which contained 235, 187, 184, 82, 79, 77, 49, and 31 unigenes, respectively, which indicated that protein degradation and cell death were associated with bud abortion.

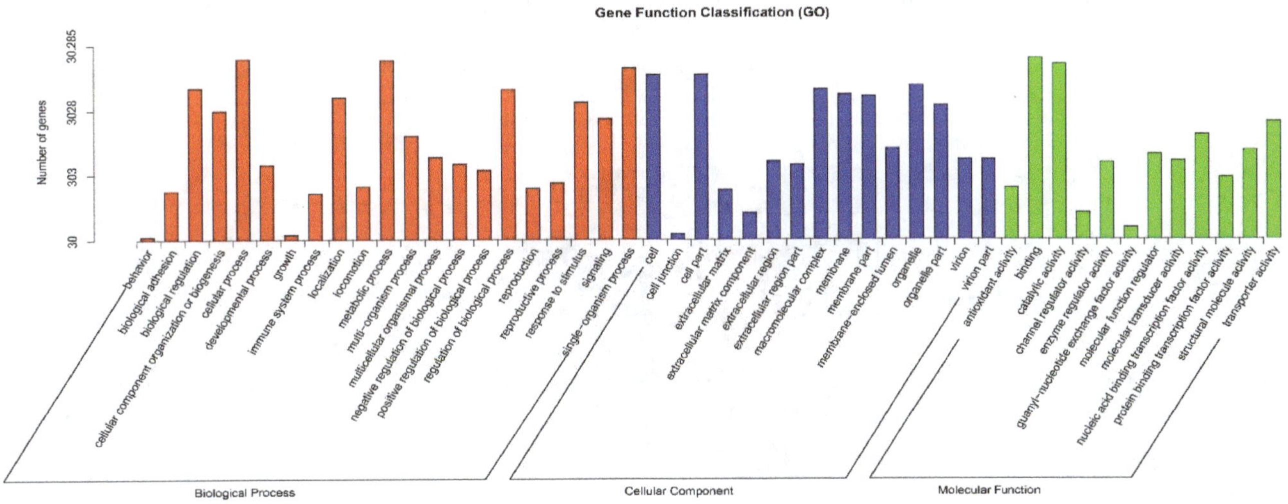

Figure 3. Gene Ontology (GO) classifications of the broccoli bud transcriptome.

2.3. Genes Express Differences and DEGs Clustering

The 97,347 assembly unigenes sequences obtained from de novo assemble of a merged set of 663,701,918 clean reads were used as the reference. We acquired the number of readcounts that aligned to each unigene and calculated the expected number of Fragments Per Kilobase of transcript sequence per Millions base pairs sequenced (FPKM) value to estimate the expression levels of the unigenes. The FPKM distribution and heatmap indicated the presence of many DEGs among the four samples, and the expression patterns of aborted buds of ML93219 (ML_AB) and aborted buds of *ogu* CMS93219 (CMS_AB), and normal buds of ML93219 (ML_NB) and normal buds of *ogu* CMS93219 (CMS_NB) were more similar, respectively (Figures 4 and 5). Changes in the DEGs indicated that the abundances of transcripts were different at the same stage of normal and abortive buds development between ML93219 and CMS93219.

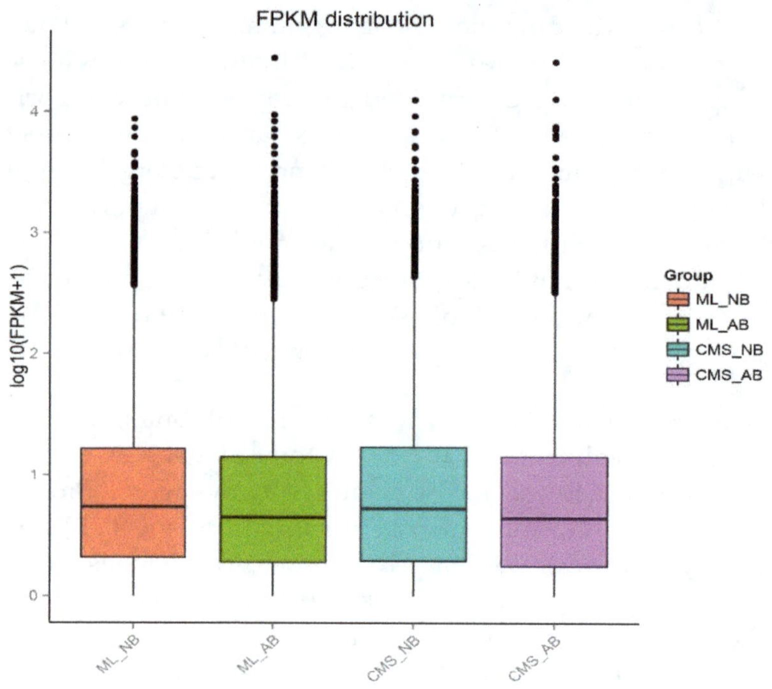

Figure 4. Boxplot of Fragments per kb per million fragments (FPKM) distribution for the four samples. Five statistics are represented by different regions of the Boxplot; from the top down they are the maximum, upper quartile, median, lower quartile, and minimum, respectively.

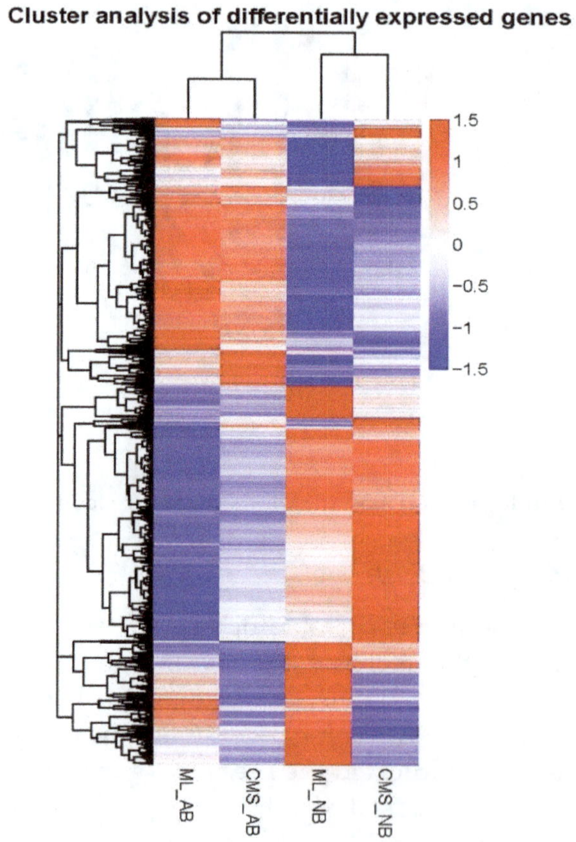

Figure 5. Cluster analysis of differentially expressed genes among the four samples. Heatmap of differentially expressed genes among the four samples. Red indicates high expression, and blue indicates low expression. Color from red to blue represents descending log10 (FPKM + 1).

2.4. DEGs in Normal and Abortive Buds of ogu CMS

To explore the reference sample CMS_AB gene expression levels, the gene expression variations were determined using three comparisons, between CMS_AB and CMS_NB, between CMS_AB and ML_NB and between CMS_AB and ML_AB. Compared with CMS_NB, there were 6575 and 5482 genes up- and downregulated in CMS_AB, respectively. Compared with ML_NB, in CMS_AB, there were 6192 and 6321 genes that were up- and downregulated, respectively. Compared with ML_AB, there were 182 and 825 genes that were up- and downregulated in CMS_AB, respectively (Figure 6a–c).

To determine the key genes involved in *ogu* CMS line bud abortion process, we compared the DEGs of normal and abortive buds from ML93219 and CMS93219 to define the differentially expressed genes in common. In CMS_AB, 74 and 21 genes were upregulated and downregulated, respectively, compared with CMS_NB, ML_NB and ML_AB (Figure 7; Tables S2–S5), which are most likely associated with *ogu* CMS bud abortion in broccoli.

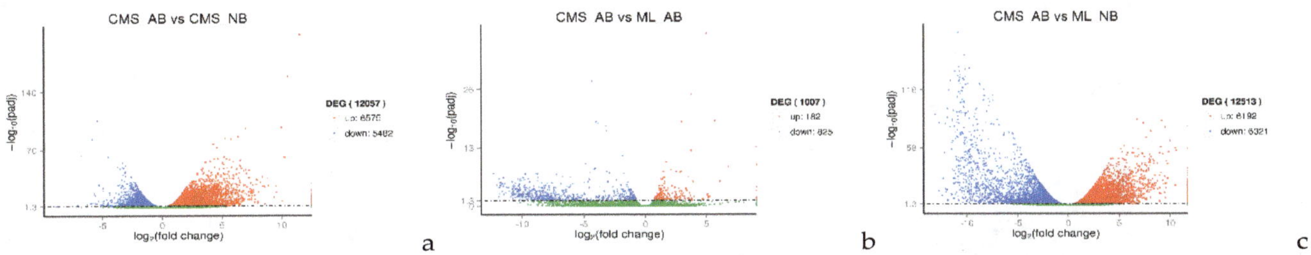

Figure 6. Differentially expressed genes (DEGs) between normal and abortive buds from ML93219 and CMS93219. (**a**) DEGs between CMS AB and CMS NB. (**b**) DEGs between CMS AB and ML AB. (**c**) DEGs between CMS AB and ML NB. The *x*-axis indicates the $\log_2$ (fold change) between the two samples. The *y*-axis indicates $-\log_{10}$ (padj) (*p*-adjusted). The scatter points in the figure represent individual genes, the green dots indicate genes with no significant differences, the red dots indicate up-regulated genes with significant differences, and the blue dots indicate down-regulated genes with significant differences. The screening condition for DEGs is padj < 0.05.

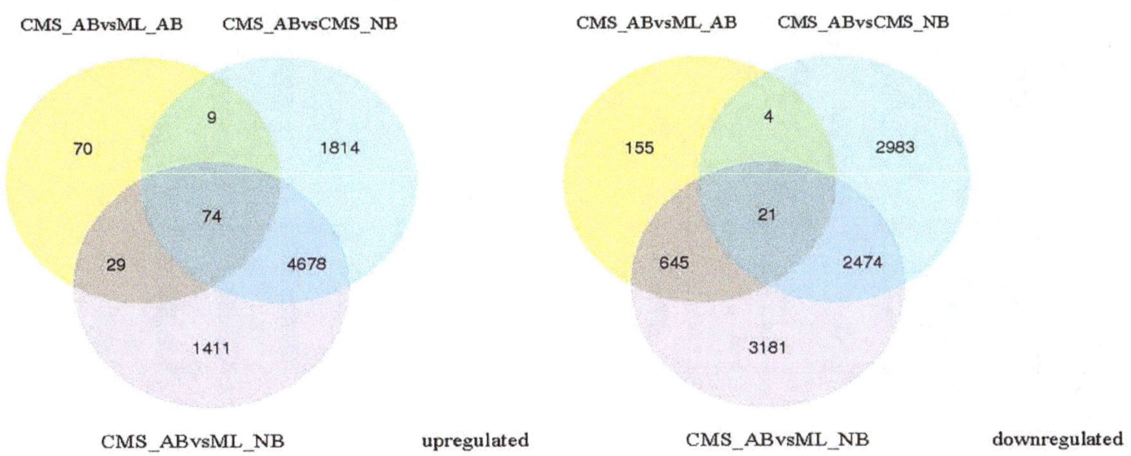

Figure 7. DEGs between normal and abortive buds from broccoli maintainer and *ogu* cytoplasmic male sterile (CMS) lines.

2.5. Validation of RNA-Seq Data by Quantitative Real-Time Reverse Transcription PCR (qRT-PCR)

We performed qRT-PCR to verify the DEGs identified by RNA-Seq, using the same samples that were used for the RNA-Seq analysis. Among the 21 randomly selected DEGs based on the expression level fold-change and differences in expression, 11 genes displayed higher expression quantification and 10 genes showed lower expression quantification in CMS_AB, compared with CMS_NB, ML_NB, and ML_AB. All 21 genes showed the same expression patterns in qRT-PCR as they did in the RNA-Seq databases experiment, indicating the high reliability of our RNA-Seq data (Figure 8).

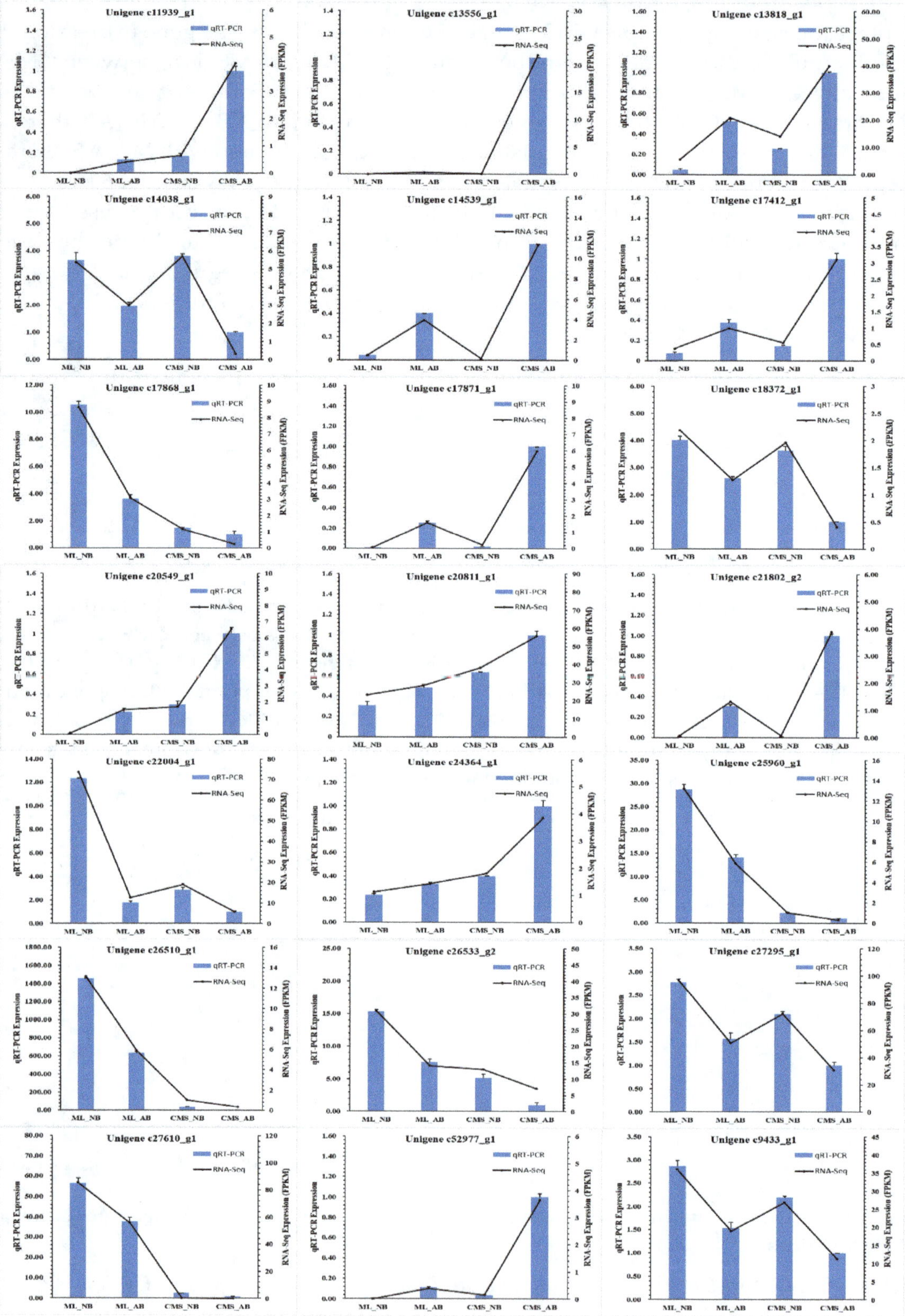

Figure 8. Verification of the DEGs by qRT-PCR. Eleven DEGs with higher expression and ten DEGs with lower expression in CMS_AB were selected for qRT-PCR validation. The relative expression level of each gene was expressed as the FPKM among four samples in the RNA-Seq data (black line) and qRT-PCR data (blue bar). To normalize the expression data, the broccoli *β-actin* gene was used as the internal control. The bars represent the standard deviation.

2.6. *Ogu* CMS Bud Abortion-Related Genes in Broccoli

To further understand the function and biological process of the DEGs, GO term (corrected $p < 0.05$) and KEGG pathway (corrected $p < 0.05$) enrichment were performed to analyse the 95 differentially coexpressed genes. Nineteen were enriched and the most significantly enriched GO terms were "cell wall organization" and "external encapsulating structure organization" (corrected $p = 3.26 \times 10^{-5}$) in the biological process (BP) group, "extracellular region" (corrected $p = 0.003937$) in the cellular component (CC) group and "polygalacturonase activity" (corrected $p = 0.015471$) in the molecular function (MF) group (Table 4). The GO terms of "cell wall organization or biogenesis" and "extracellular region" contained 10 and 13 genes, respectively, and were the biggest categories in the biological process and cellular component groups (Table 4), respectively. Only "polygalacturonase activity" GO terms in molecular function group was enriched and contained 4 genes (Table 4). Furthermore, there were 8 same genes were enriched in "sucrose metabolic process," "starch metabolic process," "disaccharide metabolic process," "cellular glucan metabolic process," "glucan metabolic process" and "oligosaccharide metabolic process" (Table 4); 5 same genes were enriched between "external encapsulating structure" and "cell wall" (Table 4). "Phenylalanine metabolism" and "Phenylpropanoid biosynthesis" (K00430) contained the same three genes (Gene ID: c20440_g1, c20661_g1, c9433_g1) that were significantly enriched in KEGG pathways; meanwhile, the three genes were enriched by GO enrichment (Tables 4 and 5).

Based on the enrichment results and functional annotation of the 19 enriched differential expressed genes in common by GO term, we found that "Glycosyl hydrolases" and "Reactive oxygen species (ROS) scavenger" related genes accounted for a high proportion (Figure 9). In addition, "Inhibitor," "Plant defense," and "Cell division and expansion" related genes were also significantly upregulated in CMS_AB and "Transporter" related gene Kinesin-4 downregulated in CMS_AB (Figure 9). Furthermore, five of the differentially expressed genes in common (GI: c11939_g1, c14539_g1, c25960_g1, c52977_g1, c9433_g1) enriched by GO terms were verified by qRT-PCR, and had significantly different expression between the abortive buds and normal buds of CMS93219 (Figure 8). These results suggested that most of the differentially expressed genes in common found in our study are required for *ogu* CMS bud abortion in broccoli.

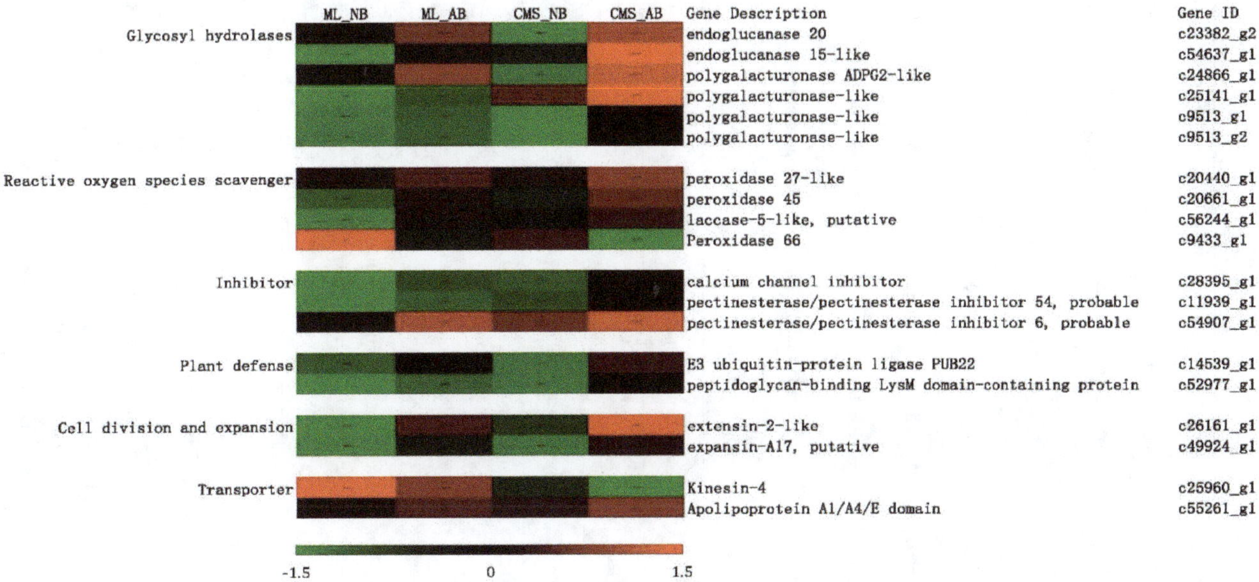

Figure 9. Heatmap of the 19 enriched common differentially expressed genes in normal and abortive broccoli buds. The bar represents the expression levels for each gene ($\log_{10}$ (FPKM + 1)) in the ML_NB, ML_AB, CMS_NB, and CMS_AB groups, as indicated by red or green rectangles. Red means upregulation of genes and green means downregulation.

Table 4. Results of DEGs enriched by Gene Ontology (GO) term.

GO Accession	Description	Term Type	*p* Value	Corrected *p* Value	DEG Item	Gene Names
GO:0071555	cell wall organization	Biological process	9.34×10^{-10}	3.25×10^{-6}	9	c23382_g2, c9513_g2, c9513_g1, c24866_g1, c25141_g1, c11939_g1, c54907_g1, c49924_g1, c26161_g1
GO:0045229	external encapsulating structure organization	Biological process	1.34×10^{-9}	3.25×10^{-6}	9	c54907_g1, c26161_g1, c49924_g1, c23382_g2, c24866_g1, c25141_g1, c11939_g1, c9513_g1, c9513_g2
GO:0071554	cell wall organization or biogenesis	Biological process	1.66×10^{-9}	3.25×10^{-6}	10	c49924_g1, c26161_g1, c52977_g1, c54907_g1, c9513_g1, c9513_g2, c24866_g1, c25141_g1, c11939_g1, c23382_g2
GO:0005985	sucrose metabolic process	Biological process	6.13×10^{-6}	0.005404	8	c54907_g1, c54637_g1, c23382_g2, c25141_g1, c11939_g1, c24866_g1, c9513_g2, c9513_g1
GO:0005982	starch metabolic process	Biological process	6.46×10^{-6}	0.005404	8	c23382_g2, c25141_g1, c11939_g1, c24866_g1, c9513_g2, c9513_g1, c54907_g1, c54637_g1
GO:0005984	disaccharide metabolic process	Biological process	8.99×10^{-6}	0.005849	8	c54907_g1, c54637_g1, c23382_g2, c9513_g2, c9513_g1, c11939_g1, c25141_g1, c24866_g1
GO:0006073	cellular glucan metabolic process	Biological process	1.2×10^{-5}	0.006395	8	c24866_g1, c11939_g1, c25141_g1, c9513_g2, c9513_g1, c23382_g2, c54637_g1, c54907_g1
GO:0044042	glucan metabolic process	Biological process	1.2×10^{-5}	0.006395	8	c23382_g2, c9513_g1, c9513_g2, c25141_g1, c11939_g1, c24866_g1, c54907_g1, c54637_g1
GO:0009311	oligosaccharide metabolic process	Biological process	1.66×10^{-5}	0.008084	8	c23382_g2, c25141_g1, c11939_g1, c24866_g1, c9513_g2, c9513_g1, c54907_g1, c54637_g1
GO:0044264	cellular polysaccharide metabolic process	Biological process	3.33×10^{-5}	0.013006	9	c54637_g1, c24364_g1, c54907_g1, c24866_g1, c11939_g1, c25141_g1, c9513_g1, c9513_g2, c23382_g2
GO:0005976	polysaccharide metabolic process	Biological process	5.27×10^{-5}	0.017134	9	c25141_g1, c11939_g1, c24866_g1, c9513_g2, c9513_g1, c23382_g2, c54637_g1, c24364_g1, c54907_g1
GO:0044723	single-organism carbohydrate metabolic process	Biological process	7.34×10^{-5}	0.022634	11	c11939_g1, c25141_g1, c24866_g1, c9513_g1, c9513_g2, c562_g1, c23382_g2, c17871_g1, c54637_g1, c54907_g1, c24364_g1
GO:0044262	cellular carbohydrate metabolic process	Biological process	0.00015	0.043887	9	c23382_g2, c9513_g1, c9513_g2, c25141_g1, c11939_g1, c24866_g1, c24364_g1, c54907_g1, c54637_g1
GO:0030312	external encapsulating structure	Cellular component	3.87×10^{-6}	0.005404	6	c23382_g2, c54907_g1, c49924_g1, c11939_g1, c26161_g1, c24866_g1
GO:0005576	extracellular region	Cellular component	7.52×10^{-6}	0.005507	11	c56244_g1, c28395_g1, c55261_g1, c14539_g1, c49924_g1, c20661_g1, c24866_g1, c25141_g1, c20440_g1, c9513_g2, c9513_g1
GO:0005618	cell wall	Cellular component	2.31×10^{-5}	0.010397	5	c11939_g1, c26161_g1, c24866_g1, c49924_g1, c54907_g1
GO:0071944	cell periphery	Cellular component	4.68×10^{-5}	0.016257	8	c49924_g1, c50518_g1, c26161_g1, c54907_g1, c11939_g1, c24866_g1, c23382_g2, c22601_g2
GO:0004650	polygalacturonase activity	Molecular function	6.12×10^{-6}	0.005404	4	c9513_g1, c9513_g2, c25141_g1, c24866_g1
GO:0004553	hydrolase activity, hydrolyzing O-glycosyl compounds	Molecular function	2.96×10^{-5}	0.012366	8	c54637_g1, c25141_g1, c24866_g1, c9513_g1, c9513_g2, c23382_g2, c562_g1, c17871_g1
GO:0016798	hydrolase activity, acting on glycosyl bonds	Molecular function	4.72×10^{-5}	0.016257	8	c17871_g1, c562_g1, c23382_g2, c9513_g1, c9513_g2, c25141_g1, c24866_g1, c54637_g1

Table 5. Results of DEGs enriched by Kyoto Encyclopedia of Genes and Genomes (KEGG) pathway.

#Term	ID	Input Number	*p* Value	Corrected *p*-Value	Input
Phenylalanine metabolism	ko00360	3	0.000337	0.006398357	c9433_g1, c20661_g1, c20440_g1
Phenylpropanoid biosynthesis	ko00940	3	0.001398	0.013279387	c9433_g1, c20661_g1, c20440_g1
Propanoate metabolism	ko00640	1	0.042547	0.179678515	c17871_g1
Cutin, suberine and wax biosynthesis	ko00073	1	0.048213	0.179678515	c24364_g1
Fatty acid elongation	ko00062	1	0.051033	0.179678515	c21267_g1
Endocrine and other factor-regulated calcium reabsorption	ko04961	1	0.060379	0.179678515	c22601_g2
alpha-Linolenic acid metabolism	ko00592	1	0.083359	0.179678515	c27465_g3
Photosynthesis	ko00195	1	0.08789	0.179678515	c28693_g1
Synaptic vesicle cycle	ko04721	1	0.090598	0.179678515	c22601_g2
Parkinson's disease	ko05012	1	0.095095	0.179678515	c25708_g1
Glycerolipid metabolism	ko00561	1	0.104024	0.179678515	c57011_g1
Peroxisome	ko04146	1	0.132037	0.187274571	c24364_g1
Pyruvate metabolism	ko00620	1	0.145727	0.187274571	c17871_g1
Cysteine and methionine metabolism	ko00270	1	0.149117	0.187274571	c17871_g1
Huntington's disease	ko05016	1	0.161716	0.187274571	c22601_g2
Glycerophospholipid metabolism	ko00564	1	0.164214	0.187274571	c57011_g1
Oxidative phosphorylation	ko00190	1	0.174958	0.187274571	c25708_g1
Glycolysis/Gluconeogenesis	ko00010	1	0.177418	0.187274571	c17871_g1
Endocytosis	ko04144	1	0.208772	0.208772251	c22601_g2

2.7. Transcription Factors Are Involved in Broccoli *ogu* CMS Bud Abortion Control

Among the 95 common differentially expressed genes, four transcription factors were identified, including predicted transcription factor basic helix-loop-helix (bHLH) 137 (GI: c13818_g1), ethylene-responsive transcription factor (ERF) 115 (GI: c21802_g2), transcription factor DUO1 (GI: c18372_g1), and PosF21/RF2a/BZIP34 (GI: c14038_g1), which were distributed in four different gene families. Compared with CMS_NB, ML_NB, and ML_AB, transcription factors bHLH137 ERF115 were upregulated in CMS_AB, suggesting that these transcription factors may function as positive regulators in *ogu* CMS bud abortion in broccoli. Conversely, transcription factors DUO1 and PosF21/RF2a/BZIP34 were downregulated in CMS_AB, suggesting that they are negative regulators. Moreover, the expression quantifications of the four transcription factors were confirmed by qRT-PCR (Figure 8), suggesting that the four transcription factors play key roles in *ogu* CMS bud abortion in broccoli.

3. Discussion

The bud abortion phenomenon is a very complex bioprocess and demands many molecular events during bud development. In this study, we used RNA-Seq technology to explore the genes involved in *ogu* CMS bud abortion process and to provide a comprehensive analysis of the genes involved in *ogu* CMS bud abortion control in broccoli. Compared with CMS_NB, ML_NB, ML_AB, there were 6575, 182, and 6192 genes that were upregulated and 5482, 825 and 6321 genes that were downregulated in CMS_AB, respectively (Figure 6a–c), among which 74 genes were significantly upregulated and 21 genes were significantly downregulated equally in CMS_AB with serious bud abortion (Figure 7; Tables S2–S5). qRT-PCR proved that our RNA-Seq data was highly reliable (Figure 8). Functional categories of the common differentially expressed genes by GO term enrichment analysis showed that gene associated with cell wall composition and metabolism, such as cell wall organization, cell wall organization or biogenesis; extracellular region, such as external encapsulating structure organization; sugar metabolism, such as sucrose metabolic process, starch metabolic process, disaccharide metabolic process, cellular glucan metabolic process, glucan metabolic process, oligosaccharide metabolic process related genes were strongly induced in the buds abortion of broccoli (Table 4). These results indicated that bud abortion may be closely related to cell wall organization, external encapsulating structures and sugar metabolism related genes play an important role in bud abortion in broccoli. Our results were consistent with the differential expression genes obtained by cDNA-AFLP technique in radish and cabbage between normal bud and dead bud [6–8].

3.1. Genes Related to Programmed Cell Death Are Involved in Bud Abortion

Programmed cell death (PCD) is an important physiological process in single cells and penetrates the whole plant life cycle, which can help plants to control and organize the destruction of non-functional or redundant damaged cells [13–15]. Although PCD is a natural result of ageing, it may be switched on by environmental stress or irregular development in plants [15]. Increases in caspase-like proteases [16] and ROS [17,18] activities of metacaspase gene family related genes [19] are the important features of PCD. In this study, several genes associated with PCD were significantly changed in *ogu* CMS abortive buds compared with normal buds and the maintainer abortive buds (CMS_NB, ML_NB and ML_AB). Among the 8833 differential expression genes between ML_AB vs. ML_NB and CMS_AB vs. CMS_NB, there was 11.75% participate in the redox process, 11.08% with redox enzyme activity and 55.38% with catalytic activity, indicating that redox process involved in bud abortion in broccoli. ROS scavengers related genes, such as: peroxidase 27-like and peroxidase 45, L-lactate dehydrogenase, laccase-5-like and fatty acyl-CoA reductase 7 (Table S5), which were all upregulated in abortive buds compared with normal buds, suggesting that higher levels of ROS were produced in abortive buds and ROS scavenger-related genes were closely related with broccoli bud abortion. Moreover, 13 caspase-like and metacaspase activity genes involved in cell apoptosis were also discovered, such as metacaspase-1, metacaspase-3, metacaspase-5, metacaspase-6, metacaspase-7, metacaspase-9, pentapeptide repeats protein, caspase recruitment domain, BTB/POZ domain-containing protein POB1. The expression of these genes in abortion buds was significantly higher than that of normal buds suggesting that bud abortion was related with PCD.

3.2. Glycosyl Hydrolases, Inhibitors and Plant Defence Related Genes Are Implicated in Bud Abortion

In this study, several glycosyl hydrolase-related genes were determined as significantly upregulated in *ogu* CMS abortive buds, such as endoglucanase 20, endoglucanase 15-like, polygalacturonase ADPG2-like and polygalacturonase-like. However, polygalacturonase plays an important role during the life cycle of cell separation, being involved in cell wall modification, abscission and dehiscence in *Arabidopsis thaliana* [20,21], and endoglucanase is involved in cell wall biogenesis or degradation, cellulose degradation and polysaccharide degradation [22], which suggested that glycosyl hydrolase-related genes were required for *ogu* CMS bud abortion in broccoli. Pectinesterase/pectinesterase inhibitors modify cell walls via demethylesterification of cell wall pectin, which negatively regulates catalytic activity. As a voltage-gated inward-rectifying Ca^{2+} channel (VDCC) across the vacuole membrane, the calcium channel is an essential components of the slow vacuolar (SV) channel and is the major ROS-responsive Ca^{2+} channel, which is the possible target of Al-dependent inhibition and is involved in the regulation of stomatal movement [23,24]. In our study, pectinesterase/pectinesterase inhibitor 6 and 54 probable and calcium channel inhibitor-related genes were highly expressed in *ogu* CMS abortive buds, indicating that these genes may be important in controlling *ogu* CMS bud abortion. Moreover, E3 ubiquitin-protein ligase and peptidoglycan-binding LysM domain-containing protein both play important roles in the plant defence response [23,25–30] and nicotianamine synthase is involved in the cellular response to ethylene stimulus [31]: these genes were distinguished significantly upregulated in *ogu* CMS abortive buds, suggesting their likely involvement in *ogu* CMS bud abortion.

3.3. Transcription Factors Associated with Bud Development

Transcription factors are critical to regulate gene expression during plant development and in response to biotic and abiotic stresses [32–38]. In the present study, predicted transcription factors bHLH137, ERF115, DUO1 and PosF21/RF2a/BZIP34 were obviously changed in abortive broccoli buds compared with normal buds. Interestingly, the genes encoding stress-responsive transcription factors ERF115 and gibberellin-responsive transcription factors bHLH137 showed similar expression patterns: both were significantly upregulated in abortive buds (ML_AB and CMS_AB) compared with normal buds (ML_NB and CMS_NB), and in *ogu* CMS abortive buds (CMS_AB), they were significantly upregulated compared with maintainer abortive buds (ML_AB) (Figure 8). Conversely, the genes encoding transcription factors DUO1 and PosF21/RF2a/BZIP34, which are positive regulators of transcription showed similar expression patterns: both were significantly downregulated in abortive buds, and in *ogu* CMS abortive buds they were significantly downregulated compared with maintainer abortive buds (Figure 8). ERF 115 as a transcriptional activator of the phytosulfokine PSK5 peptide hormone family that binds to the GCC-box pathogenesis-related promoter element and limit quiescent center cell division activity when surrounding stem cells are damaged and is also a proteolytic target of the APC/C-FZR1 complex [39]. bHLH transcription factors belong to a family of transcriptional regulators and have a range of different roles in plant cells and tissue development, as well as in plant metabolism [40]. Transcription factor DUO1 could be involved in pollen sperm cell differentiation [41]. PosF21/RF2a/BZIP34 is a transcription factor with an activatory role [42,43], which might be involved in the sporophytic control of cell wall patterning and gametophytic control of pollen development, and play a role in the control of metabolic pathways regulating cellular transport and lipid metabolism [44,45]. Therefore, transcription factors play a key role in the complex regulatory networks of *ogu* CMS bud abortion.

3.4. Molecular Mechanisms Associated with CMS

Previous studies have shown that the genes involved in reactive oxygen species (ROS) homeostasis or antioxidative system balance may be important factors contributing to pollen abortion in cotton [46] and wheat [47]. Carbohydrate and energy metabolisms, oxidation-reduction system and phenylpropanoid metabolism pathways related genes may be important factor for CMS in soybean [48], rapeseed [49], cabbage [50], onion [51] and wheat [52]. Male sterility might be related to energy metabolism turbulence, excessive ethylene synthesis, and suffocation of starch synthesis in pepper [53]. In addition, pentatricopeptide repeat proteins, heat shock proteins, stress proteins, MYB, bHLH and heat shock transcription factors and anther development related genes may be the candidates for pollen abortion in *Brassica* crops [49,50,54]. In this study, we found that genes related to polygalacturonase metabolism, glycosyl hydrolases, oxidation reduction process, phenylalanine metabolism, phenylpropanoid biosynthesis were significantly changed in *ogu* CMS abortive buds compared with the other groups. Ethylene-responsive transcription factor 115 and transcriptional factor basic helix-loop-helix 137 were both significantly upregulated in *ogu* CMS abortive buds. Therefore, our results were basically consistent with the results of previous studies [46–54], and the genes discovered related to energy metabolism, oxidation reduction process and phenylpropanoid biosynthesis, ethylene-responsive transcription factor 115 and transcriptional factor basic helix-loop-helix 137 may be important factors contributing to *Broccoli ogu* CMS pollen abortion and bud abortion. Further experiments are needed to elucidate the molecular mechanisms of these genes that lead to broccoli CMS and bud abortion.

4. Materials and Methods

4.1. Plant Materials

Broccoli (*Brassica oleracea* var. *italica*) maintainer ML93219 (showing slight bud abortion) and *ogu* CMS93219 (serious levels of bud abortion) were bred by the Institute of Vegetables and Flowers, Chinese Academy of Agricultural Sciences. The backcross paternal line of *ogu* CMS93219 was ML93219 and the number of backcross generations was sixteen. In the spring of 2015, the plants were grown in an experimental greenhouse at the Institute of Vegetable and Flowers, Chinese Academy of Agricultural Sciences, Changping (Beijing, China). We handled the main bouquet using the approach proposed by Shu et al. [55]. When the plants began to flower, four kinds of bud samples, ML_NB, ML_AB, CMS_NB and CMS_AB, were collected and labelled with three biological replicates. To ensure the integrity of the sample RNA, isolated buds were immediately frozen in liquid nitrogen and stored at −80 °C before RNA extraction.

4.2. RNA Extraction and Quality Testing

Total RNA was extracted using an EASYspin Plus Plant RNA-38 Kit, according to the manufacturer's instructions (Juhuatech Co., Ltd., Beijing, China). The integrity and purity of the RNA samples were determined by 1% agarose gels electrophoresis, and the RNA concentration was measured by Qubit® RNA Assay Kit in Qubit® 2.0 Flurometer (Life Technologies Corporation, Carlsbad, CA, USA) and the integrity of the RNA was assessed by an RNA Nano 6000 Assay Kit of the Agilent Bioanalyzer 2100 system (Agilent Technologies Inc., Santa Clara, CA, USA).

4.3. RNA-Seq Library Construction and Illumina Sequencing

Twelve strand-specific RNA-Seq libraries were constructed with cDNA fragments of 250–300 bp in length. An Illumina TruSeq PE Cluster Kit v3-cBot-HS on a cBot Cluster Generation System was then used to cluster the samples, according to the manufacturer's instructions. After cluster generation, the libraries were sequenced on an Illumina Hiseq™ 4000 system and reads were generated.

4.4. RNA-Seq Data Quality Control and Transcriptome de Novo Assembly

We obtained clean reads by removing the adaptor reads, unknown sequences "N" (reads containing more than 10% unknown nucleotides), low quality reads (reads containing more than 50% bases with Q-value ≤ 5) from the raw data. The Q20, Q30 and GC-content were then calculated based on the clean reads. The high quality clean reads were used for downstream analyses. Transcriptome *de novo* assembly was executed using Trinity [56] with min_kmer_cov set to 2 by default and other parameters set at their defaults. After assembly, the longest transcripts of each gene were selected as the unigenes.

4.5. Unigene Function Annotation

We annotated the unigenes based on seven databases, NCBI blast (2.2.28+) was used to search against the Nr (E-value = 1×10^{-5}), Nt (E-value = 1×10^{-5}), Swiss-Prot (E-value = 1×10^{-5}) and KOG databases (E-value = 1×10^{-3}). The unigenes were divided into 26 groups and their participation in different metabolic pathways based on KOG annotation was assessed. Pfam annotated was determined using the HMMER 3.0 package [57], hmmscan (e-value = 0.01). GO annotations for the unigenes were determined by Blast2GO v2.5 [12] with the self-write script (e-value = 1×10^{-6}) based on the annotation result of Nr and Pfam, which has three ontologies: molecular function, cellular component and biological process [58]. KEGG [59] related annotations were identified by the KAAS and KEGG Automatic Annotation Server [60] (E-value = 1×10^{-10}) to determine the metabolic pathway of unigenes.

4.6. Analysis of DEGs

Alignment results of bowtie were counted by RSEM [61]. FPKM [62] values were used to calculate the gene expression levels of the four groups of normal and abortive buds from the maintainer and *ogu* CMS lines. FPKM has become the most commonly used method to estimate the level of gene expression and takes into account the effects of sequencing depth and gene length on the calculation of gene expression [62]. There were three biological replicates; therefore, the calculated gene expression could be used directly to compare the gene expression between samples. Referring to the statistical method of Storey and Tibshirani [63], $|\log_2\text{Fold change}| > 1$ and p-adjusted < 0.05 were set as the threshold for significantly differential expression. p-Values were adjusted to control the false discovery rate, referring to Benjamini and Hochberg's approach [64]. Then, based on the Wallenius non-central hyper-geometric distribution [65], GO and KEGG functional enrichment analysis of the DEGs was executed by the GOseq [66] and KOBAS software [60], respectively.

4.7. qRT-PCR Validation

qRT-PCR analyses with the three biological replicates samples used for RNA-Seq were performed to verify the DGEs results. Twenty-one common differentially expressed genes were randomly selected that accounted for about 22.1% of the 95 common differentially expressed genes. Specific primers were designed using the Primer-BLAST tool (available online: http://www.ncbi.nlm.nih.gov/tools/primer-blast/index.cgi?LINK_LOC=BlastHome) in NCBI and synthesized by Sangon Biotech Co., Ltd. (Shanghai, China). cDNAs were reverse transcribed from total RNA using a PrimeScript RT reagent Kit (Takara, Dalian, China). qRT-PCR was carried out according to the SYBR PrimeScript RT-PCR Kit manufacturer specifications (Takara) on an ABI Prism®7900 Real-Time PCR System (Applied Biosystems, Foster City, CA, USA). To normalize the gene expression data, we used the broccoli *β-actin* gene as an internal standard [67]. The $2^{-\Delta\Delta Ct}$ method [68] was used to determine the relative expression of genes. The standard deviation was calculated based on the three biological replicates. The specific primers sequences are listed in Table S6.

5. Conclusions

In this study, we found that buds abortion was related with polygalacturonase metabolism, glycosyl hydrolases, oxidation reduction process, phenylalanine metabolism, and phenylpropanoid biosynthesis. Moreover, 19 common differentially expressed genes associated glycosyl hydrolases, reactive oxygen species scavenging, inhibitor, plant defense, cell division and expansion, transporter, and four transcriptional factors—ethylene-responsive transcription factor 115, transcriptional factor basic helix-loop-helix 137, transcription factors DUO1, and PosF21/RF2a/BZIP34—may be associated with *ogu* CMS abortive buds. In conclusion, our results not only increased our understanding of *ogu* CMS bud abortive mechanisms and provided a valuable resource for further functional characterization of *ogu* CMS bud abortion, but also laid the foundation for molecular breeding to overcoming bud abortion in broccoli, as well as other *Brassica* crops in the future.

Author Contributions: Conceptualization, J.S. and Y.L.; Data curation, J.S.; Formal analysis, J.S., L.Z., Z.L. and H.L.; Funding acquisition, Y.L.; Investigation, J.S.; Methodology, J.S.; Resources, Y.L., Z.F., L.Y., M.Z. and Y.Z.; Supervision, Y.L.; Writing—original draft, J.S. and L.Z.; Writing—review & editing, J.S. and Y.L. All authors read and approved the final manuscript.

Acknowledgments: This work was supported by the National Natural Science Foundation of China (Grant No. 31372067), the China Agriculture Research System (Grant No. CARS-25-A), Key Projects in the National Science and Technology Pillar Program of China (Grant No. 2013BAD01B04), the Key Laboratory of Biology and Genetic Improvement of Horticultural Crops, Ministry of Agriculture, China, and the Science and Technology Innovation Program of the Chinese Academy of Agricultural Sciences (Grant No. CAAS-ASTIP-IVFCAAS). We thank Liwen Bianji, Edanz Group China (available online: www.liwenbianji.cn/ac), for editing the English text of a draft of this manuscript.

References

1. Zhang, J.; Li, Q.F.; Huang, W.W.; Xu, X.Y.; Zhang, X.L.; Hui, M.X.; Zhang, M.K.; Zhang, L.G. A vacuolar processing enzyme *RsVPE1* gene of radish is involved in floral bud abortion under heat stress. *Int. J. Mol. Sci.* **2013**, *14*, 13346–13359. [CrossRef] [PubMed]
2. Zhang, J.; Sun, X.L.; Zhang, L.G.; Hui, M.X.; Zhang, M.K. Analysis of differential gene expression during floral bud abortion in radish (*Raphanus sativus* L.). *Genet. Mol. Res.* **2013**, *12*, 2507–2516. [CrossRef] [PubMed]
3. Liu, J.; Zhang, L.G.; Wang, F.M.; Hui, M.X.; Zhang, M.K. Observation of histocytological feature on radish flower bud during aborting. *Acta Agric. Bor. Occid. Sin.* **2008**, *5*, 272–276.
4. Wang, Q.B.; Zhang, Y.Y.; Liu, Y.M.; Yang, L.M.; Zhuang, M.; Sun, P.T. The floral and seed setting characteristics in two types of male sterile lines of cabbage (*Brassica oleracea* L. var. *capitata*). *Acta Hortic. Sin.* **2011**, *38*, 61–68.
5. Shu, J.S.; Liu, Y.M.; Li, Z.S.; Zhang, L.L.; Fang, Z.Y.; Yang, L.M.; Zhuang, M.; Zhang, Y.Y.; Li, Z.S.; Sun, P.T. Study on the floral characteristics and structure in two types of male sterile lines of broccoli (*Brassica oleracea* var. *italica*). *J. Plant. Genet. Resour.* **2014**, *15*, 113–119.
6. Jia, J.; Zhang, L.G. mRNA differential display and EST sequence analysis of aborted bud and normal bud in radish (*Raphanus sativus*). *Acta Agric. Nucl. Sin.* **2008**, *22*, 426–431.
7. Zhang, L.G.; Jia, J.; Zhang, S.L.; Zhang, Y. cDNA-AFLP differential expression analysis of genes related with aborting bud in Chinese cabbage. *J. Agric. Biotechnol.* **2010**, *18*, 489–492.
8. Wang, Q.; Liu, Z.; Feng, H. cDNA-AFLP differential expression analysis of genes about bud aborting in genetic male sterile line of *Brassica campestris* L. ssp. *chinensis* (L.) *Makino* var. *rosularis* Tsen *et* Lee. *Mol. Plant. Breed.* **2014**, *12*, 118–126.
9. Kensler, T.W.; Chen, J.G.; Egner, P.A.; Fahey, J.W.; Jacobson, L.P.; Stephenson, K.K.; Ye, L.X.; Coady, J.L.; Wang, J.B.; Wu, Y.; et al. Effects of glucosinolate-rich broccoli sprouts on urinary levels of aflatoxin-DNA adducts and phenanthrene tetraols in a randomized clinical trial in He Zuo township, Qidong, People's Republic of China. *Cancer Epidemiol. Biomark. Prev.* **2005**, *14*, 2605–2613. [CrossRef] [PubMed]
10. Canene-Adams, K.; Lindshield, B.L.; Wang, S.; Jeffery, E.H.; Clinton, S.K.; Erdman, J.W., Jr. Combinations of

tomato and broccoli enhance antitumor activity in dunning r3327-h prostate adenocarcinomas. *Cancer Res.* **2007**, *67*, 836–843. [CrossRef] [PubMed]
11. Jeffery, E.H.; Araya, M. Physiological effects of broccoli consumption. *Phytochem. Rev.* **2009**, *8*, 283–298. [CrossRef]
12. Götz, S.; García-Gómez, J.M.; Terol, J.; Williams, T.D.; Nagaraj, S.H.; Nueda, M.J.; Robles, M.; Talon, M.; Dopazo, J.; Conesa, A. High-throughput functional annotation and data mining with the Blast2GO suite. *Nucleic Acids Res.* **2008**, *36*, 3420–3435.
13. Thomas, S.G.; Franklin-Tong, V.E. Self-incompatibility triggers programmed cell death in Papaver pollen. *Nature* **2004**, *429*, 305–309. [CrossRef] [PubMed]
14. Reape, T.J.; Molony, E.M.; McCabe, P.F. Programmed cell death in plants: Distinguishing between different modes. *J. Exp. Bot.* **2008**, *59*, 435–444. [CrossRef] [PubMed]
15. Rybaczek, D.; Musiałek, M.W.; Balcerczyk, A. Caffeine-induced premature chromosome condensation results in the apoptosis-like programmed cell death in root meristems of *Vicia faba*. *PLoS ONE* **2015**, *10*, e0142307. [CrossRef] [PubMed]
16. Bosch, M.; Franklin-Tong, V.E. Temporal and spatial activation of caspase-like enzymes induced by self-incompatibility in Papaver pollen. *Proc. Natl. Acad. Sci. USA* **2007**, *104*, 18327–18332. [CrossRef] [PubMed]
17. Wilkins, K.A.; Bancroft, J.; Bosch, M.; Ings, J.; Smirnoff, N.; Franklin-Tong, V.E. Reactive oxygen species and nitric oxide mediate actin reorganization and programmed cell death in the self-incompatibility response of papaver. *Plant Physiol.* **2011**, *156*, 404–416. [CrossRef] [PubMed]
18. Zhang, F.J.; Wang, Z.Q.; Dong, W.; Sun, C.Q.; Wang, H.B.; Song, A.P.; He, L.Z.; Fang, W.M.; Chen, F.D.; Teng, N.J. Transcriptomic and proteomic analysis reveals mechanisms of embryo abortion during chrysanthemum cross breeding. *Sci. Rep.* **2014**, *4*, 6536. [CrossRef] [PubMed]
19. Fagundes, D.; Bohn, B.; Cabreira, C.; Leipelt, F.; Dias, N.; Bodanese-Zanettini, M.H.; Cagliari, A. Caspases in plants: Metacaspase gene family in plant stress responses. *Funct. Integr. Genom.* **2015**, *15*, 639–649. [CrossRef] [PubMed]
20. Gonzalez-Carranza, Z.H.; Elliott, K.A.; Roberts, J.A. Expression of polygalacturonases and evidence to support their role during cell separation processes in *Arabidopsis thaliana*. *J. Exp. Bot.* **2007**, *58*, 3719–3730. [CrossRef] [PubMed]
21. Ogawa, M.; Kay, P.; Wilson, S.; Swain, S.M. ARABIDOPSIS DEHISCENCE ZONE POLYGALACTURONASE1 (ADPG1), ADPG2, and QUARTET2 are polygalacturonases required for cell separation during reproductive development in *Arabidopsis*. *Plant Cell* **2009**, *21*, 216–233. [CrossRef] [PubMed]
22. Libertini, E.; Li, Y.; McQueen-Mason, S.J. Phylogenetic analysis of the plant endo-β-1, 4-glucanase gene family. *J. Mol. Evol.* **2004**, *58*, 506–515. [CrossRef] [PubMed]
23. Kawano, T.; Kawano, T.; Kadono, T.; Fumoto, K.; Lapeyrie, F.; Kuse, M.; Isobe, M.; Furuichi, T.; Muto, S. Aluminum as a specific inhibitor of plant TPC1 Ca^{2+} channels. *Biochem. Biophys. Res. Commun.* **2004**, *324*, 40–45. [CrossRef] [PubMed]
24. Peiter, E.; Maathuis, F.J.; Mills, L.N.; Knight, H.; Pelloux, J.; Hetherington, A.M.; Sanders, D. The vacuolar Ca^{2+}-activated channel TPC1 regulates germination and stomatal movement. *Nature* **2005**, *434*, 404–408. [CrossRef] [PubMed]
25. Zhao, J.W.; Wang, J.L.; An, L.L.; Doerge, R.W.; Chen, Z.J.; Grau, C.R.; Meng, J.L.; Osborn, T.C. Analysis of gene expression profiles in response to Sclerotinia sclerotiorum in *Brassica napus*. *Planta* **2007**, *227*, 13–24. [CrossRef] [PubMed]
26. Libault, M.; Wan, J.; Czechowski, T.; Udvardi, M.; Stacey, G. Identification of 118 *Arabidopsis* transcription factor and 30 ubiquitin-ligase genes responding to chitin, a plant-defense elicitor. *Mol. Plant Microbe Interact.* **2007**, *20*, 900–911. [CrossRef] [PubMed]
27. Seo, D.H.; Ryu, M.Y.; Jammes, F.; Hwang, J.H.; Turek, M.; Kang, B.G.; Kwak, J.M.; Kim, W.T. Roles of four *Arabidopsis* U-Box E3 ubiquitin ligases in negative regulation of abscisic acid-mediated drought stress responses. *Plant Physiol.* **2012**, *160*, 556–568. [CrossRef] [PubMed]
28. Wan, J.R.; Zhang, X.C.; Neece, D.; Ramonell, K.M.; Clough, S.; Kim, S.Y.; Stacey, M.G.; Stacey, G. A LysM receptor-like kinase plays a critical role in chitin signaling and fungal resistance in *Arabidopsis*. *Plant Cell* **2008** *20*, 471–481. [CrossRef] [PubMed]

29. Brotmana, Y.; Landau, U.; Pnini, S.; Lisec, J.; Balazadeh, S.; Mueller-Roeber, B.; Zilberstein, A.; Willmitzer, L.; Chet, I.; Viterbo, A. The LysM receptor-like kinase LysM RLK1 is required to activate defense and abiotic-stress responses induced by overexpression of fungal chitinases in *Arabidopsis* plants. *Mol. Plant.* **2012**, *5*, 1113–1124. [CrossRef] [PubMed]
30. Tanaka, K.; Nguyen, C.T.; Liang, Y.; Cao, Y.; Stacey, G. Role of LysM receptors in chitin-triggered plant innate immunity. *Plant. Signal. Behav.* **2013**, *8*, e22598. [CrossRef] [PubMed]
31. Garcia, M.J.; Lucena, C.; Romera, F.J.; Alcantara, E.; Perez-Vicente, R. Ethylene and nitric oxide involvement in the up-regulation of key genes related to iron acquisition and homeostasis in *Arabidopsis*. *J. Exp. Bot.* **2010**, *61*, 3885–3899. [CrossRef] [PubMed]
32. Xu, Z.S.; Chen, M.; Li, L.C.; Ma, Y.Z. Functions and application of the AP2/ERF transcription factor family in crop improvement. *J. Integr. Plant Biol.* **2011**, *53*, 570–585. [CrossRef] [PubMed]
33. Chen, M.K.; Shpak, E.D. *ERECTA* family genes regulate development of cotyledons during embryogenesis. *FEBS Lett.* **2014**, *588*, 3912–3917. [CrossRef] [PubMed]
34. Nakano, T.; Fujisawa, M.; Shima, Y.; Ito, Y. The AP2/ERF transcription factor SlERF52 functions in flower pedicel abscission in tomato. *J. Exp. Bot.* **2014**, *65*, 3111–3119. [CrossRef] [PubMed]
35. Ito, Y.; Nakanot, T. Development and regulation of pedicel abscission in tomato. *Front. Plant Sci.* **2015**, *6*, 442. [CrossRef] [PubMed]
36. Jisha, V.; Dampanaboina, L.; Vadassery, J.; Mithöfer, A.; Kappara, S.; Ramanan, R. Overexpression of an AP2/ERF type transcription factor *OsEREBP1* confers biotic and abiotic stress tolerance in rice. *PLoS ONE* **2015**, *10*, e0127831. [CrossRef] [PubMed]
37. Mishra, S.; Phukan, U.J.; Tripathi, V.; Singh, D.K.; Luqman, S.; Shukla, R.K. PsAP2 an AP2/ERF family transcription factor from *Papaver somniferum* enhances abiotic and biotic stress tolerance in transgenic tobacco. *Plant Mol. Biol.* **2015**, *89*, 173–186. [CrossRef] [PubMed]
38. Babitha, K.C.; Vemanna, R.S.; Nataraja, K.N.; Udayakumar, M. Overexpression of *EcbHLH57* transcription factor from *Eleusine coracana* L. in tobacco confers tolerance to salt, oxidative and drought stress. *PLoS ONE* **2015**, *10*, e0137098. [CrossRef] [PubMed]
39. Heyman, J.; Cools, T.; Vandenbussche, F.; Heyndrickx, K.S.; Van Leene, J.; Vercauteren, I.; Vanderauwera, S.; Vandepoele, K.; De Jaeger, G.; Van Der, S.D.; et al. ERF115 controls root quiescent center cell division and stem cell replenishment. *Science* **2013**, *342*, 860–863. [CrossRef] [PubMed]
40. Heim, M.A.; Jakoby, M.; Werber, M.; Martin, C.; Weisshaar, B.; Bailey, P.C. The basic helix–loop–helix transcription factor family in plants: A genome-wide study of protein structure and functional diversity. *Mol. Biol. Evol.* **2003**, *20*, 735–747. [CrossRef] [PubMed]
41. Borg, M.; Brownfield, L.; Khatab, H.; Sidorova, A.; Lingaya, M.; Twel, D. The R2R3 MYB transcription factor DUO1 activates a male germline-specific regulon essential for sperm cell differentiation in *Arabidopsis*. *Plant Cell* **2011**, *23*, 534–549. [CrossRef] [PubMed]
42. Aeschbacher, R.A.; Schrott, M.; Potrykus, I.; Saul, M.W. Isolation and molecular characterization of PosF21, an *Arabidopsis thaliana* gene which shows characteristics of a b-Zip class transcription factor. *Plant J.* **1991**, *1*, 303–316. [CrossRef] [PubMed]
43. Yin, Y.H.; Zhu, Q.; Dai, S.H.; Lamb, C.; Beachy, R.N. RF2a, a bZIP transcriptional activator of the phloem-specific rice tungro bacilliform virus promoter, functions in vascular development. *EMBO J.* **1997**, *16*, 5247–5259. [CrossRef] [PubMed]
44. Shen, H.; Cao, K.; Wang, X. A conserved proline residue in the leucine zipper region of AtbZIP34 and AtbZIP61 in *Arabidopsis thaliana* interferes with the formation of homodimer. *Biochem. Biophys. Res. Commun.* **2007**, *362*, 425–430. [CrossRef] [PubMed]
45. Gibalová, A.; Renák, D.; Matczuk, K.; Dupl'áková, N.; Cháb, D.; Twell, D.; Hongys, D. AtbZIP34 is required for *Arabidopsis* pollen wall patterning and the control of several metabolic pathways in developing pollen. *Plant Mol. Biol.* **2009**, *70*, 581–601.
46. Yang, P.; Han, J.; Huang, J. Transcriptome sequencing and de novo analysis of cytoplasmic male sterility and maintenance in JA-CMS cotton. *PLoS ONE* **2014**, *9*, e112320. [CrossRef] [PubMed]
47. Liu, Z.; Shi, X.; Li, S.; Hu, G.; Zhang, L.; Song, X. Tapetal-Delayed programmed cell death (PCD) and oxidative stress-induced male sterility of *Aegilops uniaristata* cytoplasm in wheat. *Int. J. Mol. Sci.* **2018**, *19*, 1708. [CrossRef] [PubMed]
48. Li, J.; Han, S.; Ding, X.; He, T.; Dai, J.; Yang, S.; Gai, J. Comparative transcriptome analysis between the

cytoplasmic male sterile line NJCMS1A and its maintainer NJCMS1B in soybean (*Glycine max* (L.) Merr.). *PLoS ONE* **2015**, *10*, e0126771. [CrossRef] [PubMed]

49. Du, K.; Liu, Q.; Wu, X.; Jiang, J.; Wu, J.; Fang, Y.; Li, A.; Wang, Y. Morphological structure and transcriptome comparison of the cytoplasmic male sterility line in Brassica napus (SaNa-1A) derived from somatic hybridization and its maintainer line SaNa-1B. *Front. Plant Sci.* **2016**, *7*, 1313. [CrossRef] [PubMed]
50. Wang, S.; Wang, C.; Zhang, X.X.; Chen, X.; Liu, J.J.; Jia, X.F.; Jia, S.Q. Transcriptome de novo assembly and analysis of differentially expressed genes related to cytoplasmic male sterility in cabbage. *Plant Physiol. Biochem.* **2016**, *105*, 224–232. [CrossRef] [PubMed]
51. Liu, Q.; Lan, Y.; Wen, C.; Zhao, H.; Wang, J.; Wang, Y. Transcriptome sequencing analyses between the cytoplasmic male sterile line and its maintainer line in Welsh onion (*Allium fistulosum* L.). *Int. J. Mol. Sci.* **2016**, *17*, 1058. [CrossRef] [PubMed]
52. Zhang, G.; Ye, J.; Jia, Y.; Zhang, L.; Song, X. ITRAQ-based proteomics analyses of sterile/fertile anthers from a thermo-sensitive cytoplasmic male-sterile wheat with *Aegilops kotschyi* cytoplasm. *Int. J. Mol. Sci.* **2018**, *19*, 1344. [CrossRef] [PubMed]
53. Wu, Z.; Cheng, J.; Qin, C.; Hu, Z.; Yin, C.; Hu, K. Differential proteomic analysis of anthers between cytoplasmic male sterile and maintainer lines in *Capsicum annuum* L. *Int. J. Mol. Sci.* **2013**, *14*, 22982–22996. [CrossRef] [PubMed]
54. Mei, S.; Liu, T.; Wang, Z. Comparative transcriptome profile of the cytoplasmic male sterile and fertile floral buds of radish (*Raphanus sativus* L.). *Int. J. Mol. Sci.* **2016**, *17*, 42. [CrossRef] [PubMed]
55. Shu, J.; Liu, Y.; Li, Z.; Zhang, L.; Fang, Z.; Yang, L.; Zhuang, M.; Zhang, Y.; Lv, H. Organelle simple sequence repeat markers help to distinguish carpelloid stamen and normal cytoplasmic male sterile sources in broccoli. *PLoS ONE* **2015**, *10*, e0138750. [CrossRef] [PubMed]
56. Grabherr, M.G.; Haas, B.J.; Yassour, M.; Levin, J.Z.; Thompson, D.A.; Amit, I.; Adiconis, X.; Fan, L.; Raychowdhury, R.; Zeng, Q.; et al. Full-length transcriptome assembly from RNA-Seq data without a reference genome. *Nat. Biotechnol.* **2011**, *29*, 644–652. [CrossRef] [PubMed]
57. Zhang, Z.; Wood, W.I. A profile hidden Markov model for signal peptides generated by HMMER. *Bioinformatics* **2003**, *19*, 307–308. [CrossRef] [PubMed]
58. Ashburner, M.; Ball, C.A.; Blake, J.A.; Botstein, D.; Butler, H.; Cherry, J.M.; Davis, A.P.; Dolinski, K.; Dwight, S.S.; Eppig, J.T.; et al. Gene Ontology: Tool for the unification of biology. *Nat. Genet.* **2000**, *25*, 25–29. [CrossRef] [PubMed]
59. Kanehisa, M.; Araki, M.; Goto, S.; Hattori, M.; Hirakawa, M.; Itoh, M.; Katayama, T.; Kawashima, S.; Okuda, S.; Tokimatsu, T.; et al. KEGG for linking genomes to life and the environment. *Nucleic Acids Res.* **2008**, *36*, D480–D484. [CrossRef] [PubMed]
60. Mao, X.; Cai, T.; Olyarchuk, J.G.; Wei, L. Automated genome annotation and pathway identification using the KEGG Orthology (KO) as a controlled vocabulary. *Bioinformatics* **2005**, *21*, 3787–3793. [CrossRef] [PubMed]
61. Li, B.; Dewey, C. RSEM: Accurate transcript quantification from RNA-Seq data with or without a reference genome. *BMC Bioinform.* **2011**, *12*, 323. [CrossRef] [PubMed]
62. Trapnell, C.; Williams, B.A.; Pertea, G.; Mortazavi, A.; Kwan, G.; van Baren, M.J.; Salzberg, S.L.; Wold, B.J.; Pachter, L. Transcript assembly and quantification by RNA-Seq reveals unannotated transcripts and isoform switching during cell differentiation. *Nat. Biotechnol.* **2010**, *28*, 511–515. [CrossRef] [PubMed]
63. Storey, J.D.; Tibshirani, R. Statistical significance for genomewide studies. *Proc. Nat. Acad. Sci. USA* **2003**, *100*, 9440–9445. [CrossRef] [PubMed]
64. Benjamini, Y.; Hochberg, Y. Controlling the false discovery rate: A practical and powerful approach to multiple testing. *J. R. Stat. Soc. Ser. B Methodol.* **1995**, *57*, 289–300.
65. Wallenius, K.T. Biased Sampling: The Non-Central Hypegeometric Probability Distribution. Ph.D. Thesis, Stanford University, Stanford, CA, USA, 1963.
66. Young, M.D.; Wakefield, M.J.; Smyth, G.K.; Oshlack, A. Gene ontology analysis for RNA-Seq: Accounting for selection bias. *Genome Biol.* **2010**, *11*, R14. [CrossRef] [PubMed]
67. Gómez-Lobato, M.E.; Hasperuéb, J.H.; Civelloa, P.M.; Chaves, A.R.; Martínez, G.A. Effect of 1-MCP on the expression of chlorophyll degrading genes during senescence of broccoli (*Brassica oleracea* L.). *Sci. Hortic* **2012**, *144*, 208–211.
68. Livak, K.J.; Schmittgen, T.D. Analysis of relative gene expression data using real-time quantitative PCR and the $2^{-\Delta\Delta CT}$ Method. *Methods* **2001**, *25*, 402–408. [CrossRef] [PubMed]

8

DNA Methylation Analysis of Dormancy Release in Almond (*Prunus dulcis*) Flower Buds using Epi-Genotyping by Sequencing

Ángela S. Prudencio [1,†], Olaf Werner [2,†], Pedro J. Martínez-García [1], Federico Dicenta [1], Rosa M. Ros [2] and Pedro Martínez-Gómez [1,*]

1 Department of Plant Breeding, CEBAS-CSIC, P.O. Box 164, Espinardo, 30100 Murcia, Spain; asanchez@cebas.csic.es (A.S.P.); pjmartinez@cebas.csic.es (P.J.M.-G.); fdicenta@cebas.csic.es (F.D.)

2 Department of Plant Biology, Faculty of Biology, University of Murcia, Espinardo, 30100 Murcia, Spain; werner@um.es (O.W.); rmros@um.es (R.M.R.)

* Correspondence: pmartinez@cebas.csic.es

† These authors contributed equally to this work.

Abstract: DNA methylation and histone post-translational modifications have been described as epigenetic regulation mechanisms involved in developmental transitions in plants, including seasonal changes in fruit trees. In species like almond (*Prunus dulcis* (Mill.) D.A: Webb), prolonged exposure to cold temperatures is required for dormancy release and flowering. Aiming to identify genomic regions with differential methylation states in response to chill accumulation, we carried out Illumina reduced-representation genome sequencing on bisulfite-treated DNA from floral buds. To do this, we analyzed almond genotypes with different chilling requirements and flowering times both before and after dormancy release for two consecutive years. The study was performed using epi-Genotyping by Sequencing (epi-GBS). A total of 7317 fragments were sequenced and the samples compared. Out of these fragments, 677 were identified as differentially methylated between the almond genotypes. Mapping these fragments using the *Prunus persica* (L.) Batsch v.2 genome as reference provided information about coding regions linked to early and late flowering methylation markers. Additionally, the methylation state of ten gene-coding sequences was found to be linked to the dormancy release process.

Keywords: *Prunus*; flowering; bisulfite sequencing; genomics; epigenetics; breeding

---

1. Introduction

The almond tree (*Prunus dulcis* (Mill.) D.A. Webb), like the rest of the *Prunus* species, is a deciduous fruit tree that undergoes a cyclical process of flowering, sprouting, development, and winter rest, called dormancy. The dormancy state protects the plant from potential damage from cold weather during the winter [1,2]. The dormancy period is overcome when the tree accumulates sufficient chilling hours (the chilling requirement). After dormancy release, the tree is able to sprout and flower under favorable climatic conditions [3,4]. The study of molecular mechanisms leading to dormancy release and flowering is of great interest for almond breeding programs aiming to adapt new cultivars to specific growing areas [5,6]. The dormancy release process involves sensing environmental cues (such as temperature), signal transduction, and gene expression regulation to establish a suitable response according to the stimuli received [7,8]. Transcription reprogramming leading to dormancy release may thus be mediated by epigenetic mechanisms [9,10].

Epigenetics are chemical modifications affecting DNA or structural proteins (histones) within the chromatin. Two types of epigenetic modifications have been described: DNA methylation

(5′ Methylated Cytosine, 5mC) in plants and histone Post-Translational Modifications (PTMs), which include the acetylation and methylation of histones [11,12].

Epigenetic changes are part of the transcriptional regulation machinery of genomes. The dynamic but heritable character of such modifications makes them interesting regulators mediating adaptive responses to environmental changes, such as seasonal cycles, and in the long term, climate change [13]. DNA methylation is associated with cell status stability and regulation of expression. DNA methylation occurs in three sequence contexts: CG and CHG, which are found in promoter and coding regions, and CHH (where H = A, C or T), found in non-coding regions and transposable elements (TEs) [14].

Several works have described the role of epigenetics in the regulation of dormancy in deciduous plant species. Santamaría et al. [15], for instance, described a methylation decrease concomitant with H4 deacetylation and the progress of dormancy release in *Castanea sativa* Mill. In peach (*Prunus persica*), de la Fuente et al. [16] identified a genome-wide pattern of the PTM Trymethylation of Histone 3 on Lis (K) residue 27 (H3K27me3) during bud dormancy release, and Lloret et al. [17] found a relationship between gene expression, PTMs, and sorbitol synthesis during bud dormancy progression and release. Rothkegel et al. [18] showed that DNA methylation is one of the mechanisms participating in the regulation of MADS-box (MCM1-AGAMOUS-DEFICIENS-SRF) genes controlling bud dormancy in sweet cherry (*Prunus avium* L.). In apple (*Malus domestica* (Suckow) Borkh.), genome methylation patterns have been linked to chilling acquisition during dormancy [19]. In the case of almond, preliminary results from the transcriptome sequencing of non-dormant and dormant flower buds showed differential expression in a DNA methyltransferase gene and in the *S-ADENOSYL METHIONINE SYNTHETASE* gene responsible for the synthesis of the molecule SAM (S-adenosyl methionine), which donates the methyl group to the DNA molecule [20]. In addition, DNA methylation phenomena have also been associated with floral self-incompatibility [21] and with bud falling phenomena [22] in this species.

Genome-wide analysis of DNA methylation can be done by bisulfite sequencing, which uses Next Generation Sequencing (NGS) to analyze digested and bisulfite-treated DNA samples. The epi-Genotyping by Sequencing (epiGBS) technique was developed to represent a small part of the genome for cost-effective exploration and comparative analysis of DNA methylation and genetic variation in hundreds of de novo samples. Furthermore, this method makes it possible to genotype samples without a prior reference genome [23].

The objective of this work was to analyze the DNA methylation status of almond flower buds using epi-GBS for the first time. For this purpose, we evaluated dormant and non-dormant flower buds from two almond genotypes with different chilling requirements and flowering times using the epi-GBS protocol.

2. Results

2.1. Evaluation of the Quality of the Epi-GBS Analysis

We sequenced 9518 fragments (about a 1244 kb size) and identified 7317 methylated or unmethylated fragments. Furthermore, we were able to reconstruct the original sequence of 4377 fragments. The total length of the "mock genome" obtained by merging the reconstructed fragments was 662,458 bp. Regarding the quality of this epi-GBS analysis, the absence of a secondary peak towards the right of the read coverage histograms shows that the data do not suffer from PCR duplication bias in either year (Figure 1; data from the flower buds sampled in 2015–2016). These read coverage results show the uniformity of the reads and correct PCR amplification (good quality) around the whole genome in both contexts and seasons of study.

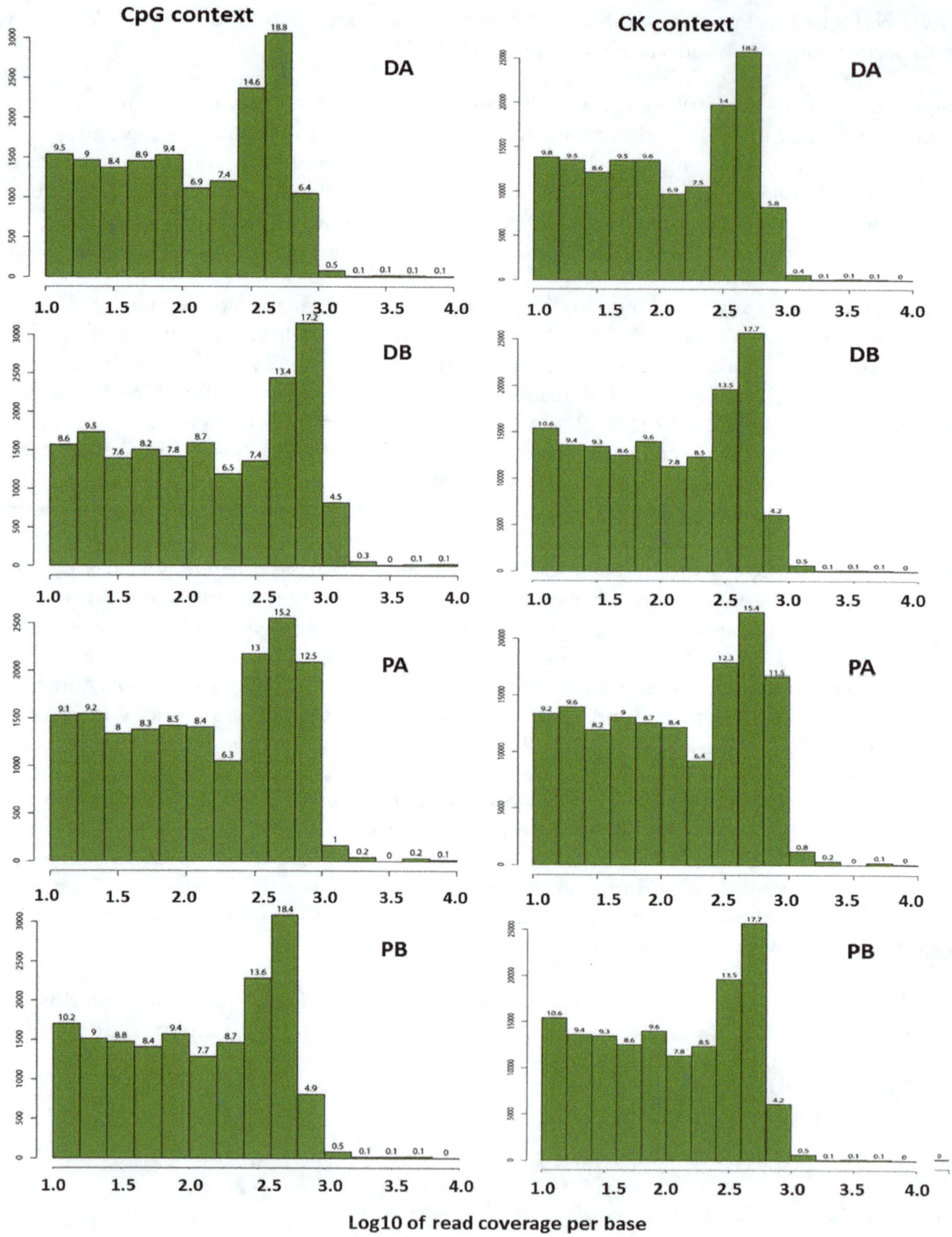

Figure 1. Read coverage of the samples tested in the CpG and CK (including CHG and CHH) contexts during the 2015–2016 season. "D" = 'Desmayo Largueta', "P" = 'Penta', "A" = dormant bud stage, "B" = non-dormant bud stage.

The histograms of CpG methylation showed that roughly 70% to 75% of the cytosine positions in the CpG context of the mock genome were unmethylated, around 10% of the positions were completely methylated and the remaining positions were partially methylated to varying degrees in both seasons of study (Figure 2; data from 2015–2016 season).

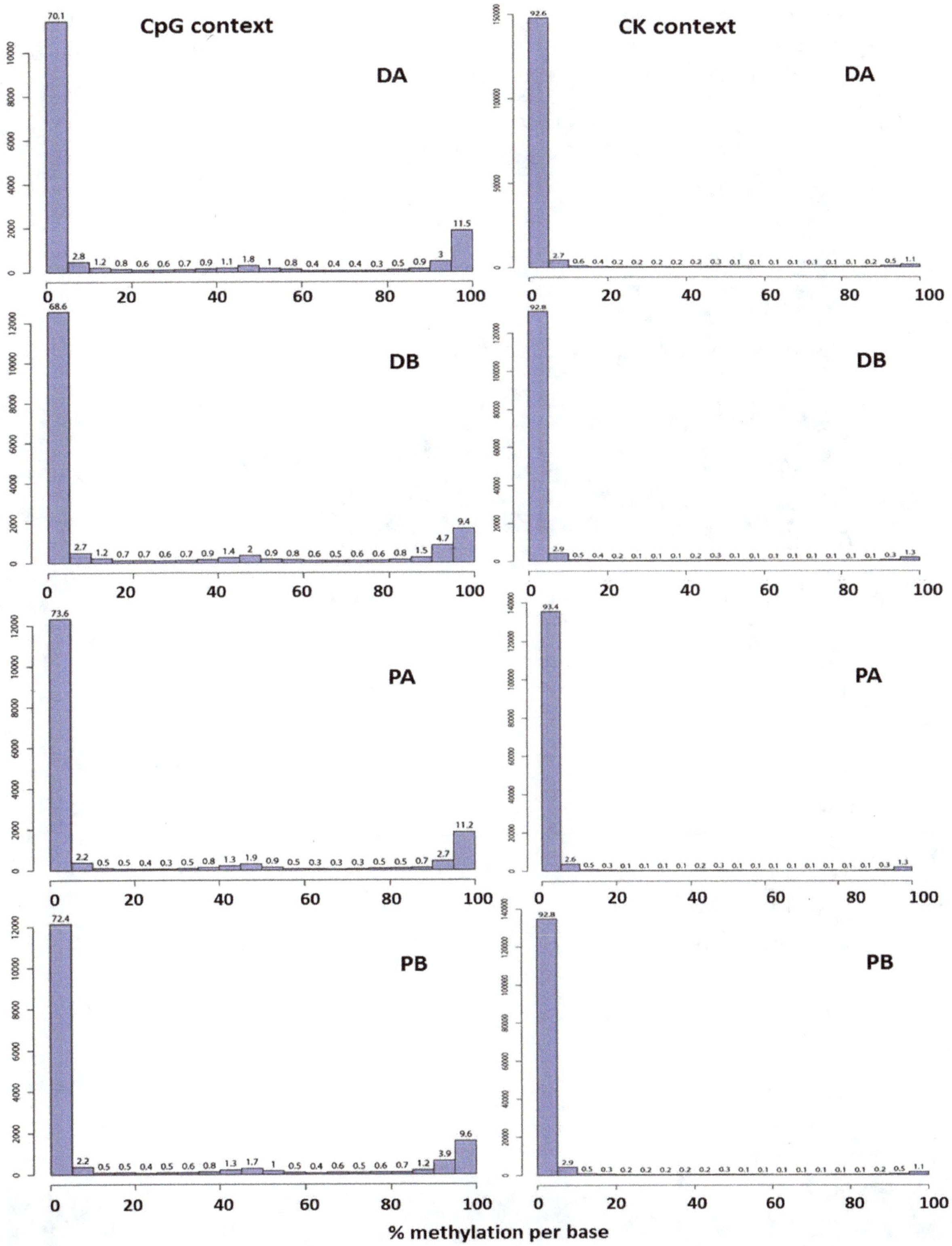

Figure 2. Percentage of DNA methylation of the samples tested in the CpG and CK (including CHG and CHH) contexts during the 2015–2016 season. "D" = 'Desmayo Largueta', "P" = 'Penta', "A" = dormant bud stage, "B" = non-dormant bud stage.

Furthermore, the correlation analyses clearly show that samples of the same variety cluster together independently of the developmental stage. The Pearson's correlation coefficient was constantly 0.99 in comparisons within each variety and in the range of 0.84 to 0.85 in comparisons between samples of different varieties (Figure 3A). These results were also corroborated by clustering analysis, in which samples belonging to the same variety were close together while the two varieties were separated by long branches (Figure 3B). The DNA methylation pattern is generally variety-dependent rather than dormancy-dependent.

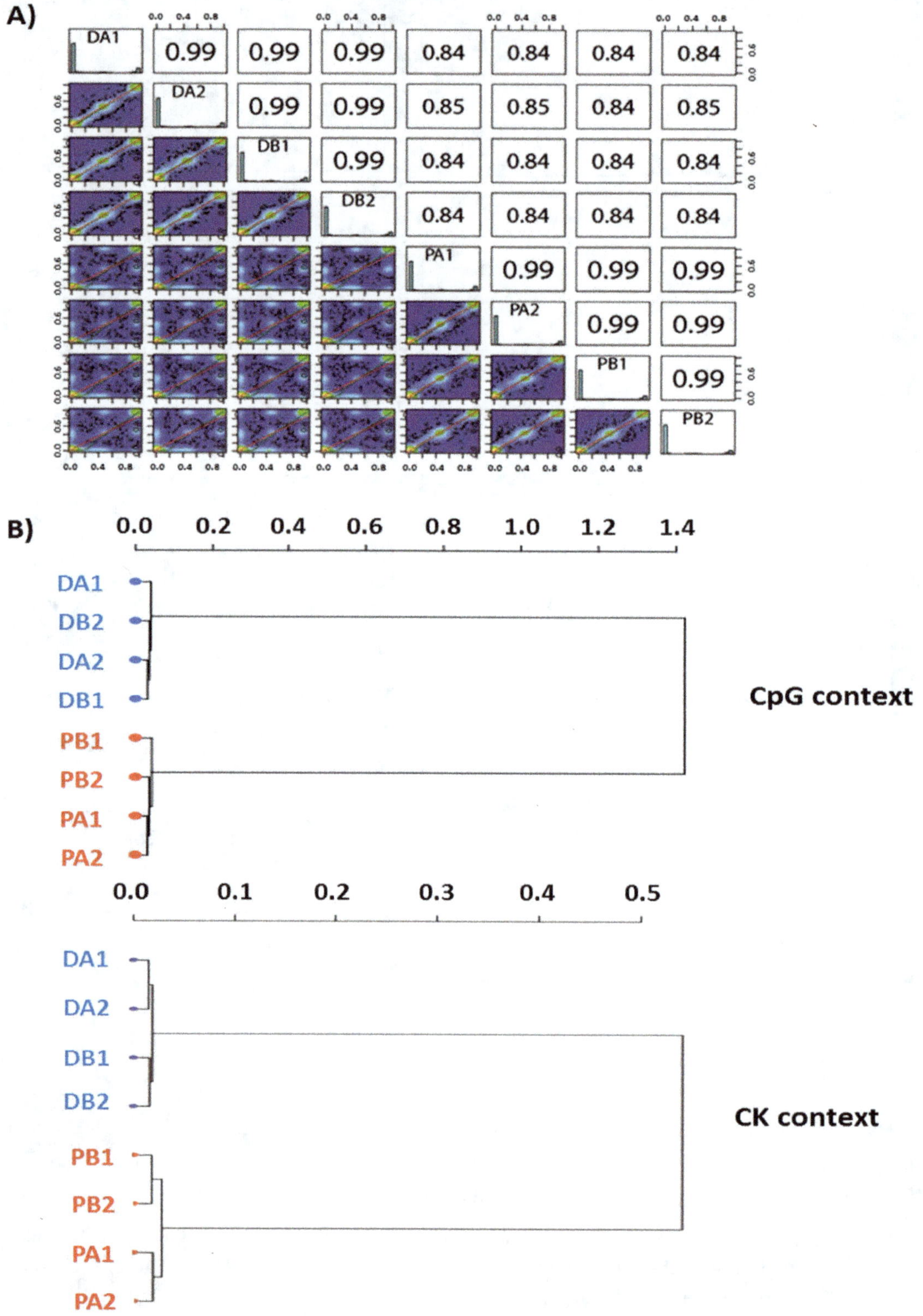

Figure 3. (**A**) Almond correlations and (**B**) clustering analysis of the methylated fragments in both CpG and CK contexts. "D" = 'Desmayo Largueta', "P" = 'Penta', "A" = dormant bud stage, "B"=non-dormant bud stage. "1" samples are from the 2015–2016 season and "2" samples are from the 2016–2017 season.

2.2. Differentyally Methylated Genes Detected

Quantitative analysis showed that 7317 different fragments were methylated in at least one sample: 5109 'Cs' were methylated in 'Desmayo Largueta' A samples; 5089 'Cs' were methylated in 'Desmayo Largueta' B samples; 4955 'Cs' were methylated in 'Penta' A samples; and 5003 'Cs' were methylated in 'Penta' B samples.

Table 1 shows that the number of differentially methylated fragments (DMFs) detected was variable depending on the comparison performed. Furthermore, a total of 677 DMFs were found between 'Desmayo Largueta' and 'Penta' genotype samples in all stages analyzed. However, when comparing dormancy state samples (A and B), 23 DMFs were found between 'Desmayo Largueta' stage A and 'Desmayo Largueta' stage B samples and 48 DMFs were found between 'Penta' stage A samples and 'Penta' stage B samples. Of those DMFs, ten were common between 'Desmayo Largueta' and 'Penta' in the A to B stage comparison. The DMFs were divided into hypermethylated or hypomethylated categories using 'Penta' or 'stage B' samples as the reference. The fragment sequences are included in Table S1.

Table 1. Number of differentially methylated fragments (DMFs) detected by epi-Genotyping by Sequencing (epi-GBS) according to sample comparisons.

Differentially Methylated Fragments		
D–P genotype comparison		
Hypo (<)	Hyper (>)	Stages
307	370	A and B
677 DMFs		
A–B stage comparison		
Hypo (<)	Hyper (>)	Genotype
3	20	D
21	27	P
3	7	D and P
10 DMFs		

"D" = 'Desmayo Largueta', "P" = 'Penta'; "A" = dormant bud stage, "B" = non-dormant bud stage.

More than 99% of the identified DMFs were mapped on the *Prunus persica* v2.1 genome (Table S2), and those located between 2 kb upstream and 1 kb downstream from the gene coding sequences were selected for subsequent annotation analysis. The number of differentially methylated genes (DMGs) thus identified is shown in Tables 2 and 3.

Table 2. The number of differentially methylated genes (DMGs) identified from sequenced fragments mapping onto the *Prunus persica* genome (v2.1).

Differentially Methylated Genes			
Methylation State	**Gene Position**	**Gene Hits**	**Genes Identified**
Hypermethylated	Upstream	36	
	Inside	291	
	Downstream	134	
Total		461	423
Hypomethylated	Upstream	19	
	Inside	201	
	Downstream	80	
Total		300	281
Equally-methylated	Upstream	6	
	Inside	41	
	Downstream	8	
Total		55	27
Total DMGs		816	731

The methylation state refers to the number of 5′ Methylated Cytosines (5mCs) in 'Desmayo Largueta' samples compared to 'Penta' samples. The category "equally methylated" refers to genes whose number of 5mCs was the same between samples but in which the 5mCs were located in different fragment positions. The gene position is based on gene orientation with respect to the fragment-mapping region ("upstream", "inside", and "downstream").

DMGs were classified according to their position with respect to the fragment-mapping region. The most frequently mapped fragments were those within gene regions ("inside" DMGs) followed by the 5′regulatory regions ("downstream" DMGs) and, finally, in the 3′ regions ("upstream" DMGs).

Data shown in Table 2 indicate that DMGs were found as hypermethylated in 'Desmayo Largueta' samples (in both the A and B stages) to a greater extent than in 'Penta' samples (423).

Table 3. The number of differentially methylated genes between the A and B dormancy states of flower buds, identified from sequenced fragments mapped to the *P. persica* genome (v2.1).

Methylation State	Gene Position	Differentially Methylated Genes		
		'Desmayo Largueta'	'Penta'	Common
Hyper-methylated	Upstream	7	3	1
	Inside	5	2	2
	Downstream	5	9	4
Total		17	14	7
Hypo-methylated	Upstream	1	-	-
Total		18	14	7

The methylation state refers to the number of 5mCs in stage A (dormant buds) samples with respect to the number found in stage B (non-dormant buds) samples. The gene position is based on gene orientation with respect to the fragment-mapping region.

We found enriched hypermethylated genes in 'Desmayo Largueta' flower bud samples in the following processes related to primary metabolism in the "biological function" GO (Gene Ontology) category: amino-acid and carbohydrate synthesis and protein phosphorylation (Figure S1). ATP binding and protein kinase and phosphatase activity were the two main "molecular function" GO terms found (Figure 4).

In a gene-level analysis, the following candidate genes appeared as hypermethylated in 'Desmayo Largueta' flower bud samples: genes related to transcription regulation, including transcription factors (Prupe.1G395600, Prupe.5G088700, Prupe.6G343100); genes linked to RNA-mediated silencing (Prupe.7G221200); genes linked to chromatin remodelling (Prupe.8G221300, and LATE ELONGATED HYPOCOTYL (LHY), encoded by Prupe.2G200400); and, especially, genes involved in the auxin response (Prupe.1G000200, Prupe.1G067400, Prupe.7G048400) and AUXIN RESPONSE FACTOR (ARF) signal transduction (Prupe.3G010900, Prupe.5G217700, Prupe.7G228800). We also identified DNA repair proteins, such as those encoded by Prupe.1G510000, Prupe.2G013900, Prupe.3G029600, Prupe.3G16000 and Prupe.5G066100. Finally, proteins participating in oxidoreduction processes, such as LATE EMBRYOGENESIS ABUNDANT (LEA) proteins encoded by Prupe.4G026900 and Prupe.4G02700, also appeared as hypermethylated in 'Desmayo Largueta' flower bud samples (Table S3).

Regarding the hypomethylated genes in the 'Desmayo Largueta' samples, we found cellular protein localization within the "biological function" GO category (Figure S2). ATP-coupled transmembrane transport and ATP-binding activity, on the other hand, appeared in the "molecular function" GO category (Figure 5).

We were able to identify a wide range of DNA-binding proteins encoded by the hypomethylated genes in the 'Desmayo Largueta' samples: histone methyltransferases (encoded by Prupe.1G050800 and Prupe.7G271600); the transcription factor NUCLEAR FACTOR-Y (NF-Y) (encoded by Prupe.2G47600); DNA topoisomerases (encoded by Prupe.1G173400 and Prupe.1G173500); and FAR-RED IMPAIRED RESPONSE 1 (FAR1) (Prupe.1G196400). Interestingly, a single gene coding for a HYDROPHOBIC SEED PROTEIN (HSP) also appeared as hypomethylated in the 'Desmayo Largueta' samples.

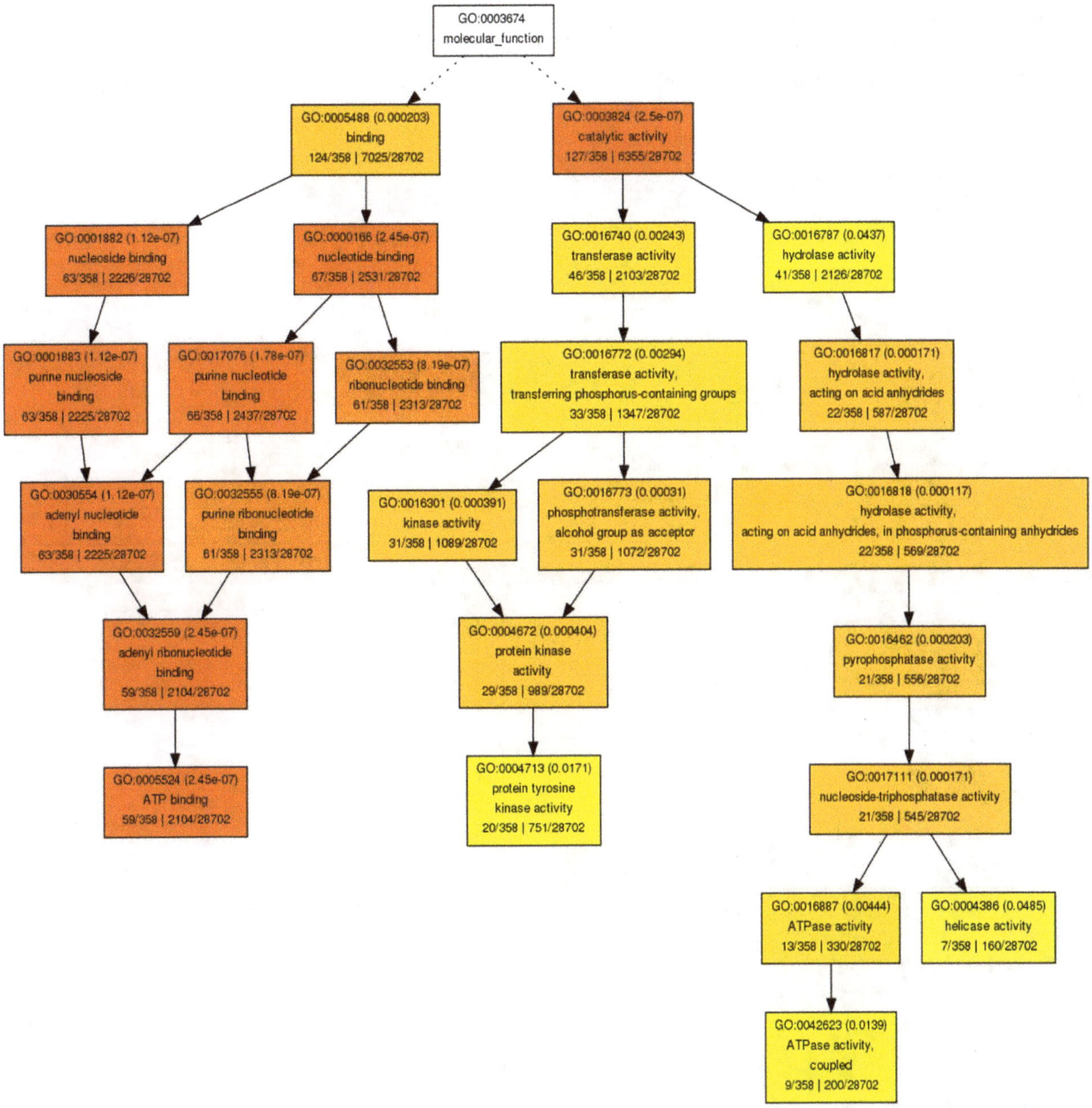

Figure 4. GO terms of the "Molecular function" category represented in genes identified as hypermethylated in 'Desmayo Largueta' flower buds in both the "A" = dormant bud stage and "B" = non dormant bud stage.

2.3. Differentyally Methylated Genes Related to Bud Dormancy

More identified DMGs were found as hypermethylated in stage A samples (dormant buds) than in stage B samples (non-dormant buds) (Table 3). Furthermore, just one hypomethylated gene could be functionally annotated, and it was mapped in the 3′ regulatory region of the gene Prupe.4G277200, which encodes for a REGULATION OF CHROMOSOME CONDENSATION (RCC1) protein (Table 4).

Common stage A hypermethylated genes coded for a MITOGEN-ACTIVATED PROTEIN (MAP)-kinase and a phosphatase (Prupe.4G270800 and Prupe.1G287200); an LEUCINE RICH REPEAT-TOLL INTERLEUKIN 1 RECEPTOR (LRR-TIR) apoptotic ATPase associated with disease resistance (Prupe.3G130700); a GDSL (Gly, Asp, Ser and Leu motif) lipase (Prupe.6G307900); an Nt-C2 family protein (Prupe.2G074400); and a Glycerophosphatidylinositol (GPI) anchor synthase (Prupe.2G019300). The Prupe.1G125600 gene was annotated, but its protein function is unknown

(Table 4). Moreover, genes coding for VACUOLAR PROTEIN SORTING 1 (VPS1) proteins were detected as hypermethylated in stage A in both 'Desmayo Largueta' and 'Penta'. VPS1 genes corresponded to Prupe.3G026400 in 'Desmayo Largueta' samples and Prupe.2G029500 in 'Penta' samples.

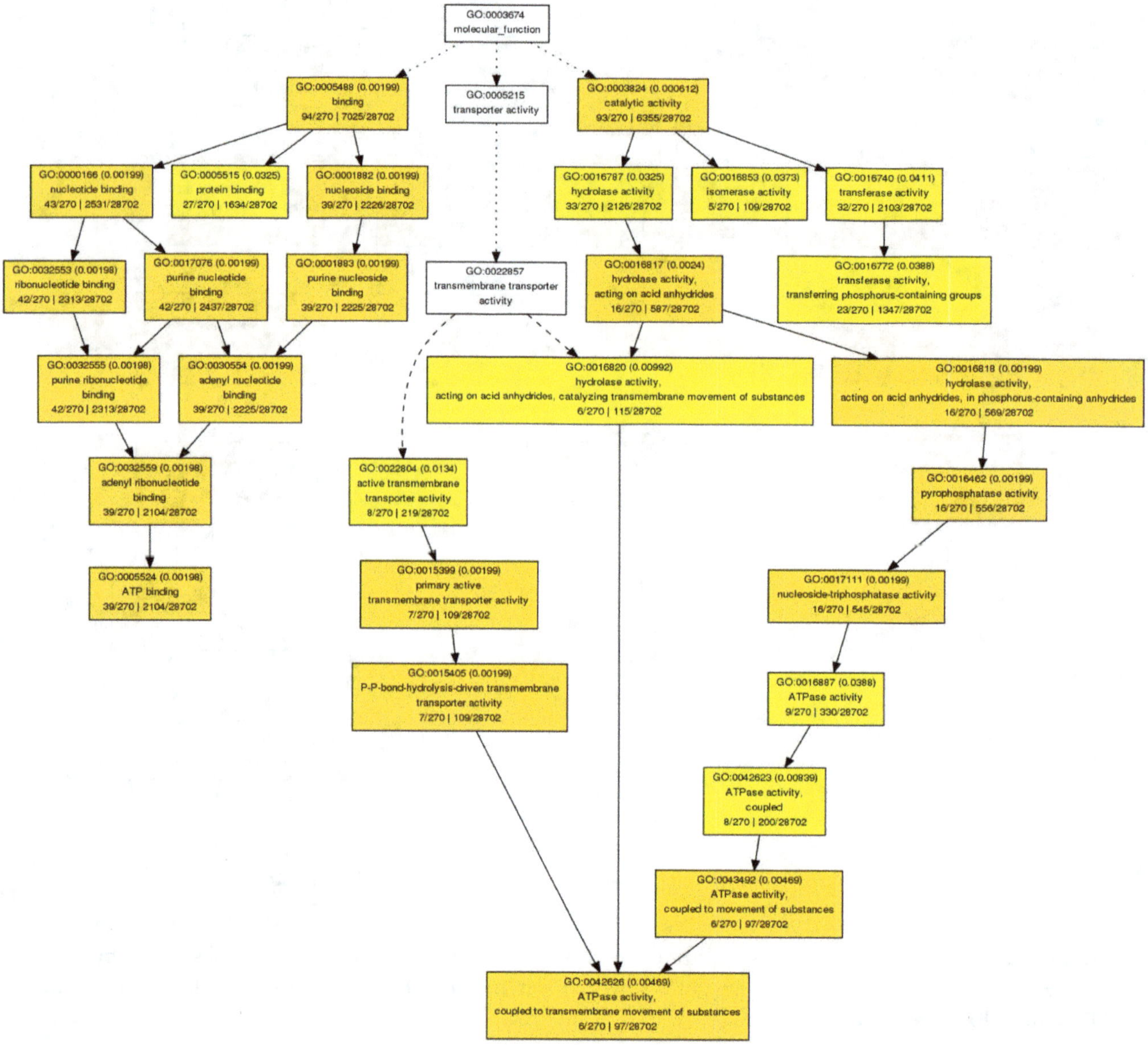

Figure 5. "Molecular function" GO terms represented in genes identified as hypomethylated in 'Desmayo Largueta' flower buds in both the "A" = dormant bud stage and "B" = non dormant bud stage.

Three additional LRR-TIR apoptotic ATPases were identified as encoded by hypermethylated genes in 'Desmayo Largueta' stage A samples (Table 4). Other genes that were found coded for defense proteins, a phosphatase, a cornichon protein associated with cell polarity and a Cytochrome P-450 (CytP450) protein member. On the other hand, hypermethylated genes in 'Penta' stage A samples coded for the ASSEMBLY PROTEIN 180 (AP180) clathrin assembly protein, the Tryptophan-Aspartic acid-Sterile Alpha Motif domain containing protein (WDSAM1) ubiquitination protein, lipases, a glycosyl hydrolase and a FCF2 rRNA processing protein recently described in yeast (Table 4).

Table 4. Hypermethylated genes identified in 'Desmayo Largueta' (D) and 'Penta' (P) stage A flower buds compared to those found in stage B flower buds (Chromosome, Fragment ID, Prupe.ID, Functional annotation).

FragmentID	Comparison	Chromosome	Gene Position	Prupe.Gene Code
fragment1289	AB	Pp01	Downstream	Prupe.1G287200
fragment1294	AB	Pp01	Upstream	Prupe.1G125600
fragment3735	PAPB	Pp01	Downstream	Prupe.1G099900
fragment32	AB	Pp03	Downstream	Prupe.3G130700
fragment341	AB	Pp06	Downstream	Prupe.6G307900
fragment797	AB	Pp02	Inside	Prupe.2G074400
fragment1708	DADB	Pp02	Inside	Prupe.2G019300
fragment4206	PAPB	Pp02	Inside	Prupe.2G019300
fragment341	DADB	Pp02	Downstream	Prupe.2G031100
fragment255	DADB	Pp02	Upstream	Prupe.2G053300
fragment92	DADB	Pp02	Inside	Prupe.2G057100
fragment92	DADB	Pp02	Inside	Prupe.2G057800
fragment797	DADB	Pp02	Upstream	Prupe.2G074300
fragment3707	PAPB	Pp04	Downstream	Prupe.4G186400
fragment4154	PAPB	Pp04	Downstream	Prupe.4G253800
fragment238	AB	Pp04	Downstream	Prupe.4G270800
fragment3299	PAPB	Pp06	Downstream	Prupe.6G307900
fragment1289	DADB	Pp02	Inside	Prupe.2G146000
fragment341	DADB	Pp03	Upstream	Prupe.3G026400
fragment797	DADB	Pp03	Downstream	Prupe.3G130700
fragment2263	DADB	Pp05	Inside	Prupe.5G036900
fragment157	PAPB	Pp06	Upstream	Prupe.6G014500
fragment1289	DADB	Pp05	Downstream	Prupe.5G038500
fragment483	PAPB	Pp02	Upstream	Prupe.2G039500
fragment507	DADB	Pp06	Inside	Prupe.6G097800
fragment1708	DADB	Pp06	Upstream	Prupe.6G331300
fragment2727	DADB	Pp06	Upstream	Prupe.6G331300
fragment849	PAPB	Pp01	Downstream	Prupe.1G105700

3. Discussion

Using our epiGBS variant as a first approach is a less expensive technique than complete GBS with highly accurate results. Furthermore, without the sequenced genome of the species, it is easier to perform bioinformatic analysis with well-defined fragments obtained by epiGBS. In this work, a conversion with bisulfite and a subsequent sequencing were performed to evaluate the 5mC variants of the samples analyzed. Subsequently, using bioinformatic analysis, these differentially methylated regions were mapped in the reference genome.

Applying epi-GBS to 'Desmayo Largueta' and 'Penta' flower bud gDNA samples provided data about methylation (5mC) variants depending on the genotype and dormancy state of the flower buds. Quality evaluation of the analysis showed that more than 90% of all cytosine positions were completely unmethylated and that only 1.0% to 1.3% of the positions were completely methylated, with higher methylation in the CpG context. These results agree with previous results in different plant species indicating that CG methylation is the typical genomic region for DNA methylation with less methylation abundance in the CHG and CHH contexts [23]. As a result of the specificity of the methylome of plants with respect to that of animals, we adapted bisulfite conversion methods to allow for correct analysis in plants for all cytosine contexts [24].

DNA methylomes have now been analyzed in many plants species, including Arabidopsis, rice, maize, and tomato. DNA methylation results in these species indicate that the distribution of methylation marks across the genomes is generally conserved, although variations can be observed between species depending on several factors, including transposon abundance and genome size [25]. In agreement with our results, polymorphism (5mC variants) can be observed by comparing different genotypes or even accessions within the same species. In addition, recent results evaluating methylomes from 1,227 different accessions of *Arabidopsis* distributed worldwide have shown important polymorphisms between accessions [26].

It is interesting to note the high degree of differential methylation that seems to be fixed between the two almond genotypes analyzed. This fact is of practical importance in cultivar improvement for developing epigenetic markers based on methylation variants and taking into account the high flexibility of methylation patterns in relation to external signals in order to identify markers based on methylation polymorphisms. In contrast to standard sequencing, bisulfite sequencing makes it possible to obtain information that conditions the phenotype. As a consequence, knowing the methylation state might help us understand the genetic determinism of important agronomic traits more deeply. Although the methylation patterns are highly variable in response to different external factors, the markers that we have detected in our almond genotypes are conserved in different stages of development and in different years and can therefore be considered as stable and conserved epigenetic marks.

In this study, data showed important differences between genotypes, which displayed different phenotypes in terms of breeding traits (chilling requirements for dormancy release, flowering and ripening times, almond production and almond characteristics) (Tables 1 and 2). It is remarkable that more hypermethylated than hypomethylated fragments were identified in stage A (dormant flower buds) almond samples in both genotypes (Table 3). This is concordant with the general decrease in 5mC during dormancy progression in *C. sativa* [15]. On the other hand, the most frequently mapped fragments were within the gene regions ("inside" DMGs), followed by the 5′regulatory regions ("downstream" DMGs) and, in last place, in the 3′ regions ("upstream" DMGs). According to Vining et al. [27], 5mC in promoters and gene body parts is related to a repressed state of chromatin, a condition that inhibits the accessibility of the transcriptional machinery.

Among the genes found as hypermethylated in 'Desmayo Largueta' with respect to 'Penta' flower buds (Table S3), ARF transcription factors were highly represented. Its known that the expression of genes like ARFs involved in the auxin response are subjected to epigenetic regulation [9,28], and ARF transcriptional regulation is required for developmental processes like germination [29]. Accordingly, Zhang et al. [30] observed a flowering delay in Arabidopsis when *ARF6* and *ARF8* were repressed. Nonetheless, the reason underlying the hypermethylated state of genes participating in the auxin response pathway in both dormant and non-dormant flower buds of the early flowering genotype 'Desmayo Largueta' has yet to be unraveled.

Another hymermethylated gene in 'Desmayo Largueta' flower bud samples was a member of the LEA family (Table S3). LEA proteins are involved in osmoprotection, which is activated in response to low temperatures [31]. When hypermethylated, this gene showed a repressed state of expression, although in low chilling requirement cultivars like 'Desmayo Largueta', osmoprotection would not be so necessary or may be regulated in a different way. The LEA gene family has been characterized by Du et al. [32] in *Prunus mume* (Siebold) Siebold & Zucc., and differential expression has been identified during bud dormancy in this species [33].

The LHY protein, on the other hand, is a well-described flowering time regulator in response to the photoperiod [34,35], and the gene network controlling this trait has been studied [36]. 'Desmayo Largueta' is a low-chill cultivar whose dormancy period takes place under short photoperiod conditions such as the experimental conditions of this work. It would be interesting to study LHY behavior during dormancy progression in different almond cultivars.

The methylation variants observed may be associated with evolutionary changes related to each genotype's features [37]. It will be interesting to distinguish which variants are related to traits of agronomic interest in order to explore adaptive mechanisms to the environment [38]. Recently, for instance, Garg et al. [39] identified conserved methylation polymorphisms distributed throughout rice varieties with different responses to drought resistance.

We found other hypermethylated genes in dormant (A) flower buds with respect to dormancy released (B) flower buds, including a MAP kinase (MAPK) and a phosphatase (Prupe.1G287200). MAP kinases and phosphatases have been found to participate in the initial response to cold induced by an increase in Ca2+ [31]. Furthermore, MAPK3 has been shown to be a central regulator of seed dormancy in barley [40]. Regarding the other genes hypermethylated in the A state, LRR-TIR apoptotic ATPases may be activated in a type of programmed cell death (PCD) called developmental cell death (DCD), leading to a differentiation of cells after dormancy release, as occurs in floral morphogenesis or in the pollen tube [41,42]. Nt-C2 and VSP1 proteins, on the other hand, are involved in vesicular trafficking from the cell membrane, and this process has been linked to cell wall differentiation and appears to be important in the dormancy release process [43,44]. Finally, GPI anchoring (a post-translational modification of proteins consisting of glycosylation) proteins are involved in intercellular signaling, as occurs in flowering transition as shown in *Populus* genus by Rinne et al. [45].

4. Materials and Methods

4.1. Plant Material and Experimental Design

We used flower buds from 'Desmayo Largueta', a traditional almond cultivar with very low chilling requirements and an extra-early flowering time, and 'Penta', a cultivar released from the CEBAS-CSIC Almond Breeding Program (Murcia, South-East Spain) with high chilling requirements and an extra-late flowering time. The plant material consisted of flower buds at stages A (dormancy phase) and B (after dormancy release) that were referenced to the phenological stages described by Felipe [46] (Figure 6). Dormancy release evaluation was performed by the forcing method according to Prudencio et al. [6]. Almond flower buds were picked from the experimental field of CEBAS-CSIC during two seasons of study: 2015–2016 and 2016–2017.

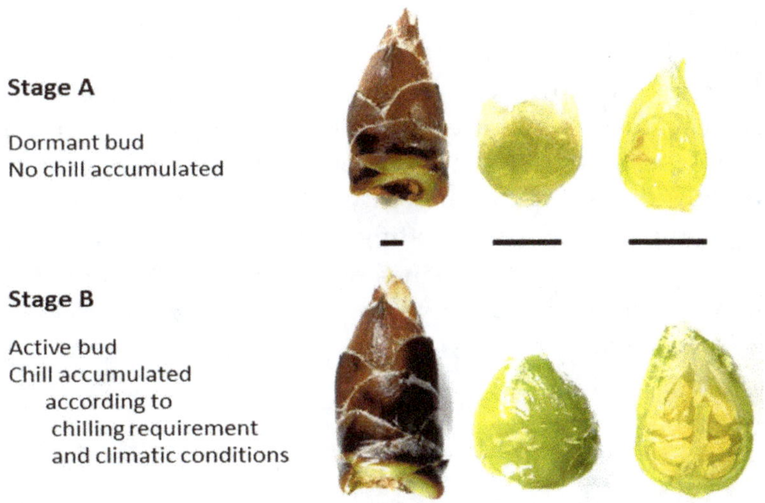

Figure 6. Plant material from 'Desmayo Largueta' and 'Penta' almond cultivars. Flower buds in the dormant state (**A**) and after dormancy release (**B**) according to Felipe [41]. Scale bars represent 1 mm in each case.

4.2. Epi-GBS Protocol

Every sample ('Desmayo Largueta' stage A, 'Penta' stage A, 'Desmayo Largueta' stage B, and 'Penta' stage B from the first and second season of study) consisted of a pool of ten flower buds. Genomic DNA was extracted from each sample following the method described by Doyle and Doyle [47]. The DNA samples were quantified using Qubit (Thermo Fisher Scientific, Alcobendas, Spain) and diluted to 1 μg in 100 μL. A total of 20 μL was digested using *PstI* restriction enzyme. Adaptors consisting of barcoded oligos were ligated to every sample (Table S4). Non-phosphorylated hemimethylated adapters were used to reduce costs. Fragmented samples (libraries generated by restriction) were pooled and purified and subsequently subjected to nick translation with C-dNTPs (Zymo Research, Irvine, CA, USA) and 7.5 μL of DNA PolI (NEB, Ipswich, MA, USA) in NEB buffer 2. An EZ DNA Methylation-Lightning kit (Zymo Research) was used for bisulfite treatment, and fragments were selected by size with a Thermo Scientific Size Selection kit (Thermo Fisher Scientific). Libraries were amplified using the Kapa HiFi HotStart Uracil+ ReadyMix (Roche, Barcelona, Spain) and purified with Magjet NGS Cleanup (Thermo Fisher Scientific). Paired-end Illumina 2500 reads (2 × 100 bases) were generated by Macrogen (Seoul, Korea) [23].

4.3. Bioinformatic Analysis of DNA Methylation

The process_radtags program of the Stacks 1.48 pipeline [48]. Checking the integrity of the restriction site was disabled with the "-disable_rad_check" option and quality filtering with the default settings was disabled with the exception the rad_check. This was necessary because the bisulfite treatment changes unmethylated cytosines in the recognition sequence of *PstI*, and, as a result, checking the restriction cut site would filter out all fragments. The ustacks program of the pipeline was used to align the fragments into perfectly matching stacks. The default settings were used with the exception of -M, which was set to 4 in order to increase the maximum distance (in nucleotides) between stacks. Finally, cstacks was used to build a catalog of consensus loci. A custom C program was used for the reconstruction of the original sequences of the fragments by comparing the reads with origins in the "Watson" and "Crick" strands of the genomic DNA.

The reconstructed DNA fragments were merged by another custom C program to produce one continuous "mock genome". Bismark_v0.19.0 [49] was used to align the original fragments to the mock genome and to extract the methylation information. The Bismark coverage reports were used as input for the methylKit R package [50]. A methyl kit was used to elaborate histograms of C-methylation and coverage and to assess sample similarity and correlation using the default settings. For the hierarchical clustering of the samples, dist was set to "correlation" and method to "ward". Finally, we used the calculate DiffMeth function of a MethylKit to search for differentially methylated cytosines with the settings difference = 25, qvalue = 0.01. We looked for both hypermethylated and hypomethylated bases setting type = hyper and = hypo, respectively. The positions of the differentially methylated cytosines were extracted from the MethylKit files. Another custom made C-program was used to identify the original fragments where these differentially methylated cytosines were located.

4.4. Gene Finding and Annotation

The sequence of each fragment was mapped against the *P. persica* reference genome (v2.0) [51] with Gmap [52]. Two different output files, in the gff3 and SAM format, were obtained. The gff3 ouput files were processed to extract the boundary coordinates (start and end positions) of each hit using command line tools. After that, the boundary coordinates were used by a second custom python script to retrieve three different categories of annotations based on gene locations on the *P. persica* reference genome: upstream and downstream genes (in a size window of 10,000 bp) and "inside genes" (fragments within gene sequence). Finally, SAM format files were processed using a custom python script to extract the alignment information (number of exons, percentage of coverage, percentage of identity, and amino acid changes). Functional annotation of genes selected by distance to the mapped fragment was carried out using AgriGO software using Singular Enrichment Analysis (SEA), and Fisher's test [53] (Figure 7).

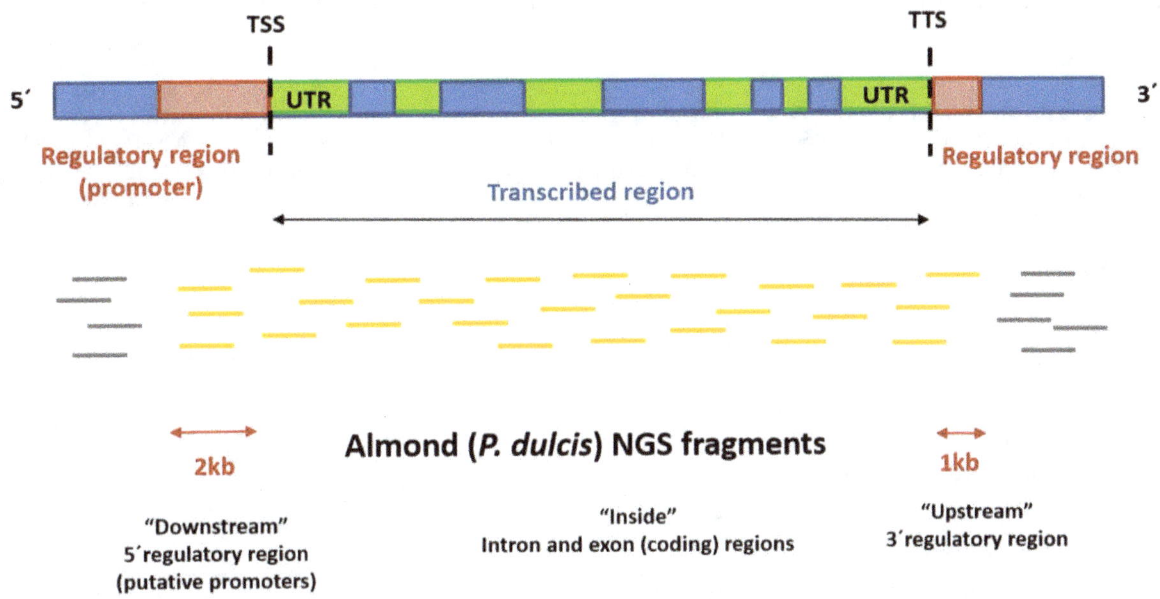

Figure 7. Schematic representation of a plant gene and classification of almond (*Prunus dulcis*) DMFs mapped in the *P. persica* genome. DMFs mapping from 2 kb upstream TSS to 1 kb downstream TTS were selected for functional annotation (fragments colored in orange). DMFs were classified according to the gene position—downstream, inside, or upstream—with respect to the fragment-mapping region. Fragments mapped in intergenic regions (colored in grey) were discarded as putative gene regulatory regions. **TSS:** Transcriptional Start Site, **UTR:** Untranslated Transcription Region, **TTS:** Transcriptional Terminal Site. Exons and introns within the transcribed region are colored in blue and green, respectively.

5. Conclusions

In this study, we applied the epi-GBS protocol to almond (*P. dulcis*) DNA samples for the first time. The technical potential is evident in the discovery of epigenetic variants, based on 5mC, that are genotype-dependent. According to the results obtained, the DNA methylation (5mC) pattern is generally genotype-dependent rather than dormancy state-dependent. Comparative DNA methylation studies of both almond varieties released from breeding programs and traditional varieties will surely contribute to our knowledge of methylation variants and provide candidate epialleles linked to agronomic traits. Such polymorphisms can be screened in large populations using NGS to confirm the locus methylation state associated with a given character of interest. In spite of coverage limitation, we were able to identify genes whose DNA methylation state changed between the dormant and active state of the flower buds. This was possible in both the traditional early-flowering genotype 'Desmayo Largueta' and the extra-late-flowering genotype 'Penta' from the CEBAS-CSIC Almond Breeding Program. Furthermore, common genes arose from the analysis. In the future, it would be interesting to improve the technique coverage to obtain a greater representation of the genome. Ultimately, the results will be an essential complement to RNAseq experiments in bud dormancy progression.

Supplementary Materials

Figure S1. GO terms of the "Biological Function" category represented in genes identified as hypermethylated in 'Desmayo Largueta' flower buds, in both the A and B dormancy states, compared to 'Penta' flower buds. Figure S2. GO terms of the "Biological Function" category represented in genes identified as hypomethylated in 'Desmayo Largueta' flower buds, in the both A and B dormancy states, compared to 'Penta' flower buds. Table S1. Total NGS fragments identified by Fragment ID. Table S2. Percentage of DMFs mapped in the *Prunus persica* genome (v2.1), Mapping region, Coverage, Percent identity, Number of exons, Amino acid changes, Comparison and Category. Table S3. Hypermethylated, Hypomethylated and Equally-methylated genes detected in 'Desmayo Largueta' flower buds with respect to 'Penta' flower buds in both the A and B stages (Chromosome, Fragment ID, Prupe.ID, Functional annotation). Gene list for GO annotation. Table S4. Sample barcodes. "D" = 'Desmayo Largueta', "P" = 'Penta', "A" = dormant bud stage; "B" = non-dormant bud stage.

DNA samples with number "1" correspond to the 2015–2016 season, and those with number "2" correspond to the 2016–2017 season.

Author Contributions: A.S.P. and O.W. participated in the design and coordination of the study. A.S.P. and O.W. carried out the epiGBS protocol. A.S.P., O.W. and P.J.M.-G. carried out data analysis. A.S.P., O.W., P.J.M.-G., R.M.R., P.M.-G. and F.D. participated in the manuscript elaboration and discussion.

Abbreviations

5mC	5′ Methylated Cytosine
AP180	ASSEMBLY PROTEIN 180
ARF	AUXIN RESPONSE FACTOR
CytP450	Cytochrome P-450
DMF	Differentially Methylated Fragment
DMG	Differentially Methylated Gene
epiGBS	Epi-Genotyping By Sequencing
FAR1	FAR-RED IMPAIRED RESPONSE 1
GDSL	Gly, Asp, Ser, and Leu motif
GO	Gene Ontology
GPI	Glycerophosphatidylinositol
H3K27me3	Trymethylation of Histone 3 on Lis (K) residue 27
HSP	HYDROPHOBIC SEED PROTEIN
LEA	LATE EMBRYOGENESIS ABUNDANT
LHY	LATE ELONGATED HYPOCOTYL
LRR-TIR	LEUCINE RICH REPEAT-TOLL INTERLEUKIN 1 RECEPTOR
MADS	MCM1-AGAMOUS-DEFICIENS-SRF
MAPK	MITOGEN-ACTIVATED PROTEIN KINASE
NGS	Next Generation Sequencing
NF-Y	NUCLEAR FACTOR-Y
PTMs	Post-Translational Modifications
SAM	S-ADENOSYL METHIONINE
VPS1	VACUOLAR PROTEIN SORTING 1
WDSAM1	Tryptophan-Aspartic acid-Sterile Alpha Motif 1

References

1. Vitasse, Y.; Lenz, A.; Körner, C. The interaction between freezing tolerance and phenology in temperate deciduous trees. *Front. Plant Sci.* **2014**, *5*. [CrossRef] [PubMed]
2. Beauvieux, R.; Wenden, B.; Dirlewanger, E. Bud Dormancy in Perennial Fruit Tree Species: A Pivotal Role for Oxidative Cues. *Front. Plant Sci.* **2018**, *9*, 657. [CrossRef] [PubMed]
3. Lang, B.A.; Early, J.D.; Martin, G.C.; Darnell, R.L. Endo-, para-, and eco-dormancy physiological terminology and classification for dormancy research. *HortScience* **1987**, *22*, 371–377.
4. Egea, J.; Ortega, E.; Martínez-Gómez, P.; Dicenta, F. Chilling and heat requirements of almond cultivars for flowering. *Environ. Exp. Bot.* **2003**, *50*, 79–85. [CrossRef]
5. Martínez-Gómez, P.; Prudencio, A.S.; Gradziel, T.M.; Dicenta, F. The delay of flowering time in almond: A review of the combined effect of adaptation, mutation and breeding. *Euphytica* **2017**, *213*, 197. [CrossRef]
6. Prudencio, A.S.; Martínez-Gómez, P.; Dicenta, F. Evaluation of breaking dormancy, flowering and productivity of extra-late and ultra-late flowering almond cultivars during cold and warm seasons in South-East of Spain. *Sci. Hort.* **2018**, *235*, 39–46. [CrossRef]

7. Cooke, J.E.K.; Eriksson, M.E.; Juntilla, O. The dynamic nature of bud dormancy in trees: Environmental control and molecular mechanisms. *Plant Cell Environ.* **2012**, *35*, 1707–1728. [CrossRef] [PubMed]
8. Abbott, A.G.; Zhebentyayeva, T.; Barakat, A.; Liu, Z. The genetic control of bud-break in trees. *Adv. Bot. Res.* **2015**, *74*, 201–228.
9. Yaish, M.W.; Colasanti, J.; Rothstein, S.J. The role of epigenetics processes in controlling flowering time exposed to stress. *J. Exp. Bot.* **2011**, *62*, 3727–3735. [CrossRef] [PubMed]
10. Ríos, G.; Leida, C.; Conejero, C.; Badenes, M.L. Epigenetic regulation of bud dormancy events in perennial plants. *Front. Plant Sci.* **2014**, *5*, 247. [PubMed]
11. Saze, H. Epigenetic memory transmission through mitosis and meiosis in plants. *Semin. Cell Dev. Biol.* **2008**, *19*, 527–536. [CrossRef] [PubMed]
12. Feng, S.; Jacobsen, S.E. Epigenetic modifications in plants: An evolutionary perspective. *Curr. Opin. Plant Biol.* **2011**, *14*, 179–186. [CrossRef] [PubMed]
13. Lämke, J.; Bäurle, I. Epigenetic and chromatin-based mechanisms in environmental stress adaptation and stress memory in plants. *Genome Biol.* **2017**, *18*, 124. [CrossRef] [PubMed]
14. Pascual, J.; Cañal, M.J.; Correia, B.; Escandon, M.; Hasbún, R.; Meijón, M.; Pinto, G.; Valledor, L. Can Epigenetics Help Forest Plants to Adapt to Climate Change? In *Epigenetics in Plants of Agronomic Importance: Fundamentals and Applications: Transcriptional Regulation and Chromatin Remodelling in Plants*; Alvarez-Venegas, R., De la Peña, C., Casas-Mollano, J.A., Eds.; Springer: Cham, Switzerland, 2014; pp. 125–146.
15. Santamaría, M.; Hasbún, R.; Valera, M.; Meijón, M.; Valledor, L.; Rodríguez, J.L.; Rodríguez, R. Acetylated H4 histone and genomic DNA methylation patterns during bud set and bud burst in *Castanea sativa*. *J. Plant Physiol.* **2009**, *166*, 1360–1369. [CrossRef] [PubMed]
16. de la Fuente, L.; Conesa, A.; Lloret, A.; Badenes, M.L.; Ríos, G. Genome-wide changes in histone H3 lysine 27 trimethylation associated with bud dormancy release in peach. *Tree Genet. Gen.* **2015**, *11*, 45. [CrossRef]
17. Lloret, A.; Martínez-Fuentes, A.; Agustí, M.; Badenes, M.L.; Ríos, G. Chromatin-associated regulation of sorbitol synthesis in flower buds of peach. *Plant Mol. Biol.* **2017**, *95*, 507–517. [CrossRef] [PubMed]
18. Rothkegel, K.; Sánchez, E.; Montes, C.; Greve, M.; Tapia, S.; Bravo, S.; Almeida, A.M. DNA methylation and small interference RNAs participate in the regulation of MADS-box genes involved in dormancy in sweet cherry (*Prunus avium* L.). *Tree Physiol.* **2017**, *37*, 1739–1751. [CrossRef] [PubMed]
19. Kumar, G.; Rattan, U.K.; Singh, A.K. Chilling-Mediated DNA Methylation Changes during Dormancy and Its Release Reveal the Importance of Epigenetic Regulation during Winter Dormancy in Apple (*Malus* × *domestica* Borkh.). *PLoS ONE* **2016**, *11*, e0149934. [CrossRef] [PubMed]
20. Prudencio, A.S.; Dicenta, F.; Martínez-Gómez, P. Gene expression analysis of flower bud dormancy breaking in almond using RNA-Seq. In Proceedings of the VII International Symposium on Almonds & Pistachios, Adelaida, Australia, 5–9 November 2017.
21. Fernández i Martí, A.; Gradziel, T.M.; Socias i Company, R. Methylation of the Sf locus in almond is associated with S-RNase loss of function. *Plant Mol. Biol.* **2014**, *86*, 681–689. [CrossRef] [PubMed]
22. Fresnedo-Ramírez, J.; Chan, H.M.; Parfitt, D.E.; Crisosto, C.H.; Gradziel, T.M. Genome-wide DNA-(de) methylation is associated with Noninfectious Bud-failure exhibition in Almond (*Prunus dulcis* [Mill.] D.A.Webb). *Sci. Rep.* **2017**, *7*, 42686.
23. van Gurp, T.P.; Wagemaker, N.V.; Wouters, B.; Vergeer, P.; Ouborg, J.N.; Werhoeben, K.J.V. epiGBS: Reference-free reduced representation bisulfite sequencing. *Nat. Meth.* **2016**, *13*, 322–329. [CrossRef] [PubMed]
24. Niederhuth, C.E.; Bewick, A.J.; Ji, L. Widespread natural variation of DNA methylation within angiosperms. *Gen. Biol.* **2016**, *17*, 194. [CrossRef] [PubMed]
25. How-Kit, A.; Emeline, T.; Deleuze, J.F.; Gallusci, P. Locus-Specific DNA Methylation Analysis and Applications to Plants. In *Plant Epigenetics*; Rajewsky, N., Jurga, S., Barciszewski, J., Eds.; Springer: Berlin, Germany, 2017; pp. 303–328.
26. Kawakatsu, T.; Huang, S.S.; Jupe, F. Epigenomic diversity in a global collection of *Arabidopsis thaliana* accesions. *Cell* **2016**, *166*, 492–505. [CrossRef] [PubMed]
27. Vining, K.J.; Pomraning, K.R.; Wilhelm, L.J.; Priest, H.D.; Pellegrini, M.; Mockler, T.C. Dynamic DNA cytosine methylation in the *Populus trichocarpa* genome: Tissue-level variation and relationship to gene expression. *BMC Gen.* **2012**, *13*, 27. [CrossRef] [PubMed]

28. Xiao, W.; Custard, K.D.; Brown, R.C.; Lemmon, B.E.; Harada, J.J.; Goldberg, R.B.; Fischer, R.L. DNA methylation is critical for Arabidopsis embryogenesis and seed viability. *Plant Cell* **2006**, *18*, 805–814. [CrossRef] [PubMed]
29. Liu, P.P.; Montgomery, T.A.; Fahlgren, N.; Kasschau, K.D.; Nonogaki, H.; Carrington, J.C. Repression of AUXIN RESPONSE FACTOR10 by microRNA160 is critical for seed germination and post-germination stages. *Plant J.* **2007**, *52*, 133–146. [CrossRef] [PubMed]
30. Zhang, G.-Z.; Jin, S.-H.; Li, P.; Jiang, X.-Y.; Li, Y.-J.; Hou, B.K. Ectopic expression of UGT84A2 delayed flowering by indole-3-butyric acid-mediated transcriptional repression of ARF6 and ARF8 genes in *Arabidopsis*. *Plant Cell Rep.* **2017**, *36*, 1995–2006. [CrossRef] [PubMed]
31. Bañuelos, M.L.G.; Moreno, L.V.; Winzerling, J.; Orozco, J.A.; Gardea, A.A. Winter metabolism in deciduous trees: Mechanisms, genes and associated proteins. *Rev. Fitotec. Mex.* **2008**, *31*, 295–308.
32. Du, D.; Zhang, Q.; Cheng, T.; Pan, H.; Yang, W.; Sun, L. Genome-wide identification and analysis of late embryogenesis abundant (LEA) genes in *Prunus mume*. *Mol. Biol. Rep.* **2013**, *40*, 1937–1946. [CrossRef] [PubMed]
33. Yamane, H.; Kashiwa, Y.; Kakehi, E.; Yonemori, K.; Mori, H.; Hayashi, K.; Iwamoto, K.; Tao, R.; Kataoka, I. Differential expression of dehydrin in flower buds of two Japanese apricot cultivars requiring different chilling requirements for bud break. *Tree Physiol.* **2006**, *26*, 1559–1563. [CrossRef] [PubMed]
34. Fujiwara, S.; Oda, A.; Yoshida, R.; Niinuma, K.; Miyata, K.; Tomozoe, Y.; Tajima, T.; Nakagawa, M.; Hayashi, K.; Coupland, G.; et al. Circadian Clock Proteins LHY and CCA1 Regulate SVP Protein Accumulation to Control Flowering in Arabidopsis. *Plant Cell* **2008**, *20*, 2960–2971. [CrossRef] [PubMed]
35. Park, B.S.; Eo, H.J.; Jang, I.-C.; Kang, H.-G.; Song, J.T.; Seo, H.S. Ubiquitination of LHY by SINAT5 regulates flowering time and is inhibited by DET1. *Biochem. Biophys. Res. Commun.* **2010**, *398*, 242–246. [CrossRef] [PubMed]
36. Park, M.-J.; Kwon, Y.-J.; Gil, K.-E.; Park, C.-M. LATE ELONGATED HYPOCOTYL regulates photoperiodic flowering via the circadian clock in Arabidopsis. *BMC Plant Biol.* **2016**, *16*, 114. [CrossRef] [PubMed]
37. Varriale, A. DNA Methylation in Plants and Its Implications in development, Hybrid Vigour, and Evolution. In *Plant Epigenetics*; Rajewsky, N., Jurga, S., Barciszewski, J., Eds.; Springer: Berlin, Germany, 2017; pp. 263–280.
38. Viggiano, L.; de Pinto, M.C. Dynamic DNA Methylation Patterns in Stress Response. In *Plant Epigenetics*; Rajewsky, N., Jurga, S., Barciszewski, J., Eds.; Springer: Berlin, Germany, 2017; pp. 281–302.
39. Garg, R.; Narayana Chevala, V.V.S.; Shankar, R.; Jain, M. Divergent DNA methylation patterns associated with gene expression in rice cultivars with contrasting drought and salinity stress response. *Sci. Rep.* **2015**, *5*, 14922. [CrossRef] [PubMed]
40. Nakamura, S.; Pourkheirandish, M.; Morishige, H.; Kubo, Y.; Nakamura, M.; Ichimura, K.; Seo, S.; Kanamori, H.; Wu, J.; Ando, T.; et al. Mitogen-Activated Protein Kinase Kinase 3 Regulates Seed Dormancy in Barley. *Curr. Biol.* **2016**, *26*, 775–781. [CrossRef] [PubMed]
41. Koonin, E.V.; Aravind, L. Origin and evolution of eukaryotic apoptosis: The bacterial connection. *Cell Death Differ.* **2002**, *9*, 394. [CrossRef] [PubMed]
42. Del Duca, S.; Serafini-Fracassini, D.; Cai, G. Senescence and programmed cell death in plants: Polyamine action mediated by transglutaminase. *Front. Plant Sci.* **2014**, *5*, 120. [CrossRef] [PubMed]
43. Ebine, K.; Ueda, T. Roles of membrane trafficking in plant cell wall dynamics. *Front. Plant Sci.* **2015**, *6*, 878. [CrossRef] [PubMed]
44. Kim, S.J.; Brandizzi, F. The plant secretory pathway: An essential factory for building the plant cell wall. *Plant Cell Physiol.* **2014**, *55*, 687–693. [CrossRef] [PubMed]
45. Rinne, P.L.H.; Welling, A.; Vahala, J.; Ripel, L.; Ruonala, R.; Kangasjärvi, J.; van der Schoot, C. Chilling of Dormant Buds Hyperinduces FLOWERING LOCUS T and Recruits GA-Inducible 1,3-β-Glucanases to Reopen Signal Conduits and Release Dormancy in Populus. *Plant Cell* **2011**, *23*, 130–146. [CrossRef] [PubMed]
46. Felipe, A.J. Phenological states of almond. In Proceedings of the Third GREMPA Colloquium, Bari, Italy, 3–7 October 1977; pp. 101–103. (In Italian)
47. Doyle, J.J.; Doyle, M. A rapid DNA isolation procedure for small quantities of fresh leaf tissue. *Phytochem. Bull.* **1987**, *19*, 11–15.

48. Catchen, J.; Hohenlohe, P.A.; Bassham, S.; Amores, A.; Cresko, W.A. Stacks: An analysis tool set for population genomics. *Mol. Ecol.* **2013**, *22*, 3124–3140. [CrossRef] [PubMed]
49. Krueger, F.; Andrews, S.R. Bismark: A flexible aligner and methylation caller for Bisulfite-Seq applications. *Bioinformatics* **2011**, *27*, 1571–1572. [CrossRef] [PubMed]
50. Akalin, A.; Kormaksson, M.; Li, S.; Garrett-Bakelman, F.E.; Figueroa, M.E.; Melnick, A.; Mason, C.E. methylKit: A comprehensive R package for the analysis of genome-wide DNA methylation profiles. *Genome Biol.* **2012**, *13*, R87. [CrossRef] [PubMed]
51. Verde, I.; Jenkins, J.; Dondini, L.; Micali, S.; Pagliarani, G.; Vendramin, E. The Peach v2.0 release: High-resolution linkage mapping and deep resequencing improve chromosome-scale assembly and contiguity. *BMC Gen.* **2017**, *18*, 225. [CrossRef] [PubMed]
52. Wu, T.D.; Watanabe, C.K. GMAP: A genomic mapping and alignment program for mRNA and EST sequences. *Bioinformatics* **2005**, *21*, 1859–1875. [CrossRef] [PubMed]
53. Tian, T.; Yue, L.; Hengyu, Y.; Qi, Y.; Xin, Y.; Zhou, D.; Wenying, X.; Zhen, S. agriGO v2.0: A GO analysis toolkit for the agricultural community, 2017 update. *Nucleic Acids Res.* **2017**, *45*, W122–W129. [CrossRef] [PubMed]

9

Genome-Wide Association Study Reveals Novel Genomic Regions for Grain Yield and Yield-Related Traits in Drought-Stressed Synthetic Hexaploid Wheat

Madhav Bhatta [1], Alexey Morgounov [2], Vikas Belamkar [1] and P. Stephen Baenziger [1,*]

[1] Department of Agronomy and Horticulture, University of Nebraska-Lincoln, Lincoln, NE 68583, USA; madhav.bhatta@huskers.unl.edu (M.B.); vikas.belamkar@unl.edu (V.B.)

[2] International Maize and Wheat Improvement Center (CIMMYT), 06511 Emek, Ankara, Turkey; a.morgounov@cgiar.org

* Correspondence: pbaenziger1@unl.edu

Abstract: Synthetic hexaploid wheat (SHW; $2n = 6x = 42$, AABBDD, *Triticum aestivum* L.) is produced from an interspecific cross between durum wheat ($2n = 4x = 28$, AABB, *T. turgidum* L.) and goat grass ($2n = 2x = 14$, DD, *Aegilops tauschii* Coss.) and is reported to have significant novel alleles-controlling biotic and abiotic stresses resistance. A genome-wide association study (GWAS) was conducted to unravel these loci [marker–trait associations (MTAs)] using 35,648 genotyping-by-sequencing-derived single nucleotide polymorphisms in 123 SHWs. We identified 90 novel MTAs (45, 11, and 34 on the A, B, and D genomes, respectively) and haplotype blocks associated with grain yield and yield-related traits including root traits under drought stress. The phenotypic variance explained by the MTAs ranged from 1.1% to 32.3%. Most of the MTAs (120 out of 194) identified were found in genes, and of these 45 MTAs were in genes annotated as having a potential role in drought stress. This result provides further evidence for the reliability of MTAs identified. The large number of MTAs (53) identified especially on the D-genome demonstrate the potential of SHWs for elucidating the genetic architecture of complex traits and provide an opportunity for further improvement of wheat under rapidly changing climatic conditions.

Keywords: marker–trait association; haplotype block; genes; root traits; D-genome; genotyping-by-sequencing; single nucleotide polymorphism; durum wheat; bread wheat; complex traits

---

1. Introduction

Drought is one of the most important abiotic stresses that reduce crop productivity and is expected to increase with the change in climate [1]. Erratic rainfall patterns caused by climate change may aggravate drought stress and will have a major impact on agriculture [2,3]. The most prominent example of the impact of drought stress on agriculture was the 2012 drought stress in the United States, where moderate to extreme drought stress occurred across the central agricultural states that resulted in crop harvest failure for corn (*Zea mays* L.), sorghum (*Sorghum bicolor* L.), and soybean (*Glycine max* L.), and the agriculture loss due to drought was estimated to be $30 billion [4]. To cope with the challenges of drought stress, plant breeders have been focusing on improving drought tolerance since several decades [2,3,5]. However, the drought tolerance is a complex phenomenon as most of the traits associated with drought tolerance are polygenic in nature, and understating the genetic architecture of drought tolerance is still underway [3] including in wheat (*Triticum sps.*) [2]. Wheat is one of the most important staple cereal crops mainly grown under rainfed conditions [3,6] and is expected to suffer from drought stress [3]. Therefore, breeding for drought tolerance and identifying genomic regions and underlying candidate genes associated with drought tolerance are important for wheat improvement.

Bread wheat (*T. aestivum* L.) has limited genetic and phenotypic diversity available for breeding for drought tolerance [2]. This is mainly due to the genetic bottleneck experienced during its origin and subsequent domestication [7,8]. Diversity can be increased through the production of synthetic hexaploid wheat (SHW) and its utilization in breeding programs [9–11]. Synthetic hexaploid wheat ($2n = 6x = 42$, AABBDD) is produced from an interspecific cross between durum wheat ($2n = 4x = 28$, AABB, *T. turgidum* L.) and goat grass ($2n = 2x = 14$, DD, *Aegilops tauschii* Coss.).

The SHWs are reported to have significant genetic variation for biotic [12,13] and abiotic stresses resistance [2,10,14]. However, previous studies focused mainly on biotic stresses including leaf rust (incited by *Puccinia triticina*) [13,15,16], stem rust (incited by *P. graminis*) [15,16], stripe rust (incited by *P. striiformis*) [12,15,16], Fusarium head blight (incited by *Fusarium graminearum*) [13], yellow spot (incited by *Pyrenophora tritici-repentis*) [15,16], septoria nodorum (incited by *Parastagonospora nodorum*) [15,16], Septoria tritici blotch (incited by *Mycosphaerella graminicola*) [13,15], cereal cyst nematode (incited by *Heterodera avenae*) [15], crown rot (incited by *F. pseudograminearum*) [16], and root-lesion nematode (incited by *Pratylenchus thornei* and *P. neglectus*) [15]. Therefore, exploiting genetic variation under abiotic stresses such as drought is needed to further utilize the potential of SHWs.

About 800 quantitative trait loci (QTLs) and marker–trait associations (MTAs) have been reported for drought tolerant traits (agronomic, physiological, root, and yield-related traits) using bi-parental mapping (~691 QTLs) and genome-wide association studies (GWASs; ~109 MTAs) in wheat [3]. However, only 68 QTLs are major QTLs that explain more than 19% of phenotypic variation [3]. This study was conducted to identify novel genomic regions associated with grain yield (GY) and yield-related traits using GWAS performed using 35,648 genotyping-by-sequencing (GBS)-derived single nucleotide polymorphisms (SNPs) in 123 SHWs grown under two drought-stressed growing seasons (2016 and 2017) in Konya, Turkey. Subsequently, the underlying genes for the MTAs identified were investigated for their potential role in drought stress using the functional annotations. To the best of our knowledge, this is the first report on GWAS on GY and yield-related traits under drought stress in SHWs. The results from this study will be a valuable resource for the genetic improvement of GY and yield-related traits in drought stress, introgression of desirable genes from SHWs into elite wheat germplasm, genomic selection, and marker-assisted selection in the breeding program.

2. Results and Discussion

2.1. Weather Conditions

The mean monthly air temperatures were similar at Konya in both growing seasons with 13 °C in 2015–2016 and 12 °C in 2016–2017 compared to the 25-year mean monthly air temperature (11 °C) in Turkey (Table 1). The total rainfall during 2016–2017 (243 mm) was slightly higher than that during 2015–2016 (222.4 mm) growing season. Total rainfalls during the wheat-growing season (September–July) were 48.9% lower in 2015–2016 and 44.2% lower in 2016–2017 compared to the 25-year mean total rainfall (435.1 mm) in Turkey.

Although winter wheat water requirements are higher from mid-March to mid-June (from the spring tillering period to the mid-grain filling period), rainfalls were lower in both years compared to the 25-year mean rainfall. The plants were exposed to drought stress from tillering through grain filling. Hence, the results from the present study can be used to understand the effects and genetics of drought in SHW.

Table 1. Mean monthly temperatures and total monthly rainfalls in two growing seasons (2016 and 2017) and 25-year averaged data in Konya, Turkey.

Month	Konya, 2015–2016	Konya, 2016–2017	Turkey, 1991–2015	Konya, 2015–2016	Konya, 2016–2017	Turkey, 1991–2015
	Temperature (temp) (°C) [a]	Temp (°C)	Temp (°C) [b]	Rainfall (mm) [c]	Rainfall (mm)	Rainfall (mm) [d]
September	22.8	17.1	19.0	35.8	11.2	23.1
October	15.3	13.2	13.7	34.4	0.0	48.3
November	7.5	4.9	7.0	5.8	16.6	58.0
December	−0.1	0.5	2.1	8.0	26.8	73.0
January	1.6	0.2	0.1	37.0	9.0	65.6
February	4.9	3.4	1.3	0.4	69.2	60.0
March	8.5	8.2	5.3	37.8	31.0	61.6
April	15.2	12.7	10.4	9.6	33.2	62.7
May	18.4	16.7	15.2	38.4	41.2	54.6
June	23.7	24.4	19.5	15.0	4.8	34.7
July	26.6	27.7	22.9	0.2	0.0	15.1
Total/average	13.1	11.7	10.6	222.4	243.0	435.1

2.2. Phenotypic Variation for Yield and Yield-Related Traits

A combined analysis of variance (ANOVA) across years identified significant cross-over genotype × year interaction for all traits except for flag leaf width (FLW) and stem diameter (STMDIA) (Table S4). Therefore, analysis of variance was computed for both years separately and the results indicated that the SHWs showed significant variation for GY and yield-related traits in each year (Table 2). For instance, GY ranged from 200 $g \cdot m^{-2}$ to 341 $g \cdot m^{-2}$ with an average yield of 259 $g \cdot m^{-2}$ in 2016 and from 241 $g \cdot m^{-2}$ to 392 $g \cdot m^{-2}$ with an average yield of 290 $g \cdot m^{-2}$ in 2017 (Table 2). The large variation among the traits in each year can be attributed to the collection of diverse accessions of SHWs from different countries and different genetic backgrounds [11,14].

Table 2. Phenotypic variation for grain yield and yield-related traits with best linear unbiased predictor values, range, percentage of coefficient of variation (CV), and broad sense heritability (H^2) of 123 synthetic hexaploid wheat grown in two seasons (2016 and 2017) in Konya, Turkey.

Trait	2016				2017			
	Mean	Range	CV	H^2	Mean	Range	CV	H^2
Grain yield ($g \cdot m^{-2}$)	259	200–341	9.7	0.32	290	241–392	9.9	0.56
Harvest index	0.4	0.24–0.66	10.9	0.63	0.34	0.27–0.41	6.3	0.64
Biomass weight ($g \cdot m^{-2}$)	671	537–827	9.1	0.39	865	684–1098	8.9	0.63
Thousand kernel weight (g)	32.1	24–42	10.5	0.75	41	33–50	8	0.90
Grain volume weight ($Kg \cdot hL^{-1}$)	65.6	52–77	7.2	0.91	74	68–77	2.3	0.76
Awn length (cm)	6	2.3–8.6	24.3	0.61	5.6	0.5–8.0	28.3	0.95
Flag leaf length (cm)	22.4	21.8–22.8	0.8	0.91	12	9.9–16.4	7.6	0.53
Flag leaf width (cm)	1	0.96–1.13	2.8	0.67	1	0.9–1.3	6.1	0.49
Flag leaf area (cm^2)	18.9	17.6–19.7	2.2	0.85	10.1	7.7–14	11.6	0.52
Stem diameter (mm)	2.9	2.4–3.5	6.9	0.57	2.9	2.5–4.0	7.4	0.63
Root length (cm)	393	392–395	0.20	0.6	192.2	72–375	20	0.31

Broad-sense heritability (H^2) ranged from low to high (Table 2). Low to moderate H^2 was observed for GY (0.32–0.56), biomass weight (BMWT; 0.39–0.63), FLW (0.49–0.67), and root length (RTLN; 0.31–0.60); moderate H^2 was observed for harvest index (HI; 0.63–0.64) and STMDIA (0.57–0.63); moderate to high H^2 was observed for flag leaf length (FLLN; 0.53–0.91), flag leaf area (FLA; 0.52–0.85) and awn length (AWNLN; 0.61–0.95); and high H^2 was observed for thousand kernel weight (TKW; 0.75–0.90) and grain volume weight (GVWT; 0.76–0.91), indicating the genetic instability of these traits across years under drought stress. Similar H^2 for most of these traits have been observed in previous studies [17–26].

2.3. Principal Component Analysis and Phenotypic Correlation

Principal component (PC) bi-plot analysis showed the association between GY and yield-related traits based on correlation matrix (Figure 1). The first two PCs that explained 43.4% (2016) and 44.9% (2017) of variation better explained the relationship between traits in the two-dimensional space. In the PC biplot, we observed two distinct groupings. The first one comprised of GY, HI, BMWT, TKW, GVWT, AWNLN, and RTLN whereas the second one had FLLN, FLW, and FLA (Figure 1). The traits grouping with GY are the more important traits for improving GY in drought-stressed conditions. The association observed in the PC biplot was supported by the significant correlations of GY with BMWT, HI, TKW, and GVW in both years (Figure S1). Similar correlations for these traits were observed in the previous studies [18,27–32].

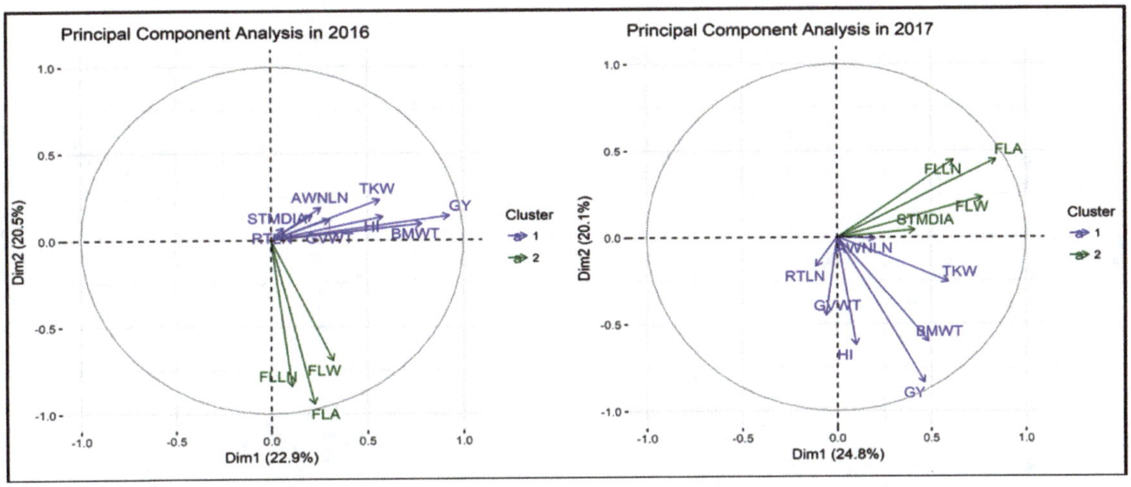

Figure 1. Principal component bi-plot analysis of 123 drought-stressed synthetic hexaploid wheat grown in two seasons (2016 and 2017) in Konya, Turkey. AWLN, awn length; BMWT, biomass weight; FLA, flag leaf area; FLLN, flag leaf length; FLW, flag leaf width; GVWT, grain volume weight; GY, grain yield; HI, harvest index; RTLN, root length; STMDIA, stem diameter; and TKW, thousand kernel weight.

2.4. Population Structure and Genome-Wide Association Study

Population structure analysis of 123 SHWs was performed using 35,648 SNPs [filtered for minor allele frequency (MAF) > 0.05 and missing data < 20%] using the Bayesian clustering algorithm implemented in Structure software and the results showed that these lines were divided into three subgroups (Figure S2 and Table S1). The details of the population structure and genetic diversity of these SHWs have been previously reported in Bhatta et al. [11].

The GWAS identified novel genomic regions for GY and yield-related traits and the MTA explained the high phenotypic variance. The Fixed and random model Circulating Probability Unification algorithm (FarmCPU), with kinship, population structure (Q) or PC, best linear unbiased predictors (BLUPs) for each trait, and 35,648 GBS-derived SNPs was used to identify MTAs. The GBS-derived SNPs were well distributed across each of the chromosome (Figure S3). We identified 194 MTAs distributed across 21 chromosomes for GY and yield-related traits with phenotypic variance explained (PVE) ranging from 1.1% to 32.3% (Figure 2 and Figure S4, Table S5). The highest number of MTAs was observed for GY (29), followed by STMDIA (23), FLA (20), and TKW (20) while the lowest MTAs were observed for HI (10) (Figure 2). Of the 194 MTAs, 75 MTAs were detected on the A genome, with 66 MTAs on the B genome, and 53 MTAs on the D-genome. The highest MTAs were present on chromosome 7A (26 MTAs) and the lowest MTAs were present on chromosome 3D (four MTAs)

(Figure 2). Most of the MTAs identified in the present study were year-specific, suggesting the influence of genotype × environment interaction on the phenotype of the traits measured in two years. However, 120 of the 194 significant SNPs were in 83 genes and 45 of these MTAs were present within genes and their annotations suggested their potential role in drought stress. This result further provided confidence that the MTAs identified in the study are likely reliable MTAs (Table S5).

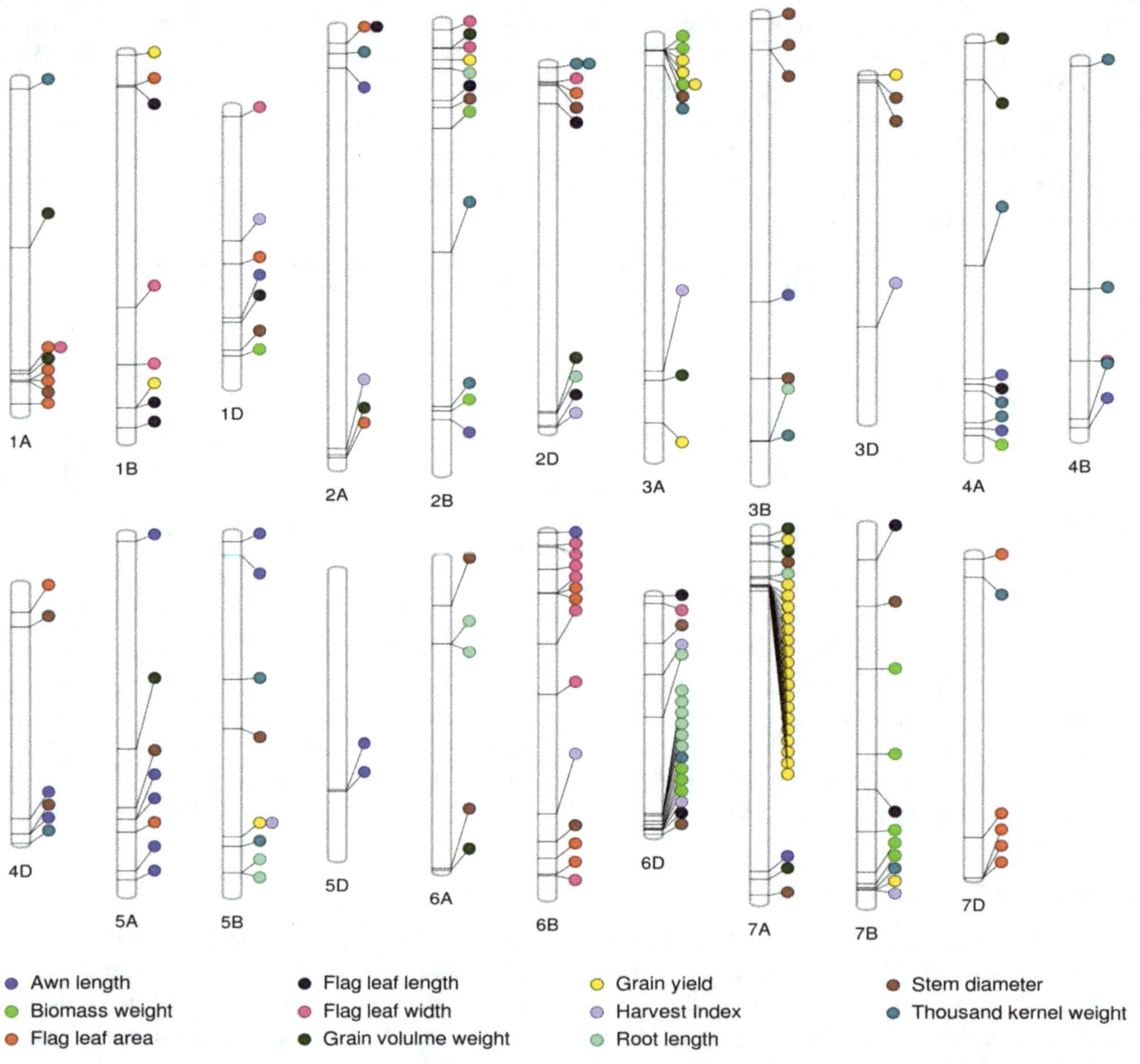

Figure 2. Significant markers trait associations identified on each chromosome for grain yield and yield-related traits obtained from the genome-wide association study of 123 synthetic hexaploid wheats grown in 2016 and 2017 in Konya, Turkey.

2.4.1. Grain Yield

The 29 MTAs for GY were observed in 29 different genomic regions on seven chromosomes including 1B, 2B, 3A, 3D, 5B, 7A, and 7B with PVE ranging from 7.6% to 17.9% (Figure 2, and Table S5). Earlier studies have reported QTLs/MTAs for GY on wheat chromosomes 1B [5,17,19,33], 2B [17,19,29,33,34], 3A [17,30,33,34], 3D [33], 5B [5,28,30,31,33,35], 7A [20,27,30,33], and 7B [17,30,33].

However, it is difficult to align our findings with previous studies due to the use of different marker systems [90K SNP, short sequence repeat (SSR), diversity arrays technology (DART) marker vs. GBS-derived SNP marker], absence of precise location information in published studies, or the use of a different version of the reference wheat genome in previous studies than the International Wheat Genome Sequencing Consortium (IWGSC) RefSeq v1.0. However, identification of several MTAs on the same chromosome as earlier studies provided increased confidence on these associations.

The present study identified four major haplotype blocks (from 19 bp to 433 kb) on chromosome 7A with two to six SNPs associated with GY in 2016 (Figure 3). First haplotype block consisted of six MTAs within the 433 kb range, second haplotype block consisted of four MTAs within the 81 bp range, third haplotype block consisted of two MTAs within the 19 bp range, and fourth haplotype block consisted of three MTAs within the 314 kb range. The PVE on GY by the first, second, third, and fourth haplotype blocks were 17.2%, 24.6%, 21.9%, and 8.2%, respectively.

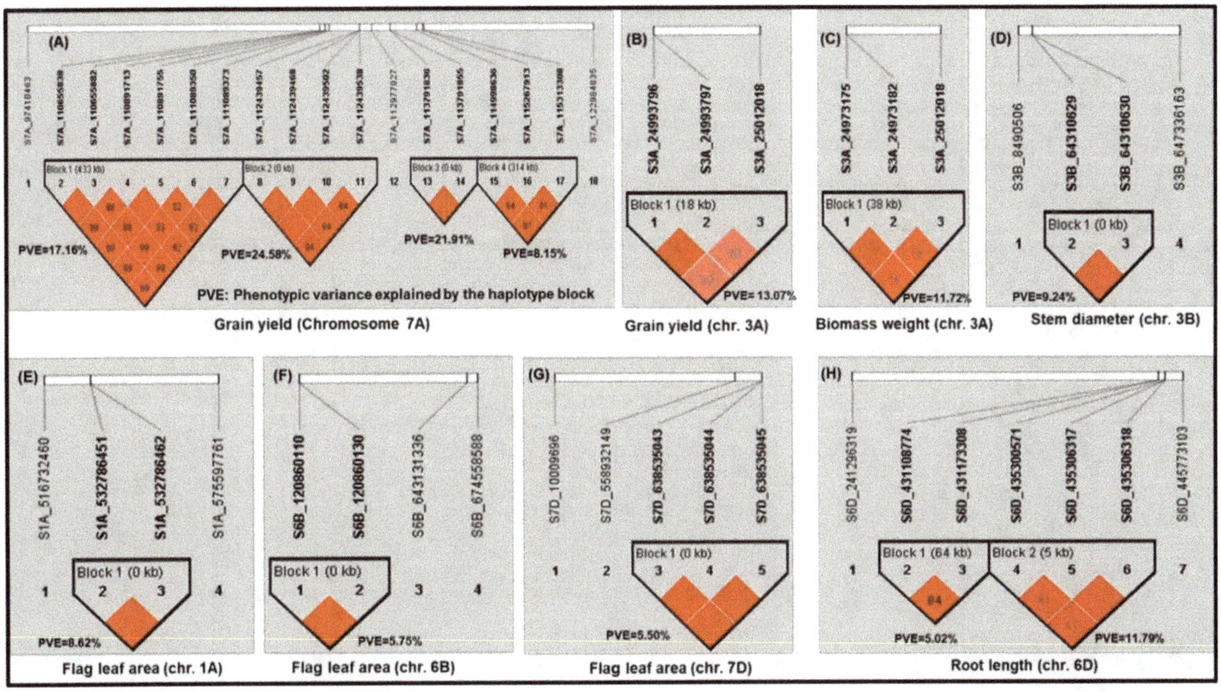

Figure 3. Linkage disequilibrium (LD) values (R^2) and haplotype blocks with significant marker–trait associations (MTAs; $\geq$2) observed (**A**) on chromosome 7A for GY, (**B**) on chromosome 3A for GY, (**C**) on chromosome 3A for BMWT, (**D**) on chromosome 3B for STMDIA, (**E**) on chromosome 1A for FLA, (**F**), on chromosome 6B for FLA, (**G**) on chromosome 7D for FLA, and (**H**) on chromosome 6D for RTLN and phenotypic variance explained (PVE) by each haplotype block. Dark red color represents the strong LD whereas light red color represents the weak LD between pairs of MTAs.

One MTA (S7A_112977027; 112977027 bp) present in-between the second (537 kb away) and third (837 kb) haplotype blocks was within the gene, TraesCS7A01G158200.1, and PVE on GY was 12.8% (Table S5). This gene was annotated as a member of sentrin-specific protease of Ubiquitin-like Protease 1 (*Ulp1*) gene family (Table 3). The *Ulp1* is a small ubiquitin related modifier (SUMO)-specific protease that affects several important biological processes in plants including response to abiotic stress [21]. It has been shown to play a role in drought tolerance in Arabidopsis (*Arabidopsis thaliana*) [36] and rice (*Oryza sativa*) [22,37]. This makes this MTA interesting and a stronger candidate for future functional validation studies.

Table 3. List of significant markers associated with GY and yield-related traits, favorable alleles (underlined), SNP effects, and drought-related putative genes from genome-wide association study of 123 drought stressed synthetic hexaploid wheats grown in 2016 in Konya, Turkey.

Trait	SNP [a]	$-\log_{10}(p)$	Alleles	SNP Effect	PVE (%)	Gene ID	Annotation
GY	S3A_686179591	4.08	A/G	−14.28	10.7	TraesCS3A01G445100	F-box family protein
GY	S7A_112977027	5.24	A/T	−19.17	12.8	TraesCS7A01G158200.1	Sentrin-specific protease
HI	S3A_593313534	13.56	T/C	0.08	16.0	TraesCS3A01G343700	WRKY transcription factor
HI	S6D_157451060	4.01	A/G	−0.03	6.2	TraesCS6D01G170900.1	Cytochrome P450, putative
HI	S6D_462272376	12.01	G/A	0.02	14.5	TraesCS6D01G382600.1	LOB-domain protein-like
BMWT	S1D_441309135	4.82	C/G	−105.34	14.4	TraesCS1D01G357500.1	Protein DETOXIFICATION
BMWT	S7B_450630784	4.06	A/G	−25.88	10.7	TraesCS7B01G242600.1	F-box family protein
TKW	S2A_47781717	4.52	G/A	0.93	4.2	TraesCS2A01G093500	F-box family protein
TKW	S4A_625466381	4.12	T/G	1.21	15.3	TraesCS4A01G347600	Protein kinase family protein
TKW	S4D_509427923	4.91	C/G	−1.72	10.1	TraesCS4D01G364700	Cytochrome P450 family protein
TKW	S6D_452410667	8.16	A/G	−1.54	17.7	TraesCS6D01G360800	Protein kinase family protein
AWNLN	S4D_461573496	5.71	T/C	0.32	9.0	TraesCS4D01G290700.1	60S ribosomal protein L18a
AWNLN	S5A_562540562	11.67	C/T	−1.71	11.3	TraesCS5A01G361300.1	Guanine nucleotide exchange family protein
FLLN	S1B_667135914	4.38	C/T	−0.16	20.8	TraesCS1B01G447400	Disease resistance protein RPM1
FLW	S6D_16376439	4.85	C/T	−0.02	13.3	TraesCS6D01G040100.1	Mitochondrial transcription termination factor-like
FLA	S1D_278097355	4.74	G/C	0.21	11.5	TraesCS1D01G197200.1	P-loop containing nucleoside triphosphate hydrolases superfamily protein
FLA	S6B_120860110	4.01	G/A	0.17	9.3	TraesCS6B01G125800	Cytochrome P450 family protein, expressed
FLA	S6B_120860130	4.01	A/T	−0.17	9.3	TraesCS6B01G125900	Cytochrome P450 family protein, expressed
STMDIA	S1D_431523575	6.58	A/G	−0.06	10.3	TraesCS1D01G341500	Disease resistance protein (NBS-LRR class) family
STMDIA	S3D_10133372	9.83	G/T	−0.11	8.6	TraesCS3D01G028500.1	Leucine-rich repeat receptor-like protein kinase family protein
STMDIA	S6A_94238211	6.9	T/G	0.06	7.5	TraesCS6A01G122200.1	Protein kinase, putative
RTLN	S5B_669373985	4.62	T/C	0.27	6.9	TraesCS5B01G502200	GRAM domain-containing protein/ABA-responsive
RTLN	S5B_669374027	4.62	T/C	0.27	6.9	TraesCS5B01G502200	GRAM domain-containing protein/ABA-responsive
RTLN	S6D_431108774	4.01	A/G	−0.27	5.8	TraesCS6D01G332800.1	Protein DETOXIFICATION
RTLN	S7A_94404310	4.01	G/A	0.52	7.5	TraesCS7A01G143200.2	Phosphatase 2C family protein

PVE: phenotypic variance explained; GY, grain yield; HI, harvest index; BMWT, biomass weight; TKW, thousand kernel weight; GVWT, grain volume weight; AWNLN, awn length; FLLN, flag leaf length; FLW, flag leaf width; FLA, flag leaf area; STMDIA, stem diameter; RTLN, root length. [a] S+chromosome_chromosome position in bp.

Table 4. List of significant markers associated with GY and yield-related traits, favorable alleles (underlined), SNP effects, and drought-related putative genes obtained from genome-wide association study of 123 drought stressed synthetic hexaploid wheats grown in 2017 in Konya, Turkey.

Trait	SNP [a]	$-\log_{10}(p)$	Alleles	SNP Effect	PVE (%)	Gene-ID	Annotation
GY	S3A_25012018	4.81	A/G	−20.02	12.7	TraesCS3A01G047300	F-box-domain-containing protein
GY	S3D_1203058	4.12	T/G	14.32	12.8	TraesCS3D01G002700	Disease resistance protein RPM1
BMWT	S3A_25012018	6.08	A/G	−59.44	14.4	TraesCS3A01G047300	F-box-domain-containing protein
TKW	S4B_11905230	8.94	C/G	−1.11	3.9	TraesCS4B01G016200.1	LOB domain-containing protein, putative
TKW	S4B_637722874	5.17	T/C	0.86	2.0	TraesCS4B01G344200.1	Zinc finger (C3HC4-type RING finger) family protein
GVWT	S1A_522189599	4.11	A/G	−0.55	2.5	TraesCS1A01G334800	Cytochrome P450
GVWT	S4A_73454791	5.64	C/T	−0.63	5.5	TraesCS4A01G074200.2	Microtubule associated protein family protein, putative, expressed
AWNLN	S5B_43896804	7.31	C/T	−1.13	6.0	TraesCS5B01G038700	F-box family protein
AWNLN	S6B_643657	10.55	C/T	−1.04	5.7	TraesCS6B01G001000	F-box family protein
FLLN	S1B_631203243	5.26	A/G	−0.21	9.8	TraesCS1B01G400600.1	Rp1-like protein
FLLN	S2B_140752747	4.15	G/C	0.29	1.6	TraesCS2B01G167500.1	Cytochrome P450, putative
FLLN	S2D_642055122	4.12	T/C	0.25	5.9	TraesCS2D01G579800	protein kinase family protein
FLLN	S4A_612662321	5.3	C/T	−0.23	5.3	TraesCS4A01G325200	F-box family protein
FLLN	S6D_463762312	5.72	G/A	0.25	6.0	TraesCS6D01G386300	Cytochrome P450, putative
FLW	S1A_516732460	6.99	A/G	−0.03	9.5	TraesCS1A01G326700.1	Citrate-binding protein
FLW	S6B_26200560	7.13	C/A	0.03	12.3	TraesCS6B01G042800	F-box family protein
FLA	S1A_516732460	6.9	A/G	−0.42	8.0	TraesCS1A01G326700.1	Citrate-binding protein
FLA	S2A_764065400	4.18	G/T	−0.19	3.8	TraesCS2A01G563200	NBS-LRR resistance-like protein
STMDIA	S6B_610963076	5.7	T/G	0.06	7.9	TraesCS6B01G346900-TraesCS6B01G347000	NBS-LRR disease resistance protein and F-box protein-like
RTLN	S2D_620326979	4.22	T/C	192.21	9.9	TraesCS2D01G541000.1	Disease resistance protein RPM1

PVE: phenotypic variance explained; GY, grain yield; HI, harvest index; BMWT, biomass weight; TKW, thousand kernel weight; GVWT, grain volume weight; AWNLN, awn length; FLLN, flag leaf length; FLW, flag leaf width; FLA, flag leaf area; STMDIA, stem diameter; RTLN, root length. [a] S+chromosome_chromosome position in bp.

Another major haplotype block (18 kb) of three MTAs was observed on chromosome 3A in 2017 (Figure 3) and the PVE on GY by this haplotype block was 13.1% (Figure 3). The chromosome 3A is known to be an important chromosome that contains useful QTLs for GY and yield-related traits [17–22,30–38] and the haplotype block identified will have a significance in the crop improvement program. All three MTAs present in this haplotype block of chromosome 3A were found in the gene, TraesCS3A01G047300 (Table S5), which was annotated as a member of the F-box gene family (Table 4). These three SNPs were indicated as having a moderate impact on the protein as they resulted in a missense mutation and caused an amino acid change. Such changes may alter the function of the protein [39], which makes this F-box gene a strong candidate for future functional characterization studies under drought tolerance in wheat. The F-box proteins are known to regulate many important biological processes, such as embryogenesis, floral development, plant growth and development, biotic and abiotic stresses, hormonal responses, and senescence [39]. Two other MTAs observed on chromosome 3A and 3D were present within genes (F-box family protein: TraesCS3A01G445100 and disease resistance protein RPM1: TraesCS3D01G002700) that have been previously reported to be involved in drought tolerance [39,40] (Table 4).

The GY haplotype blocks and other MTAs identified in the present study have not been mapped to date and four MTAs were in the genes, of which functional annotations suggested that they are likely involved in drought tolerance. This result implied that haplotype blocks observed on chromosome 3A (3 MTAs) and 7A (16 MTAs), and one MTA on chromosome 3D (1) for GY are novel and may potentially be used in a marker-assisted breeding program, focusing on improving drought tolerance in wheat after validating them in different populations and environments.

2.4.2. Harvest Index

A total of 10 SNPs significantly associated with HI were identified on chromosomes 1D, 2A, 2D, 3A, 3D, 5B, 6B, 6D, and 7B (Figure 2) with PVE ranging from 2.2% to 18.7% (Table S5). Previous studies have reported QTLs/MTAs responsible for HI on chromosomes 2D [27], 3A [27,30], 6B [41], and 7B [30]. To the best of our knowledge, the six MTAs identified for HI on chromosomes 1D, 2A, 3D, 5B, and 6D have not been reported and they are potentially novel MTAs responsible for HI.

Six MTAs for HI detected on chromosomes 2A, 3A, 6B, 6D, and 7B were found in genes and two of these genes have annotations suggesting their involvement in drought stress (Table 3 and Table S5). The two genes are WRKY transcription factor (TraesCS3A01G343700) found on chromosome 3A and cytochrome P450 (TraesCS6D01G170900.1) found on chromosome 6D. The role of WRKY transcription factor is well known in abiotic stresses including drought tolerance [42,43]. The cytochrome P450 genes are a large superfamily of enzymes and are involved in many metabolic pathways including drought tolerance in rice [44] and Arabidopsis [45,46]. The multi-trait marker associated with GY and HI was located on chromosome 5B (S5B_598463062) with PVE ranging from 15.9% to 18.7% (Table S5).

2.4.3. Biomass Weight

The 15 MTAs responsible for BMWT were identified on chromosomes 1D, 2B, 3A, 4A, 6D, and 7B (Figure 2) with PVE ranging from 4.9% to 14.4% (Table S5). Previous studies have reported QTLs/MTAs responsible for BMWT on chromosomes 1D [30,41], 2B [27], 6D [27], and 7B [30]. The four MTAs identified for BMWT on chromosome 3A and 4A have not been reported and they are potentially novel MTAs responsible for BMWT.

A novel haplotype block (38 kb) of three SNPs on chromosome 3A associated with BMWT was identified in 2017 (Figure 3) with PVE by the haplotype block of 11.7%. This MTA (S3A_25012018) was also associated with GY and PVE ranged from 12.7% (GY) to 14.4% (BMWT).

All three MTAs present in this haplotype block were within genes (Table S5) and one of the genes had annotations suggesting its involvement in drought tolerance was an F-box family protein (TraesCS3A01G047300) (Table 4) [39]. Excluding MTAs on haplotype block, eight MTAs for BMWT detected on chromosomes 1D, 2B, 6D, and 7B were found in genes (Table S5) and two of the genes had annotations suggesting its involvement in drought stress (Tables 3 and 4). The genes associated with two MTAs are F-box family protein (TraesCS7B01G242600) [39] and protein DETOXIFICATION containing multi-antimicrobial extrusion protein (MatE) (TraesCS1D01G357500) [23].

2.4.4. Thousand Kernel Weight

A total of 20 MTAs responsible for TKW were detected in 19 different genomic regions on chromosomes 1A, 2A, 2B, 2D, 3A, 3B, 4A, 4B, 4D, 5B, 6D, 7B, and 7D (Figure 2) with PVE ranging from 1.6% to 22.2% (Table S5). Earlier studies have reported QTLs/MTAs for TKW on chromosomes 1A [19,24,29,30], 2A [20], 2B [20,29,30], 2D [19], 3A [24,25,29], 3B [20,26], 4B [5], 5B [24], 7B [30], and 7D [20]. In the present study, only one MTA (S2D_7309581) responsible for TKW was detected on chromosome 2D in both years and assumed to be a stable MTA, because this MTA was detected despite significant genotype x year interaction. The five MTAs identified for TKW on chromosomes 4A, 4D, and 6D have not been previously reported and they are potentially novel MTAs responsible for TKW.

Twelve MTAs responsible for TKW detected on chromosomes 2A, 2B, 3A, 3B, 4A, 4B, 5B, and 6D were found in genes (Table S5) and five of these genes had annotations suggesting their involvement in drought stress (Tables 3 and 4). The genes associated with MTAs involved in drought tolerance are F-box family protein (chromosome 3A; TraesCS3A01G047300) [39], protein kinase family protein (chromosome 4A; TraesCS4A01G347600 and chromosome 6D; TraesCS6D01G360800) [47], cytochrome P450 family protein (chromosome 4D; TraesCS4D01G364700) [44–46], and zinc finger (C3HC4-type RING finger) family protein (chromosome 4B; TraesCS4B01G344200.1) [48–50]. The SNP S4D_509427923 was annotated as a missense variant and thus may have a moderate impact on the protein function (Table S5).

2.4.5. Grain Volume Weight

Thirteen MTAs responsible for GVWT were identified on chromosomes 1A, 2A, 2B, 2D, 3A, 4A, 5A, 6A, and 7A (Figure 2) with PVE ranging from 1.3% to 16.2% (Table S5). Earlier studies have reported QTLs/MTAs for GVWT on chromosomes 1A [33], 2A [33,51], 2B [33,51], 2D [33,51], 5A [33] and 7A [33,51]. Four MTAs identified for GVWT on chromosomes 3A, 4A, and 6A have not been previously reported and they are potentially novel MTAs responsible for GVWT.

Eight MTAs responsible for GVWT detected on chromosomes 1A, 2A, 2B, 4A, 6A, and 7A were found in genes (Table S5) and two of these genes had annotations suggesting their involvement in drought stress (Table 4). The genes associated with two MTAs involved in drought tolerance are cytochrome P450 (TraesCS1A01G334800) [44–46] on chromosome 1A and microtubule-associated protein family protein (TraesCS4A01G074200.2) [52] on chromosome 4A.

2.4.6. Awn Length

Twenty MTAs responsible for AWNLN were observed on chromosomes 1D, 2A, 2B, 3B, 4A, 4B, 4D, 5A, 5B, 5D, 6B, and 7A (Figure 2) with PVE ranging from 1.1% to 20.1% (Table S5). Earlier studies have reported QTLs/MTAs for AWNLN on chromosomes 2A [53,54], 4A [54], 4B [54], 5A [54], and 6B [53,54]. The nine MTAs identified for AWNLN on chromosomes 2B, 3B, 4D, 5B, 5D, and 7A have not been previously reported and they are potentially novel MTAs responsible for AWNLN.

Eleven MTAs responsible for AWNLN detected on chromosomes 1D, 2A, 4D, 5A, 5B, 6B, and 7A were found in genes (Table S5) and four of these genes had annotations suggesting their involvement in drought stress (Tables 3 and 4). The genes associated with four MTAs involved in drought tolerance are 60S ribosomal protein L18a (chromosome 4D; TraesCS4D01G290700) [55], guanine nucleotide exchange family protein (chromosome 5A; TraesCS5A01G361300) [56], and F-box family protein (chromosome 5B; TraesCS5B01G038700 and chromosome 6B; TraesCS6B01G001000) [39]. It has been reported that the putative 60S ribosomal protein L18a is an upregulated transcript in response to drought stress in ears and silks during the flowering stage in maize [55].

2.4.7. Flag Leaf Length

Thirteen MTAs responsible for FLLN were detected on chromosomes 1B, 1D, 2A, 2B, 2D, 4A, 6D, and 7B (Figure 2) with PVE ranging from 1.58% to 32.3% (Table S5). Previous studies have reported QTLs for FLLN on chromosomes 1B [57,58], 2B [57–60], 2D [57,61], 4A [57–59], and 7B [61]. The four MTAs identified for FLLN on chromosomes 1D, 2A, and 6D have not been previously reported and they are potentially novel MTAs responsible for FLLN.

Eleven MTAs responsible for FLLN detected on chromosomes 1B, 1D, 2B, 2D, 4A, 6D, and 7B were found in genes (Table S5) and four of these genes had annotations suggesting their involvement in drought stress (Tables 3 and 4). The genes associated with four MTAs involved in drought stress are F-box family protein (chromosome 4A: TraesCS4A01G325200) [39], cytochrome P450 (chromosome 2B; TraesCS2B01G167500 and chromosome 6D; TraesCS6D01G386300) [44–46], and Rp1-like protein (chromosome 1B; TraesCS1B01G400600) [40].

2.4.8. Flag Leaf Width

Sixteen MTAs responsible for FLW were detected on chromosomes 1A, 1B, 1D, 2B, 2D, 4B, 6B, and 6D (Figure 2) with PVE ranging from 1.6% to 15.2% (Table S5). Previous studies have found QTLs for FLW on chromosomes 1B [57,60,61], 1D [57,59], 2B [57,59], 2D [57,59,61], 4B [59,60], and 6B [57–59]. The two MTAs identified for FLW on chromosomes 1A and 6D have not been previously reported and they are potentially novel MTAs responsible for FLW.

Thirteen MTAs responsible for FLW detected on chromosomes 1A, 1B, 1D, 2D, 4B, 6B, and 6D were found in genes (Table S5) and three of these genes had annotations suggesting their involvement in drought stress (Tables 3 and 4). The genes associated with three MTAs involved in drought stress are citrate-binding protein (chromosome 1A; TraesCS1A01G326700) [62], F-box family protein (chromosome 6B; TraesCS6B01G042800) [39], and mitochondrial transcription termination factor-like (chromosome 6D; TraesCS6D01G040100) [63]. The SNPs S1A_516732460 and S6D_16376439 were annotated as a missense variant and thus may impact the function of the proteins that are annotated as citrate-binding protein and mitochondrial transcription termination factor-like protein, respectively (Table S5).

2.4.9. Flag Leaf Area

Twenty MTAs responsible for FLW were detected on chromosomes 1A, 1B, 1D, 2A, 2D, 4D, 5A, 6B, and 7D (Figure S2) with PVE ranging from 8.1% to 23.1% (Table S5). Previous studies have reported QTLs for FLA on chromosomes 1B [58,59], 1D [57,59,61], 2A [57,59,61], 2D [57,59,61], 4D [58], 5A [57,58,60,61], 6B [58], and 7D [61]. The four MTAs identified for FLA on chromosome 1A have not been previously reported and they are potentially novel MTAs responsible for FLA.

Three novel haplotype blocks were observed for FLA on chromosomes 1A (two MTAs), 6B (two MTAs) and 7D (three MTAs) with PVE by these haplotype block ranging from 5.5% to 8.6% (Figure 3). Fourteen MTAs responsible for FLA detected on chromosomes 1A, 1B, 1D, 2A, 4D, 5A, 6B, and 7D were found in genes (Table S5) and five of these genes had annotations suggesting their involvement in drought stress (Tables 3 and 4). The genes associated with five MTAs involved in drought stress are citrate-binding protein (TraesCS1A01G326700), P-loop containing nucleoside triphosphate hydrolases superfamily protein (TraesCS1D01G197200) [64], cytochrome P450 (TraesCS6B01G125900) [44–46], and NBS-LRR resistance-like protein [40].

The multi-trait marker associated with FLW and FLA was located on chromosome 1A (S1A_516732460) with PVE ranging from 8.0% to 9.5% (Table S4). Another multi-trait marker associated with FLLN and FLA was located on chromosome 2A (S2A_29874199) with PVE ranging from 23.1% to 32.3% (Table S4). The multi-trait MTA indicates that the related candidate gene may affect multiple traits.

2.4.10. Stem Diameter

In the present study, 23 MTAs responsible for STMDIA were identified on chromosomes 1A, 1D, 2B, 2D, 3A, 3B, 3D, 4D, 5A, 5B, 6A, 6B, 6D, 7A, and 7B (Figure 2) with PVE ranging from 2.7% to 28.8% (Table S5). Earlier study has identified one minor QTL (QSd-3B) for STMDIA on chromosome 3B that explained 8.7% of the phenotypic variance [65]. It means that, 19 MTAs detected on chromosomes 1A, 1D, 2B, 2D, 3A, 3D, 4D, 5A, 5B, 6A, 6B, 6D, 7A, and 7B except chromosome 3B in the present study may potentially be a novel MTAs controlling STMDIA under drought stress.

Four MTAs were detected on chromosome 3B and two of them (1 bp apart) were observed in one haplotype block in 2016 with PVE was 9.2% (Figure 3). Fifteen MTAs for STMDIA detected on chromosomes 1D, 2B, 3A, 3B, 6B, 6D, 7A, and 7B were found in genes (Table S5) and four of these genes had annotations suggesting their involvement in drought stress (Tables 3 and 4). The genes associated with four MTAs involved in drought stress are leucine-rich repeat receptor-like protein kinase family protein (TraesCS3D01G028500) [66], protein kinase (TraesCS6A01G122200.1) [47], disease resistance protein (NBS-LRR class) family (TraesCS1D01G341500 and TraesCS6B01G346900) [40], and F-box protein family (TraesCS6B01G347000) [39].

2.4.11. Root Length

RTLN is one of the most important traits under drought stress. We have measured RTLN 3–4 days after anthesis (Zadoks 60 growth stage) under the drought-stressed field condition using WinRhizo® (WinRhizo reg. 2009c, Regent Instruments Inc., Quebec City, QC, Canada). This trait is very unique compared to previous studies where they focused on the roots of seedlings [67–70] rather than direct field-based measurements (labor intensive, time consuming, and expensive). Identification of QTL governing RTLN is very important in wheat, especially for the wheat grown under drought stress. Limited information is available on QTL related to root traits in wheat [67,69–72].

In the present study, 15 MTAs responsible for RTLN were identified on chromosomes 2B, 2D, 3B, 5B, 6A, 6D, and 7A (Figure 2) with PVE ranging from 5.3% to 18.5% (Table S5). Earlier studies have reported QTLs for RTLN on chromosomes 2B [68], 3B [69], 5B [67,69], 6A [68,69], and 6D [67,70] in hexaploid wheat and on chromosomes 2B [71,72], 3B [72], 6A [72], and 7A [72] in tetraploid wheat. The MTA identified for RTLN on chromosome 2D has not been previously reported and it is potentially novel MTAs responsible for RTLN under drought stress. Furthermore, previous studies identified very few QTLs for RTLN on the D-genome of wheat [67,70]. Therefore, the MTAs (eight MTAs) for RTLN detected on the D-genome of SHWs in the present study are potentially novel.

Seven out of eight MTAs responsible for RTLN were present on chromosome 6D. Two haplotype blocks (the haplotype block1 with a size of 64 kb and the haplotype block2 with a size of 5kb) were identified from five out of seven MTAs for RTLN on chromosome 6D with PVE ranging from 5.0% to 11.8% (Figure 3). One SNP (S6D_435300571) present in the haplotype block2 was found in the gene (TraesCS6D01G332800) with PVE of 13.0%. The gene associated with this SNP is protein detoxification gene-containing multi-antimicrobial extrusion protein (MatE) (Table 3) and has been reported to be expressed mainly in the root than shoots under drought stress [73]. For instance, *MatE* family genes such as *HvAACT1* in barley [74] and *TaMate* in wheat [75], encode proteins that are primarily localized to root epidermis cells [74] and required for external resistance [23]. In the present study, this gene was also significantly associated with BMWT on chromosome 1D. This result implied that that this gene plays an important role for RTLN and BMWT in drought-stressed conditions.

Excluding MTAs on haplotype block, eight MTAs for RTLN detected on chromosomes 2D, 3B, 5B, 6A, and 7A were found in genes (Table S5) and four of these genes had annotations suggesting their involvement in drought stress (Tables 3 and 4). The genes associated with four MTAs involved in drought stress are GRAM domain-containing protein/ABA-responsive (TraesCS5B01G502200) [76–81], phosphatase 2C family protein (TraesCS7A01G143200.2) [82], and disease resistance protein RPM1 (TraesCS2D01G541000.1) [40]. The SNPs S7A_94404310 was annotated as a missense variant and thus may have a moderate impact on the protein function (Table S5).

2.5. Potential Candidate Gene Annotations Affecting Yield and Yield-related Traits under Drought Stress

This study identified ~194 MTAs present on different chromosomes and associated with multiple traits. These 62 MTAs were associated with either the same trait in multiple years (MTA stability in different environments) or multiple traits within the same year or across years (suggesting epistasis) (Table S5). Additionally, ~45 of the MTAs were present in genes with annotations relevant to the respective trait under drought stress (Tables 3 and 4). Interestingly, we noticed MTAs associated with the same or related traits were located within genes that had the exact same annotation (Figure 4; and Table S6). For instance, some of the MTAs for GY (2 MTAs), BMWT (2), TKW (1), AWNLN (2), FLLN (1), FLW (1), and STMDIA (1) were located within genes annotated as F-box family protein (Table S6). Similarly, the genes annotated as cytochrome P450 harbored MTAs for HI (1), TKW (1), FLA (2), GVWT (1), and FLLN (2). Additional examples are provided in Figure 4 and Table S6 in Supplementary Materials. This result indicated the likely gene families that are important for GY and yield-related traits under drought stress.

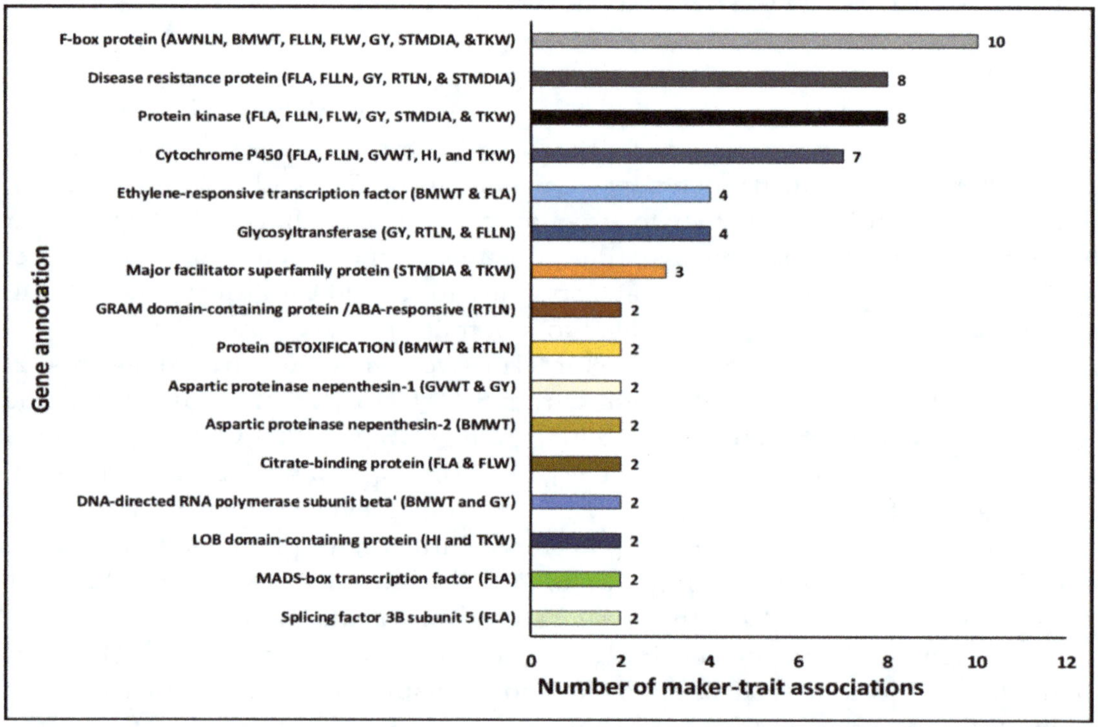

Figure 4. Potential candidate gene functions harboring SNPs affecting yield and yield-related traits under drought stress. The count of marker–trait associations (for either single or multiple traits) located within genes that have the same gene annotation is shown. AWLN, awn length; BMWT, biomass weight; FLA, flag leaf area; FLLN, flag leaf length; FLW, flag leaf width; GVWT, grain volume weight; GY, grain yield; HI, harvest index; RTLN, root length; STMDIA, stem diameter; and TKW, thousand kernel weight.

3. Materials and Methods

3.1. Site Description

A field experiment was conducted during two growing seasons (2016 and 2017) under drought stressed conditions (rainfed) at the research farm located at the Bahri Dagdas International Agricultural Research Institute in Konya, Turkey (37°51′15.894″ N, 32°34′3.936″ E; Elevation = 1021 m). This site was characterized by a low precipitation (below 300 mm), low humidity, and slightly alkaline clay loam soil [83].

3.2. Plant Materials and Experimental Design

One hundred twenty-three SHWs developed from two introgression programs were used (Table S1). The first group was developed by Kyoto University, Japan from one spring durum ('Langdon') parent crossed with 14 different *Ae. tauschii* accessions resulting in 14 different lines (Table S2). The remaining 109 lines were the second group of synthetics that were developed by International Maize and Wheat Improvement Center (CIMMYT) from the six durum parents crossed with 11 different *Ae. tauschii* accessions mainly from the Caspian Sea Basin area. Initially, 13 crosses among six winter durum wheats were involved in the creation of 13 different winter-type synthetics. However, due to the partial sterility, segregation, and continuous selection in the early generation, 109 lines were selected as unique lines because of their differences in phenotype [14] and their kinship values [11]. The synthetic genotypes used in the present study are unique and have been developed recently and tested for their agronomic traits [14], genetic diversity, and population structure [11]. The detailed information of these SHWs were provided in Bhatta et al. [11].

The experimental design was an augmented design (plot size: 1.2 m × 5 m) with replicated checks ('Gerek' and 'Karahan') in the 2016 growing season and modified alpha-lattice design (plot size: 1.2 m × 5 m) including 123 SHWs and replicated checks ('Gerek' and 'Karahan') with two replications in the 2017 growing season. The SHWs were planted on 20 September in 2015 and harvested on 18 July 2016 for the 2016 growing season, whereas the SHWs were planted on 15 September in 2016 and harvested on 21 July 2017 for the 2017 growing season.

3.3. Trait Measurements

The GY was obtained by harvesting four middle rows of 1.008 m^2 (i.e., 84 m × 120 m) and reported in $g{\cdot}m^{-2}$. The HI, BMWT, TKW, GVWT, AWNLN, FLLN, FLW, and FLA (0.8 × FLLN × FLW) were measured using previously reported protocols [18,32,59,61]. The STMDIA was measured from five randomly selected plants per plot using a digital Vernier caliper at the second internode from the soil surface at physiological maturity. The RTLN was measured from randomly selected three plants per plot after 3–4 days of flowering (Zadoks 60 growth stage) using WinRhizo software (WinRhizo reg. 2009c, Regent Instruments Inc., Quebec City, QC, Canada).

3.4. Phenotypic Data Analysis

A combined ANOVA was performed using the following model:

$$y_{\mathrm{ijklmn}} = \mu + Yr_{\mathrm{i}} + R(Yr)_{\mathrm{ji}} + B(R(Yr))_{\mathrm{kji}} + C_{\mathrm{l}} + G_{\mathrm{m(kji)}} + GXYr_{\mathrm{mi}} + e_{\mathrm{ijklmn}} \quad (1)$$

where y_{ijklm} is the GY and yield-related trait; μ is overall mean; Yr_{i} is the effect of ith year; $R(Yr)_{\mathrm{ji}}$ is the effect of jth replication within the ith year; $B(R(Yr))_{\mathrm{kji}}$ is the effect of kth incomplete block within jth replication of ith environment; C_{l} is the lth checks; G_{mkji} is the effect of mth genotypes (new variable, where check is coded as 0 and entry is coded is 1 and the genotype is taken as a new variable × entry) within the kth incomplete block of jth replication in the ith year; $GXYr_{\mathrm{mi}}$ is the interaction effect of mth genotype and ith year; and e_{ijklmn} is the residual. In the combined ANOVA, year and check were assumed as fixed effects, whereas genotype, genotype × year interaction, replication nested within a year, and incomplete block nested within replications were assumed as random effects.

Individual analyses of variance were performed because most of the traits had highly significant genotype × year interaction and therefore will be discussed hereafter. An augmented design was analyzed using the following model for the estimation of BLUPs in the year 2016:

$$y_{\mathrm{ijkl}} = \mu + B_{\mathrm{i}} + C_{\mathrm{j}} + G_{\mathrm{k(i)}} + e_{\mathrm{ijkl}} \quad (2)$$

where y_{ijkl} is the trait, μ is the overall mean; B_{i} is the effect of ith incomplete block; C_{j} is the jth check; $G_{\mathrm{k(i)}}$ (new variable, where check is coded as 0 and entry is coded as 1 and genotype is taken as a new variable × entry) is the effect of kth genotypes within the ith block; e_{ijkl} is the residual. In ANOVA calculated for the 2016 datasets, the check was assumed as a fixed effect, whereas genotype and incomplete block were assumed as random effects.

An alpha (α) lattice design with two replications was analyzed using the following model for the estimation of BLUPs in the year 2017:

$$y_{\mathrm{ijkl}} = \mu + R_{\mathrm{i}} + B(R)_{\mathrm{ji}} + C_{\mathrm{k}} + G_{\mathrm{l(ji)}} + e_{\mathrm{ijkl}} \tag{3}$$

where y_{ijk} is the trait, μ is the overall mean; R_i is the effect of ith replication; $B(R)_{ji}$ is the effect of jth block within the ith replication; C_k is the kth checks; G_{lji} (new variable, where check is coded as 0 and entry is coded as 1 and the genotype is taken as a new variable × entry) is the effect of kth genotypes within jth incomplete block of ith replication; e_{ijkl} is the residual. In ANOVA calculated for the 2017 datasets, the check was assumed as a fixed effect, whereas genotype, replication, and incomplete block nested within replication were assumed as random effects.

All phenotypic data were analyzed using PROC MIXED in SAS 9.4 (SAS Institute Inc., Cary, NC) [84] using the restricted maximum likelihood (REML) approach unless mentioned otherwise.

Broad-sense heritability for each trait in each year was calculated based on entry mean basis using Equations (4–6) for 2016, 2017, and combine0d experiments, respectively:

$$H^2 = \frac{\sigma^2 g}{\sigma^2 g + \sigma^2 e} \tag{4}$$

$$H^2 = \frac{\sigma^2 g}{\sigma^2 g + \frac{\sigma^2 e}{r}} \tag{5}$$

$$H^2 = \frac{\sigma^2 g}{\sigma^2 g + \frac{\sigma^2 yr}{n} + \frac{\sigma^2 gxyr}{nr}} \tag{6}$$

where $\sigma^2{}_{g}$, $\sigma^2{}_{yr}$, $\sigma^2{}_{gxyr}$, and $\sigma^2{}_{e}$ are the variance components for genotype, year, genotype × year, and error, respectively, and n and r are the numbers of years and replications, respectively.

Pearson's correlation of GY and yield-related traits was calculated based on BLUPs for each trait in each year using PROC CORR in SAS. The PC biplot analysis (PCA-biplot) was performed based on the correlation matrix to avoid any variation due to the different scales of the measured variables using 'factoextra' package in R software [85].

3.5. Genotyping and SNP Discovery

Genomic DNA was extracted from two to three fresh young leaves of 14-day-old seedlings using BioSprint 96 Plant Kits (Qiagen, Hombrechtikon, Switzerland), as described in Bhatta et al. [11]. The GBS libraries were constructed in 96-plex following digestion with two restriction enzymes, PstI and MspI [86] and pooled libraries were sequenced using Illumina, Inc. (San Diego, CA, USA) next-generation sequencing platforms at the Wheat Genetics Resource Center at Kansas State University (Manhattan, KS, USA). SNP calling was performed using TASSEL v. 5.2.40 GBS v2 Pipeline (available online: https://bitbucket.org/tasseladmin/tassel-5-source/wiki/Tassel5GBSv2Pipeline) [87] with physical alignment to the Chinese Spring genome sequence (RefSeq v1.0) made available by the IWGSC [88] using default settings with the one exception that the number of times for a GBS tag to be present and included for SNP calling was changed from the default value of 1 to 5 to increase the stringency in SNP calling. The identified SNPs with MAF less than 5% and missing data more than 20% were removed from the analysis. All lines had missing data less than 20% and none of them were dropped due to missing percentage-filtering criterion. The GBS-derived SNPs are provided in Table S3 in Supplementary Materials.

3.6. Population Structure and Genome-Wide Association Study Analysis

Population structure of 123 genotypes was assessed using the Bayesian clustering algorithm in the program STRUCTURE v 2.3.4 (available online: https://web.stanford.edu/group/pritchardlab/structure_software/release_versions/v2.3.4/html/structure.html) [89] and principal component (PC) analysis using TASSEL (available online: http://www.maizegenetics.net/tassel) [90], as described in Bhatta et al. [11].

Many GWASs were previously performed using the mixed linear model (MLM), where the population structure (Q) or PC was set as a fixed effect and kinship (K) as a random effect to control false positives [91,92]. However, the MLM may lead to confounding between population structure, kinship, and quantitative trait nucleotides (QTNs) that results in false negatives due to model overfitting [78]. Recently, the multilocus mixed model (MLMM), which tests multiple markers simultaneously by fitting pseudo QTNs, in addition to testing markers in stepwise MLM, has been proposed, which is advantageous over conventional GLM and MLM testing one marker at a time [78]. One of the examples of recently popular GWAS analysis algorithm that is based on MLMM is FarmCPU [78,79]. The FarmCPU uses a fixed effect model (FEM) and a random effect model (REM) iteratively to remove the confounding between testing markers and kinship that results in false negatives, prevents model overfitting, and control false positives simultaneously [78]. Therefore, GWAS was performed on the adjusted BLUPs for each trait in each year to identify SNPs associated with GY and yield-related traits in SHWs using FarmCPU with population structures (Q_1 and Q_2) or first three principal components (PC_1, PC_2, and PC_3) as covariates by looking at the model fit using Quantile-Quantile (Q-Q) plots and FarmCPU-calculated kinship [78] implemented in MVP R software package (available online: https://github.com/XiaoleiLiuBio/MVP). A uniform suggestive genome-wide significance threshold level of p-value = 9.99×10^{-5} ($-\log_{10}p = 4.00$) was selected for MTAs considering the deviation of the observed test statistics values from the expected test statistics values in the Q-Q plots [28,80] from the two-year results of the present study.

3.7. Haplotype Block Analysis

Haplotype blocks with linkage disequilibrium (LD) values (squared correlation coefficient between locus allele frequency; $R^2 > 0.2$) in adjacent regions (<500 kb) of significant MTAs were visualized and plotted using default parameters (Hardy–Weinberg p-value cut off at 1% and MAF > 0.001) of Haploview software (available online: https://www.broadinstitute.org/haploview/haploview) [81]. PVE by each haplotype block on the trait of interest was calculated using multiple regression analysis that accounted for the population structure by removing the haplotype allele of less than 5% in SAS using PROC REG.

3.8. Putative Candidate Gene Analysis

The genes underlying the MTAs and subsequently their annotations were retrieved using a Perl script and the IWGSC RefSeq v1.0 annotations [88] provided for the Chinese Spring. The underlying genes were further examined for their association with GY and yield-related traits under drought stress using previously published literature. Additionally, the SnpEff program (available online: http://snpeff.sourceforge.net/) was used for SNP annotation and predicting the effects of SNPs on the protein function. The MTAs present within genes or flanked (5 kb) by genes were investigated [93].

4. Conclusions

The present study showed SHWs have large amounts of genetic variation for GY and yield-related traits. The GWAS in 123 SHWs using 35,648 SNPs identified several novel (90 MTAs: 45 MTAs on the A genome, 11 on the B genome, and 34 on the D-genome) genomic regions or haplotype blocks associated with GY and yield-related traits in drought-stressed conditions. Most of the MTAs (120 MTAs) were present in genes ad several of them (45 MTAs) were annotated with functions related to drought stress.

This provided further evidence for the reliability of the MTAs identified. We also identified MTAs on different chromosomes associated with multiple traits but within genes having the same annotation. This resulted in the identification of candidate genes belonging to the same gene family that likely have a major role in affecting GY and yield-related traits under drought stress in SHWs. The large number of MTAs, especially on the D-genome (53 MTAs with 34 MTAs being novel) identified in the present study, demonstrate the potential of SHWs for elucidating the genetic architecture of complex traits and provide an opportunity for further improvement of wheat under rapidly growing drought-stressed environment worldwide.

Author Contributions: Conceptualization, M.B., A.M., and P.S.B.; methodology, M.B.; validation, M.B.; formal analysis, M.B.; investigation, M.B.; resources, M.B., A.M. and P.S.B.; data curation, M.B. and V.B.; writing of the original draft preparation, M.B.; writing of review and editing, M.B., A.M., V.B., and P.S.B.; supervision, A.M. and P.S.B.; project administration, A.M. and P.S.B.

Acknowledgments: We would like to thank CIMMYT-Turkey for providing seeds of synthetic hexaploid wheat and Jesse Poland for providing the GBS data. We would like to acknowledge CIMMYT-Turkey staffs including Adem Urglu and Ibrahim Ozturk, Seregey Schepelew from Omsk, Russia, Bahri Dagdas International Agricultural Research Institute staffs including Emel Özer, Fatih Özdemir, and Enes Yakişir.

Abbreviations

ANOVA	analysis of variance
AWNLN	awn length
BLUP	best linear unbiased predictor
BMWT	biomass weight
FarmCPU	fixed and random model circulating probability unification
FLA	flag leaf area
FLLN	flag leaf length
FLW	flag leaf width
GBS	genotyping-by-sequencing
GVWT	grain volume weight
GWAS	genome wide association study
GY	grain yield
HI	harvest index
IWGC	international wheat genome sequencing consortium
MAF	minor allele frequency
QTN	quantitative trait nucleotide
MLM	mixed linear model
MLMM	multilocus mixed model
MTA	marker trait association
QTL	quantitative trait loci
RTLN	root length
SHW	synthetic hexaploid wheat
SNP	single nucleotide polymorphism
STMDIA	stem diameter
TKW	thousand kernel weight

References

1. Kang, Y.; Khan, S.; Ma, X. Climate change impacts on crop yield, crop water productivity and food security—A review. *Prog. Nat. Sci.* **2009**, *19*, 1665–1674. [CrossRef]
2. Becker, S.R.; Byrne, P.F.; Reid, S.D.; Bauerle, W.L.; McKay, J.K.; Haley, S.D. Root traits contributing to drought tolerance of synthetic hexaploid wheat in a greenhouse study. *Euphytica* **2016**, *207*, 213–224. [CrossRef]

3. Gupta, P.; Balyan, H.; Gahlaut, V.; Gupta, P.K.; Balyan, H.S.; Gahlaut, V. QTL analysis for drought tolerance in wheat: Present status and future possibilities. *Agronomy* **2017**, *7*, 5. [CrossRef]
4. Smith, A.B.; Matthews, J.L. Quantifying uncertainty and variable sensitivity within the US billion-dollar weather and climate disaster cost estimates. *Nat. Hazards* **2015**, *77*, 1829–1851. [CrossRef]
5. Pinto, R.S.; Reynolds, M.P.; Mathews, K.L.; McIntyre, C.L.; Olivares-Villegas, J.-J.; Chapman, S.C. Heat and drought adaptive QTL in a wheat population designed to minimize confounding agronomic effects. *Theor. Appl. Genet.* **2010**, *121*, 1001–1021. [CrossRef] [PubMed]
6. Gummadov, N.; Keser, M.; Akin, B.; Cakmak, M.; Mert, Z.; Taner, S.; Ozturk, I.; Topal, A.; Yazar, S.; Morgounov, A. Genetic gains in wheat in Turkey: Winter wheat for irrigated conditions. *Crop J.* **2015**, *3*, 507–516. [CrossRef]
7. Cavanagh, C.R.; Chao, S.; Wang, S.; Huang, B.E.; Stephen, S.; Kiani, S.; Forrest, K.; Saintenac, C.; Brown-Guedira, G.L.; Akhunova, A.; et al. Genome-wide comparative diversity uncovers multiple targets of selection for improvement in hexaploid wheat landraces and cultivars. *Proc. Natl. Acad. Sci. USA* **2013**, *110*, 8057–8062. [CrossRef] [PubMed]
8. Cox, T. Deepening the wheat gene pool. *J. Crop Prod.* **1997**, *1*, 145–168. [CrossRef]
9. Dreisigacker, S.; Kishii, M.; Lage, J.; Warburton, M. Use of synthetic hexaploid wheat to increase diversity for CIMMYT bread wheat improvement. *Aust. J. Agric Res.* **2008**, *59*, 413–420. [CrossRef]
10. Ogbonnaya, F. C.; Abdalla, O.; Mujeeb-Kazi, A.; Kazi, A.G.; Xu, S.S.; Gosman, N.; Lagudah, E.S.; Bonnett, D.; Sorrells, M.E.; Tsujimoto, H. Synthetic hexaploids: Harnessing species of the primary gene pool for wheat improvement. *Plant Breed. Rev.* **2013**, *37*, 35–122. [CrossRef]
11. Bhatta, M.; Morgounov, A.; Belamkar, V.; Poland, J.; Baenziger, P.S. Unlocking the novel genetic diversity and population structure of synthetic hexaploid wheat. *BMC Genom.* **2018**, *19*, 591. [CrossRef] [PubMed]
12. Zegeye, H.; Rasheed, A.; Makdis, F.; Badebo, A.; Ogbonnaya, F.C. Genome-wide association mapping for seedling and adult plant resistance to stripe rust in synthetic hexaploid wheat. *PLoS ONE* **2014**, *9*, e105593. [CrossRef] [PubMed]
13. Das, M.K.; Bai, G.; Mujeeb-Kazi, A.; Rajaram, S. Genetic diversity among synthetic hexaploid wheat accessions (*Triticum aestivum*) with resistance to several fungal diseases. *Genet. Resour. Crop Evol.* **2016**, *63*, 1285–1296. [CrossRef]
14. Morgounov, A.; Abugalieva, A.; Akan, K.; Akın, B.; Baenziger, S.; Bhatta, M.; Dababat, A.A.; Demir, L.; Dutbayev, Y.; El Bouhssini, M.; et al. High-yielding winter synthetic hexaploid wheats resistant to multiple diseases and pests. *Plant Genet. Resour.* **2018**, *16*, 273–278. [CrossRef]
15. Ogbonnaya, F.C.; Imtiaz, M.; Bariana, H.S.; McLean, M.; Shankar, M.M.; Hollaway, G.J.; Trethowan, R.M.; Lagudah, E.S.; Van Ginkel, M. Mining synthetic hexaploids for multiple disease resistance to improve bread wheat. *Aust. J. Agric Res.* **2008**, *59*, 421–431. [CrossRef]
16. Jighly, A.; Alagu, M.; Makdis, F.; Singh, M.; Singh, S.; Emebiri, L.C.; Ogbonnaya, F.C. Genomic regions conferring resistance to multiple fungal pathogens in synthetic hexaploid wheat. *Mol. Breed.* **2016**, *36*, 1–19. [CrossRef]
17. Sehgal, D.; Autrique, E.; Singh, R.; Ellis, M.; Singh, S.; Dreisigacker, S. Identification of genomic regions for grain yield and yield stability and their epistatic interactions. *Sci. Rep.* **2017**, *7*. [CrossRef] [PubMed]
18. Bhatta, M.; Eskridge, K.M.; Rose, D.J.; Santra, D.K.; Baenziger, P.S.; Regassa, T. Seeding Rate, genotype, and topdressed nitrogen effects on yield and agronomic characteristics of winter wheat. *Crop Sci.* **2017**, *57*, 951–963. [CrossRef]
19. Ogbonnaya, F.C.; Rasheed, A.; Okechukwu, E.C.; Jighly, A.; Makdis, F.; Wuletaw, T.; Hagras, A.; Uguru, M.I.; Agbo, C.U. Genome-wide association study for agronomic and physiological traits in spring wheat evaluated in a range of heat prone environments. *Theor. Appl. Genet.* **2017**, *130*, 1819–1835. [CrossRef] [PubMed]
20. Sukumaran, S.; Lopes, M.; Dreisigacker, S.; Reynolds, M. Genetic analysis of multi-environmental spring wheat trials identifies genomic regions for locus-specific trade-offs for grain weight and grain number. *Theor. Appl. Genet.* **2018**, *131*, 985–998. [CrossRef] [PubMed]
21. Murtas, G. A Nuclear Protease Required for Flowering-Time Regulation in Arabidopsis Reduces the Abundance of SMALL UBIQUITIN-RELATED MODIFIER Conjugates. *Plant Cell* **2003**, *15*, 2308–2319. [CrossRef] [PubMed]
22. Choudhary, M.K.; Basu, D.; Datta, A.; Chakraborty, N.; Chakraborty, S. Dehydration-responsive nuclear proteome of rice (*Oryza sativa* L.) illustrates protein network, novel regulators of cellular adaptation, and evolutionary perspective. *Mol. Cell. Proteomics* **2009**, *8*, 1579–1598. [CrossRef] [PubMed]

23. Yokosho, K.; Yamaji, N.; Fujii-Kashino, M.; Ma, J.F. Functional Analysis of a MATE gene OsFRDL2 revealed its involvement in Al-induced secretion of citrate, but a lower contribution to Al tolerance in rice. *Plant Cell Physiol.* **2016**, *57*, 976–985. [CrossRef] [PubMed]
24. Sun, C.; Zhang, F.; Yan, X.; Zhang, X.; Dong, Z.; Cui, D.; Chen, F. Genome-wide association study for 13 agronomic traits reveals distribution of superior alleles in bread wheat from the Yellow and Huai Valley of China. *Plant Biotechnol. J.* **2017**, *15*, 953–969. [CrossRef] [PubMed]
25. Zanke, C.D.; Ling, J.; Plieske, J.; Kollers, S.; Ebmeyer, E.; Korzun, V.; Argillier, O.; Stiewe, G.; Hinze, M.; Neumann, K.; et al. Whole genome association mapping of plant height in winter wheat (*Triticum aestivum* L.). *PLoS ONE* **2014**, *9*, e113287. [CrossRef] [PubMed]
26. Golabadi, M.; Arzani, A.; Mirmohammadi Maibody, S.A.M.; Sayed Tabatabaei, B.E.; Mohammadi, S.A. Identification of microsatellite markers linked with yield components under drought stress at terminal growth stages in durum wheat. *Euphytica* **2011**, *177*, 207–221. [CrossRef]
27. Kumar, N.; Kulwal, P.L.; Balyan, H.S.; Gupta, P.K. QTL mapping for yield and yield contributing traits in two mapping populations of bread wheat. *Mol. Breed.* **2007**, *19*, 163–177. [CrossRef]
28. Sukumaran, S.; Dreisigacker, S.; Lopes, M.; Chavez, P.; Reynolds, M.P. Genome-wide association study for grain yield and related traits in an elite spring wheat population grown in temperate irrigated environments. *Theor. Appl. Genet.* **2015**, *128*, 353–363. [CrossRef] [PubMed]
29. Lozada, D.N.; Mason, R.E.; Babar, M.A.; Carver, B.F.; Guedira, G.B.; Merrill, K.; Arguello, M.N.; Acuna, A.; Vieira, L.; Holder, A.; et al. Association mapping reveals loci associated with multiple traits that affect grain yield and adaptation in soft winter wheat. *Euphytica* **2017**, *213*. [CrossRef]
30. Neumann, K.; Kobiljski, B.; Denčić, S.; Varshney, R.K.; Börner, A. Genome-wide association mapping: A case study in bread wheat (*Triticum aestivum* L.). *Mol. Breed.* **2011**, *27*, 37–58. [CrossRef]
31. Edae, E.A.; Byrne, P.F.; Haley, S.D.; Lopes, M.S.; Reynolds, M.P. Genome-wide association mapping of yield and yield components of spring wheat under contrasting moisture regimes. *Theor. Appl. Genet.* **2014**, *127*, 791–807. [CrossRef] [PubMed]
32. Bhatta, M.; Regassa, T.; Rose, D.J.; Baenziger, P.S.; Eskridge, K.M.; Santra, D.K.; Poudel, R. Genotype, environment, seeding rate, and top-dressed nitrogen effects on end-use quality of modern Nebraska winter wheat. *J. Sci. Food Agric.* **2017**, *97*, 5311–5318. [CrossRef] [PubMed]
33. Bordes, J.; Goudemand, E.; Duchalais, L.; Chevarin, L.; Oury, F.X.; Heumez, E.; Lapierre, A.; Perretant, M.R.; Rolland, B.; Beghin, D.; et al. Genome-wide association mapping of three important traits using bread wheat elite breeding populations. *Mol. Breed.* **2014**, *33*, 755–768. [CrossRef]
34. Hoffstetter, A.; Cabrera, A.; Sneller, C. Identifying quantitative trait loci for economic traits in an elite soft red winter wheat population. *Crop Sci.* **2016**, *56*, 547. [CrossRef]
35. Wang, S.-X.; Zhu, Y.-L.; Zhang, D.-X.; Shao, H.; Liu, P.; Hu, J.-B.; Zhang, H.; Zhang, H.-P.; Chang, C.; Lu, J.; et al. Genome-wide association study for grain yield and related traits in elite wheat varieties and advanced lines using SNP markers. *PLOS ONE* **2017**, *12*, e0188662. [CrossRef] [PubMed]
36. Catala, R.; Ouyang, J.; Abreu, I.A.; Hu, Y.; Seo, H.; Zhang, X.; Chua, N.-H. The Arabidopsis E3 SUMO Ligase SIZ1 Regulates Plant Growth and Drought Responses. *Plant Cell* **2007**, *19*, 2952–2966. [CrossRef] [PubMed]
37. Park, H.J.; Kim, W.-Y.; Park, H.C.; Lee, S.Y.; Bohnert, H.J.; Yun, D.-J. SUMO and SUMOylation in plants. *Mol. Cells* **2011**, *32*, 305–316. [CrossRef] [PubMed]
38. Ali, M.L.; Baenziger, P.S.; Ajlouni, Z.A.; Campbell, B.T.; Gill, K.S.; Eskridge, K.M.; Mujeeb-Kazi, A.; Dweikat, I. Mapping QTL for agronomic traits on wheat chromosome 3a and a comparison of recombinant inbred chromosome line populations. *Crop Sci.* **2011**, *51*, 553–566. [CrossRef]
39. Lechner, E.; Achard, P.; Vansiri, A.; Potuschak, T.; Genschik, P. F-box proteins everywhere. *Curr. Opin. Plant Biol.* **2006**, *9*, 631–638. [CrossRef] [PubMed]
40. Lee, H.-A.; Yeom, S.-I. Plant NB-LRR proteins: Tightly regulated sensors in a complex manner. *Brief. Funct. Genomics* **2015**, *14*, 233–242. [CrossRef] [PubMed]
41. Ain, Q.; Rasheed, A.; Anwar, A.; Mahmood, T.; Imtiaz, M.; Mahmood, T.; Xia, X.; He, Z.; Quraishi, U.M. Genome-wide association for grain yield under rainfed conditions in historical wheat cultivars from Pakistan. *Front. Plant Sci.* **2015**, *6*. [CrossRef] [PubMed]
42. He, G.-H.; Xu, J.-Y.; Wang, Y.-X.; Liu, J.-M.; Li, P.-S.; Chen, M.; Ma, Y.-Z.; Xu, Z.-S. Drought-responsive WRKY

transcription factor genes TaWRKY1 and TaWRKY33 from wheat confer drought and/or heat resistance in Arabidopsis. *BMC Plant Biol.* **2016**, *16*, 116. [CrossRef] [PubMed]

43. Ning, P.; Liu, C.; Kang, J.; Lv, J. Genome-wide analysis of WRKY transcription factors in wheat (*Triticum aestivum* L.) and differential expression under water deficit condition. *PeerJ* **2017**, *5*, e3232. [CrossRef] [PubMed]
44. Tamiru, M.; Undan, J. R.; Takagi, H.; Abe, A.; Yoshida, K.; Undan, J. Q.; Natsume, S.; Uemura, A.; Saitoh, H.; Matsumura, H.; Urasaki, N.; Yokota, T.; Terauchi, R. A cytochrome P450, OsDSS1, is involved in growth and drought stress responses in rice (*Oryza sativa* L.). *Plant Mol. Biol.* **2015**, *88*, 85–99. [CrossRef] [PubMed]
45. Seki, M.; Narusaka, M.; Ishida, J.; Nanjo, T.; Fujita, M.; Oono, Y.; Kamiya, A.; Nakajima, M.; Enju, A.; Sakurai, T.; et al. Monitoring the expression profiles of 7000 Arabidopsis genes under drought, cold and high-salinity stresses using a full-length cDNA microarray: Expression profiling under abiotic stresses. *Plant J.* **2002**, *31*, 279–292. [CrossRef] [PubMed]
46. Kushiro, T.; Okamoto, M.; Nakabayashi, K.; Yamagishi, K.; Kitamura, S.; Asami, T.; Hirai, N.; Koshiba, T.; Kamiya, Y.; Nambara, E. The Arabidopsis cytochrome P450 CYP707A encodes ABA 8′-hydroxylases: key enzymes in ABA catabolism. *EMBO J.* **2004**, *23*, 1647–1656. [CrossRef] [PubMed]
47. Yan, J.; Su, P.; Wei, Z.; Nevo, E.; Kong, L. Genome-wide identification, classification, evolutionary analysis and gene expression patterns of the protein kinase gene family in wheat and Aegilops tauschii. *Plant Mol. Biol.* **2017**, *95*, 227–242. [CrossRef] [PubMed]
48. Sakamoto, H.; Maruyama, K.; Sakuma, Y.; Meshi, T.; Iwabuchi, M.; Shinozaki, K.; Yamaguchi-Shinozaki, K. Arabidopsis Cys2/His2-Type zinc-finger proteins function as transcription repressors under drought, cold, and high-salinity stress conditions. *Plant Physiol.* **2004**, *136*, 2734–2746. [CrossRef] [PubMed]
49. Ciftci-Yilmaz, S.; Morsy, M.R.; Song, L.; Coutu, A.; Krizek, B.A.; Lewis, M.W.; Warren, D.; Cushman, J.; Connolly, E.L.; Mittler, R. The EAR-motif of the Cys2/His2-type zinc finger protein Zat7 plays a key role in the defense response of Arabidopsis to salinity stress. *J. Biol. Chem.* **2007**, *282*, 9260–9268. [CrossRef] [PubMed]
50. Ma, K.; Xiao, J.; Li, X.; Zhang, Q.; Lian, X. Sequence and expression analysis of the C3HC4-type RING finger gene family in rice. *Gene* **2009**, *444*, 33–45. [CrossRef] [PubMed]
51. Cabral, A.L.; Jordan, M.C.; Larson, G.; Somers, D.J.; Humphreys, D.G.; McCartney, C.A. Relationship between QTL for grain shape, grain weight, test weight, milling yield, and plant height in the spring wheat cross RL4452/'AC Domain'. *PLoS ONE* **2018**, *13*, e0190681. [CrossRef] [PubMed]
52. Zhou, J.; Wang, X.; Jiao, Y.; Qin, Y.; Liu, X.; He, K.; Chen, C.; Ma, L.; Wang, J.; Xiong, L.; et al. Global genome expression analysis of rice in response to drought and high-salinity stresses in shoot, flag leaf, and panicle. *Plant Mol. Biol.* **2007**, *63*, 591–608. [CrossRef] [PubMed]
53. Sourdille, P.; Cadalen, T.; Gay, G.; Gill, B.; Bernard, M. Molecular and physical mapping of genes affecting awning in wheat. *Plant Breed.* **2002**, *121*, 320–324. [CrossRef]
54. Echeverry-Solarte, M.; Kumar, A.; Kianian, S.; Mantovani, E.E.; McClean, P.E.; Deckard, E.L.; Elias, E.; Simsek, S.; Alamri, M.S.; Hegstad, J.; et al. Genome-wide mapping of spike-related and agronomic traits in a common wheat population derived from a supernumerary spikelet parent and an elite parent. *Plant Genome* **2015**, *8*, 0. [CrossRef]
55. Li, H.Y.; Wang, T.Y.; Shi, Y.S.; Fu, J.J.; Song, Y.C.; Wang, G.Y.; Li, Y. Isolation and characterization of induced genes under drought stress at the flowering stage in maize (*Zea mays*): Full Length Research Paper. *DNA Seq* **2007**, *18*, 445–460. [CrossRef] [PubMed]
56. Pandey, S.; Assmann, S.M. The Arabidopsis putative g protein–coupled receptor GCR1 interacts with the G protein α subunit GPA1 and regulates abscisic acid signaling. *Plant Cell* **2004**, *16*, 1616–1632. [CrossRef] [PubMed]
57. Wu, Q.; Chen, Y.; Fu, L.; Zhou, S.; Chen, J.; Zhao, X.; Zhang, D.; Ouyang, S.; Wang, Z.; Li, D.; et al. QTL mapping of flag leaf traits in common wheat using an integrated high-density SSR and SNP genetic linkage map. *Euphytica* **2016**, *208*, 337–351. [CrossRef]
58. Yang, D.; Liu, Y.; Cheng, H.; Chang, L.; Chen, J.; Chai, S.; Li, M. Genetic dissection of flag leaf morphology in wheat (*Triticum aestivum* L.) under diverse water regimes. *BMC Genet* **2016**, *17*. [CrossRef] [PubMed]
59. Fan, X.; Cui, F.; Zhao, C.; Zhang, W.; Yang, L.; Zhao, X.; Han, J.; Su, Q.; Ji, J.; Zhao, Z.; et al. QTLs for flag leaf size and their influence on yield-related traits in wheat (*Triticum aestivum* L.). *Mol. Breed.* **2015**, *35*, 24. [CrossRef]
60. Liu, K.; Xu, H.; Liu, G.; Guan, P.; Zhou, X.; Peng, H.; Yao, Y.; Ni, Z.; Sun, Q.; Du, J. QTL mapping of flag

leaf-related traits in wheat (*Triticum aestivum* L.). *Theor. Appl. Genet.* **2018**, *131*, 839–849. [CrossRef] [PubMed]

61. Hussain, W.; Baenziger, P.S.; Belamkar, V.; Guttieri, M.J.; Venegas, J.P.; Easterly, A.; Sallam, A.; Poland, J. Genotyping-by-sequencing derived high-density linkage map and its application to QTL mapping of flag leaf traits in bread wheat. *Sci. Rep.* **2017**, *7*, 16394. [CrossRef] [PubMed]
62. Ros, B.; Thümmler, F.; Wenzel, G. Analysis of differentially expressed genes in a susceptible and moderately resistant potato cultivar upon Phytophthora infestans infection. *Mol. Plant Pathol.* **2004**, *5*, 191–201. [CrossRef] [PubMed]
63. Zhao, Y.; Cai, M.; Zhang, X.; Li, Y.; Zhang, J.; Zhao, H.; Kong, F.; Zheng, Y.; Qiu, F. Genome-Wide identification, evolution and expression analysis of mTERF gene family in maize. *PLoS ONE* **2014**, *9*, e94126. [CrossRef] [PubMed]
64. Cotsaftis, O.; Plett, D.; Johnson, A.A.T.; Walia, H.; Wilson, C.; Ismail, A.M.; Close, T.J.; Tester, M.; Baumann, U. Root-specific transcript profiling of contrasting rice genotypes in response to salinity stress. *Mol. Plant* **2011**, *4*, 25–41. [CrossRef] [PubMed]
65. Hai, L.; Guo, H.; Xiao, S.; Jiang, G.; Zhang, X.; Yan, C.; Xin, Z.; Jia, J. Quantitative trait loci (QTL) of stem strength and related traits in a doubled-haploid population of wheat (*Triticum aestivum* L.). *Euphytica* **2005**, *141*, 1–9. [CrossRef]
66. Shumayla; Sharma, S.; Kumar, R.; Mendu, V.; Singh, K.; Upadhyay, S.K. Genomic dissection and expression profiling revealed functional divergence in Triticum aestivum leucine rich repeat receptor like kinases (TaLRRKs). *Front. Plant Sci.* **2016**, *7*. [CrossRef]
67. Landjeva, S.; Neumann, K.; Lohwasser, U.; Börner, A. Molecular mapping of genomic regions associated with wheat seedling growth under osmotic stress. *Biol. Plant.* **2008**, *52*, 259–266. [CrossRef]
68. Ren, Y.; He, X.; Liu, D.; Li, J.; Zhao, X.; Li, B.; Tong, Y.; Zhang, A.; Li, Z. Major quantitative trait loci for seminal root morphology of wheat seedlings. *Mol. Breed.* **2012**.
69. Bai, C.; Liang, Y.; Hawkesford, M.J. Identification of QTLs associated with seedling root traits and their correlation with plant height in wheat. *J. Exp. Bot.* **2013**, *64*, 1745–1753. [CrossRef] [PubMed]
70. Atkinson, J.A.; Wingen, L.U.; Griffiths, M.; Pound, M.P.; Gaju, O.; Foulkes, M.J.; Le Gouis, J.; Griffiths, S.; Bennett, M.J.; King, J.; et al. Phenotyping pipeline reveals major seedling root growth QTL in hexaploid wheat. *J. Exp. Bot.* **2015**, *66*, 2283–2292. [CrossRef] [PubMed]
71. Canè, M.A.; Maccaferri, M.; Nazemi, G.; Salvi, S.; Francia, R.; Colalongo, C.; Tuberosa, R. Association mapping for root architectural traits in durum wheat seedlings as related to agronomic performance. *Mol. Breed.* **2014**, *34*, 1629–1645. [CrossRef] [PubMed]
72. Maccaferri, M.; El-Feki, W.; Nazemi, G.; Salvi, S.; Canè, M.A.; Colalongo, M.C.; Stefanelli, S.; Tuberosa, R. Prioritizing quantitative trait loci for root system architecture in tetraploid wheat. *J. Exp. Bot.* **2016**, *67*, 1161–1178. [CrossRef] [PubMed]
73. Oono, Y.; Yazawa, T.; Kanamori, H.; Sasaki, H.; Mori, S.; Handa, H.; Matsumoto, T. Genome-wide transcriptome analysis of cadmium stress in rice. *BioMed Res. Int.* **2016**, *2016*. [CrossRef] [PubMed]
74. Furukawa, J.; Yamaji, N.; Wang, H.; Mitani, N.; Murata, Y.; Sato, K.; Katsuhara, M.; Takeda, K.; Ma, J.F. An aluminum-activated citrate transporter in barley. *Plant Cell Physiol.* **2007**, *48*, 1081–1091. [CrossRef] [PubMed]
75. Ryan, P.R.; Raman, H.; Gupta, S.; Horst, W.J.; Delhaize, E. A second mechanism for aluminum resistance in wheat relies on the constitutive efflux of citrate from roots. *Plant Physiol.* **2009**, *149*, 340–351. [CrossRef] [PubMed]
76. Baron, K.N.; Schroeder, D.F.; Stasolla, C. GEm-related 5 (GER5), an ABA and stress-responsive GRAM domain protein regulating seed development and inflorescence architecture. *Plant Sci.* **2014**, *223*, 153–166. [CrossRef] [PubMed]
77. Liu, L.; Li, N.; Yao, C.; Meng, S.; Song, C. Functional analysis of the ABA-responsive protein family in ABA and stress signal transduction in *Arabidopsis*. *Chin. Sci. Bull.* **2013**, *58*, 3721–3730. [CrossRef]
78. Liu, X.; Huang, M.; Fan, B.; Buckler, E.S.; Zhang, Z. Iterative usage of fixed and random effect models for powerful and efficient genome-wide association studies. *PLoS Genet.* **2016**, *12*, e1005767. [CrossRef] [PubMed]
79. Arora, S.; Singh, N.; Kaur, S.; Bains, N.S.; Uauy, C.; Poland, J.; Chhuneja, P. Genome-wide association study of grain architecture in wild wheat Aegilops tauschii. *Front. Plant Sci.* **2017**, *8*. [CrossRef] [PubMed]
80. Sukumaran, S.; Xiang, W.; Bean, S.R.; Pedersen, J.F.; Kresovich, S.; Tuinstra, M.R.; Tesso, T.T.; Hamblin, M.T.;

Yu, J. Association mapping for grain quality in a diverse sorghum collection. *Plant Genome J.* **2012**, *5*, 126. [CrossRef]
81. Barrett, J.C.; Fry, B.; Maller, J.; Daly, M.J. Haploview: analysis and visualization of LD and haplotype maps. *Bioinformatics* **2005**, *21*, 263–265. [CrossRef] [PubMed]
82. Singh, A.; Giri, J.; Kapoor, S.; Tyagi, A.K.; Pandey, G.K. Protein phosphatase complement in rice: Genome-wide identification and transcriptional analysis under abiotic stress conditions and reproductive development. *BMC Genom.* **2010**, *11*, 435. [CrossRef] [PubMed]
83. Evaluation of deficit irrigation for efficient sheep production from permanent sown pastures in a dry continental climate. *Agric. Water Manag.* **2013**, *119*, 135–143. [CrossRef]
84. SAS 9.4 Product Documentation. Available online: https://support.sas.com/documentation/94/ (accessed on 16 August 2018).
85. Kassambara, A.; Mundt, F. Factoextra: Extract and visualize the results of multivariate data analyses. 2017. Available online: http://www.sthda.com/english/rpkgs/factoextra/ (accessed on 16 August 2018).
86. Poland, J.A.; Brown, P.J.; Sorrells, M.E.; Jannink, J. Development of high-density genetic maps for barley and wheat using a novel two-enzyme genotyping-by-sequencing approach. *PLoS ONE* **2012**, *7*. [CrossRef] [PubMed]
87. Glaubitz, J.C.; Casstevens, T.M.; Lu, F.; Harriman, J.; Elshire, R.J.; Sun, Q.; Buckler, E.S. TASSEL-GBS: A high capacity genotyping by sequencing analysis pipeline. *PLoS ONE* **2014**, *9*. [CrossRef] [PubMed]
88. International Wheat Genome Sequencing Consortium (IWGSC). Shifting the limits in wheat research and breeding using a fully annotated reference genome. *Science* **2018**, *361*, 7191. [CrossRef] [PubMed]
89. Pritchard, J.K.; Stephens, M.; Donnelly, P. Inference of Population Structure Using Multilocus Genotype Data. *Genetics* **2000**, *155*, 945–959. [PubMed]
90. Bradbury, P.J.; Zhang, Z.; Kroon, D.E.; Casstevens, T.M.; Ramdoss, Y.; Buckler, E.S. TASSEL: Software for association mapping of complex traits in diverse samples. *Bioinformatics* **2007**, *23*, 2633–2635. [CrossRef] [PubMed]
91. Yu, J.; Pressoir, G.; Briggs, W.H.; Vroh Bi, I.; Yamasaki, M.; Doebley, J.F.; McMullen, M.D.; Gaut, B.S.; Nielsen, D.M.; Holland, J.B.; et al. A unified mixed-model method for association mapping that accounts for multiple levels of relatedness. *Nat. Genet.* **2006**, *38*, 203–208. [CrossRef] [PubMed]
92. Zhang, Z.; Ersoz, E.; Lai, C.-Q.; Todhunter, R.J.; Tiwari, H.K.; Gore, M.A.; Bradbury, P.J.; Yu, J.; Arnett, D.K.; Ordovas, J.M.; et al. Mixed linear model approach adapted for genome-wide association studies. *Nature Genetics* **2010**, *42*, 355–360. [CrossRef] [PubMed]
93. Cingolani, P.; Platts, A.; Wang, L.L.; Coon, M.; Nguyen, T.; Wang, L.; Land, S.J.; Lu, X.; Ruden, D.M. A program for annotating and predicting the effects of single nucleotide polymorphisms, SnpEff. *Fly (Austin)* **2012**, *6*, 80–92. [CrossRef] [PubMed]

GmBRC1 is a Candidate Gene for Branching in Soybean (*Glycine max* (L.) Merrill)

Sangrea Shim [1,2], Jungmin Ha [1,2], Moon Young Kim [1,2], Man Soo Choi [3], Sung-Taeg Kang [4], Soon-Chun Jeong [5], Jung-Kyung Moon [6] and Suk-Ha Lee [1,2,*]

1 Department of Plant Science and Research Institute of Agriculture and Life Sciences, Seoul National University, Seoul 08826, Korea; sangreashim@gmail.com (S.S.); saga16@snu.ac.kr (J.H.); moonykim@snu.ac.kr (M.Y.K.)
2 Plant Genomics and Breeding Institute, Seoul National University, Seoul 08826, Korea
3 National Institute of Crop Sciences, Rural Development Administration, Wanju-gun, Jeollabuk-do 55365, Korea; mschoi73@korea.kr
4 Department of Crop Science & Biotechnology, Dankook University, Cheonan-si, Chungcheongnam-do 31116, Korea; kangst@dankook.ac.kr
5 Bio-Evaluation Center, Korea Research Institute of Bioscience and Biotechnology, Cheongju-si, Chungcheongbuk-do 28116, Korea; scjeong@kribb.re.kr
6 National Institute of Agricultural Sciences, Rural Development Administration, Jeonju-si, Jeollabuk-do 54874, Korea; moonjk2@korea.kr
* Correspondence: sukhalee@snu.ac.kr

Abstract: Branch number is one of the main factors affecting the yield of soybean (*Glycine max* (L.)). In this study, we conducted a genome-wide association study combined with linkage analysis for the identification of a candidate gene controlling soybean branching. Five quantitative trait nucleotides (QTNs) were associated with branch numbers in a soybean core collection. Among these QTNs, a linkage disequilibrium (LD) block *qtnBR6-1* spanning 20 genes was found to overlap a previously identified major quantitative trait locus *qBR6-1*. To validate and narrow down *qtnBR6-1*, we developed a set of near-isogenic lines (NILs) harboring high-branching (HB) and low-branching (LB) alleles of *qBR6-1*, with 99.96% isogenicity and different branch numbers. A cluster of single nucleotide polymorphisms (SNPs) segregating between NIL-HB and NIL-LB was located within the *qtnBR6-1* LD block. Among the five genes showing differential expression between NIL-HB and NIL-LB, *BRANCHED1* (*BRC1*; Glyma.06G210600) was down-regulated in the shoot apex of NIL-HB, and one missense mutation and two SNPs upstream of *BRC1* were associated with branch numbers in 59 additional soybean accessions. *BRC1* encodes TEOSINTE-BRANCHED1/CYCLOIDEA/PROLIFERATING CELL FACTORS 1 and 2 transcription factor and functions as a regulatory repressor of branching. On the basis of these results, we propose *BRC1* as a candidate gene for branching in soybean.

Keywords: soybean; branching; genome-wide association study (GWAS); near-isogenic line (NIL); *BRANCHED1* (*BRC1*); TCP transcription factor

1. Introduction

Soybean (*Glycine max* (L.)) is a major food crop and a rich source of protein and oil in human diet and animal feed. Two cultivation methods with different planting densities are used to maximize soybean yield. The high-density planting method is mainly practiced in the USA [1], where the yield of soybean increases with planting density until saturation. The low-density planting method is practiced in Asia not only to avoid disease and lodging, but also to reduce seed and labor cost; however, the productivity in low-density planting is lower than that in high-density planting [2]. An important factor of low-density planting is branching plasticity, which offsets yield losses [3]. Branching pattern and branch number are dependent on environmental factors such as planting density and light quality [3–5]. These factors have obstructed the identification of genes regulating branch development. However, variation in branch number among soybean cultivars of diverse origins under low-density planting suggests the presence of genetic differences [6,7]. Thus, breeding for soybean genotypes with optimal plant architecture adapted to a specific planting density is necessary for improving yield and enabling mechanical harvesting.

To date, a dozen quantitative trait loci (QTLs) regulating branch development have been identified using recombinant inbred line (RIL) or F_2 populations in soybean [7–11]. However, the identified QTLs span a large number of plausible genes because of the low resolution of genetic linkage maps and low recombination frequency in mapping populations. Recently, three known QTLs have been narrowed down to identify candidate genes based on a high-density linkage map using the 6K single nucleotide polymorphism (SNP) chip (BARCSoySNP6K), and positive phenotypic correlation has been demonstrated between branch number and total pod number [8]. Among the QTLs, a major one (*qBR6-1*) on chromosome (Chr) 6, with a logarithm of odds (LOD) score of 10.3 and 14.5% of the phenotypic variation in branch numbers, was shown to contain 13 genes [8]. One of these genes was a gene encoding TEOSINTE-BRANCHED1/CYCLOIDEA/PROLIFERATING CELL FACTORS 1 and 2 (TCP) transcription factor, also known as *BRANCHED1* (*BRC1*), which is involved in gene networks of axillary branching via interactions with the auxin hormone network [8]. In soybean, the *BRC1* gene has not been genetically identified.

A genome-wide association study (GWAS) is used to identify associations between genetic loci and traits. It is a powerful method, as it provides high genetic resolution derived from all recombination events that occurred during the evolution of a natural population [12]. GWAS has been used successfully to identify the genetic basis of complex agronomic traits in Arabidopsis (*Arabidopsis thaliana* (L.)), rice (*Oryza sativa* (L.)), and maize (*Zea mays* (L.)) [13–15]. In soybean, GWAS has been performed to identify the association of genetic regions with agronomic traits, including flowering time, maturity date, plant height, and seed oil content [16,17]. However, the detection of false-positive associations with phenotypes is a weakness of GWAS, which is caused by the population structure and kinship relatedness in natural populations [16]. Utilizing linkage analysis in combination with GWAS has been previously used to overcome the limitations of QTL mapping and to validate the results of GWAS [16].

The objective of this study was to narrow down a genomic region of *qBR6-1* to identify a candidate gene responsible for soybean branching. To this end, GWAS was conducted to identify quantitative trait nucleotides (QTNs) associated with branch number in a soybean core collection comprising 400 soybean genotypes grown in three locations. To validate and narrow down a QTN that was found to be overlapped with the previously identified major QTL, *qBR6-1*, we developed and analyzed a set of near-isogenic lines (NILs) carrying high-branching (HB) and low-branching (LB) alleles of *qBR6-1* derived from an F_6 residual heterozygous line (RHL) heterozygous for *qBR6-1*. Using these NILs, we detected differential expression of genes within the linkage disequilibrium (LD) block of the QTN on Chr 6. In addition, we examined nucleotide variations in a selected candidate gene between NIL-HB and NIL-LB, and confirmed their allelic associations with branch numbers in soybean accessions obtained from the United States Department of Agriculture Germplasm Resources Information Network (USDA-GRIN). This study will provide a better understanding of genetic basis underlying the branch development in soybean and valuable information to improve plant architecture for soybean cultivars with high yields.

2. Results

2.1. Variation and Heritability in Branch Number of the Soybean Core Collection

We present variations in branch number of the soybean core collection according to the location of cultivation (Wanju, Cheonan, and Ochang) and best linear unbiased predictor (BLUP) values in Figure S1. The number of branches varied from 2.0–20.7 in Wanju, 1.0–21.3 in Cheonan, and 0.0–14.3 in Ochang (Table S1). The broad-sense heritability (H^2) of branch numbers in the soybean core collection was 57.7% (Table S1). Analysis of variance (ANOVA) showed significant effects of genotype as well as genotype-by-environment (G × E) interaction on branch numbers ($p < 0.0001$). Additionally, a continuous distribution of branch numbers was observed in all three locations, indicating that branch numbers were regulated by multiple genetic factors. To identify reliable QTNs associated with branch numbers, the BLUP values were calculated for each genotype and used in GWAS (Figure S1).

2.2. Population Structure and LD

Covariates from the population structure were stratified using 81,078 SNP markers. The log likelihood (LnP(D)) from STRUCTURE [18] analysis showed a continuous increase with the number of sub-populations (K). Therefore, we determined the optimum K value using the ΔK method. The ΔK value classified the core collection into two groups (Figure S2). In a scree plot, the proportion of variance drastically decreased until the number of principal components (PCs) reached two (Figure 1A). The extent of LD estimated by PLINK [19] showed that the average pair-wise squared correlation coefficient (r^2) between alleles was dropped to half of its maximum value at 160–170 kb for the entire genome, with 140–150 kb for euchromatin and 460–470 kb for heterochromatin (Figure 1B).

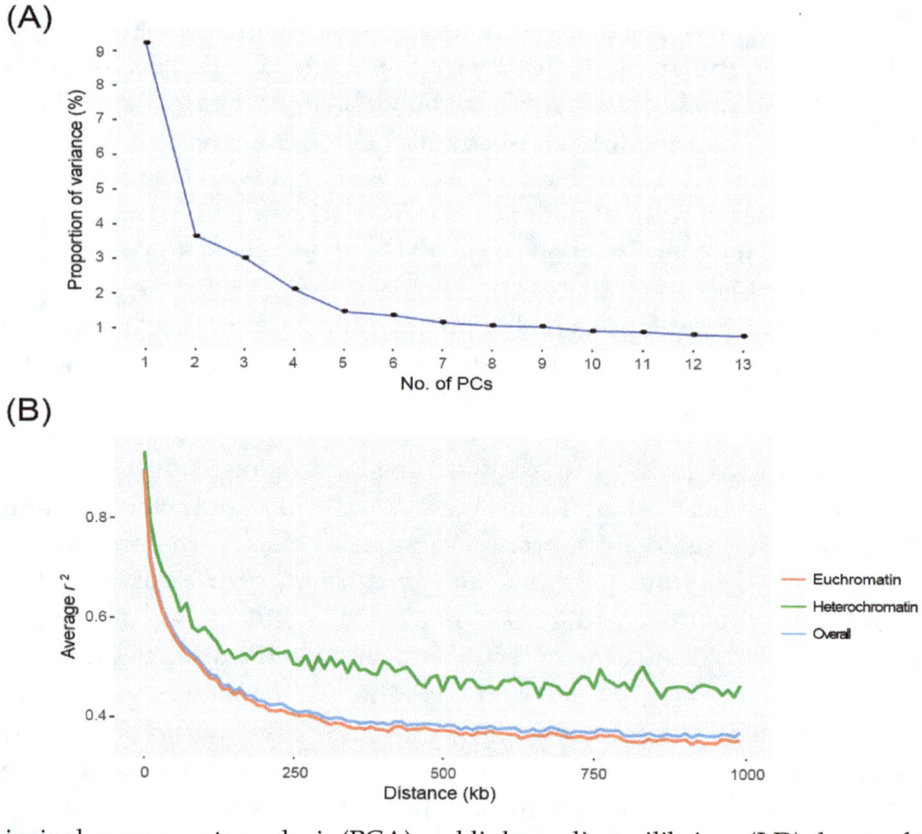

Figure 1. Principal components analysis (PCA) and linkage disequilibrium (LD) decay of the soybean core collection. (**A**) Scree plot showing the proportion of variance explained by principal components (PCs). (**B**) LD decay based on pairwise r^2 values. Blue line represents the overall rate of LD decay. Red and green lines indicate LD decay patterns of euchromatin and heterochromatin, respectively.

2.3. Determination of Genetic Association with Branch Numbers Using GWAS

To determine the best fitted GWAS model for branch numbers, quantile–quantile (QQ) plots generated using the generalized linear model (GLM) + population structure (Q) were compared with those generated using the mixed linear model (MLM) + Q (from principal component analysis; PCA) and kinship matrix (K) (Figure S3). The GLM + Q model showed a strong inflation of *p*-value (blue dots) compared with the MLM + Q (PCA) and K model (red dots), indicating erroneous inflation of false-positive signal. Thus, MLM + Q (PCA) and K model was more appropriate for the identification of QTNs associated with branch numbers in this study.

A total of five significant QTNs showing significant association with branch numbers were identified on Chr 6, 11, 12, and 20 (Table 1 and Figure 2A). The most highly significant QTN, *qtnBR12-1* at 38,057,780 base pair (bp) in the euchromatic region of Chr 12, explained 6.4% of the phenotypic variation in branch numbers. The second most highly significant QTN, *qtnBR6-1* at 20,663,101 bp in heterochromatin of Chr 6, accounted for 5.8% of the phenotypic variation (Figure 2B). Phenotypic variation in branch numbers explained by two other QTNs, *qtnBR11-1* and *qtnBR11-2* at 16,074,992 and 28,613,118 bp, respectively, in heterochromatin of Chr 11 was 5.0% and 5.6%, respectively. The last QTN, *qtnBR20-1*, located at 42,471,316 bp in euchromatin of Chr 20, accounted for 4.9% of the phenotypic variation in branch numbers.

On the basis of the rate of LD decay, we extended the chromosomal regions both upstream and downstream of the QTN positions, up to 140 kb for euchromatin and 460 kb for heterochromatin (Table 1). The LD blocks of *qtnBR6-1*, *qtnBR11-1*, and *qtnBR11-2* were adjacent to a major QTL *qBR6-1* and a minor QTL *qBR11-1* reported previously [8]. The *qtnBR6-1*, spanning 20 protein-coding genes (Figure 2C), overlapped with the QTL *qBR6-1*, which has been shown to play a major role in branch development. Using the publicly available soybean RNA-seq data [20], in silico expression profiling of all 20 genes in *qtnBR6-1* revealed that 13 genes, including the TCP transcription factor gene (*BRC1*; Glyma.06G210600), were expressed in the shoot apical meristem (SAM) (Figure 3A).

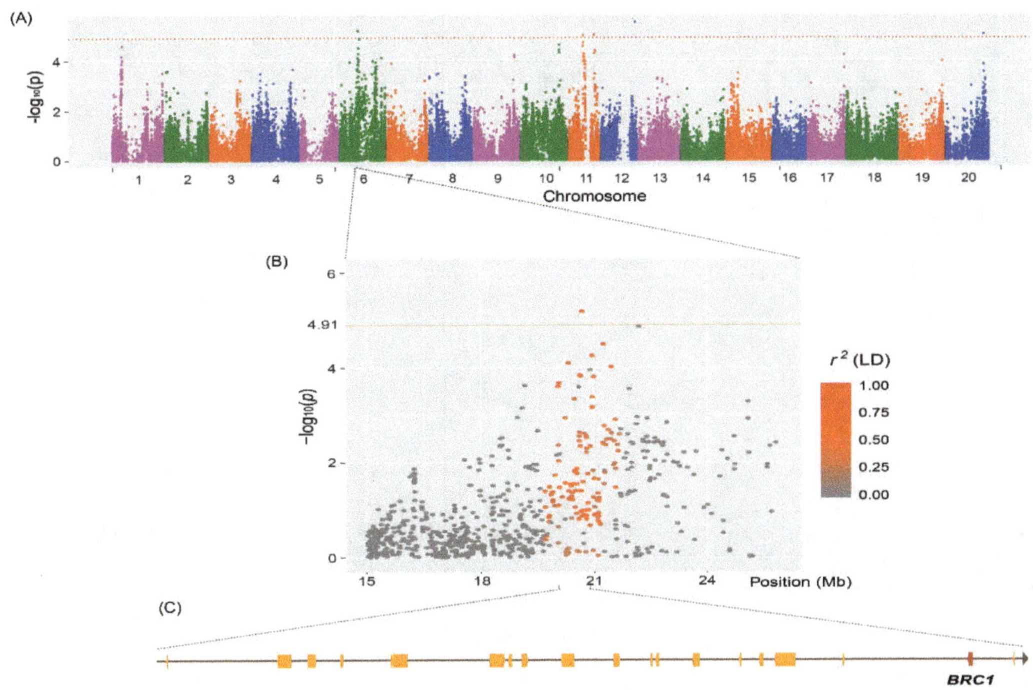

Figure 2. Genome-wide association study (GWAS) of branch numbers in the soybean core collection. (**A**) Genome-wide Manhattan plot of branch numbers. (**B**) LD region harboring *qtnBR6-1* on Chr 6. The pair-wise r^2 values between markers in LD are presented along a color gradient ranging from gray to red. The *p*-value threshold is indicated with red horizontal lines in Manhattan plots. (**C**) Protein-coding genes located in the LD block of *qtnBR6-1*.

Table 1. Quantitative trait nucleotides (QTNs) associated with branch numbers in the soybean core collection identified by genome-wide association study (GWAS).

QTN ID	Marker ID	Chr [a]	Marker Position (bp)	*p*-Value	Phenotypic R^2 (%)	Chromosomal Location	Linkage Disequilibrium Block		No. of Genes	Known QTL	Reference [b]
qtnBR6-1	AX-90305605	6	20,663,101	6.43×10^{-6}	5.8	Heterochromatin	20,433,101	20,893,101	20	*qBR6-1*	[8]
qtnBR11-1	AX-90512426	11	16,074,992	9.98×10^{-6}	5.0	Heterochromatin	15,844,992	16,304,992	13	*qBR11-1*	[8]
qtnBR11-2	AX-90472718	11	28,613,118	9.51×10^{-6}	5.6	Heterochromatin	28,383,118	28,843,118	23	*qBR11-1*	[8]
qtnBR12-1	AX-90419363	12	38,057,780	2.89×10^{-6}	6.4	Euchromatin	37,987,780	38,127,780	14		
qtnBR20-1	AX-90519199	20	42,471,316	8.88×10^{-6}	4.9	Euchromatin	42,401,316	42,541,316	13		

[a] Chr represents the soybean chromosome. [b] Reference represented reference article for previously reported quantitative trait loci (QTL).

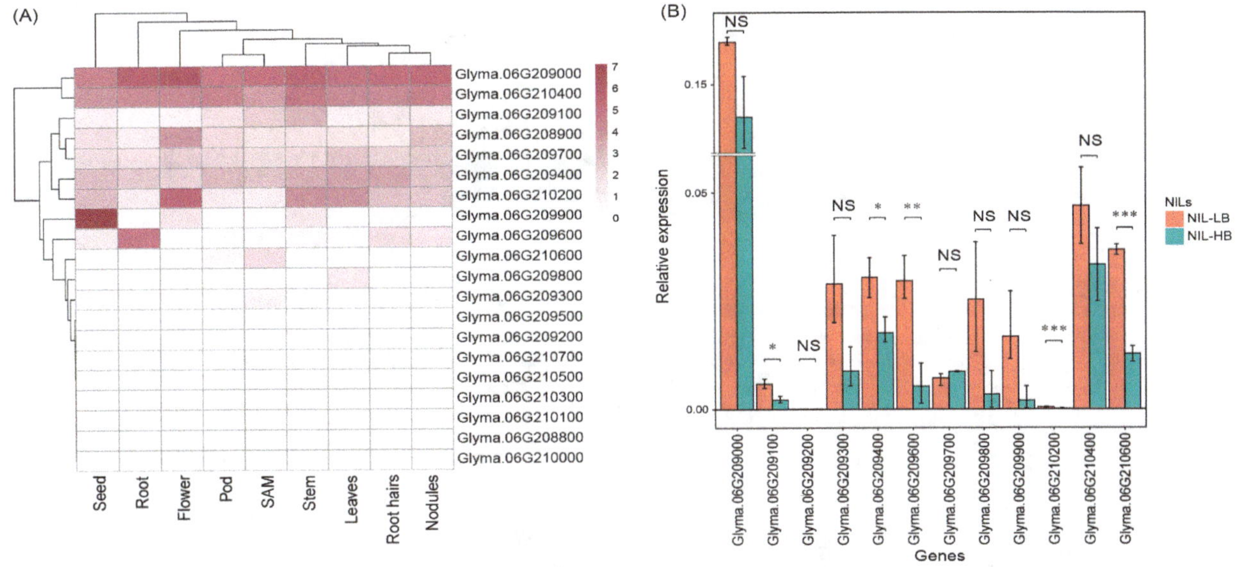

Figure 3. Expression patterns of genes located in the LD block of *qtnBR6-1*. (**A**) Heatmap showing the expression patterns of 20 genes obtained from the publicly available RNA-seq data. Gene expression values of $\log_2$(FPKM+1) were used to generate the heatmap. (**B**) Comparison of expression levels of selected genes in the shoot apical meristem between NILs with high-branching (HB) and low-branching (LB) alleles (NIL-HB and NIL-LB, respectively). The red and green bars indicate mean expression level of genes for nine samples (three biological replicates x three technical replicates) of each NIL (NIL-LB and NIL-HB), respectively. Black variance bars represent standard error of the mean (SEM). Statistically significant differences in relative gene expression between NIL-HB and NIL-LB are indicated with asterisks (*, $p < 0.05$; **, $p < 0.01$; ***, $p < 0.001$). NS, not significant.

2.4. Isogenicity and Phenotypic Differences Between NIL-HB and NIL-LB Associated with qBR6-1

To validate and narrow down *qtnBR6-1*, we explored a set of NILs carrying HB and LB alleles of *qBR6-1*. NIL-HB with more branches contained HB allele derived from paternal genotype SS0404-T5-76. NIL-LB carried LB allele from maternal genotype Jiyu69, showing fewer branches. The NILs and parental genotypes, SS0404-T5-76 and Jilyu69, were resequenced at an average depth of 31.6X (Table S2). On average, 92.4% of the paired-end reads were mapped to the soybean reference genome sequence. Within 895 Mb of consensus genome sequence of the NILs with at least 10X mapping depth, a total of 286,467 nucleotide variants were identified on all chromosomes except Chr 6 (carrier chromosome) between NIL-HB and NIL-LB, resulting in 99.97% isogenicity (Table S3). In addition, the non-QTL region on the carrier Chr 6 showed 99.89% isogenicity. The QTL region of *qBR6-1* on Chr 6 harbored 3798 nucleotide variants between NIL-HB and NIL-LB, of which 96.2% were shared with polymorphisms between parental genotypes SS0404-T5-76 and Jilyu69. Within the LD block of *qtnBR6-1*, SNPs segregating between NIL-HB and NIL-LB clustered in a 343 kb interval (20,555–20,898 kb) harboring 16 protein-coding genes (Figure 4A).

Additionally, we evaluated the effect of planting density on the branching performance of NIL-HB and NIL-LB in the field and greenhouse. The results showed a significant difference in the number of branches between NIL-HB and NIL-LB grown under low planting density (Table 2 and Figure S4). On average, NIL-HB plants had two more branches than NIL-LB plants. High planting density displayed no difference in branch numbers between NIL-HB and NIL-LB.

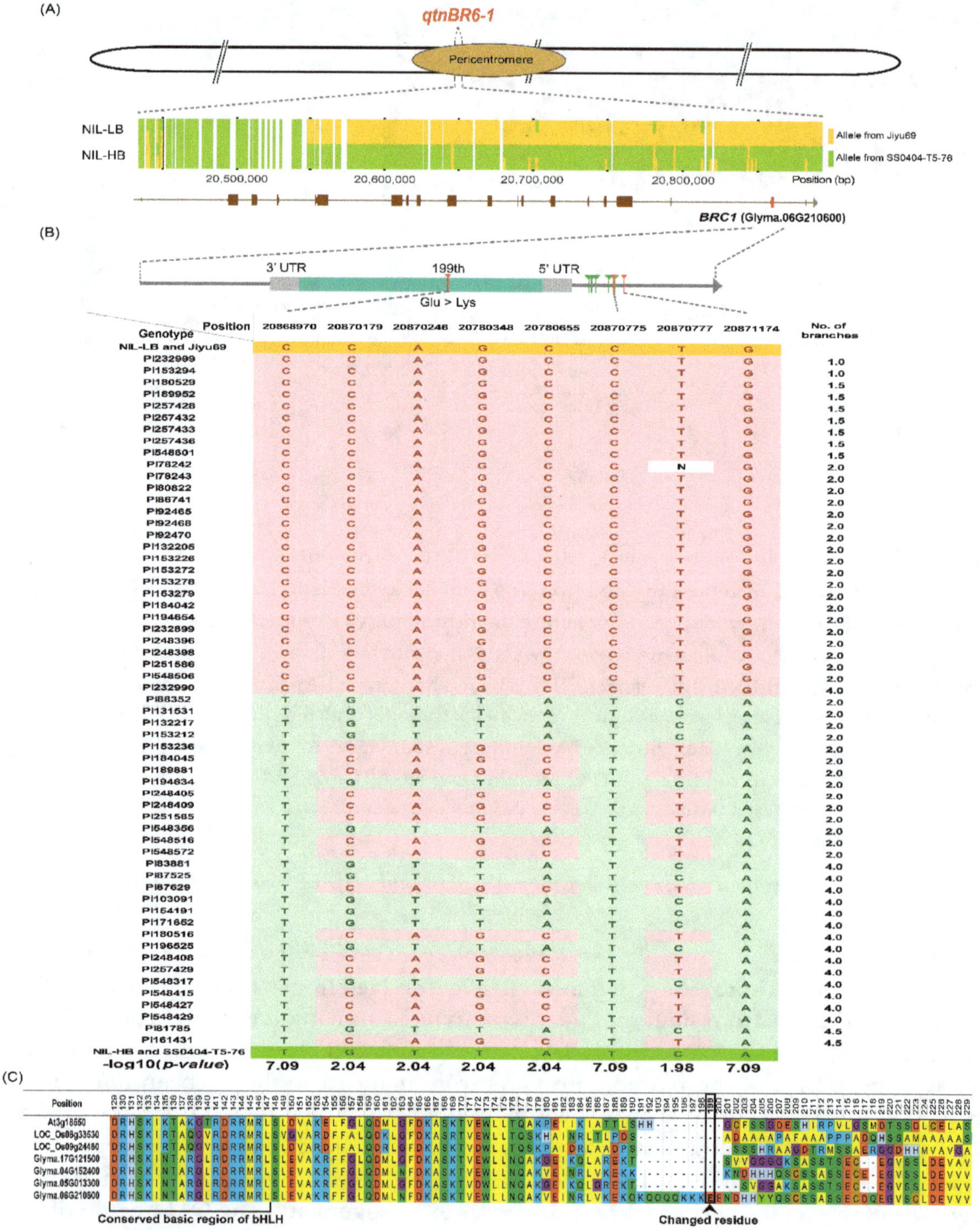

Figure 4. Allelic association of nucleotide variants in *BRC1* with branch numbers and protein sequence alignment of *BRC1*. (**A**) Physical location of *qtnBR6-1* in the pericentromeric region of Chr 6, and a cluster of single nucleotide polymorphism (SNPs) segregating between NIL-HB and NIL-LB. The SNPs between NIL-HB and NIL-LB within the LD block of *qtnBR6-1* are originated from SS0404-T5-76 (high-branching, green) and Jiyu69 (low-branching, orange), respectively. The X-axis indicates the genomic position (bp). (**B**) Allelic association of SNPs in *BRC1* with branch numbers among 59 United States Department of Agriculture (USDA) soybean accessions. Among eight SNPs, one missense mutation and two SNPs upstream of *BRC1*, indicated by red bars, were associated with branch numbers in 59 USDA soybean accessions. (**C**) Protein sequence alignment of *BRC1* orthologues. The missense mutation (glutamate to lysine) at amino acid position 199 is outlined with a black box.

Table 2. Effect of planting density on the branch number of soybean near-isogenic lines (NILs) carrying high-branching (HB) and low-branching (LB) alleles (NIL-HB and NIL-LB, respectively) at the quantitative trait loci (QTL) *qBR6-1*.

Year	Planting Density	Growth Condition	NIL-HB	NIL-LB	*p*-Value
2017	Low	Greenhouse	5.8 ± 1.7	3.4 ± 0.7	0.0001
		Field	14.5 ± 1.1	12.2 ± 2.1	0.019
2018	Low	Field	7.1 ± 1.4	5.1 ± 1.6	0.0003
	High		3.3 ± 1.9	2.6 ± 0.8	0.057

2.5. Candidate Gene Identification and Allelic Association Analysis

To narrow down the 13 candidate genes identified in the LD block of *qtnBR6-1*, cDNA was isolated from the shoot apex of NILs and used for expression analysis by qRT-PCR. Among these 13 genes, a multi-copy gene (Glyma.06G208900) encoding ATPase E1-E2 type family protein could not be analyzed by qRT-PCR because the primers were not sequence-specific. Of the remaining 12 genes, five were significantly down-regulated in NIL-HB; the genes encode MIZU-KUSSEI-like protein (Glyma.06G209100), P-loop containing nucleoside triphosphate hydrolases superfamily protein (Glyma.06G209400), adenine nucleotide alpha hydrolases-like superfamily protein (Glyma.06G209600), unknown protein (Glyma.06G210200), and TCP transcription factor (*BRC1*; Glyma.06G210600) (Figure 3B). On the basis of molecular genetic data available in Arabidopsis [21,22], *BRC1* was identified as the most promising candidate gene responsible for branch development in soybean.

Analysis of the nucleotide sequence of *BRC1* in NILs, Jily69, and SS0404-T5-76 revealed one SNP in the coding sequence, resulting in a missense mutation at amino acid position 199 (glutamate to lysine), and seven additional SNPs in the 2 kb upstream sequence (Figure 4B). Furthermore, the SNP in *BRC1* coding sequence and two SNPs in the upstream sequence were significantly associated with branch number in 59 USDA-GRIN soybean accessions (available online: http://www.ars-grin.gov).

3. Discussion

Shoot branching influences seed yield in soybean and is regulated by a complex mechanism of axillary bud outgrowth following axillary meristem initiation [21,23]. The fate of axillary buds, that is, whether to outgrow into a branch or to remain as a bud, is determined by an orchestrated regulatory process induced by endogenous hormonal and developmental signals [24]. Such a regulatory process is affected by environmental factors such as planting density, shading, light quality, soil nitrogen content, and soil water content [3,4,25–27]. The soybean core collection used in this study showed a more significant effect on branch numbers than locations or interactions between genotypes and locations. We identified five QTNs showing association with branch numbers based on BLUP values from three different locations (Table 1 and Figure 2). The LD block of one of these five QTNs, *qtnBR6-1*, co-localized with a previously reported major QTL, *qBR6-1*, which was validated based on the clustering of SNPs segregating between NIL-HB and NIL-LB (Figure 4A).

The outgrowth of axillary buds is inhibited by the active shoot apex; this phenomenon is referred to as apical dominance. Decapitation abolishes apical dominance and triggers the growth of one or more axillary buds because auxin, which is synthesized in the shoot apex, is mobilized to the lower parts of plants and inhibits branch outgrowth [28]. Considering the relevance of shoot apex to branch development, we compared the expression levels of 13 genes showing transcriptional activity in SAM using publicly available RNA-seq data (Figure 3A). Among the five significantly down-regulated genes in NIL-HB (Figure 3B), we identified *BRC1* (Glyma.06G210600) as the most promising candidate gene responsible for soybean shoot branching because *BRC1* functions as a regulatory hub that integrates hormonal signals and external stimuli for determining the fate of axillary buds [21,22]. In Arabidopsis, *BRC1* is expressed in axillary buds and SAM [21,29]. In pea (*Pisum sativum* L.), *PsBRC1* shows the highest expression level in lateral buds and is also expressed in the shoot apex [30]. In this study,

NIL-LB showed higher transcriptional activity of *BRC1* in the shoot apex than NIL-HB, although the expression of *BRC1* was not examined in axillary buds. Our data are consistent with previous studies showing that *BRC1* orthologues in rice (*Ostb1*) and maize (*tb1*) negatively regulate branch development [21,22,31,32].

The effects of planting density and shading on branch development are well-established in plants [3,5,21,33,34]. High planting density and shading have the same effect on light quality as they both reduce the ratio of red to far red light (R/FR) [24]. Shoot branching increases under high R/FR ratios caused by low planting density, but decreases under low R/FR ratios [3,5,21,33,34]. Similarly, in this study, NILs grown under high planting density produced fewer branches than those grown under low planting density, and showed no difference in branch numbers (Table 2). However, a significant difference in branch numbers was observed between NIL-HB and NIL-LB cultivated at low planting density. This indicates that the gene regulating branch number responds to low planting density. The light signal perceived by phytochrome B (PHYB), a photoreceptor, is transduced to the endogenous signal via *BRC1*, which functions as a molecular mediator [34]. Under shade or high planting density, an elevated FR signal inactivates PHYB by converting the active form of phytochrome (Pr) to the inactive form (Pfr) [35]. The inactive Pfr form of PHYB up-regulates *BRC1*, resulting in the inhibition of branch development [34,35]. These findings link the exogenous light signal with the endogenous molecular regulator in branch development, and suggest *GmBRC1* as the causal gene responsible for branch development in soybean. The function of *BRC1* is highly conserved across other plant species [21,30–32].

The *BRC1* gene and its orthologues are characterized by a highly conserved basic region of basic helix–loop–helix (bHLH) on a functional domain of TCP genes [21,31,32,36], and belong to CYCLODEA/TEOSINTE BRANCHED1 (CYC/TB1)-type TCP [37]. In this study, a missense mutation (glutamate to lysine) in the CYC/TB1-type TCP domain was identified between NIL-HB and NIL-LB, which showed a tight association with branch numbers in 59 USDA soybean accessions (Figure 4B). However, the altered amino acid residue was not located in the highly conserved basic region of bHLH, and was not conserved among other CYC/TB1-type TCP orthologues in Arabidopsis, rice, and soybean (Figure 4C). Thus, it is not clear if the difference in branch numbers between NIL-HB and NIL-LB could be attributed to the amino acid change at position 199. In addition to the SNP in *BRC1* coding sequence, two out of seven SNPs within the 2 kb upstream sequence of *BRC1* showed tight association with branch numbers. As both these SNPs are located in the putative promoter region of *BRC1*, they are predicted to affect *BRC1* expression. This is consistent with previous reports that *BRC1* regulates branching at the transcription level [21,31,32]; *BRC1* expression was down-regulated in NIL-HB (Figure 3B). An example similar to our results is of maize *tb1;* sequence variation in the upstream region of *tb1*, resulting in low expression, is associated with increased branch development in maize [31,38,39].

In conclusion, we propose *BRC1* (Glyma.06G210600) as the candidate gene regulating branch development in soybean. Further functional validation of these results by overexpression or knockout of *GmBRC1* is required for a thorough understanding of the regulatory mechanism of branch development in soybean, which will provide key insights into the complex genetic modules mediating branch development in soybean. Agronomically, soybean cultivars with optimal plant architecture depending on cultivation methods can be developed based on the allelic information of *BRC1* gene. In western countries including USA, soybean cultivars for high yield and mechanical harvesting can be improved by selecting genotypes with alleles contributing low branching phenotype. Besides, introgression of alleles responsible for high branching phenotype to other elite cultivars will enable breeding of high yielding soybean cultivars with high branch number and contribute to labor saving cultivation practice in Asian countries.

4. Materials and Methods

4.1. Plant Materials

A soybean core collection comprising 400 soybean genotypes with diverse origins was obtained from the National Agrobiodiversity Center in the Rural Development Administration (RDA, Jeonju, Korea) for GWAS (Table S4). To validate and narrow down the locus for branching, a set of NILs carrying HB and LB alleles of *qBR6-1* was developed from an RHL selected from the F_6 RIL population of Jiyu69 (low-branching) × SS0404-T5-76 (high-branching). Among five individuals in the progeny of RHL, plants showing the highest and lowest number of branches were selected as NILs carrying the HB and LB alleles, respectively, in 2016, and were designated as NIL-HB and NIL-LB, respectively. These NILs were genotyped using simple sequence repeat markers flanking *qBR6-1* (Table S5). We also used 59 USDA-GRIN soybean accessions with known branch numbers to confirm the allelic association of SNPs within a candidate gene between NIL-HB and NIL-LB.

The soybean core collection was grown in three different locations, namely, Wanju (35°50′27.384″ N, 127°2′46.1826″ E), Cheonan (36°49′49.2816″ N, 127°10′1.9122″ E), and Ochang (36°43′14.0982″ N, 127°26′1.1148″ E), in Korea in 2017. NIL-HB and NIL-LB were planted in a greenhouse and experimental field of Seoul National University, Suwon, Korea (37°16′12.094″ N, 126°59′20.756″ E). In the greenhouse, three plants of each line (NIL-HB and NIL-LB) were grown in a rectangular pot (64.3 cm × 23.0 cm × 16.9 cm). The plant-to-plant and row-to-row spacing was 20 and 80 cm, respectively, in low-density planting, and 10 and 40 cm, respectively, in high-density planting. All experiments were performed in triplicate, and field evaluation of branching in NILs was conducted in 2017 and 2018.

4.2. Phenotyping of Branch Number and Statistical Analysis

The number of branches generated on the main stem of the soybean core collection genotypes and NILs was evaluated in three biological replicates. Phenotypic differences between NILs were examined by analysis of variance (ANOVA) using the R software (available online: http://www.R-project.org). To minimize the effect of environmental factors on branch numbers at three different locations in GWAS, the BLUP value was predicted by the lme4 package of R, considering the variation among genotypes and locations [40]. The BLUP values were calculated according to the following equation:

$$Y_{ik} = \mu + G_i + L_k + GL_{ik} + e_{ik}$$

where Y_{ik} represents the phenotypic measurement, μ is the total mean, G_i is the genotypic effect of the ith genotype, L_k is the effect of the kth location, GL_{ik} represents interaction between genotype and location, and e_{ik} is the residual error. The BLUP values of each soybean genotype were calculated with random effect and used as phenotypes for GWAS. Broad-sense heritability (H^2) of branch numbers was calculated using the following equation:

$$H^2 = \frac{\sigma_g^2}{(\sigma_g^2 + \frac{\sigma_{gl}^2}{n} + \frac{\sigma_e^2}{nr})}$$

where ${\sigma_g}^2$ represents genotypic variance, ${\sigma_{gi}}^2$ is the variance of interaction between genotype and location, ${\sigma_e}^2$ represents the variance of error components, n represents the number of locations, and r represents the number of replications.

4.3. Population Structure and LD Analysis

SNP genotypic data of the soybean core collection previously produced from the 180K Axiom® SoyaSNP array were explored for GWAS [41]. SNPs with minor allele frequency <0.05 and missing genotype >10% were excluded. The remaining 81,078 SNPs were used in STRUCTURE [18], PCA, kinship analysis, LD analysis, and GWAS. To stratify covariates (Q) from the population structure,

STRUCTURE and PCA analyses were applied. In STRUCTURE, burn-in and Markov chain Monte Carlo (MCMC) values were set at 10,000 and 100,000, respectively. The STRUCTURE analysis was carried out for number of sub-populations (K) values ranging from 1 to 13. To evaluate the optimum K for this population, the ΔK method was applied. The use of MLM with covariates from STRUCTURE produced erroneous inflation and false-positive signals. Therefore, population stratification was analyzed using PCA, which has been previously proposed as an alternative method for investigating relatedness [42]. Therefore, covariates implemented by PCA analysis in TASSEL v5.2 [43] were adopted for MLM analysis. To determine the optimal number of PCs, a scree plot was generated based on the proportion of variance explained by PCs. The kinship matrix (K) for soybean core collection was analyzed using TASSEL v5.2 [43].

LD was analyzed using PLINK software [19] with LD window length of 1 Mb and an unlimited number of variants within LD window (–r2 –ld-window-kb 1000 –ld-window 99999). Considering the different patterns of LD decay in heterochromatin and euchromatin, genomic regions specified as pericentromeric regions in Soybase (available online: http://soybase.org) were downloaded and used in the LD analysis. The rate of LD decay for the soybean core collection was measured in physical distance, where the average pair-wise r^2 value between alleles dropped to half of its maximum value.

4.4. GWAS

GWAS was conducted using TASSEL v5.2 [43]. Two statistical models, GLM + Q (from STRUCTURE) and MLM + Q (from PCA) and K, were considered. The Q and K were regarded as fixed and random effects in GLM and MLM models, respectively. Quantile–quantile plots of both models were compared for determining the best fit. The threshold p-value ($1/n$, where n is the number of SNPs (81,078)) was used for the identification of QTNs significantly associated with branch number.

4.5. Expression Patterns of Genes in LD Block of *qtnBR6-1*

The identified QTNs were extended based on the rate of LD decay, depending on the chromosomal region (euchromatin vs. heterochromatin). To investigate the expression patterns and levels of genes located within the LD block of the *qtnBR6-1*, RNA-seq data (fragments per kilobase of exon model per million mapped reads (FPKM) values) for nine tissues, including flower, leaf, nodule, pod, root, root hair, SAM, seed, and stem, of soybean cv. Williams 82 were obtained from Phytozome v12.0 (available online: https://phytozome.jgi.doe.gov/pz/portal.html) [20]; RNA-seq data for axillary buds were not available in the public database. A heatmap with hierarchical clustering of genes was constructed using the R package pheatmap for visualizing gene expression levels in nine tissues, based on the $\log_2$(FPKM + 1) values. Only genes expressed in SAM were selected for further analysis.

4.6. Resequencing of NIL-HB, NIL-LB, and Parental Genotypes

The two NILs (NIL-HB and NIL-LB) and their parental genotypes, SS0404-T5-76 and Jiyu69, were resequenced. Raw sequence reads were mapped to the soybean reference genome (Wm82.a2) downloaded from Phytozome [20] using BWA [44], Samtools [45], and Vcftools [46]. Annotation of SNPs was conducted using SnpEff [47]. Nucleotide positions with more than ten supporting reads per genotype were analyzed further. SNPs segregating between NIL-HB and NIL-LB in the LD block of *qtnBR6-1* were compared with the sequence of SS0404-T5-76 and Jiyu69.

4.7. qRT-PCR Analysis of Candidate Genes in NILs

To determine the expression levels of selected genes in NILs, total RNA was extracted from the shoot apex (<3 mm) of NILs at the R1 stage using Ribospin™ Plant (GeneAll, Seoul, Korea), and cDNA was synthesized using Bio-Rad iScript™ cDNA Synthesis Kit (Hercules, CA, USA). Next, qRT-PCR was performed on a LightCycler® 480 (Roche Diagnostics, Laval, QC, Canada) using Bio-Rad iQ™ SYBR Green Supermix Kit. Primer sequences were designed using PRIMER3plus (available online: http://www.bioinformatics.nl/cgi-bin/primer3plus/primer3plus.cgi) [48]. Appropriate primer pairs that did not amplify orthologues of target genes were selected using a stand-alone version of electronic PCR [49] (Table S5). Each qRT-PCR reaction mixture (20 μL volume) contained 100 ng cDNA template, and 300 μM each of forward and reverse primer. Amplification was performed using the following conditions: initial denaturation at 95 °C for 5 min, followed by 40 cycles of denaturation at 95 °C for 10 s, and annealing and extension at 60 °C for 1 min. Three biological samples of each NIL were analyzed in triplicate to increase statistical power. The *ACTIN11* (*ACT11*) gene was used as a reference for data normalization. Normalized data were analyzed using the method of Livak and Schmittgen [50]. Statistical significance was analyzed using Fisher's least significant difference ($p < 0.05$) in R.

4.8. Analysis of BRC1 SNPs and Amino Acid Sequence

SNPs identified in the candidate gene *BRC1* based on a comparison between NIL-HB and NIL-LB were tested for association with branch number in a collection of 59 soybean accessions obtained from USDA-GRIN. Genomic DNA of each soybean accession was extracted using Exgene™ Plant SV mini DNA extraction kit (GeneAll, Seoul, Korea). On the basis of the identified SNPs between NIL-HB and NIL-LB, primers were designed using PRIMER3Plus (Table S5) [48] and validated using an electronic PCR algorithm [49]. PCR products were sequenced using ABI 3730XL DNA analyzer (Applied Biosystems, Foster, CA, USA). Branch numbers of soybean accessions were downloaded from the GRIN website. Association analysis between branch numbers and allelic variations was performed using ANOVA. Amino acid sequences of *BRC1* orthologues from Arabidopsis, rice, and soybean were aligned using MEGA7 [51].

Author Contributions: Conceptualization, S.S., J.H., M.Y.K., and S.-H.L.; Methodology, S.S.; Validation, S.S.; Formal Analysis, S.S.; Investigation, S.S.; Resources, M.S.C., S.-T.K., S.-C.J., and J.-K.M.; Data Curation, S.S.; Writing—Original Draft Preparation, S.S.; Writing—Review and Editing, M.Y.K., J.H., and S.-H.L.; Visualization, S.S.; Supervision, M.Y.K., S.-C.J., and S.-H.L.; Project Administration, S.-H.L.; Funding Acquisition, S.-H.L.

Abbreviations

BLUP	Best linear unbiased predictor/prediction
BRC1	*BRANCHED1*
GLM	Generalized linear model
GWAS	Genome-wide association study
LD	Linkage disequilibrium
MLM	Mixed linear model
NIL	Near-isogenic line
QTL	Quantitative trait locus
QTN	Quantitative trait nucleotide
RHL	Residual heterozygous line
RIL	Recombinant inbred line
SAM	Shoot apical meristem
SNP	Single nucleotide polymorphism
TCP	TEOSINTE BRANCHED1/CYCLODEA/PROLIFERATING CELL FACTOR 1 and 2

References

1. Heatherly, L.G.; Elmore, R.W. Managing Inputs for Peak Production. In *Soybeans: Improvement, Production, and Uses*; Agronomy Monograph 16; American Society of Agronomy, Inc.; Crop Science Society of America, Inc.; Soil Science Society of America, Inc.: Madison, WI, USA, 2004; pp. 451–536. [CrossRef]
2. Cho, Y.; Kim, S. Growth parameters and seed yield compenets by seeding time and seed density of non-/few branching soybean cultivars in drained paddy field. *Asian J. Plant Sci.* **2010**, *9*, 140–145.
3. Agudamu; Yoshihira, T.; Shiraiwa, T. Branch development responses to planting density and yield stability in soybean cultivars. *Plant Prod. Sci.* **2016** *19*, 331–339. [CrossRef]

4. Board, J. Light Interception Efficiency and Light Quality Affect Yield Compensation of Soybean at Low Plant Populations. *Crop Sci.* **2000**, *40*, 1285–1294. [CrossRef]
5. Cox, W.J.; Cherney, J.H.; Shields, E. Soybeans Compensate at Low Seeding Rates but not at High Thinning Rates. *Agron. J.* **2010**, *102*, 1238–1243. [CrossRef]
6. Board, J.E.; Kahlon, C.S. Morphological Responses to Low Plant Population Differ Between Soybean Genotypes. *Crop Sci.* **2013**, *53*, 1109–1119. [CrossRef]
7. Sayama, T.; Hwang, T.-Y.; Yamazaki, H.; Yamaguchi, N.; Komatsu, K.; Takahashi, M.; Suzuki, C.; Miyoshi, T.; Tanaka, Y.; Xia, Z.; et al. Mapping and comparison of quantitative trait loci for soybean branching phenotype in two locations. *Breeding Sci.* **2010**, *60*, 380–389. [CrossRef]
8. Shim, S.; Kim, M.Y.; Ha, J.; Lee, Y.-H.; Lee, S.-H. Identification of QTLs for branching in soybean (*Glycine max* (L.) Merrill). *Euphytica* **2017**, *213*, 225. [CrossRef]
9. Yao, D.; Liu, Z.Z.; Zhang, J.; Liu, S.Y.; Qu, J.; Guan, S.Y.; Pan, L.D.; Wang, D.; Liu, J.W.; Wang, P.W. Analysis of quantitative trait loci for main plant traits in soybean. *Genet. Mol. Res.* **2015**, *14*, 6101–6109. [CrossRef]
10. Chen, Q.; Zhang, Z.; Liu, C.; Xin, D.; Qiu, H.; Shan, D.; Shan, C.; Hu, G. QTL Analysis of Major Agronomic Traits in Soybean. *Agr. Sci. China* **2007**, *6*, 399–405. [CrossRef]
11. Li, W.; Zheng, D.; Van, K.; Lee, S. QTL Mapping for Major Agronomic Traits across Two Years in Soybean (*Glycine max* L. Merr.). *J. Crop Sci. Biotech.* **2008**, *11*, 171.
12. Rafalski, J.A. Association genetics in crop improvement. *Curr. Opin. Plant Biol.* **2010**, *13*, 174–180. [CrossRef] [PubMed]
13. Zhao, K.; Tung, C.-W.; Eizenga, G.C.; Wright, M.H.; Ali, M.L.; Price, A.H.; Norton, G.J.; Islam, M.R.; Reynolds, A.; Mezey, J.; et al. Genome-wide association mapping reveals a rich genetic architecture of complex traits in *Oryza sativa*. *Nat. Commun.* **2011**, *2*, 467. [CrossRef] [PubMed]
14. Tian, F.; Bradbury, P.J.; Brown, P.J.; Hung, H.; Sun, Q.; Flint-Garcia, S.; Rocheford, T.R.; McMullen, M.D.; Holland, J.B.; Buckler, E.S. Genome-wide association study of leaf architecture in the maize nested association mapping population. *Nat. Genet.* **2011**, *43*, 159–162. [CrossRef] [PubMed]
15. Zhao, K.; Aranzana, M.J.; Kim, S.; Lister, C.; Shindo, C.; Tang, C.; Toomajian, C.; Zheng, H.; Dean, C.; Marjoram, P.; et al. An *Arabidopsis* Example of Association Mapping in Structured Samples. *PLoS Genet.* **2007**, *3*, e4. [CrossRef] [PubMed]
16. Cao, Y.; Li, S.; Wang, Z.; Chang, F.; Kong, J.; Gai, J.; Zhao, T. Identification of Major Quantitative Trait Loci for Seed Oil Content in Soybeans by Combining Linkage and Genome-Wide Association Mapping. *Front. Plant Sci.* **2017**, *8*. [CrossRef]
17. Zhang, J.; Song, Q.; Cregan, P.B.; Nelson, R.L.; Wang, X.; Wu, J.; Jiang, G.-L. Genome-wide association study for flowering time, maturity dates and plant height in early maturing soybean (*Glycine max*) germplasm. *BMC Genomics* **2015**, *16*, 217. [CrossRef] [PubMed]
18. Pritchard, J.K.; Stephens, M.; Donnelly, P. Inference of population structure using multilocus genotype data. *Genetics* **2000**, *155*, 945–959.
19. Purcell, S.; Neale, B.; Todd-Brown, K.; Thomas, L.; Ferreira, M.A.R.; Bender, D.; Maller, J.; Sklar, P.; de Bakker, P.I.W.; Daly, M.J.; et al. PLINK: A Tool Set for Whole-Genome Association and Population-Based Linkage Analyses. *Am. J. Hum. Genet.* **2007**, *81*, 559–575. [CrossRef]
20. Schmutz, J.; Cannon, S.B.; Schlueter, J.; Ma, J.; Mitros, T.; Nelson, W.; Hyten, D.L.; Song, Q.; Thelen, J.J.; Cheng, J.; et al. Genome sequence of the palaeopolyploid soybean. *Nature* **2010**, *463*, 178–183. [CrossRef]
21. Aguilar-Martínez, J.A.; Poza-Carrión, C.; Cubas, P. *Arabidopsis BRANCHED1* Acts as an Integrator of Branching Signals within Axillary Buds. *Plant Cell* **2007**, *19*, 458–472. [CrossRef]
22. Poza-Carrión, C.; Aguilar-Martínez, J.A.; Cubas, P. Role of TCP Gene *BRANCHED1* in the Control of Shoot Branching in Arabidopsis. *Plant Signal Behav.* **2007**, *2*, 551–552. [CrossRef] [PubMed]
23. Ghodrati, G. Study of genetic variation and broad sense heritability for some qualitative and quantitative traits in soybean (*Glycine max* L.) genotypes. *Curr. Opin. Agr.* **2013**, *2*.
24. Rameau, C.; Bertheloot, J.; Leduc, N.; Andrieu, B.; Foucher, F.; Sakr, S. Multiple pathways regulate shoot branching. *Front Plant Sci.* **2015**, *5*. [CrossRef] [PubMed]
25. Board, J.E.; Settimi, J.R. Photoperiod Effect before and after Flowering on Branch Development in Determinate Soybean. *Agron. J.* **1986**, *78*, 995–1002. [CrossRef]
26. Frederick, J.R.; Camp, C.R.; Bauer, P.J. Drought-Stress Effects on Branch and Mainstem Seed Yield and Yield Components of Determinate Soybean. *Crop Sci.* **2001**, *41*, 759–763. [CrossRef]

27. Linkemer, G.; Board, J.E.; Musgrave, M.E. Waterlogging effects on growth and yield components in late-planted soybean. *Crop Sci.* **1998**, *38*, 1576–1584. [CrossRef] [PubMed]
28. Thimann, K.V.; Skoog, F. Studies on the growth hormone of plants the inhibiting action of the growth substance on bud development. *Proc. Natl. Acad. Sci. USA* **1933**, *19*, 714–716. [CrossRef] [PubMed]
29. Niwa, M.; Daimon, Y.; Kurotani, K.; Higo, A.; Pruneda-Paz, J.L.; Breton, G.; Mitsuda, N.; Kay, S.A.; Ohme-Takagi, M.; Endo, M.; et al. BRANCHED1 Interacts with FLOWERING LOCUS T to Repress the Floral Transition of the Axillary Meristems in *Arabidopsis*[C][W][OA]. *Plant Cell* **2013**, *25*, 1228–1242. [CrossRef]
30. Braun, N.; Germain, A. de S.; Pillot, J.-P.; Boutet-Mercey, S.; Dalmais, M.; Antoniadi, I.; Li, X.; Maia-Grondard, A.; Signor, C.L.; Bouteiller, N.; et al. The Pea TCP Transcription Factor PsBRC1 Acts Downstream of Strigolactones to Control Shoot Branching. *Plant Physiol.* **2012**, *158*, 225–238. [CrossRef]
31. Doebley, J.; Stec, A.; Hubbard, L. The evolution of apical dominance in maize. *Nature* **1997**, *386*, 485–488. [CrossRef]
32. Takeda, T.; Suwa, Y.; Suzuki, M.; Kitano, H.; Ueguchi-Tanaka, M.; Ashikari, M.; Matsuoka, M.; Ueguchi, C. The *OsTB1* gene negatively regulates lateral branching in rice. *Plant J.* **2003**, *33*, 513–520. [CrossRef] [PubMed]
33. Finlayson, S.A. Arabidopsis TEOSINTE BRANCHED1-LIKE 1 Regulates Axillary Bud Outgrowth and is Homologous to Monocot TEOSINTE BRANCHED1. *Plant Cell Physiol.* **2007**, *48*, 667–677. [CrossRef] [PubMed]
34. Kebrom, T.H.; Brutnell, T.P.; Finlayson, S.A. Suppression of sorghum axillary bud outgrowth by shade, phyB and defoliation signalling pathways. *Plant Cell Environ.* **2010**, *33*, 48–58. [CrossRef]
35. Minakuchi, K.; Kameoka, H.; Yasuno, N.; Umehara, M.; Luo, L.; Kobayashi, K.; Hanada, A.; Ueno, K.; Asami, T.; Yamaguchi, S.; et al. *FINE CULM1* (*FC1*) Works Downstream of Strigolactones to Inhibit the Outgrowth of Axillary Buds in Rice. *Plant Cell Physiol.* **2010**, *51*, 1127–1135. [CrossRef] [PubMed]
36. Cubas, P.; Lauter, N.; Doebley, J.; Coen, E. The TCP domain: a motif found in proteins regulating plant growth and development. *Plant J.* **1999**, *18*, 215–222. [CrossRef] [PubMed]
37. Sorefan, K.; Booker, J.; Haurogné, K.; Goussot, M.; Bainbridge, K.; Foo, E.; Chatfield, S.; Ward, S.; Beveridge, C.; Rameau, C.; et al. *MAX4* and *RMS1* are orthologous dioxygenase-like genes that regulate shoot branching in *Arabidopsis* and pea. *Genes Dev.* **2003**, *17*, 1469–1474. [CrossRef]
38. Clark, R.M.; Linton, E.; Messing, J.; Doebley, J.F. Pattern of diversity in the genomic region near the maize domestication gene *tb1*. *Proc. Natl. Acad. Sci. USA* **2004**, *101*, 700–707. [CrossRef]
39. Clark, R.M.; Wagler, T.N.; Quijada, P.; Doebley, J. A distant upstream enhancer at the maize domestication gene *tb1* has pleiotropic effects on plant and inflorescent architecture. *Nat. Genet.* **2006**, *38*, 594–597. [CrossRef]
40. Bates, D.; Sarkar, D. lme4: Linear mixed-effects models using S4 classes. 2007.
41. Lee, Y.-G.; Jeong, N.; Kim, J.H.; Lee, K.; Kim, K.H.; Pirani, A.; Ha, B.-K.; Kang, S.-T.; Park, B.-S.; Moon, J.-K.; et al. Development, validation and genetic analysis of a large soybean SNP genotyping array. *Plant J.* **2015**, *81*, 625–636. [CrossRef]
42. Price, A.L.; Patterson, N.J.; Plenge, R.M.; Weinblatt, M.E.; Shadick, N.A.; Reich, D. Principal components analysis corrects for stratification in genome-wide association studies. *Nat. Genet.* **2006**, *38*, 904–909. [CrossRef]
43. Bradbury, P.J.; Zhang, Z.; Kroon, D.E.; Casstevens, T.M.; Ramdoss, Y.; Buckler, E.S. TASSEL: software for association mapping of complex traits in diverse samples. *Bioinformatics* **2007**, *23*, 2633–2635. [CrossRef] [PubMed]
44. Li, H.; Durbin, R. Fast and accurate short read alignment with Burrows-Wheeler transform. *Bioinformatics* **2009**, *25*, 1754–1760. [CrossRef] [PubMed]
45. Li, H.; Handsaker, B.; Wysoker, A.; Fennell, T.; Ruan, J.; Homer, N.; Marth, G.; Abecasis, G.; Durbin, R. 1000 Genome Project Data Processing Subgroup The Sequence Alignment/Map format and SAMtools. *Bioinformatics* **2009**, *25*, 2078–2079. [CrossRef] [PubMed]
46. Danecek, P.; Auton, A.; Abecasis, G.; Albers, C.A.; Banks, E.; DePristo, M.A.; Handsaker, R.E.; Lunter, G.; Marth, G.T.; Sherry, S.T.; et al. The variant call format and VCFtools. *Bioinformatics* **2011**, *27*, 2156–2158. [CrossRef]
47. Cingolani, P.; Platts, A.; Wang, L.L.; Coon, M.; Nguyen, T.; Wang, L.; Land, S.J.; Lu, X.; Ruden, D.M. A program for annotating and predicting the effects of single nucleotide polymorphisms, SnpEff: *SNPs in the genome of Drosophila melanogaster strain w1118; iso-2; iso-3. Fly (Austin)* **2012**, *6*, 80–92. [CrossRef] [PubMed]

48. Untergasser, A.; Nijveen, H.; Rao, X.; Bisseling, T.; Geurts, R.; Leunissen, J.A.M. Primer3Plus, an enhanced web interface to Primer3. *Nucleic Acids Res.* **2007**, *35*, W71–W74. [CrossRef] [PubMed]
49. Rotmistrovsky, K.; Jang, W.; Schuler, G.D. A web server for performing electronic PCR. *Nucleic Acids Res.* **2004**, *32*, W108–W112. [CrossRef]
50. Livak, K.J.; Schmittgen, T.D. Analysis of Relative Gene Expression Data Using Real-Time Quantitative PCR and the 2−ΔΔCT Method. *Methods* **2001**, *25*, 402–408. [CrossRef]
51. Kumar, S.; Stecher, G.; Tamura, K. MEGA7: Molecular Evolutionary Genetics Analysis Version 7.0 for Bigger Datasets. *Mol. Biol. Evol.* **2016**, *33*, 1870–1874. [CrossRef]

11

RrGT2, A Key Gene Associated with Anthocyanin Biosynthesis in *Rosa rugosa*, Was Identified Via Virus-Induced Gene Silencing and Overexpression

Xiaoming Sui †, Mingyuan Zhao †, Zongda Xu, Lanyong Zhao * and Xu Han *

Flower Research Laboratory, College of Forestry, Shandong Agricultural University, Taian 271018, China; suixiaomingjiayou@163.com (X.S.); zhaomingy9@163.com (M.Z.); xuzoda@163.com (Z.X.)

* Correspondence: sdzly369@163.com (L.Z.); hanxusdau@163.com (X.H.)

† These authors contributed equally.

Abstract: In this study, a gene with a full-length cDNA of 1422 bp encoding 473 amino acids, designated *RrGT2*, was isolated from *R. rugosa* 'Zizhi' and then functionally characterized. *RrGT2* transcripts were detected in various tissues and were proved that their expression patterns corresponded with anthocyanins accumulation. Functional verification of *RrGT2* in *R. rugosa* was performed via VIGS. When *RrGT2* was silenced, the *Rosa* plants displayed a pale petal color phenotype. The detection results showed that the expression of *RrGT2* was significantly downregulated, which was consistent with the decrease of all anthocyanins; while the expression of six key upstream structural genes was normal. Additionally, the in vivo function of *RrGT2* was investigated via its overexpression in tobacco. In transgenic tobacco plants expressing *RrGT2*, anthocyanin accumulation was induced in the flowers, indicating that *RrGT2* could encode a functional GT protein for anthocyanin biosynthesis and could function in other species. The application of VIGS in transgenic tobacco resulted in the treated tobacco plants presenting flowers whose phenotypes were lighter in color than those of normal plants. These results also validated and affirmed previous conclusions. Therefore, we speculated that glycosylation of *RrGT2* plays a crucial role in anthocyanin biosynthesis in *R. rugosa*.

Keywords: *Rosa rugosa*; *RrGT2* gene; Clone; VIGS; Overexpression; Tobacco; Flower color; Anthocyanin

1. Introduction

Rosa rugosa is an important ornamental plant species that belongs to the genus *Rosa* in the family *Rosaceae*. This species is native to China and is widely distributed worldwide. Because of its unique fragrance, color, cold resistance and drought resistance, there is great potential for the development of this species for use in garden applications. Many varieties of roses exist but most of them are traditional colors such as pink and purple. A few varieties are white and some lack yellow, bright red, orange and compound colors and so forth. [1]. Therefore, the development of innovative rose colors has become the main goal of breeders. At present, scientific research on *R. rugosa* in China and abroad has focused mainly on the development and protection of wild *R. rugosa* resources [2], the analysis of the genetic diversity of *R. rugosa* [3], the optimization of *R. rugosa* essential oil extraction [4], the nutritional value of *R. rugosa* [5], the cultivation and propagation technology of *R. rugosa* [6,7] and strategies for attaining high *R. rugosa* yields. Innovating new *R. rugosa* flower colors has occurred mainly by improving cultivation and management techniques or by trying to cross different varieties during *R. rugosa* breeding; relatively less molecular biology technology has been used in the innovation of new flower colors. However, due to the decrease in wild *R. rugosa* resources and the lack of natural variation in recent years, the existing *R. rugosa* varieties can no longer meet the various needs of gardening applications [1]. However, there is still much room for development in the breeding of new *R. rugosa* varieties via molecular biology. Therefore, studying the mechanism of *R. rugosa* flower color formation and enriching flower color during breeding are urgently needed. Analysis of the pigment composition of rose varieties and the study of the expression characteristics of the key genes encoding enzymes that catalyze the synthesis of rose pigments are important prerequisites for the molecular breeding of rose color traits [8]. Anthocyanins determine the color of higher plant organs. Structural genes (*CHS, CHI, F3H, F3'H, DFR, ANS, 3GT*, etc.) and regulatory genes (*MYB*, mostly those of the *R2R3-MYB, BHLH* and *WD40* classes) [9,10] related to the anthocyanin biosynthesis pathway have been cloned and sequenced and related protein functional studies have been performed in many plant species, such as petunia, maize, snapdragon and so on [11,12]. However, less anthocyanin-related research has been conducted in rose than in those species.

Flower coloration is caused by the accumulation of pigments comprising mainly flavonoids, carotenoids and betalains. Among them, flavonoids, which comprise chalcones, flavones, flavonols, flavandiols, anthocyanins and proanthocyanidins, are the most important pigments [13,14]. Anthocyanins, which are derived from the anthocyanin biosynthesis pathway, are the largest group of water-soluble plant flavonoids found in the organs of plants, including crop species [15–19]. Anthocyanins are unstable in plants; they exist mainly in the form of glycosides within the vacuole [20]. The flavonoid 3-*O*-glycosyltransferase (*3GT*) gene lies downstream in the anthocyanin synthesis pathway. The enzyme encoded by this gene can catalyze the glycosylation of UDP-glucose to replace the 3 hydroxyl groups of anthocyanin molecules and cause anthocyanin glycosylation to produce colored and stable anthocyanins. Glycosylation can alter the hydrophilicity, biochemical activity and subcellular localization of anthocyanins, which is beneficial to their transport and storage in cells and organisms [21–23].

For a long time, GT genes had failed to be cloned and their functions in plant secondary metabolism were unclear. However, since the first cDNA sequence of a GTase was isolated by transposon tagging in maize, an increasing number of GT genes have been cloned and identified [24]. Studies have shown that the *3GT* gene is expressed only in red grape (*Vitis vinifera*) varieties and not in white grape varieties. When a *3GT* transgene was transformed into a colorless embryo, a pale-red bud was produced [25,26]. Studies by Afifi et al. on the expression of five key genes involved in anthocyanin synthesis in grape cell tissue indicate that the presence of the cytotoxic factor eutypine inhibits the expression of *3GT* and thus reduces anthocyanin contents [27]. This finding indicates that *3GT* is the key gene involved in grape skin color (from white to red) and is also a key gene in the anthocyanin biosynthesis pathway [28]. In *Gentiana triflora*, *3GT* expression occurs mostly in blue petals and rarely in white flowers [29]. Expression of the *3GT* gene is very important for anthocyanin accumulation in many plant species and its expression intensity is positively correlated with anthocyanin synthesis [30].

In this study, we cloned and identified the *RrGT2* gene from the petals of *R. rugosa* for the first time. We carried out detailed bioinformatic and homology analyses of the *RrGT2* gene. VIGS results in perennial *Rosa* plants under field conditions suggested that *RrGT2* is related to the biosynthesis of anthocyanins in *R. rugosa*. Stable transformation of the *RrGT2* gene in tobacco showed that its overexpression was positively correlated with the accumulation of anthocyanins. The results of VIGS in transgenic tobacco also confirmed this conclusion. We verified the functions of the *RrGT2* gene in anthocyanin metabolism in both the positive and negative directions to provide useful information for subsequent color-improvement projects in *R. rugosa*.

2. Results

2.1. Cloning of RrGT2 and Sequence Analysis

In the early stages, we screened the *RrGT2* gene by comparing the differentially expressed genes in the anthocyanin pathway within the *R. rugosa* transcriptome data of our laboratory. The full-length CDS of *RrGT2* (MK034141) was cloned and then confirmed by sequencing [31]. The complete open reading frame from the ATG start codon to the TAA termination codon encodes a 473 amino acid protein (Figure 1A). Multiple sequence alignment (Figure 2) revealed that the *RrGT2* protein, which belongs to the GTB superfamily, displays strong species specificity in the N-terminal region and PSPG conserved domains that consist of 44 amino acid residues in the C-terminal region. A phylogenetic tree (Figure 1B) was constructed from the amino acid sequences of 21 plants, including the sequence of *RrGT2*, using MEGA 5.0 software. The results showed that the *RrGT2* gene shared the highest homology percentages with *FaUGT* (*Fragaria* × *ananassa*) and *FvGT* (*Fragaria vesca* subsp.), both of which were 89%.

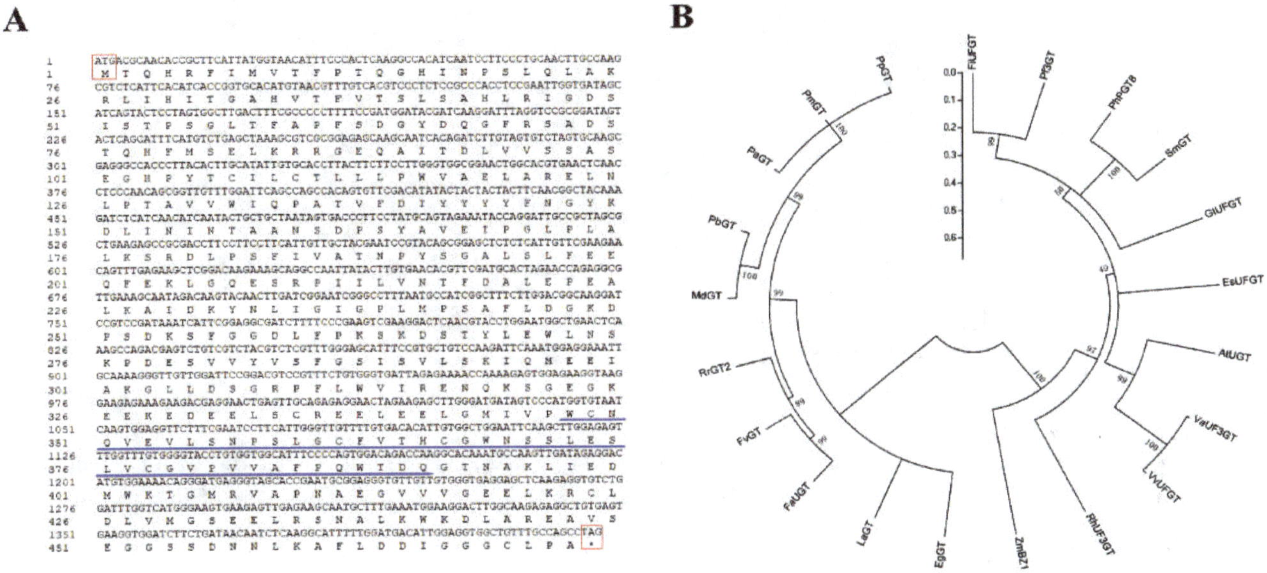

Figure 1. cDNA sequence analysis and phylogenetic tree analysis of the *RrGT2* gene. (**A**) cDNA sequence of *RrGT2* and its deduced amino acids. The red box shows the start codon and the termination codon as well as the amino acids they encode. The PSPG domains are underlined by the blue line. (**B**) Phylogenetic tree of amino acid sequences of *RrGT2* and *GT* members from other plant species. The tree was constructed by the neighbor-joining method using MEGA 5.0 software. The branch numbers represent the percentage of bootstrap values from 1000 sampling replicates and the scale indicates the branch lengths. The gene names from various plant species and the NCBI GenBank accession numbers for the sequences are as follows: *RhUF3GT* (AB599928.1) from a *Rosa* hybrid cultivar; *AtUGT* (UGTNM-121711) from *Arabidopsis thaliana*; *EsUFGT* (KJ648620) from *Epimedium sagittatum*; *FaUGT* (KP337600.1) from *Fragaria* × *ananassa*; *FiUFGT* (AF127218.1) from *Forsythia* × *intermedia*; *FvGT* (XM_004298174.2) from *Fragaria vesca* subspecies; *GtUFGT* (D85186.1) from *Gentiana triflora*; *LaGT* (XM_019560329.1) from *Lupinus angustifolius*; *MdGT* (XM_008350196.2) from *Malus* × *domestica*; *PaGT* (XM_021963118.1) from *Prunus avium*; *PbGT* (XM_009339472.2) from *Pyrus* × *bretschneideri*; *Pf3GT* (AB002818) from *Perilla frutescens*; *PhPGT8* (AB027454) from *Petunia* × *hybrida*; *PmGT* (XM_008224513.2) from *Prunus mume*; *PpGT* (XM_007221257.2) from *Prunus persica*; *SmGT* (X77369.1) from *Solanum melongena*; *VaUF3GT* (FJ169463.1) from *Vitis amurensis*; *VvUFGT* (AF000371) from *Vitis vinifera*; *ZmBZ1* (NM_001112416.1) from *Zea mays*; and *EgGT* (XM_012988948.1) from *Erythranthe guttatus*.

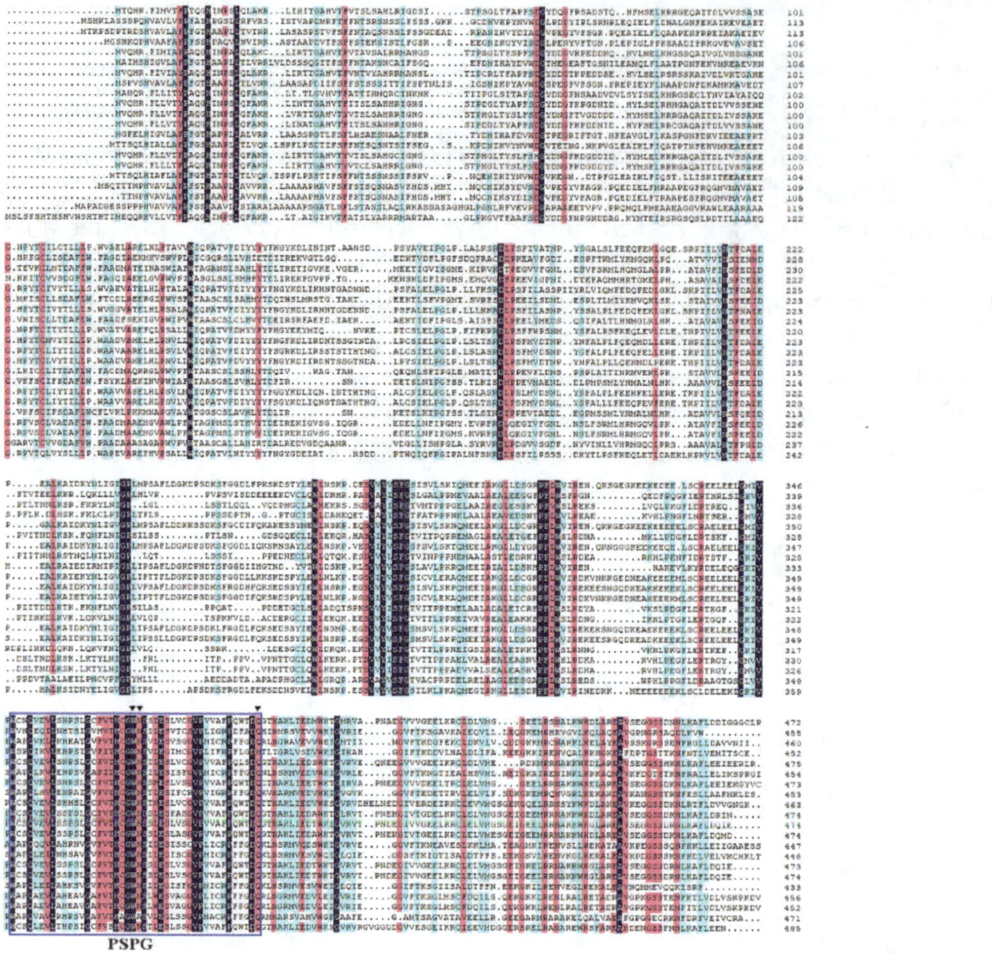

Figure 2. Amino acid sequence homology analysis of the *RrGT2* gene and GTs from other species. Alignments were performed using DNAMAN (version 6.0). The blue box shows the PSPG domains. The black triangles (from left to right) indicate the 22nd, 23rd and 44th amino acids in the PSPG box. The accession numbers are the same as those in Figure 1B.

2.2. *Temporal and Spatial Expression Patterns of RrGT2 in Rosa*

Before analyzing the expression patterns of *RrGT2*, we cloned a gene from the cDNA of *R. davurica* with the full-length primers of *RrGT2* and sequenced the gene sequence which was identical to that of *RrGT2*. So, we named it *RdGT2*.

The expression levels of the *RrGT2* and *RdGT2* gene, which significantly differed, were assessed during five flowering stages. In *R. rugosa* 'Zizhi' (Figure 3A), the highest expression level of *RrGT2* was observed during the full opening stage and the lowest was observed during the budding stage. In *Rosa davurica* (Figure 3B), the expression level of *RdGT2* was also highest during the full opening stage but lowest during the half opening stage. The expression patterns of the *RrGT2* gene in *R. rugosa* 'Zizhi' and *R. davurica* exhibited approximately the same trend.

The expression levels of the *RrGT2* and *RdGT2* gene, which also significantly differed, were assessed in seven different tissue types. The expression level in the leaves, stems and sepals was relatively high but was relatively low in the other tissues in both *R. rugosa* 'Zizhi' (Figure 3C) and *R. davurica* (Figure 3D).

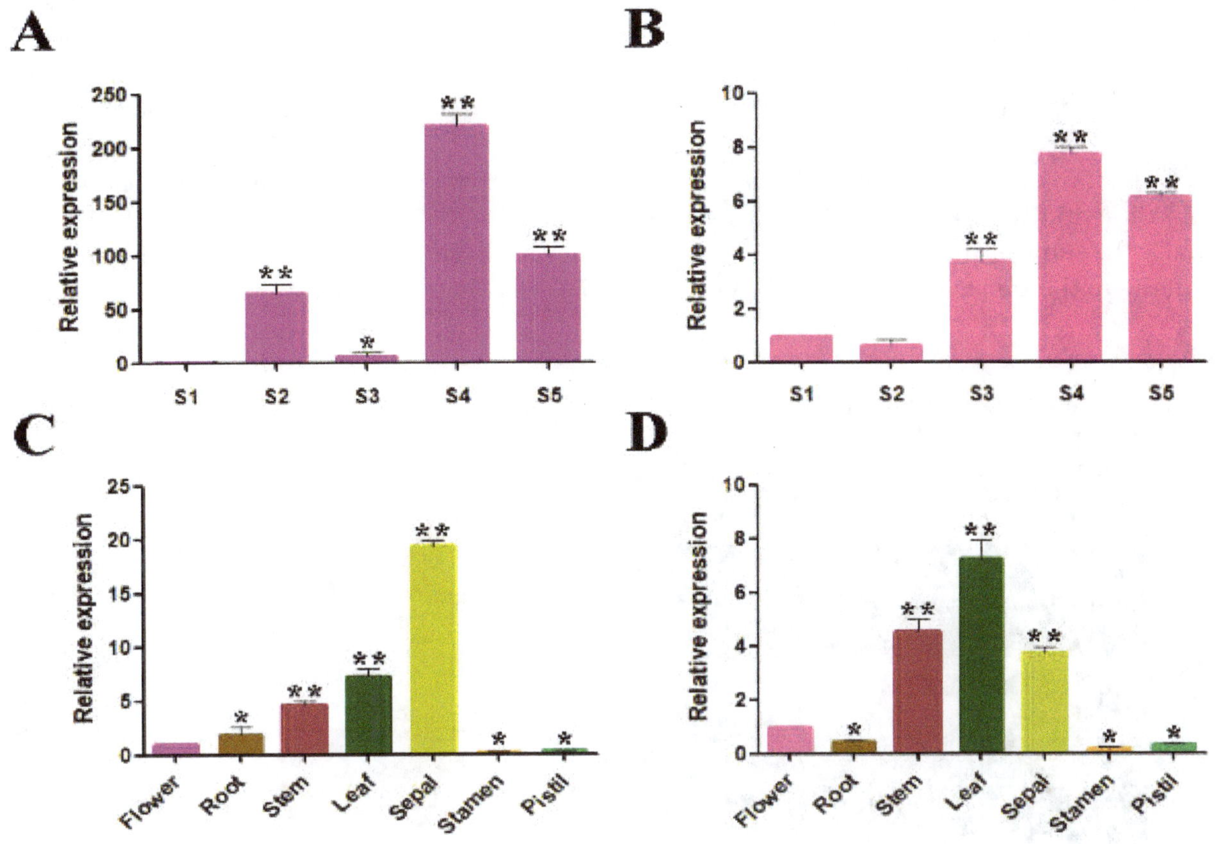

Figure 3. Temporal and spatial expression patterns of *RrGT2*. Relative expression of the *RrGT2* and *RdGT2* gene during the five flowering stages of *R. rugosa* 'Zizhi' (**A**) and *R. davurica* (**B**). S1, budding stage; S2, initial opening stage; S3, half opening stage; S4, full opening stage; S5, wilting stage. Relative expression of the *RrGT2* and *RdGT2* gene in seven different tissues of *R. rugosa* 'Zizhi' (**C**) and *R. davurica* (**D**). The error bars represent the SDs of triplicate reactions. The experiment was repeated three times and each yielded similar results. * $p < 0.05$ and ** $p < 0.01$ indicate significant differences between different flowering stages and between different tissue types.

2.3. VIGS of RrGT2 Reduced the Transcript Abundance of the Endogenous RrGT2 Gene

At 14 days after infection, GFP detection was performed on the newly grown leaves of the infected plants (TRV-GFP and TRV-GFP-*RrGT2*) and on the untreated leaves of both *Rosa* species. GFP imaging (Figures 4A and 5A) showed that the leaves treated with VIGS (TRV-GFP and TRV-GFP-*RrGT2*) showed green fluorescence under longwave ultraviolet light, while the untreated leaves in the control group showed red fluorescence. The corresponding leaves were collected for qRT-PCR detection and the *RrGAPDH* gene was used as an internal control [32] to confirm the efficiency of VIGS. The results (Figures 4D and 5D) showed that the abundance of the *RrGT2* transcript significantly decreased only in the leaves treated with TRV-GFP-*RrGT2* but was expressed normally in the leaves of the plants in the control group and TRV-GFP group.

At 30–35 days after infection, the flowers of *R. rugosa* 'Zizhi' changed from the budding stage to the initial opening stage, by which time notable differences in flower color could be observed with the naked eye. The petals in the control group and TRV-GFP group showed no definitive changes in color but the petals in the TRV-GFP-*RrGT2* group were clearly lighter in color (Figure 4B).

At 35–40 days after infection, GFP imaging and qRT-PCR were performed on the blossoming petals of untreated plants and infected plants (TRV-GFP and TRV-GFP-*RrGT2*) of both *Rosa* species. The results were as expected: the petals treated with VIGS (TRV-GFP and TRV-GFP-*RrGT2*) showed green fluorescence under longwave ultraviolet light, while the petals in the control group showed red fluorescence (Figures 4C and 5B). The results of qRT-PCR (Figures 4E and 5E) were consistent with those of the above detection. In both *Rosa* species, the relative expression trends of the *RrGT2* gene were also essentially consistent: the relative expression of the *RrGT2* gene in the TRV-GFP group was essentially the same as that in the control group. However, the transcript abundance of the endogenous *RrGT2* gene in the petals treated with TRV-GFP-*RrGT2* was significantly downregulated.

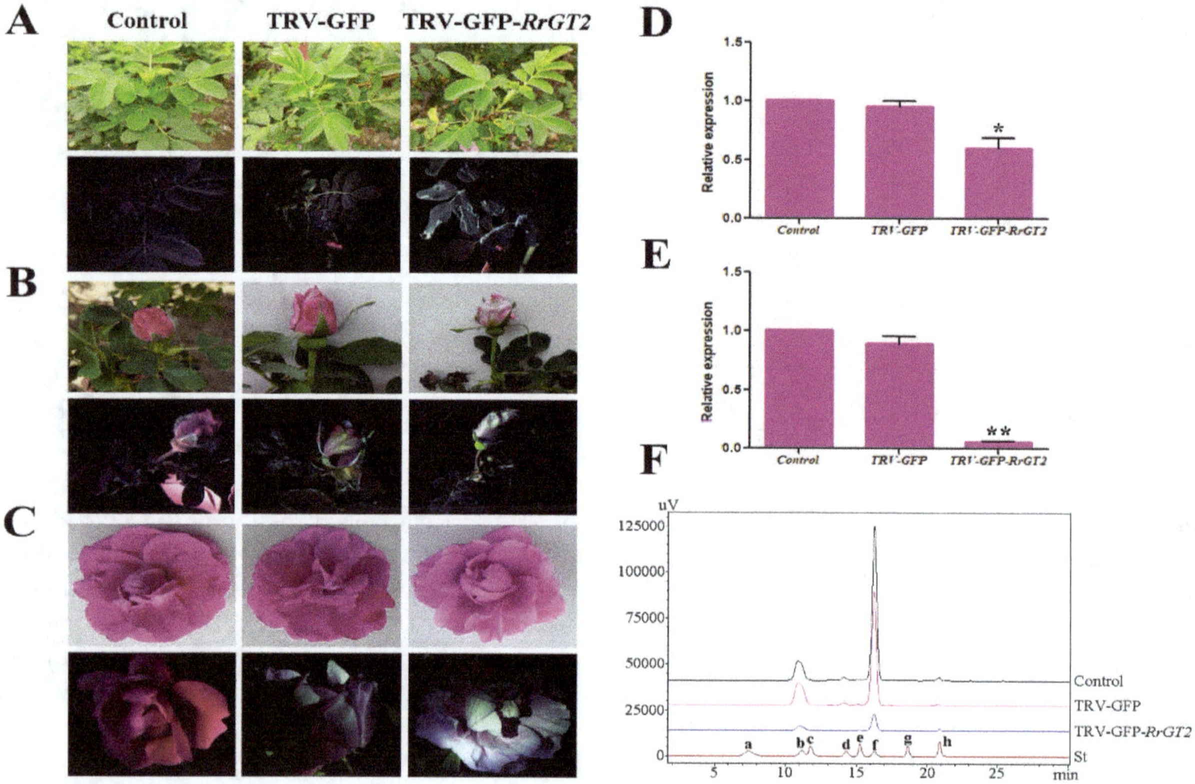

Figure 4. Validation of VIGS in *R. rugosa* 'Zizhi.' Comparisons between the control group and VIGS-treated groups (TRV-GFP and TRV-GFP-*RrGT2*) with respect to leaves (**A**), flowers between budding stage and initial opening stage (**B**) and flowers at the full opening stage (**C**). The plants were imaged under normal light and ultraviolet illumination. qRT-PCR detection of leaves (**D**) and flowers at the full opening stage (**E**) in the control and VIGS-treated groups. *RrGAPDH* was used as an internal control. The error bars represent the SDs of triplicate reactions. The experiment was repeated three times and each yielded similar results. * and ** indicate a significant difference from the control at $p < 0.05$ and $p < 0.01$, respectively, according to Student's *t*-test. (**F**) HPLC-derived chromatograms of flowers at the full opening stage. Eight kinds of anthocyanin standards (St) were used for detection: (a) Dp3G5G; (b) Cy3G5G; (c) Dp3G; (d) Pg3G5G; (e) Cy3G; (f) Pn3G5G; (g) Pg3G; and (h) Pn3G.

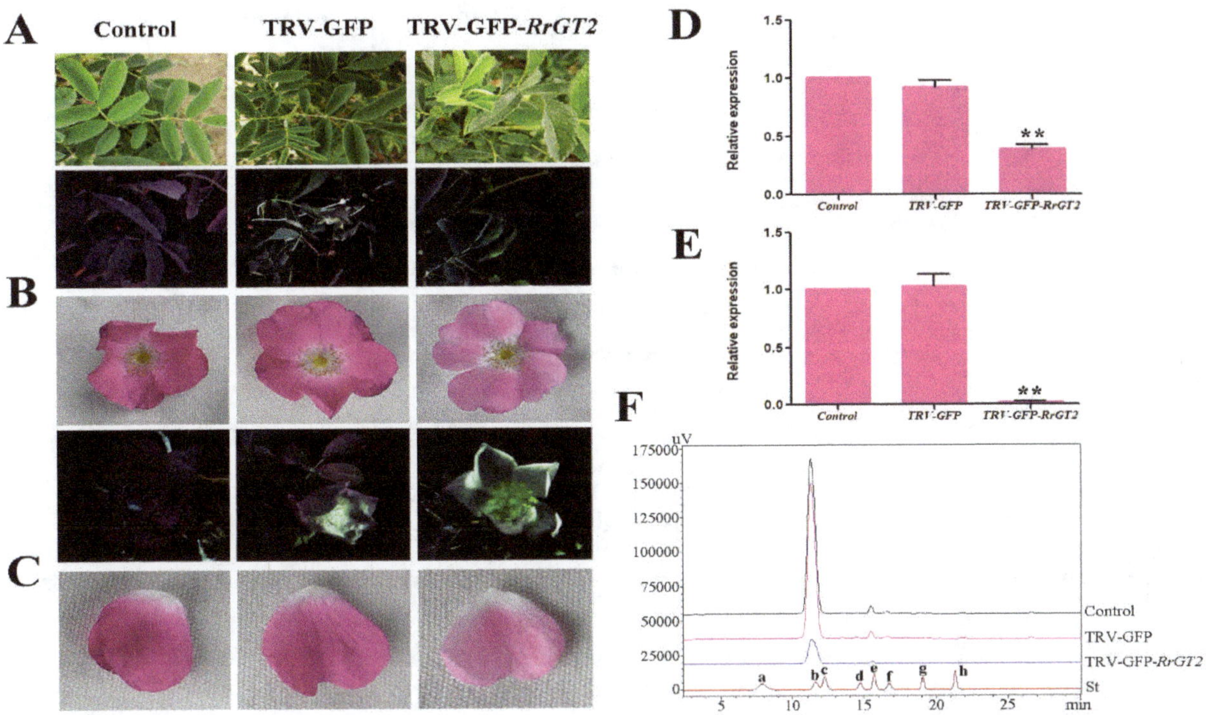

Figure 5. Validation of VIGS in *R. davurica*. Comparisons between the control group and VIGS-treated groups (TRV-GFP and TRV-GFP-*RrGT2*) with respect to leaves (**A**) and flowers at the full opening stage (**B**). The plants were imaged under normal light and ultraviolet illumination. (**C**) The color contrast of a single petal. qRT-PCR detection of leaves (**D**) and flowers at the full opening stage (**E**) in the control and VIGS-treated groups. *RrGAPDH* was used as an internal control. The error bars represent the SDs of triplicate reactions. The experiment was repeated three times and each yielded similar results. ** indicates a significant difference from the control at $p < 0.01$ according to Student's *t*-test. (**F**) HPLC-derived chromatograms of flowers at the full opening stage. Eight kinds of anthocyanin standards (St) were used for detection: (a) Dp3G5G; (b) Cy3G5G; (c) Dp3G; (d) Pg3G5G; (e) Cy3G; (f) Pn3G5G; (g) Pg3G; and (h) Pn3G.

2.4. Reduced Anthocyanin Accumulation in Rosa Petals Was Related to the VIGS of RrGT2

To clarify the role of the posttranscriptional silencing of the *RrGT2* gene in the outcome of the current color change, we compared the relative expression of the six key structural genes, *RrCHS* (KT809351), *RrCHI* (KT809352), *RrF3H* (KT809354), *RrF3'H* (MG735186), *RrDFR* (KT809350) and *RrANS* (KT809353), upstream of the *RrGT2* gene in the anthocyanin pathway (Figure 6). The results showed that the relative expression of the six genes in the control group, TRV-GFP group and TRV-GFP-*RrGT2* group of *R. rugosa* 'Zizhi' (Figure 7A) and *R. davurica* (Figure 7B) did not clearly change. Therefore, the effects of these upstream structural genes can be excluded. It is speculated that the change in flower color after VIGS treatment was related to the posttranscriptional silencing of the *RrGT2* gene.

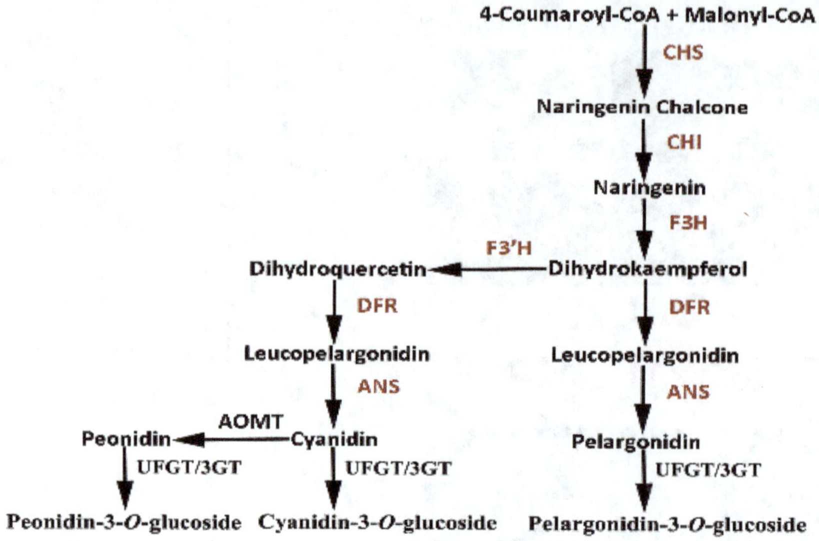

Figure 6. The metabolic pathway analysis of structural genes involved in anthocyanin biosynthesis of *R. rugosa*.

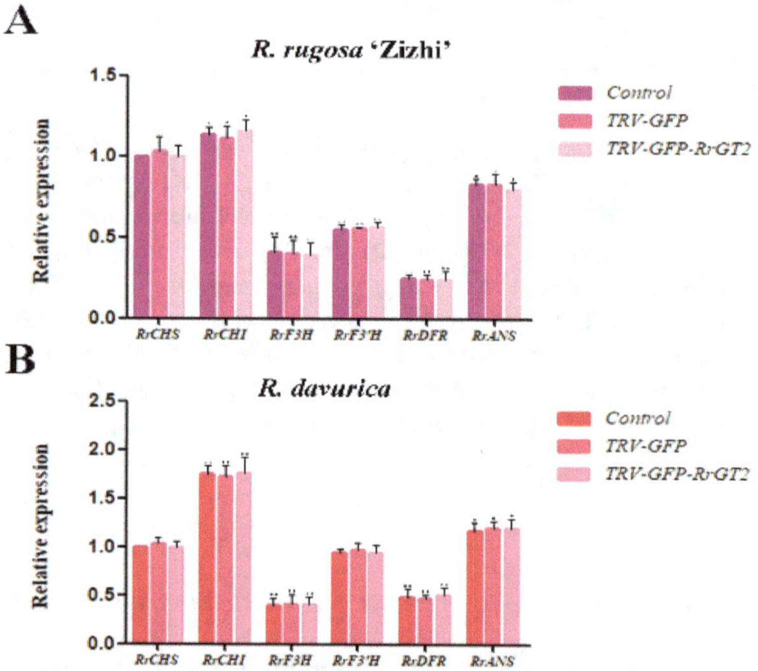

Figure 7. Relative expression levels of six key structural genes (*RrCHS*, *RrCHI*, *RrF3H*, *RrF3'H*, *RrDFR* and *RrANS*) upstream of *RrGT2* in the anthocyanin pathway of *R. rugosa* 'Zizhi' (**A**) and *R. davurica* (**B**) at the full opening stage. *RrGAPDH* was used as the internal control. The error bars represent the SDs of triplicate reactions. The experiment was repeated three times and each yielded similar results. * and ** indicate a significant difference from the relative expression levels of *RrCHS* in the control group at $p < 0.05$ and $p < 0.01$, respectively, according to Student's *t*-test.

2.5. HPLC Analysis of Rosa

The anthocyanin HPLC-generated chromatograms for *R. rugosa* 'Zizhi' (Figure 4F) and *R. davurica* (Figure 5F) showed that the components were well separated. Comparisons with standards allowed the contents of different substances to be calculated by their peak area (Table S2). In 'Zizhi,' six kinds of anthocyanins were detected: Cy3G5G, Pg3G5G, Cy3G, Pn3G5G, Pg3G and Pn3G. Pn3G5G had the highest content, while Cy3G5G had the second highest; the contents of the other four anthocyanins were relatively low. Compared with those in the control group and TRV-GFP group, the contents of several anthocyanins in response to the VIGS treatment were clearly reduced. Pn3G5G exhibited the greatest decrease in content, followed by Cy3G5G; the content of Pg3G was no longer detectable (Figure 8A). In *R. davurica*, the six anthocyanins listed above were also detected. However, Cy3G5G had the highest content and Cy3G had the second highest content; the contents of the other four anthocyanins were relatively low. Compared with those in the control group and TRV-GFP group, the contents of the six anthocyanins in response to the VIGS treatment were clearly reduced. Cy3G5G exhibited the greatest decrease in content, followed by Cy3G; no detection of Pn3G was observed (Figure 8B).

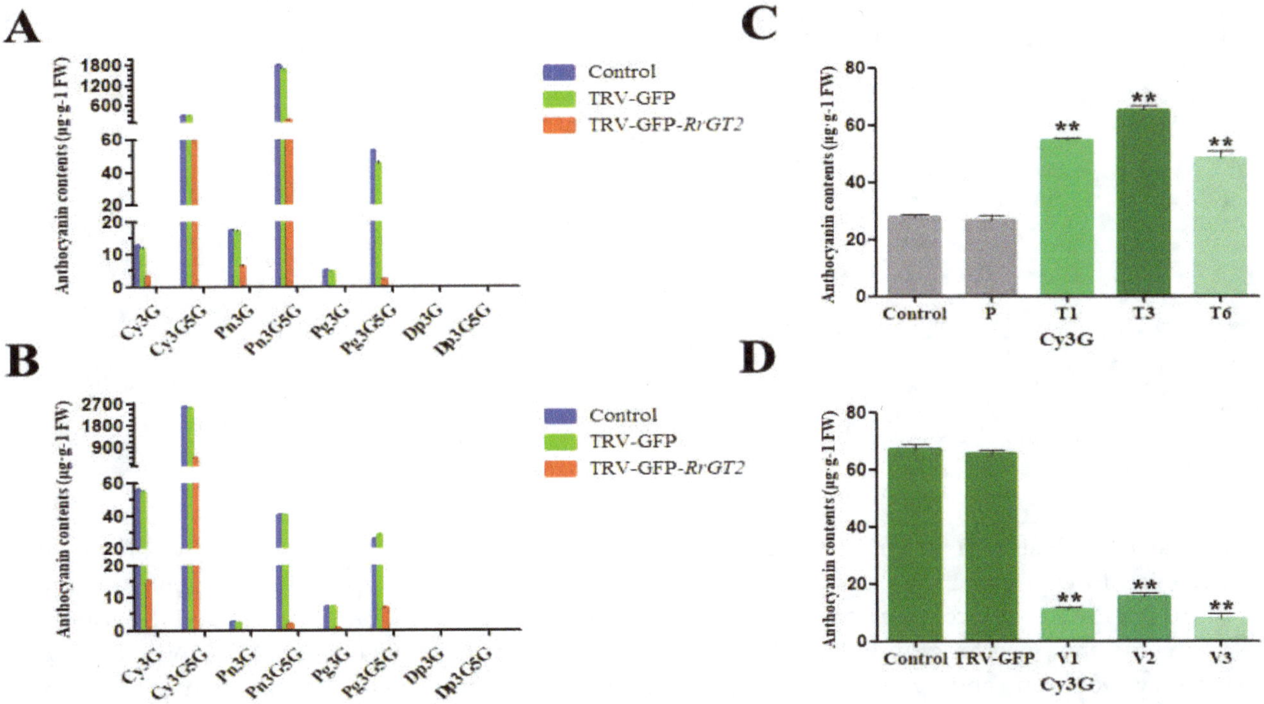

Figure 8. Comparative analysis of anthocyanin components and contents. Anthocyanin components and contents in the flowers of *R. rugosa* 'Zizhi' (**A**) and *R. davurica* (**B**) subjected to different VIGS treatments. (**C**) Anthocyanin component and contents in the flowers of transgenic tobacco. (**D**) Anthocyanin component and contents in the flowers of transgenic tobacco subjected to different VIGS treatments. ** indicates a significant difference from the control at $p < 0.01$ according to Student's t-test.

2.6. Overexpression of RrGT2 Increased the Anthocyanin Accumulation in Tobacco

The *RrGT2* gene was ectopically expressed in tobacco using the binary vector pCAMBIA1304-*RrGT2*. Six independent transgenic tobacco lines that overexpressed the *RrGT2* gene were obtained from Hyg-resistance selection and were cultured under the same conditions. PCR analysis confirmed the presence of the transformed *RrGT2* gene in all the transgenic lines as well as the absence of endogenous *RrGT2* in the tobacco plants of the control group (wild type) and empty vector group (the plants were transformed with an empty pCAMBIA1304 vector) (Figure 9A). qRT-PCR analysis revealed that the expression level of *RrGT2* was significantly higher in the transgenic plants, especially T1, T3 and T6, than in the plants of the control group and empty vector group (Figure 9B). Therefore, those three lines were used for further experiments.

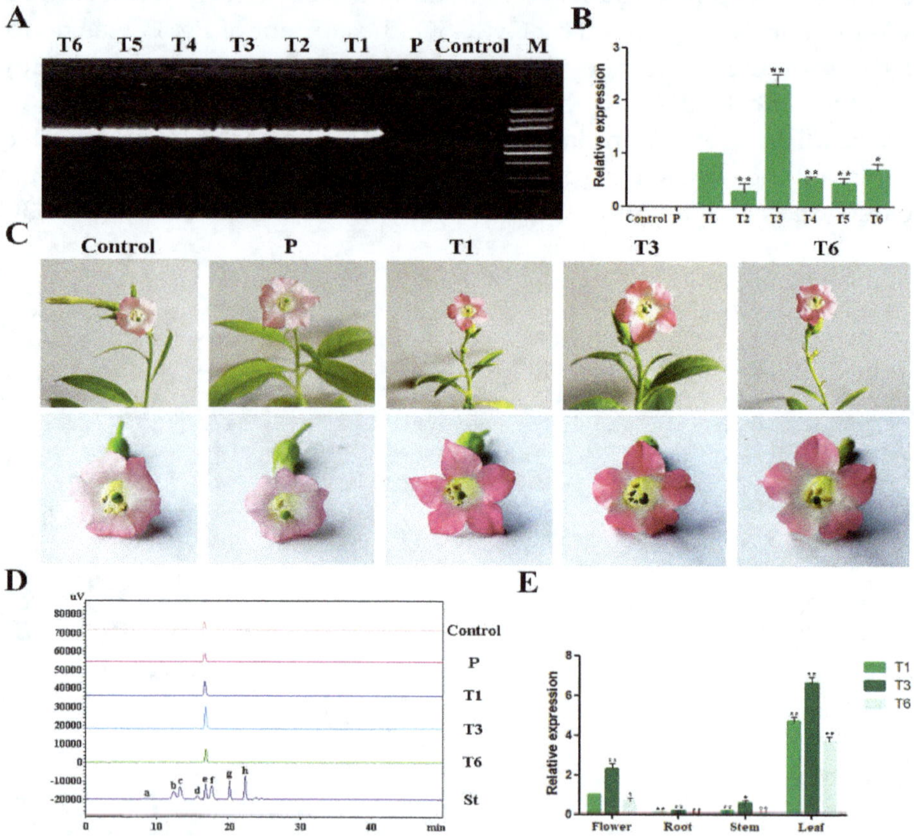

Figure 9. Analysis of tobacco lines overexpressing the *RrGT2* gene. (**A**) The results of positive PCR detection in transgenic tobacco lines. (**B**) The results of qRT-PCR detection in transgenic tobacco lines. (**C**) Phenotypic comparison of flowers between the transgenic tobacco group and the control group and empty vector group. (**D**) HPLC-derived chromatograms of the flowers of the control group, empty vector group and three transgenic tobacco lines. Eight kinds of anthocyanin standards (St) were used for detection: (a) Dp3G5G; (b) Cy3G5G; (c) Dp3G; (d) Pg3G5G; (e) Cy3G; (f) Pn3G5G; (g) Pg3G; and (h) Pn3G. (**E**) Temporal and spatial expression patterns of *RrGT2* in the three transgenic tobacco lines. The error bars represent the SDs of triplicate reactions. The experiment was repeated three times and each yielded similar results. * and ** indicate a significant difference in the relative expression levels at $p < 0.05$ and $p < 0.01$, respectively, according to Student's *t*-test.

Interestingly, there was no substantial difference in morphology between the transgenic plants and wild-type plants. However, the flower color of the transgenic tobacco lines harboring *RrGT2* was affected; compared with that of the tobacco plants in the control group and empty vector group, the petal pigmentation of the transgenic tobacco plants harboring *RrGT2* was markedly deeper (Figure 9C). This change in corolla color of the transgenic tobacco was already visible prior to anthesis. To confirm that the deeper flower color was attributed to increased pigment levels synthesized from the anthocyanin pathway, the total anthocyanins were determined qualitatively and quantitatively via HPLC. Previous studies have reported that cyanidin-3-*O*-rutinoside mainly exists in the petals of wild-type tobacco [33–35]. Even in our study, Cy3G was detected mainly in the flowers of OE-*RrGT2*. Compared with those of the plants in the control group and empty vector group, the contents of anthocyanins in the flowers of T1, T3 and T6 were significantly greater (Figures 8C and 9D and Table S3). The anthocyanin contents in the three transgenic lines were basically consistent with the trends of the relative expression of *RrGT2*. The expression of *RrGT2* in the transgenic tobacco was clearly correlated with the increased pigmentation observed in the petals.

2.7. Expression Patterns of RrGT2 in Transgenic Tobacco

The expression patterns of the *RrGT2* gene in three transgenic tobacco lines were analyzed by qRT-PCR (Figure 9E). The expression levels of *RrGT2*, which significantly differed, were assessed in four different tissue types. The expression levels in the leaves and flowers were relatively high but were relatively low in the stems and roots. Of all the tissue types of the three transgenic lines, the highest gene expression occurred in T3, followed by T1 and the lowest was in T6.

2.8. VIGS of RrGT2 Reduced the Anthocyanin Accumulation in Transgenic Tobacco

Previous tests showed that the *RrGT2* transgenic tobacco line T3 had the highest level of gene expression, so it was used as the experimental object of VIGS. At 14 days after infection, GFP detection was performed on the newly grown leaves of tobacco plants in the control group, TRV-GFP group and TRV-GFP-*RrGT2* group. GFP imaging (Figure 10A) showed that the leaves treated with VIGS (TRV-GFP and TRV-GFP-*RrGT2*) showed green fluorescence under longwave ultraviolet light, while the untreated leaves in the control group showed red fluorescence. The corresponding leaves were collected for qRT-PCR detection and the *NtACT* gene was used as an internal control. The results (Figure 10C) showed that the abundance of the *RrGT2* transcript significantly decreased only in the leaves treated with TRV-GFP-*RrGT2* but was expressed normally in the control group and TRV-GFP group. In addition, interestingly, compared with those of the control group and TRV-GFP group, the phenotypes of the leaves of the tobacco plants in the group treated with TRV-GFP-*RrGT2* were similar to those in response to photobleaching.

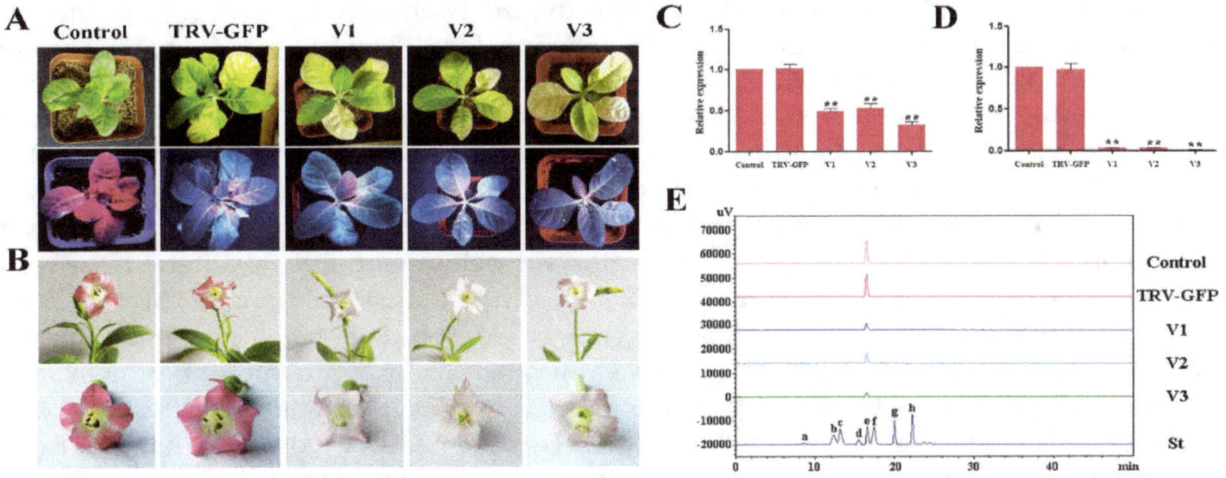

Figure 10. Analysis of OE-*RrGT2* tobacco plants treated with VIGS. (**A**) Comparisons between the control group and the VIGS-treated groups (TRV-GFP and TRV-GFP-*RrGT2*) with respect to leaves at 14 days after infection. The plants were imaged under normal light and ultraviolet illumination. V1, VIGS-*RrGT2*-1; V2, VIGS-*RrGT2*-2; V3, VIGS-*RrGT2*-3. (**B**) Phenotypic comparison of the flowers between the OE-*RrGT2*-VIGS tobacco group and the control group and empty vector group. qRT-PCR detection in the leaves (**C**) and flowers (**D**) of the control and VIGS-treated groups. The error bars represent the SDs of triplicate reactions. The experiment was repeated three times and each yielded similar results. ** indicates a significant difference from the control at $p < 0.01$ according to Student's *t*-test. (**E**) HPLC-derived chromatograms of the flowers in the control group and VIGS-treated groups. Eight kinds of anthocyanin standards (St) were used for detection: (a) Dp3G5G; (b) Cy3G5G; (c) Dp3G; (d) Pg3G5G; (e) Cy3G; (f) Pn3G5G; (g) Pg3G; and (h) Pn3G.

At 60–75 days after infection, notable differences in flower color could be observed with the naked eye. The flowers in the control group and TRV-GFP group showed no definitive changes in color but the flowers in the TRV-GFP-*RrGT2* group were clearly lighter in color (Figure 10B). The corresponding flowers were collected for qRT-PCR detection and the results (Figure 10D) showed that the abundance of the *RrGT2* transcript significantly decreased only in the flowers treated with TRV-GFP-*RrGT2* but was expressed normally in the control group and TRV-GFP group. To confirm that the lighter flower color was attributed to decreased levels of pigments synthesized from the anthocyanin pathway, the total anthocyanins were determined qualitatively and quantitatively via HPLC. Compared with those in the flowers of the control group and TRV-GFP group, the anthocyanin contents in the VIGS-*RrGT2*-1, -2 and -3 flowers were significantly lower (Figures 8D and 10E and Table S4), which was essentially consistent with the qRT-PCR results.

3. Discussion

At present, research on the flower color of *R. rugosa* requires very innovative and practical studies. Although many genes have been reported to regulate the formation of flower color, few reports on downstream structural genes such as GTs exist. The final formation of anthocyanins depends on the glycosylation of GTs, so it is very important to determine the function and influence of the *RrGT2* gene in *Rosa* color formation. In this study, we successfully cloned the *RrGT2* gene, which had a full-length cDNA of 1422 bp and encoded 473 amino acids, from the petals of *R. rugosa* 'Zizhi.'

The amino acid sequence alignment between *RrGT2* and GTs from 21 other species indicated that *RrGT2* has a common PSPG motif of the GT superfamily (Figure 2). Previous studies have shown that the conserved PSPG region is related to the substrate recognition and catalytic activity of protein-based enzymes [36–42]. If the 44 amino acids of the PSPG domain were numbered, those at positions 22, 23 and 44 would play an important role in the selection of enzyme proteoglycan donors. The twenty-second position of tryptophan (Trp, W) can correctly bind UDP-glucose, while arginine (Arg, R) can make UDP-glucuronic acid bind correctly; the twenty-third position of serine (Ser, S) is highly conserved among UDP-glucuronosyltransferases [43–45] and the forty-fourth position of glutamine (Gln, Q) and histidine (His, H) is strongly conserved among glucosyltransferases and galactotransferases, respectively [38]. Within the PSPG domain of the *RrGT2* gene, the amino acids at positions 22, 23 and 44 are tryptophan (Trp, W), asparagine (Asn, N) and glutamine (Gln, Q), respectively. Therefore, we speculate that the *RrGT2* gene uses UDP-glucose as the main glycosyl donor and has no glucuronyltransferase activity [45].

The expression of the *RrGT2* gene during floral development and in different tissues was investigated. The expression trends of the *RrGT2* gene differed during different flowering periods, indicating that the expression of the *RrGT2* gene was developmentally regulated during the anthocyanin biosynthesis process. Studies have shown that the accumulation of anthocyanins in red-skinned sand pear, strawberries and litchi is positively correlated with the activity of *UF3GT*. Boss et al. [28] also detected the expression of *UF3GT* in the peels of red grape that accumulated anthocyanin but not in other tissues of red grape or white grape without anthocyanin accumulation. The tissue-specific anthocyanin expression was similar to that of *F3GT* genes in peach, in which expression levels were greatest in tissues with pigment accumulation but relatively low in unpigmented organs [46]. Notably, the stems of both *Rosa* species were purple, which is consistent with the relatively high expression level of the *RrGT2* gene in that tissue. Interestingly, *R. davurica* is one of the parents of *R. rugosa* 'Zizhi,' so we speculate that this reason might explain the similar expression patterns between both *Rosa* species. However, with respect to the relatively low expression levels in flowers, this did not mean that the *RrGT2* gene had no effect on flower color formation. We believe that this was the result of using flowers at the budding stage as one of the tissue types. During the budding stage, the expression of the *RrGT2* gene was lowest but during the other stages, it was very high. In addition, *RrGT2* was highly expressed in the leaves and sepals of both *Rosa* species, so we infer that *RrGT2* is also involved in the glycosylation of secondary metabolites in leaves and sepals and plays an important role.

To clarify the role of *RrGT2* in the formation of *R. rugosa* flower color, the VIGS technique was used to specifically silence the *RrGT2* gene in both *Rosa* species as well as to detect and analyze the phenotypes of the flowers. The VIGS system, which involves TRV1 and TRV2, is a powerful tool for use in the functional characterization of genes in vivo [47]. At present, few reports exist about the use of the VIGS system in plant floral organs and most of the tested species thus far have been members of the Solanaceae family. For example, VIGS technology was used to study the genes controlling floral fragrance in *Petunia hybrida* [48] and the roles of the *SlMADSI*, *NbMADS4-1* and *NbMADS4-2* genes in tobacco flowers were also determined via VIGS [49]. Furthermore, the TRV recombinant virus vector was successfully used to induce the silencing of the *CHS* and *GLO1* genes in *Gerbera jamesonii* [50]. In the present study, we developed a VIGS system for use with perennial *Rosa* plants grown naturally in the field as experimental materials for the first time and we used this system to study key genes of *Rosa* color and obtained a preliminary result of gene silencing efficiency (Table S5) and other relevant results. Compared with the control conditions, the conditions resulting from the established optimal VIGS system resulted in clearly lighter petal color of both *Rosa* species, which was consistent with the significantly downregulated transcript abundance of the endogenous *RrGT2* gene. The relative expression of the six key upstream structural genes (*RrCHS*, *RrCHI*, *RrF3H*, *RrF3'H*, *RrDFR* and *RrANS*) remained unchanged. In the anthocyanin biosynthesis pathway, the upstream genes are precursors for anthocyanin biosynthesis [29]. Therefore, silencing of the *RrGT2* gene might lead to such a change.

The contents of eight anthocyanins, Cy3G, Cy3G5G, Pg3G, Pg3G5G, Pn3G, Pn3G5G, Dp3G and Dp3G5G, were analyzed qualitatively and quantitatively via HPLC. The results showed that the most abundant anthocyanin in the petals of *R. rugosa* 'Zizhi' was Pn3G5G, which is consistent with the results of Zhang et al. [51]. The content of Cy3G5G was the second highest and the other anthocyanin contents were relatively low; no presence of Dp3G or Dp3G5G was detected. With respect to *R. davurica*, this is the first time different kinds and contents of anthocyanins were detected in the petals. The Cy3G5G content was predominant; that is, the coloration of the *R. davurica* petals was affected mainly by Cy3G5G, while the other anthocyanins contributed little to flower color. After performing the VIGS treatment, we again analyzed both *Rosa* species via HPLC. The results showed that the contents of all the different kinds of anthocyanins decreased to some extent and that the decrease in the contents of several major anthocyanins was clear in both *Rosa* species. These results are in agreement with the lighter flower color phenotypes and the relatively downregulated expression level of the endogenous *RrGT2* gene in response to VIGS treatment. Therefore, it can be inferred that *RrGT2* is a key structural gene that directly affects the formation of anthocyanins in *R. rugosa*.

In plant secondary metabolism, numerous glycosides have already been isolated as biologically active compounds and some of them have been widely used as important medicines. GTs usually act in the final stages of plant secondary metabolism and are used for stabilizing and solubilizing various low-molecular-mass compounds, such as flower pigments [52,53] and for regulating the action of functional compounds such as plant hormones [54–56]. To date, most studies on the characteristics of GT enzymes have been derived from recombinant proteins produced in bacterial cells and characterized in vitro. However, very few published studies exist on the characterization of GTs in vivo. To investigate the function of the *RrGT2* gene in anthocyanin biosynthesis in vivo, *RrGT2* was first transferred into tobacco, which enhanced the flower coloration of the transgenic tobacco plants. In addition, after analyzing the tissue-specific expression of various transgenic tobacco lines, we found that the *RrGT2* gene was highly expressed not only in the flowers but also in the leaves. This phenomenon was consistent with the results of tissue-specific expression analysis in *Rosa*. Therefore, we speculated that the *RrGT2* gene may play an important role in the regulation of the growth and development of leaves, even whole plants, in addition to the anthocyanin biosynthesis pathway. However, whether in *R. rugosa* or in tobacco, the transcriptional regulatory mechanisms related to *RrGT2* expression patterns are still unclear, which requires further study in the future. Anyway, the results of the transgenic experiments proved that the exogenous *RrGT2* enzymes could also affect the synthesis of anthocyanins in different species. In other words, the function of *RrGT2* in anthocyanin biosynthesis and other aspects can be exchanged among plant species.

In this study, we used VIGS technology twice to explore the function of the *RrGT2* gene. Notably, we have applied VIGS technology to transgenic tobacco. Compared with the results of the transgenic experiments, the results of these VIGS experiments successfully verified the function of *RrGT2* in tobacco in the reverse direction. As far as we know, no such reports currently exist, so this study remains novel. In our experiments, we observed that the tobacco flowers were significantly pale in color when the *RrGT2* gene was successfully silenced. As determined via HPLC detection, this phenomenon was mostly due to decreased levels of Cy3G, which is synthesized from the anthocyanin pathway. In addition, we also observed that the phenotypes of the leaves of tobacco plants whose *RrGT2* gene was silenced were similar to those in response to photobleaching. This phenomenon was consistent with the above results of tissue-specific expression analysis of the *RrGT2* gene in *Rosa* and transgenic tobacco. This undoubtedly confirmed our previous inference about the function of the *RrGT2* gene in leaves.

4. Materials and Methods

4.1. Plant Materials

With respect to *Rosa*, *R. rugosa* 'Zizhi' and *R. davurica* plants cultivated in the *Rosa* germplasm nursery of Shandong Agricultural University were used as test materials. We collected petals at the budding stage, initial opening stage, half opening stage, full opening stage and wilting stage as well as seven different tissue samples (roots, stems, leaves, sepals, stamens, pistils and petals at the budding stage) in the mornings of sunny days from 20 April to 10 May 2017. After they were flash frozen in liquid nitrogen, all samples, which were collected in triplicate, were put into a −80 °C refrigerator for storage.

With respect to tobacco, wild-type plants were used as transgenic materials. After the tobacco seeds were disinfected by soaking in 70–75% ethanol for 2 min, rinsing with aseptic water once, soaking with 3.5% NaClO for 10–15 min and then rinsing with aseptic water 5 times, they were sown in Murashige and Skoog (MS) solid medium (without antibiotics). After 3 days of vernalization at 4 °C in darkness, the seeds were placed in a growth chamber (25 °C, 16 h/23 °C, 8 h day/night, 60% relative humidity) for approximately 30 days. The germless tobacco seedlings that grew well were selected as follow-up experimental materials.

4.2. Extraction of Total RNA and Synthesis of First-Strand cDNA

Total RNA was extracted via an EASY Spin Plant RNA Rapid Extraction Kit (Aidlab Biotech, Beijing, China) in accordance with the manufacturer's specifications. The integrity of the RNA was measured by gel electrophoresis with 1.0% nondenatured agarose, the purity and concentration of the RNA were detected by a Nanodrop 2000C ultra-microspectrophotometer (Thermo Fisher Scientific, Wilmington, DE, USA) and the qualified RNA was preserved at −80 °C. First-strand cDNA was synthesized via a 5× All-In-One RT MasterMix Reverse Transcription Kit (ABM Company, Vancouver, Canada) in accordance with both the manufacturer's protocol and the requirements of RT-PCR and qRT-PCR.

4.3. Cloning of the Full-Length CDS of RrGT2

We identified the *RrGT2* gene that contained the complete 5′ CDS from the *R. rugosa* transcriptome data in our laboratory. The cDNA 3′ terminal sequence of the target gene was then amplified by 3′-RACE technology. *RrGT2*-F and *RrGT2*-R primers (Table S1) were designed and amplified according to the full-length cDNA sequence of the *RrGT2* gene [31].

4.4. Tobacco Stable Transformation

The plasmids of *pCAMBIA1304* vectors and the *RrGT2* gene with restriction sites (*SpeI* and *BstEII*) were extracted and digested by two enzymes. The digestion products were then ligated with DNA ligase (Figure S2A) and transformed into *Agrobacterium tumefaciens*.

A. tumefaciens-mediated leaf disc transformation [57] was used to transform tobacco. First, the leaves of the cultured tobacco sterile seedlings were pruned to the appropriate size. *A. tumefaciens* infection was carried out after 2 days of preculture. Acetosyringone (AS) was added to the infective liquid and an empty *pCAMBIA1304* vector was used as a control. After infection, the plants were subjected to a dark treatment for 2 days, after which they were cultured in a growth chamber (25 °C, 16 h/23 °C, 8 h day/night, 60% relative humidity). Differentiation culture and rooting culture were carried out on MS media supplemented with relevant hormones and antibiotics. Hygromycin (Hyg) was used for screening resistant seedlings. The plantlets exhibiting good growth potential were transplanted into small flowerpots that contained substrate after seedling refining and the original growth environment was maintained.

4.5. VIGS in Rosa

On the basis of a modified TRV-GFP vector, a TRV-GFP-*RrGT2* recombinant viral vector was constructed. pTRV1 and pTRV2-GFP are two RNA strands of the TRV-GFP virus vector and the multiple cloning sites are mainly within pTRV2-GFP. For silencing *RrGT2* specifically in *R. rugosa*, a 406 bp fragment of the *RrGT2* gene was amplified and cloned into pTRV2-GFP (Figure S1A).

The pTRV1, pTRV2-GFP and pTRV2-GFP-*RrGT2* plasmids were transformed into *A. tumefaciens*, which was then cultured in YEB media that contained kanamycin, rifampicin and AS at 28 °C for 14–16 h until an OD_{600} = 1.5 was reached. Before infection, pTRV1 was added to the infection liquid that contained pTRV2-GFP and pTRV2-GFP-*RrGT2* in equal volume; the solution was subsequently mixed, forming a complete TRV-GFP and TRV-GFP-*RrGT2* virus carrier. The mixed bacterial solution was kept at room temperature in darkness for 4 h [58–63].

Perennial *Rosa* plants that grew naturally in the field were used as experimental materials and the experimental treatment time (from mid-March to mid-April) was approximately one month before *R. rugosa* flowering. And before setting an inoculation date, we will look into the weather for at least a week to avoid bad weather. In addition, we set the specific inoculation time between 14:00 and 16:00 in the afternoon, because during this time, the environment temperature is relatively high, which is more in line with the inoculation operation of VIGS. Another reason is that inoculation in the afternoon leads to a faster transition to night, making dark processing more real. Because it was difficult to inject the leaves and twigs with syringes and because vacuum infiltration could not be used in the field, we used the method that involved first scratching the leaves and twigs and then infecting them with *A. tumefaciens*. To improve the infection efficiency, 0.01% Silwet L-77 was added to the infection liquid and the plants were subjected to darkness for 24 h after infection for 10 min.

4.6. VIGS in Transgenic Tobacco

The final concentration of the virus infective fluid needed to reach OD_{600} = 1.0. In addition, the other aspects of the preparation of the virus infection solution and other preliminary preparation works were consistent with the above methods. First, we selected one of several *RrGT2* transgenic tobacco lines as the experimental object of VIGS. Tobacco infiltration was then performed as described by Liu et al. [62]. The *A. tumefaciens* cultures containing pTRV1 and pTRV2 or their derivatives (1:1, *v/v*) were injected into the lowest leaf of four-leaf stage plants by using a 1 mL needleless syringe. After infiltration, the tobacco plants were subjected to a dark treatment for 12 h, after which they were cultured in a growth chamber (25 °C, 16 h/23 °C, 8 h day/night, 60% relative humidity).

4.7. GFP Imaging

The detection and imaging of the visualized GFP in *Rosa* and transgenic tobacco plants after VIGS treatment were performed at night with a handheld high-intensity ultraviolet lamp (Model SB-100P/F; Spectronics Corporation, Westbury, NY, USA) and a Nikon D90 camera, respectively.

4.8. qRT-PCR Detection

We analyzed the gene expression by qRT-PCR on a Bio-Rad CFX96^{TM} Real-Time PCR instrument (Bio-Rad, Inc., Philadelphia, PA, USA). The qRT-PCR mixture (total volume of 20 μL) contained 10 μL of SYBR® Premix Ex Taq™ (TaKaRa, Inc., Kusatsu, Japan), 8.2 μL of ddH_2O, 0.4 μL of each primer and 1 μL of cDNA. The PCR program consisted of an initial step of 95 °C for 30 s; 40 cycles of 95 °C for 5 s and 60 °C for 30 s; and then a dissociation stage of 95 °C for 10 s, 65 °C for 5 s and 95 °C for 5 s. Each gene was assessed via three biological replicates. The relative expression levels of the genes were calculated by the $2^{-\Delta\Delta Ct}$ method [64].

4.9. Total Anthocyanin Extraction and HPLC Analysis

All samples (0.1 g fresh weight) were homogenized in liquid nitrogen, after which they were extracted with 5 mL of an acidic methanol solution (70:0.1:29.9, *v*/*v*/*v*; CH_3OH:HCl:H_2O) at 4 °C in darkness for 24 h and then sonicated for 30 min [65]. After centrifugation, each extract was passed through a membrane filter (0.22 mm).

Qualitative and quantitative analyses of anthocyanins were performed via HPLC. The chromatographic analysis was conducted using a Prominence LC-20AT series HPLC system (Shimadzu, Inc., Kyoto, Japan) with a detection wavelength of 530 nm and the column (TC-C18 column, 5 μm, 4.6 mm × 250 mm) was maintained at 30 °C. The eluent consisted of an aqueous solution A (0.1% formic acid in water) and organic solvent B (acetonitrile). The gradient elution program was modified as described previously [66]: 0 min, 10% B; 15 min, 17% B; 20 min, 23% B; 25 min, 23% B; and 30 min, 10% B. Moreover, the eluent flow rate was 1.0 mL/min, with a 10 μL injection volume. Cy3G, Cy3G5G, Pg3G, Pg3G5G, Pn3G, Pn3G5G, Dp3G and Dp3G5G (EXTRASYNTHESE Trading Company, Lille-Lezennes, France) were used as references for anthocyanin analysis. Three independent biological replicates were measured for each sample.

4.10. Statistical Analyses

Three independent biological replicates were measured for each sample and the data presented as the mean ± standard error (SE). Where applicable, data were analyzed by Student's *t* test in a two-tailed analysis. Values of $p < 0.05$ or <0.01 were considered to be statistically significant.

5. Conclusions

In conclusion, the *RrGT2* gene from *R. rugosa* was successfully cloned and characterized. Our results demonstrated that *RrGT2* has all the conserved amino acid residues that are typical of the GT enzyme. Transcript analysis revealed that *RrGT2* was expressed in specific tissues and was developmentally regulated, suggesting that *RrGT2* might act as a modified enzyme in the anthocyanin biosynthesis pathway. The functional verification of *RrGT2* in *Rosa* via VIGS revealed that *RrGT2* is a key structural gene that directly affects the formation of anthocyanins in *R. rugosa*. By overexpressing *RrGT2* in tobacco, we found an increase in anthocyanins in flowers, indicating that *RrGT2* encodes a functional GT protein for anthocyanin glucosylation and could function in other species. Furthermore, VIGS of the *RrGT2* gene in transgenic tobacco resulted in a decrease in the total content of anthocyanins that accumulated in the flowers, which further confirmed that *RrGT2* is involved in the modification of flower color.

Supplementary Materials

Table S1. Primers used in the present study. Table S2. Anthocyanin contents in the flowers of *R. rugosa* 'Zizhi' and *R. davurica* subjected to different VIGS treatments (μg·g^{-1} FW). Table S3. Anthocyanin contents in the flowers of transgenic tobacco (μg·g^{-1} FW). Table S4. Anthocyanin contents in the flowers of transgenic tobacco subjected to different VIGS treatments (μg·g^{-1} FW). Table S5. Silencing efficiency of VIGS in *R. rugosa* 'Zizhi' and *R. davurica*. Figure S1. Construction and validation of the recombinant virus vector TRV-GFP-*RrGT2*. Figure S2. Construction and validation of the recombinant expression vector pCAMBIA1304-*RrGT2*.

Author Contributions: Conceptualization, X.S. and L.Z.; Data curation, X.S., M.Z. and X.H.; Formal analysis, M.Z. and X.H.; Funding acquisition, Z.X. and L.Z.; Investigation, X.S.; Methodology, X.S. and Z.X.; Project administration, Z.X., L.Z. and X.H.; Resources, L.Z.; Software, X.S. and M.Z.; Supervision, Z.X., L.Z. and X.H.; Validation, X.S., M.Z. and X.H.; Visualization, X.S. and M.Z.; Writing—original draft, X.S.; Writing—review & editing, Z.X., L.Z. and X.H.

Acknowledgments: This project was supported by the Agricultural Seed Project of Shandong Province ([2014] No. 96) and National Science Foundation of China (NSFC) (31700622).

Abbreviations

cDNA	complementary DNA
RACE	rapid amplification of cDNA ends
CDS	coding DNA sequence
PSPG	plant secondary product glycosyltransferases
PCR	polymerase chain reaction
qRT-PCR	quantitative real-time polymerase chain reaction
VIGS	virus-induced gene silencing
TRV	tobacco rattle virus
GFP	green fluorescent protein
TRV-GFP	a modified TRV virus with GFP gene inserted into TRV virus
HPLC	high-performance liquid chromatography

References

1. Feng, L.G.; Shao, D.W.; Sheng, L.X.; Zhao, L.Y.; Yu, X.Y. Study on investigation and morphological variation of wild *Rosa rugosa* in China. *J. Shandong Agric. Univ. (Nat. Sci. Ed.)* **2009**, *40*, 484–488.
2. Yang, M.; Zhao, L.Y. Research and classification on the germplasm resources of the Pingyin *Rosa rugosa* in Shandong province. *Landsc. Archit. J. Chin.* **2003**, *7*, 61–63.
3. Yu, S.C.; Feng, Z.; Zhao, L.Y. Research on quantitative taxonomy of cultivars in Pingyin rose. *Acta Hortic. Sin.* **2005**, *32*, 327–330.
4. Feng, L.G.; Sheng, L.X.; Tao, J.; Zhao, L.Y.; Shao, D.W. Comparative studies on aroma components and contents of wild *Rosa rugosa* in China. *J. Yangzhou Univ. (Agric. Life Sci. Ed.)* **2009**, *30*, 90–94.
5. Chen, Z.J.; Zang, F.S.; Dai, Y.Q.; Yuan, X.Y.; Li, S.; Wu, Y.M.; Ni, Y.Y. Nutrient components analysis of rose fruit. *Food Res. Dev.* **2012**, *33*, 194–198.
6. Wang, Y. Natural hybridization and speciation. *Biodivers. Sci.* **2017**, *25*, 565–576. [CrossRef]
7. Henderson, I.R.; Salt, D.E. Natural genetic variation and hybridization in plants. *J. Clin. Exp. Dent.* **2017**, *9*, e1212–e1217. [CrossRef] [PubMed]
8. Chen, S.M.; Zhu, X.R.; Chen, F.D.; Luo, H.L.; Lv, G.S.; Fang, W.M.; Zhang, Z. Expression characteristics of anthocyanin structural genes in different flower color chrysanthemum cultivars. *J. Northwest Plants* **2010**, *30*, 453–458.
9. Grotewold, E. The genetics and biochemistry of floral pigments. *Annu. Rev. Plant Biol.* **2006**, *57*, 761–780. [CrossRef] [PubMed]
10. Koes, R.; Verweij, W.; Quattrocchio, F. Flavonoids: A colorful model for the regulation and evolution of biochemical pathways. *Trends Plant Sci.* **2005**, *10*, 236–242. [CrossRef] [PubMed]
11. Quattrocchio, F.; Wing, J.; van der Woude, K.; Souer, E.; de Vetten, N.; Mol, J.; Koes, R. Molecular analysis of the anthocyanin 2 gene of petunia and its role in the evolution of flower color. *Plant Cell* **1999**, *11*, 1433–1444. [CrossRef] [PubMed]
12. Paz-Ares, J.; Ghosal, D.; Wienand, U.; Peterson, P.; Saedler, H. The regulatory *c1* locus of *Zea mays* encodes a protein with homology to *myb* proto-oncogene products and with structural similarities to transcriptional activators. *EMBO J.* **1987**, *6*, 3553–3558. [CrossRef] [PubMed]
13. Winkelshirley, B. Flavonoid Biosynthesis. A Colorful Model for Genetics, Biochemistry, Cell Biology and Biotechnology. *Plant Physiol.* **2001**, *126*, 485–493. [CrossRef]

14. Lu, Y.F.; Zhang, M.L.; Meng, X.N.; Wan, H.H.; Zhang, J.; Tian, J.; Hao, S.X.; Jin, K.N.; Yao, Y.C. Photoperiod and shading regulate coloration and anthocyanin accumulation in the leaves of Malus crabapples. *Plant Cell Tissue Organ Cult.* **2015**, *121*, 619–632. [CrossRef]
15. Holton, T.A.; Cornish, E.C. Genetics and biochemistry of anthocyanin biosynthesis. *Plant Cell* **1995**, *7*, 1071. [CrossRef] [PubMed]
16. Kim, S.H.; Lee, J.R.; Hong, S.T.; Yoo, Y.K.; An, G.; Kim, S.R. Molecular cloning and analysis of anthocyanin biosynthesis genes preferentially expressed in apple skin. *Plant Sci.* **2003**, *165*, 403–413. [CrossRef]
17. Charles, B.; Imin, N.; Djordjevic, M.A. Flavonoids: New roles for old molecules. *J. Integr. Plant Biol.* **2010**, *52*, 98–111.
18. Martens, S.; Preuß, A.; Matern, U. Multifunctional flavonoid dioxygenases: Flavonol and anthocyanin biosynthesis in *Arabidopsis thaliana* L. *Phytochemistry* **2010**, *71*, 1040–1049. [CrossRef] [PubMed]
19. Zhang, Q.; Hao, R.J.; Xu, Z.D.; Yang, W.R.; Wang, J.; Cheng, T.R.; Pan, H.T.; Zhang, Q.X. Isolation and functional characterization of a R2R3-MYB regulator of *Prunus mume* anthocyanin biosynthetic pathway. *Plant Cell Tissue Organ Cult.* **2017**, *131*, 417–429. [CrossRef]
20. Ge, C.L.; Huang, C.H.; Xu, X.B. Research on anthocyanins biosynthesis in fruit. *Acta Hortic. Sin.* **2012**, *39*, 1655–1664.
21. Vogt, T.; Jones, P. Glycosyltransferases in plant natural product synthesis: Characterization of a supergene family. *Trends Plant Sci.* **2000**, *5*, 380–386. [CrossRef]
22. Lim, E.K.; Bowles, D.J. A class of plant glycosyltransferases involved in cellular homeostasis. *EMBO J.* **2004**, *23*, 2915–2922. [CrossRef] [PubMed]
23. Brazier-Hicks, M.; Edwards, R. Functional importance of the family 1 glucosyltransferase UGT72B1 in the metabolism of xenobiotics in *Arabidopsis thaliana*. *Plant J.* **2005**, *42*, 556–566. [CrossRef] [PubMed]
24. Fedoroff, N.V.; Furtek, D.B.; Nelson, O.E. Cloning of the bronze locus in maize by a simple and generalizable procedure using the transposable controlling element Activator (Ac). *Proc. Natl. Acad. Sci. USA* **1984**, *81*, 3825–3829. [CrossRef] [PubMed]
25. Kobayashi, S.; Ishimaru, M.; Ding, C.K.; Yakushiji, H.; Goto, N. Comparison of UDP-glucose: Flavonoid 3-O-glucosyltransferase (UFGT) gene sequences between white grapes (*Vitis vinifera*) and theirsports with red skin. *J. Plant Sci.* **2001**, *160*, 543–550. [CrossRef]
26. Li, X.G.; Yu, Z.Y. Research progress of anthocyanin. *J. North. Hortic.* **2003**, *4*, 6–8.
27. Afifi, M.; El-Kereamy, A.; Legrand, V.; Chervin, C.; Monje, M.-C.; Nepveu, F.; Roustan, J.-P. Control of anthocyanin biosynthesis pathway gene expression by eutypine, atoxin from Eutypa lata in grape cell tiussue cultures. *J. Plant Physiol.* **2003**, *160*, 971–975. [CrossRef] [PubMed]
28. Boss, P.K.; Davies, C.; Robinson, S.P. Anthocyanin composition and anthocyanin pathway gene expression in grapewine sports differing in berry skin colour. *Aust. J. Grape Wine Res.* **1996**, *2*, 163–170. [CrossRef]
29. Nakatsuka, T.; Nishihara, M.; Mishiba, K.; Yamamura, S. Two different mutations are involved in the formation of white-flowered gentian plants. *J. Plant Sci.* **2005**, *169*, 949–958. [CrossRef]
30. Ju, Z.G. Activities of chalcone and UDP Gal: Flavonoid-3-*O*-glycosyhransferase in relation to anthocyanin synthesis in apple. *J. Sci. Hortic.* **1995**, *63*, 175–185. [CrossRef]
31. Sui, X.M.; Wang, Y.; Zhao, M.Y.; Han, X.; Zhao, L.Y.; Xu, Z.D. Cloning and Expression Analysis of *RrGT2* Gene Related to Anthocyanin Biosynthesis in *Rosa rugosa*. *Am. J. Plant Sci.* **2018**, *9*, 2008–2019. [CrossRef]
32. Qi, Y. Cloning and Expression Analysis of Anthocyanin Biosynthesis Related Gene in *Rosa rugosa*. Master's Thesis, Shandong Agricultural University, Tai'an, China, 2016.
33. Aharoni, A.; De Vos, C.R.; Wein, M.; Sun, Z.; Greco, R.; Kroon, A.; Mol, J.N.; O'Connell, A.P. The strawberry FaMYB1 transcription factor suppresses anthocyanin and flavonol accumulation in transgenic tobacco. *Plant J.* **2001**, *28*, 319–332. [CrossRef] [PubMed]
34. Deluc, L.; Barrieu, F.; Marchive, C.; Lauvergeat, V.; Decendit, A.; Richard, T.; Carde, J.P.; Mérillon, J.M.; Hamdi, S. Characterization of a grapevine R2R3-MYB transcription factor that regulates the phenylpropanoid pathway. *Plant Physiol.* **2006**, *140*, 499–511. [CrossRef] [PubMed]
35. Nakatsuka, T.; Sato, K.; Takahashi, H.; Yamamura, S.; Nishihara, M. Cloning and characterization of the UDP-glucose: Anthocyanin 5-*O*-glucosyltransferase gene from blue-flowered gentian. *J. Exp. Bot.* **2008**, *59*, 1241–1252. [CrossRef] [PubMed]
36. Wang, C.X.; Yang, M.Z.; Pan, N.S.; Chen, Z.L. Resistance to potato virus X infection in transgenic tobacco plants with coat protein gene of virus. *J. Integr. Plant Biol.* **1993**, *35*, 819–824.

37. Zhang, X.Y.; Xue, Q.Z. Introduction of soybean glycinin gene into rice (*Oryza sativa* L.) with Agrobacterium-mediated transformation. *J. Zhejiang Agric. Univ. (Agric. Life Sci.)* **2001**, *27*, 495–499.
38. Kubo, A.; Arai, Y.; Nagashima, S.; Yoshikawa, T. Alteration of sugar donor specificities of plant glycosyltransferases by a single point mutation. *J. Arch. Biochem. Biophys.* **2004**, *429*, 198–203. [CrossRef] [PubMed]
39. Broothaerts, W.; Mitchell, H.J.; Weir, B.; Kaines, S.; Smith, L.M.; Yang, W.; Mayer, J.; Roa-Rodríguez, C.; Jefferson, R. Genetransfer to plants by diverse species of bacteria. *J. Nat.* **2005**, *433*, 629–633. [CrossRef] [PubMed]
40. Herrera-Estrella, L.; Simpson, J.; Martinez-Trujillo, M. Transgenic plants: An historical perspective. *J. Methods Mol. Biol.* **2005**, *286*, 3–32.
41. Wang, X.Q. Structure, mechanism and engineering of plant natural product glycosyltransferases. *J. FEBS Lett.* **2009**, *583*, 3303–3309. [CrossRef] [PubMed]
42. Yonekura-Sakakibara, K.; Hanada, K. An evolutionary view of functional diversity in family 1 glycosyltransferases. *J. Plant J.* **2011**, *66*, 182–193. [CrossRef] [PubMed]
43. Shao, H.; He, X.; Achnine, L.; Blount, J.W.; Dixon, R.A.; Wang, X. Crystal structures of a multifunctional triterpene/flavonoid glycosyltransferase from *Medicago truncatula*. *J. Plant Cell* **2005**, *17*, 3141–3154. [CrossRef] [PubMed]
44. Li, L.; Modolo, L.V.; Escamilla-Trevino, L.L.; Achnine, L.; Dixon, R.A.; Wang, X. Crystal structure of Medicago truncatula UGT85H2-insights into the trusctural basis of a multifunctional (iso) flavonoid glycosyltransferase. *J. Mol. Biol.* **2007**, *370*, 951–963. [CrossRef] [PubMed]
45. Modolo, L.V.; Li, L.N.; Pan, H.Y.; Blount, J.W.; Dixon, R.A.; Wang, X.Q. Crystal structures of glycosyltransferase UGT/8G1 reveal the molecular basis for glycosylation and deglycosylation of (iso) flavonoids. *J. Mol. Biol.* **2009**, *392*, 1292–1302. [CrossRef] [PubMed]
46. Cheng, J.; Wei, G.; Zhou, H.; Gu, C.; Vimolmangkang, S.; Liao, L.; Han, Y. Unraveling the mechanism underlying the glycosylation and methylation of anthocyanins in peach. *Plant Physiol.* **2014**, *166*, 1044–1058. [CrossRef] [PubMed]
47. Sun, D.; Nandety, R.S.; Zhang, Y.; Reid, M.S.; Niu, L.; Jiang, C.Z. A petunia ethylene-responsive element binding factor, PhERF2, plays an important role in antiviral RNA silencing. *J. Exp. Bot.* **2016**, *67*, 3353–3365. [CrossRef] [PubMed]
48. Spitzer, B.; Zvi, M.M.B.; Ovadis, M.; Marhevka, E.; Barkai, O.; Edelbaum, O.; Marton, I.; Masci, T.; Alon, M.; Morin, S.; et al. Reverse Genetics of floral scent: Application of Tobacco rattle virus-based gene silencing in Petunia. *Plant Physiol.* **2007**, *145*, 1241–1250. [CrossRef] [PubMed]
49. Dong, Y.Y.; Burch-Smith, T.M.; Liu, Y.L.; Mamillapalli, P.; Dinesh-Kumar, S.P. A Ligation-independent cloning Tobacco rattle virus vector for high-throughput virus-induced gene silencing identifies roles for NbMADS4-1 and -2 in floral development. *Plant Physiol.* **2007**, *145*, 1161–1170. [CrossRef] [PubMed]
50. Deng, X.B.; Elomaa, P.; Nguyen, C.X.; Hytönen, T.; Valkonen, J.P.T.; Teeri, T.H. Virus-induced gene silencing for *Asteraceae*-a reverse genetics approach for functional genomics in *Gerbera hybrida*. *Plant Biotechnol. J.* **2012**, *10*, 970–978. [CrossRef] [PubMed]
51. Zhang, L.; Xu, Z.D.; Tang, T.F.; Zhang, H.; Zhao, L.Y. Analysis of anthocyanin related compounds and metabolic pathways in the flowering process of *Rosa rugosa* 'Zizhi'. *Sci. Agric. Sin.* **2015**, *48*, 235–236.
52. Brugliera, F.; Holton, T.A.; Stevenson, T.W.; Farcy, E.; Lu, C.Y.; Cornish, E.C. Isolation and characterization of a cDNA clone corresponding to the Rt locus of *Petunia hybrida*. *Plant J.* **1994**, *5*, 81–92. [CrossRef] [PubMed]
53. Kroon, J.; Souer, E.; de Graaff, A.; Xue, Y.; Mol, J.; Koes, R. Cloning and structural analysis of the anthocyanin pigmentation locus Rt of Petunia hybrida: Characterization of insertion sequences in two mutant alleles. *Plant J.* **1994**, *5*, 69–80. [CrossRef] [PubMed]
54. Szerszen, J.B.; Szczyglowski, K.; Bandurski, R.S. A gene from Zea mays involved in conjugation of growth hormone indole-3-acetic acid. *Science* **1994**, *265*, 1699–1701. [CrossRef] [PubMed]
55. Martin, R.C.; Mok, M.C.; Mok, D.W.S. Isolation of a cytokinin gene, ZOG1, encoding zeatin *O*-glucosyltransferase from *Phaseolus lunatus*. *Proc. Natl. Acad. Sci. USA* **1999**, *96*, 284–289. [CrossRef] [PubMed]
56. Martin, R.C.; Mok, M.C.; Habben, J.E.; Mok, D.W.S. A maize cytokinin gene encoding an *O*-glucosyltransferase specific to cis-zeatin. *Proc. Natl. Acad. Sci. USA* **2001**, *98*, 5922–5926. [CrossRef] [PubMed]

57. Horsch, R.; Fry, J.E.; Hoffmann, N.L.; Eichholtz, D.Z.; Rogers, S.G.; Fraley, R.T. A simple and general method for transferring genes into plants. *Science* **1985**, *227*, 1229–1232.
58. Bachan, S.; Dinesh-Kumar, S.P. Tobacco rattle virus (TRV) based virus-induced gene silencing. *Antivir. Resist. Plants* **2012**, *894*, 83–92.
59. Burch-Smith, T.M.; Schiff, M.; Liu, Y.; Dinesh-Kumar, S.P. Efficient virus-induced gene silencing in Arabidopsis. *Plant Physiol.* **2006**, *142*, 21–27. [CrossRef] [PubMed]
60. Chen, J.C.; Jiang, C.Z.; Gookin, T.E.; Hunter, D.A.; Clark, D.G.; Reid, M.S. Chalcone synthase as a reporter in virus-induced gene silencing studies of flower senescence. *Plant Mol. Biol.* **2004**, *55*, 521–530. [CrossRef] [PubMed]
61. Jiang, C.Z.; Chen, J.C.; Reid, M. Virus-induced gene silencing in ornamental plants. *Methods Mol. Biol.* **2011**, *744*, 81–96. [PubMed]
62. Liu, Y.; Schiff, M.; Dinesh-Kumar, S.P. Virus-induced gene silencing in tomato. *Plant J.* **2002**, *31*, 777–786. [CrossRef] [PubMed]
63. Quadrana, L.; Rodriguez, M.C.; López, M.; Bermúdez, L.; Nunes-Nesi, A.; Fernie, A.R.; Descalzo, A.; Asis, R.; Rossi, M.; Asurmendi, S.; et al. Coupling virus-induced gene silencing to exogenous green fluorescence protein expression provides a highly efficient system for functional genomics in Arabidopsis and across all stages of tomato fruit development. *Plant Physiol.* **2011**, *156*, 1278–1291. [CrossRef] [PubMed]
64. Schmittgen, T.D.; Livak, K.J. Analyzing real-time PCR data by the comparative C (T) method. *Nat. Protoc.* **2008**, *3*, 1101. [CrossRef] [PubMed]
65. Yang, Q.; Yuan, T.; Sun, X.B. Preliminary studies on the changes of flower color during the flowering period in two tree peony cultivars. *Acta Hortic. Sin.* **2015**, *42*, 930–938.
66. Xie, D.Y.; Sharma, S.B.; Wright, E.; Wang, Z.Y.; Dixon, R.A. Metabolic engineering of proanthocyanidins through co-expression of anthocyanidin reductase and the PAP1 MYB transcription factor. *Plant J.* **2006**, *45*, 895–907. [CrossRef] [PubMed]

12

OsGPAT3 Plays a Critical Role in Anther Wall Programmed Cell Death and Pollen Development in Rice

Lianping Sun †, Xiaojiao Xiang †, Zhengfu Yang, Ping Yu, Xiaoxia Wen, Hong Wang, Adil Abbas, Riaz Muhammad Khan, Yingxin Zhang, Shihua Cheng * and Liyong Cao *

Key Laboratory for Zhejiang Super Rice Research and State Key Laboratory of Rice Biology, China National Rice Research Institute, Hangzhou 310006, China; sunlianping@caas.cn (L.S.); taorangongwangjimm@163.com (X.X.); cnrriyzf@sina.com (Z.Y.); pingping367@163.com (P.Y.); 18883966700@163.com (X.W.); wjiyinh@126.com (H.W.); adilabbasqau@126.com (A.A.); riazkatlang@126.com (R.M.K.); zyxrice@163.com (Y.Z.)

* Correspondence: chengshihua@caas.cn (S.C.); caoliyong@caas.cn (L.C.)

† These authors contributed equally to this work.

Abstract: In flowering plants, ideal male reproductive development requires the systematic coordination of various processes, in which timely differentiation and degradation of the anther wall, especially the tapetum, is essential for both pollen formation and anther dehiscence. Here, we show that *OsGPAT3*, a conserved glycerol-3-phosphate acyltransferase gene, plays a critical role in regulating anther wall degradation and pollen exine formation. The *gpat3-2* mutant had defective synthesis of Ubisch bodies, delayed programmed cell death (PCD) of the inner three anther layers, and abnormal degradation of micropores/pollen grains, resulting in failure of pollen maturation and complete male sterility. Complementation and clustered regularly interspaced short palindromic repeats (CRISPR)/CRISPR-associated 9 (Cas9) experiments demonstrated that *OsGPAT3* is responsible for the male sterility phenotype. Furthermore, the expression level of tapetal PCD-related and nutrient metabolism-related genes changed significantly in the *gpat3-2* anthers. Based on these genetic and cytological analyses, *OsGPAT3* is proposed to coordinate the differentiation and degradation of the anther wall and pollen grains in addition to regulating lipid biosynthesis. This study provides insights for understanding the function of *GPATs* in regulating rice male reproductive development, and also lays a theoretical basis for hybrid rice breeding.

Keywords: anther wall; tapetum; pollen accumulation; *OsGPAT3*; rice

---

1. Introduction

Rice is a key gramineous plant that is self-pollinated. Guaranteeing sufficient rice yield requires stable male sterility, which depends on normal development of anther and male gametophyte (pollen) formation [1,2]. Typical rice anthers have four lobes, and the central reproductive microsporocytes (or pollen mother cells) are surrounded by four concentrically organized somatic cell layers in each locule: the epidermis, endothecium, middle layer, and tapetum. Anther development is a multistage process involving localized cellular differentiation and degeneration, cell division and chromosomal behaviors, and synthesis and transportation of nutrients. This process is combined with changes in the structure and external environment to complete anther dehiscence and pollen maturation, and release for pollination and fertilization. All four cell layers possess specific functions and coordinate throughout the whole process to ensure normal anther development and microspore/pollen formation in rice [2–6]. The epidermis is located at the outermost layer of anthers to protect against external environmental stresses to ensure normal development of internal cells at the appropriate time for anther dehiscence and pollen release. The endothecium, with localized secondary thickening, is the second layer and is essential for anther dehiscence and pollen release. Endothecial development is concurrent with pollen maturation and degeneration of the anther tapetum and middle layer. The middle layer, which is situated between the tapetum and endothecium, undergoes programmed degeneration along with the tapetum during the pollen maturation stage [5–8]. The innermost cell layer of the anther wall, the tapetum, directly contacts developing gametophytes. The tapetum undergoes programmed cell death (PCD)-mediated degeneration to supply a series of nutritional components and structural molecules for normal pollen formation and ordinary anther development [9,10]. Proper development and timely degeneration of tapetal cells is essential for providing and supplying nutrients for sporopollenin synthesis and pollen development [1,11–13]. All developmental processes of these cell layers are dominated by specific regulators with extremely precise and systematic molecular mechanisms; mutation of these genes may result in anther development malformations and eventually lead to pollen abortion [14,15].

Programmed cell death (PCD) events, known as apoptosis, are often characterized by nuclear chromatin condensation and degeneration, membrane breakdown, and compactness of cytoplasmic organelles. PCD is usually used as a cytological feature of tapetum degradation in plants [9,12,16]. The PCD process also occurs in the most peripheral layers at the late stages of anther development. Vascular bundle cells, the cavity of dehiscence, the epidermis, and the endothecium are always prepared for apoptosis to provide nutrients for pollen mitosis and maturation [17]. Timely PCD of the anther wall, especially tapetum degradation in vivo, plays a fundamental role in male reproduction. Premature or delayed tapetal PCD and cellular degeneration can cause pollen abortion and male sterility [2,9,12,15]. Recently, multiple genes were identified to play essential roles in the process of rice tapetal PCD and pollen development, including several basic helix–loop–helix (*bHLH*) transcription factors, undeveloped tapetum 1 (*UDT1, bHLH164*) [11], tapetum degeneration retardation (*TDR, bHLH5*) [12,18], eternal tapetum 1/delayed tapetum degeneration (*EAT1/DTD, bHLH141*) [19,20], TDR-interacting protein 2 (*TIP2, bHLH142*) [21–24], MYB family transcription factor *GAMYB* [25–27], PHD-finger protein persistent tapetal cell 1 (*PTC1*) [28], and TGA transcription factor *OsTGA10* [29]. These genes were shown to regulate various aspects of anther development, especially tapetal PCD. *GAMYB* probably works upstream of *TDR1* and *PTC1* in parallel with *UDT1* to regulate rice tapetum development and pollen wall formation [27,28]. *TIP2* functions upstream of *TDR* and *EAT1*, but downstream of *UDT1*, and also binds to the promoter of *EAT1* to activate positive effects on regulation of tapetum PCD by promoting aspartic proteases *AP25*, *AP37*, and *OsCP1* [20,22]. *OsTGA10* also regulates tapetum development and pollen formation by interacting with *TIP2* and *TDR* to affect the expression of *AP25* and *MTR* [29]. In addition, some genes that encode enzymes or specific binding proteins are also essential for tapetal degeneration. Loss of function of these genes can result in anther deformity and defects in pollen exine formation. *DTC1* controls tapetal degeneration by modulating the dynamics of reactive oxygen species (ROS) with *OsMT2b* during reproduction of male rice [30]. *DEX1* regulates tapetal cell death and pollen exine formation by binding to Ca^{2+} to modulate cellular Ca^{2+} homeostasis, acting as a component required for tapetal cell death signal transduction [31]. Recessive mutation of the fasciclin glycoprotein *MTR* can dislocate its plasma membrane localization system and cause delayed tapetum PCD and reduced synthesis of Ubisch bodies, which are micron-sized particles on the inner surface of the tapetum in anthers, finally resulting in abortive pollen grains and complete male sterility [32]. The F-box protein *OsADF* is expressed in tapetal cells and microspores and works depending on *TDR* by binding E-box motifs of its promoter [33]. The two aspartic proteases, *AP25* and *AP37*, cysteine protease *OsCP1*, and apoptosis inhibitor *API5* are reported as specific regulators of the tapetal PCD process, and inhibition/mutagenesis of these genes can cause defects in pollen formation and eventually lead to male sterility [12,20,34].

Microspore cells need to undergo a series of processes after they are released from the tetrads to ultimately develop to mature fertile pollen grains, such as pollen wall formation, vacuolation, two rounds of mitosis, and starch enrichment. Normal progress of these processes requires nutritional supply and structural support from aliphatic biopolymers (sporopollenins, Ubisch bodies, epicuticular waxes, and cuticle monomers) and especially the synthesis and transport of sporopollenin. Sporopollenin is made up of polyhydroxylated aliphatic compounds and oxygenated aromatic monomers, such as phenolics, conjugated by ether and ester bonds, and acts as one of the main components of pollen exine, playing a critical role during pollen development by protecting pollen grains from abiotic and biotic stresses. Evidence from previous studies of male-sterile mutants with abnormal pollen wall formation showed that many genes are involved in these biosynthesis/transport mechanisms [1,2,15,35,36]. Two cytochrome P450 family genes, *CYP704B2* and *CYP703A3*, function as catalyzers of ω-hydroxylated fatty acids with 16- and 18-carbon chains and in-chain hydroxylase only for lauric acid to generate 7-hydroxylated lauric acid [37–39]. Defective pollen wall (*DPW*) acts as a fatty acyl-carrier protein reductase and *DPW2* as a fatty-acid acyltransferase to alter the amounts of cutin and waxes, and of lipidic and phenolic compounds, respectively, during anther development and pollen formation [40,41]. The ATP-binding cassette (ABC) transporters *ABCG15/PDA1*, *ABCG26*, and *ABCG3* work collaboratively but perform their own functions to transport different materials for anther development and pollen formation [42–45]. The lipid transfer protein *OsC6* is secreted into anther cuticle, anther locule, and the space between the tapetum and middle layer for pollen exine and orbicule formation downstream of *TDR* and *GAMYB* [46]. In addition, *WDA1* [47], *OsACOS12* [48,49], *OsNP1* [50], and *OsPKS2* [51,52] were also reported to regulate sporopollenin biosynthesis and deposition. Loss of function of these genes caused defects in not only the anther cuticle and pollen wall, but also the number of secretory, lipidic Ubisch bodies. In fact, the functions of these genes often interact; most of the recessive mutants of PCD-induced genes showed apparent defects in Ubisch body patterning and pollen exine formation. In particular, *tdr*, *ptc1*, *eat1-1*, *dex1*, and *dtc1* mutants exhibit obvious microspore collapse and pollen wall degeneration at the microspore stage [12,20,28,30]; *CYP703A3*, *ABCG15*, *ABCG26*, and *OsACOS12* also correspondingly showed high expression levels in tapetal cells and their recessive mutants due to delayed or premature degradation of the tapetum layer [42–44,49].

Acyl coenzyme A (CoA) glycerol-3-phosphate acyltransferases (GPATs), which localize to the endoplasmic reticulum (ER), are generally recognized as important catalyzers for the first step of de novo synthesis of triacylglycerol. GPATs play key roles in regulating cell growth and metabolic processes of membrane lipids, storage lipids, and extracellular lipid polyesters (cutin and suberin) by generating lysophosphatidic acids (LPAs) and acylating glycerol 3-phosphate at the sn-1 or sn-2 hydroxyl with acyl-CoA or acyl-acyl-acyl carrier protein (ACP) to alter glycerolipid triacylglycerol (TAG) biosynthesis [53–55]. The sn-1 GPATs promote acylation to produce lysophosphatidic acid (LPA) for lipid formation, while sn-2 GPATs possess a phosphatase domain to produce sn-2 monoacylglycerol (2-MAG) as the major product for cutin and suberin synthesis in plants. Most GPATs were confirmed to have sn-2 acyl transfer activity [56–60]. In *Arabidopsis*, there are ten GPATs, eight of which (*AtGPAT1* to *AtGPAT8*) belong to the sn-2 family and are divided into three sub-clades. The first clade, *AtGPAT1* to *AtGPA3*, is mainly expressed in flowers and siliques, and show sn-2 acyltransferase during the process of dicarboxylic acyl-CoA substrate utilization but not phosphatase activity. *Arabidopsis gpat1* mutant exhibits less fibrillar material, fewer vesicles in the anther locule, disrupted degeneration of the tapetum, and collapsed pollen grains [61]. *AtGPAT4*, *AtGPA6*, and *AtGPAT8* are unique bifunctional enzymes with both sn-2 acyltransferase and phosphatase activity to produce 2-MAG for cutin synthesis. *AtGPAT5* and *AtGPAt7*, which were identified as part of the suberin-associated clade, function in suberin synthesis in the wounding response for root and seed coat formation [60]. Genetic functions revealed that *AtGPAT1* and *AtGPAT6* play an important role in anther development and pollen formation. Loss of function of these two genes can result in altered ER profiles in tapetal cells, reduced pollen production, and decreased pollination; double mutation of these two genes can cause defective callose degeneration, pollen release, and complete male sterility [60,61]. Homologous sequence alignments and phenotypic analysis on other species reveal that *SlGPAT6*, *OsGPAT3*, and *ZmGPAT3* also have important roles in regulating anther cuticle biosynthesis and pollen exine formation [62–64]. However, studies on the metabolism of GPATs in regulating rice male reproductive development are still limited, particularly in tapetum PCD, and there is a lack of related mutants and phenotype analysis of relevant genes.

In this study, we further characterized *OsGPAT3* in rice male reproductive development, particularly focusing on its function in anther wall degeneration and pollen maturation. The *gpat3-2* mutant exhibited delayed tapetum PCD and Ubisch body formation and abnormal anther wall and pollen degeneration, resulting in complete male sterility. Genetic analysis and map-based cloning revealed that the mutant phenotype was caused by a single-nucleotide polymorphism (SNP) mutation in the first exon of *OsGPAT3*, a land plant sn-2 *GPAT* homolog. Two allelic mutants from our mutant library and three other mutant alleles of *OsGPAT3* generated using clustered regularly interspaced short palindromic repeats (CRISPR)/CRISPR-associated 9 (Cas9) also showed the same male-sterile phenotype. In addition, the expression pattern of many tapetum PCD-induced regulators and nutrition metabolism related genes were significantly altered resulting from recessive mutation of *OsGPAT3*. Thus, our results demonstrate that *OsGPAT3* is essential for anther wall PCD and pollen development in addition to its function during the synthesis of Ubisch bodies for anther cuticle and pollen exine formation. Our study also provides new insights into GPATs on regulating tapetum PCD and pollen maturation during plant reproductive development.

2. Results

2.1. Isolation and Phenotypic Analyses of the gpat3-2 Mutant

The *gpat3-2* mutant was first identified from the ethyl methyl sulfone (EMS)-soaked M1 progeny of an *indica* rice cultivar Zhonghui8015 (Zh8015) in Lingshui, Hainan Province. The *gpat3-2* mutant exhibited complete male sterility and was genetically stable in Hangzhou, Zhejiang Province. When the *gpat3-2* mutant was pollinated with wild-type pollens, all BC_1F_1 plants exhibited normal male fertility. Further identification of anthers in the BC_1F_2 population also indicated that the *gpat3-2* phenotype conformed to the genetic regulation of a single recessive gene ($\chi^2 = 0.17$; $p < 0.05$). Genetic analysis on the F_1 and F_2 progeny generated by the cross *gpat3-2* × 02428, a wide compatibility *japonica* cultivar, also confirmed the monofactorial recessive inheritance of the *gpat3-2* mutant (see Table S1, Supplementary Materials).

Vegetative development, including plant height, heading date, tiller number, other major agronomic traits, and general spikelet morphology of the *gpat3-2* plants did not differ from those of wild-type Zh8015 except for the anthers (Figure 1A–C). The Zh8015 anthers were normal golden yellow and had countless fertile pollen grains in the anther interior (Figure 1C–F); anther dehiscence occurred soon after glumes opened and pollens were deposited for fertilization (Figure 1C,D). Compared with wild-type anthers, the mutant anthers were small and white, without mature pollen grains (Figure 1C–F,H) and failed to dehisce, resulting in completely sterile spikelets (Figure 1C–F,H).

2.2. Defects of Anther Development and Pollen Maturation in gpat3-2 Mutant

To identify defects in the *gpat3-2* mutant, semi-thin transverse sections of anthers at different developmental stages from the wild-type plant and *gpat3-2* mutant were further examined. Rice anther development was delineated into 14 stages, which was consistent with that of *Arabidopsis thaliana* [3], based on the cellular events observed under light microscopy by semi-thin section [65,66]. By stage 7, both wild-type (WT) and *gpat3-2* anthers exhibited an obvious four-layered anther wall from surface to interior, and the wall enwrapped pollen mother cells (PMCs) within the locule. At this stage, the PMCs progressively initiated meiotic division and nestled against the tapetal layer. No obvious defects in the four somatic layers of the anther wall and microsporocytes were detected between the WT and *gpat3-2* mutant until this stage (Figure 2A,B). Subsequently, *gpat3-2* anthers began displaying obvious morphological abnormalities. At late s7 to s8a, the PMCs generated dyed cells, the middle layer became nearly invisible, and the anther wall had three layers in WT anthers. Meanwhile, the tapetal layer began having a weak point of programmed cell death (PCD) (Figure 2C). In the *gpat3-2* anthers, the middle layer was still clearly visible, the anther wall had the four-layer structure, and newly formed dyed cells were misshapen and less darkly stained; this phenomenon was also observed in tapetum cells (Figure 2D). By the end of stage 8b, tetrads were generated, tapetum PCD started, and the middle layer became nearly invisible in the wild-type anthers (Figure 2E). The *gpat3-2* anthers also formed imperfect tetrads, which exhibited unequal cleavage. However, the tapetum layer was electron-dense and expanded, and the middle layer was still apparent and showed no signs of degradation (Figure 2F).

Figure 1. Phenotype comparison between the wild-type Zhonghui8015 (Zh8015) and *gpat3-2* mutant. Plants of the wild-type Zh8015 and *gpat3-2* mutant at the anthesis stage (**A**); Mature spikelets of the wild-type Zh8015and the *gpat3-2* mutant at anthesis (**B**); The stamen morphologies of the wild-type Zh8015and *gpat3-2* mutant; lemmas and paleae were removed for clarity (**C**); Mature anther of the wild-type Zh8015 and *gpat3-2* mutant (**D**); Compressed anthers of wild-type Zh8015 (**E**) and the *gpat3-2* mutant (**F**) after I_2/KI staining (**E**,**F**); I_2/KI staining of the pollen grains of wild-type Zh8015 (**G**) and the *gpat3-2* mutant (**H**) at stage 13 (**G**,**H**). Scale bars = 1 mm in (**B**–**F**), and 100 μm in (**G**,**H**); ad, anther dehiscence.

At stage 9, microspores were released from the tetrads and tapetal cells, becoming condensed and electron-dense resulting from gradual PCD-induced degradation (Figure 2G). Although microspore cells can be formed after meiosis, most anther walls gradually became distorted and shrank, and the middle layer remained relatively discernible in this period in the mutant anthers. Furthermore, the *gpat3-2* mutant swelled and there were lightly stained tapetal cells, indicating abnormal PCD (Figure 2H). A sharp distinction between the wild-type and *gpat3-2* anthers began appearing at stage 10; normal vacuolated microspores were uniformly attached to the tapetum side as round shapes with dark-stained pollen exine, while the tapetum layer was electron-lucent after gradual PCD (Figure 2I). By comparison, internal cavities of the *gpat3-2* anthers were disordered; the tapetum layer was swollen and lightly stained with obvious degradation characteristics and the middle layer was still visible. In addition, microspores were disrupted and degraded together with the tapetum layer (Figure 2I). From the pollen mitosis stage and the mature pollen stage, wild-type pollen formed a complete double-layer exine and fertile pollen after two mitotic divisions with pollen exine deposition and starch accumulation, while the tapetum layer gradually degenerated and thinned until it almost disappeared at the end of stage 12. During this procedure, the epidermis and endothecium layer further degenerated and anther dehiscence occurred. Mature pollen grains were full of lipids, starch, and other storage nutrients, and were dark-stained with toluidine blue, indicating that the wild-type pollen grains had normal functions and were viable (Figure 2K,M,O). However, degradation of microspores and tapetal cells continued, resulting in linear pollen walls and cell detritus in the *gpat3-2* anther locule. In addition, the outer layers of the anther wall retained the original structure; epidermis, endothecium layer, and middle layer were not degraded until the end of stage 13 in *gpat3-2* anthers (Figure 2L,N,P). These observations suggested that *gpat3-2* carried defects not only in anther wall development and pollen maturation, but also in differentiation and degradation of tapetal cells and anther wall.

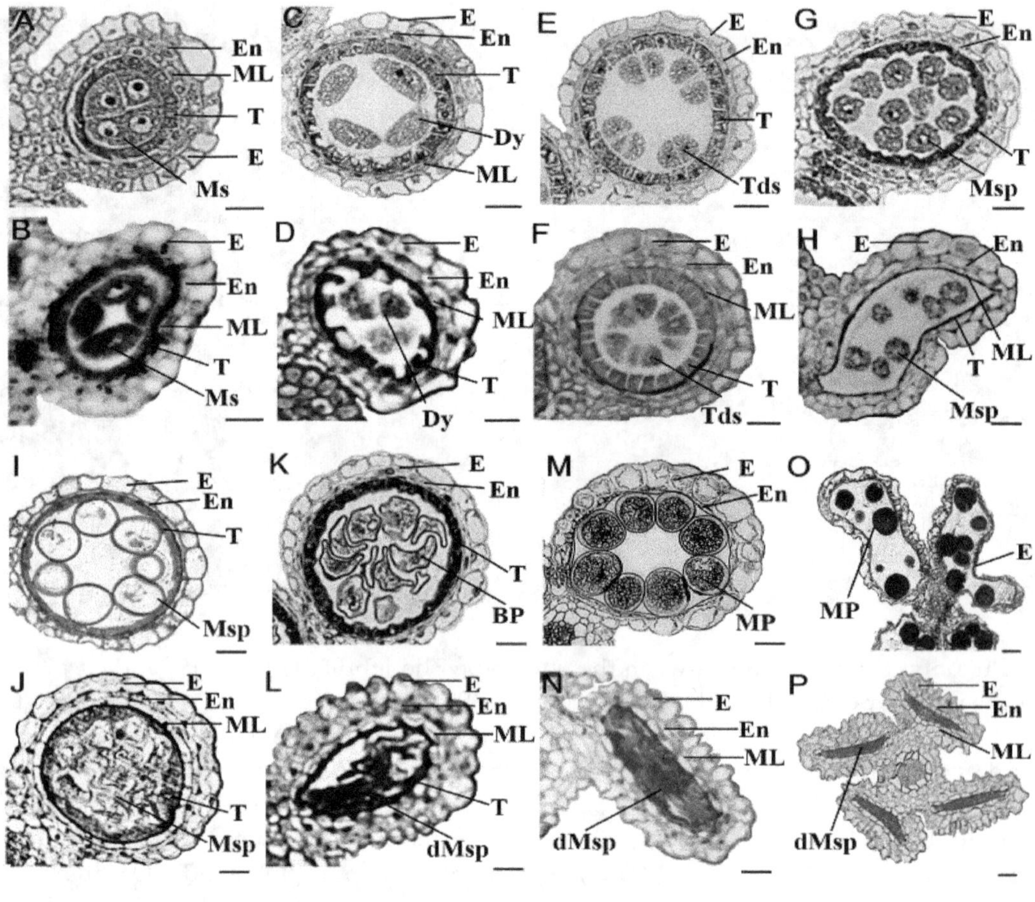

Figure 2. Transverse section analysis of anther development in wild-type Zh8015 and *Osgpat3-2* mutant. Locules from the anther section of Zh8015 (**A**,**C**,**E**,**G**,**I**,**K**,**M**,**O**) and *Osgpat3-2* (**B**,**D**,**F**,**H**,**J**,**L**,**N**,**P**) from stage 7 to stage 13 of development (stages 7, 8a, 8b, 9, 10, 11, 12, and 13). BP, bicellular pollen; dMsp, degraded microspores; Dy, dyed cell; E, epidermis; En, endothecium; ML, middle layer; Mp, mature pollen; Ms, microsporocyte; Msp, microspores; T, tapetum; Tds, tetrads. Scale bars = 20 μm.

2.3. Delayed PCD of Osgpat3-2 Tapetal Cells and Anther Wall Cells

The transverse section analysis suggested that *gpat3-2* mutation affected the differentiation and degradation of tapetal cells and anther wall. We, therefore, used a terminal deoxynucleotidyl transferase-mediated deoxyuridine triphosphate (dUTP) nick-end labeling (TUNEL) assay to test the tapetum PCD process during a range of developmental stages in WT and *gpat3-2* anthers.

Microspore mother cells were generated during meiosis and tapetal cells became condensed after meiosis; there was no detectable DNA fragmentation signal in the wild-type or *gpat3-2* tapetal cells. At stage 8a, some TUNEL-positive nuclei were detected in wild-type tapetal cells and the middle layer, indicating that normal PCD started occurring in the wild-type anthers (Figure 3A). At stage

8b, positive signals of DNA fragmentation were much stronger in wild-type tapetal cells. Positive PCD signals were also detected in both the outer layers (endothecium and middle layer) and vascular bundle cells of wild-type anthers at this stage (Figure 3C). However, no visible fragmented DNA signal was observed in the *gpat3-2* mutant anthers at both stage 8a and 8b (Figure 3B,D). At stage 9, when the microspore was released from the tetrad, PCD signals in wild-type tapetal cells became strongest, and the signal in the outer layers, vascular bundle cells, and cavity of dehiscence increased at the same time (Figure 3E). However, PCD signals were still not detected in the *gpat3-2* tapetum; only weak signals of DNA fragmentation were detected in *gpat3-2* outer layers and vascular bundle cells (Figure 3F). At stage 10, positive PCD signals in wild-type tapetal cells and other tissues became much weaker than at stage 9 (Figure 3G), while PCD signals unexpectedly became detectable and very strong in most tissues of *gpat3-2* anthers, including tapetal cells, the outer layers (endothecium and middle layer), vascular bundle cells, and even microspore cells inside the chamber (Figure 3H). At stage 12, PCD signals of wild-type tapetal cells gradually became invisible and only small, positive signals were detected in the outer layers and cavity of dehiscence (Figure 3I). Conversely, PCD signals continued increasing in all parts of *gpat3-2* anthers (Figure 3J). Nevertheless, at the dehiscence stage, none positive PCD signals were detected in wild-type tapetum, but the signals became much stronger in outer layers, vascular bundle cells, and cavity of dehiscence to meet the requirement of anther cracking and pollination (Figure 3K). However, positive PCD signals became strongest in all cell layers of *gpat3-2* anthers at this stage, including the degenerated microspores. These TUNEL assays demonstrate that the PCD of tapetum and peripheral layers were normal and orderly, and the delay and disorder of PCD in tapetum, anther wall, and microspores possibly resulted in the failure of pollen formation in the *gpat3-2* mutant.

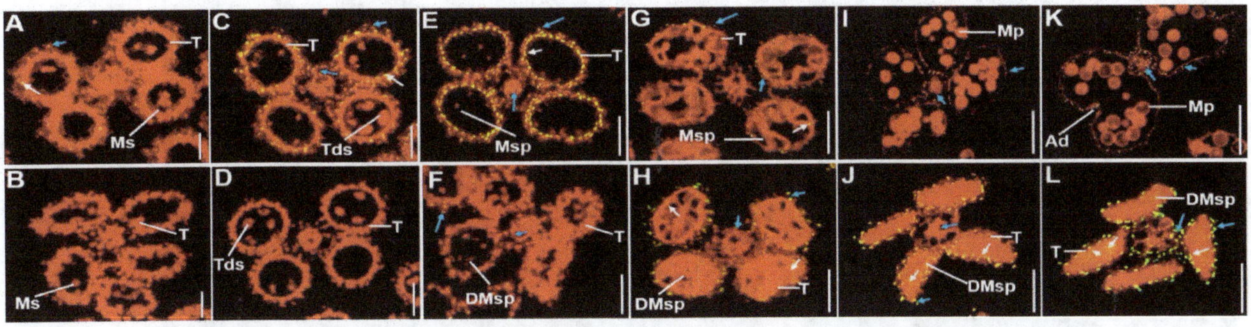

Figure 3. Detection of DNA fragmentation in wild-type and *gpat3-2* anthers using a terminal deoxynucleotidyl transferase-mediated deoxyuridine triphosphate (dUTP) nick-end labeling (TUNEL) assay. The wild-type and *gpat3-2* mutant anthers at stage 8a (**A**,**B**), stage 8b (**C**,**D**), stage 9 (**E**,**F**), late stage 10 (**G**,**H**), stage 12 (**I**,**J**), and dehiscence stage (**K**,**L**). A red signal indicates propidium iodide (PI) staining, while yellow and green fluorescence indicates a TUNEL-positive signal. TUNEL-positive signals detected in the tapetum cells of both wild-type and *gpat3-2* anthers are marked by white arrows, while TUNEL-positive signal observed in the outer cell layers (including the epidermis, endothecium, and middle layer), and vascular bundle cells are marked by blue arrows. Ad, anther dehiscence; DMsp, degenerated microspore; Mp, mature pollen; Ms, microsporocyte; Msp, microspore; Tds, tetrads; T, tapetum. Scale bars = 50 um.

2.4. Osgpat3-2 Exhibits Defects in the Formation of Ubisch Bodies and Pollen Accumulation

For further verification and understanding of abnormalities of the *gpat3-2* mutation, we performed a more detailed scanning electron microscope (SEM) observation of the surfaces of wild-type and *gpat3-2* anthers and pollen grains at different development stages. No obvious differences of the anther morphologies or outer wall structure were observed between wild-type and the *gpat3-2* anthers at stage 9; both showed a smooth epidermis and microspores were released (Figure 4A,B,D-I,D-II). However, further enlargement of the inner surface of locules and pollen exine revealed that the

Ubisch bodies were produced by the tapetum and transported to microspores in wild-type anthers, while the *gpat3-2* anthers still had a smooth inner surface and abnormal pollen exine, indicating the defects of tapetum PCD (Figure 4C,E-I,E-II). From the vacuolated pollen stage (stage 10) to the mature pollen stage (stage 12), the size of wild-type anthers gradually increased to almost double, the anther epidermis constantly thickened and was covered with a three-dimensional spaghetti-like cutin layer, the continuous synthesis and transport of Ubisch bodies resulted in a neat arrangement of inner surface, and the uninucleate microspores enlarged and developed to trinucleate pollen with regular shaped pollen exine, which was formed by secreted tapetum-produced sporopollenin precursors from Ubisch bodies (Figure 4III,V,VII of A–E). By contrast, the size of *gpat3-2* anthers was only slightly increased from stage 9 to 11 and basically no longer increased after this stage (Figure 4A-II,IV,VI,VIII). Anther epidermis was still smooth, which indicated defects in the synthesis of typical cutin and in the synthesis of fatty acids in the anther wall (Figure 4B-II,IV,VI,VIII). The inner wall of anthers gradually became irregular at stages 11 and 12 (Figure 4C-VI,VIII) and was smooth but uneven with randomly distributed flocs of Ubisch bodies and degraded microspores at stages 9 and 10 (Figure 4B-II,IV). Further observation of the microspores from different developmental periods showed that microspores in *gpat3-2* anther locules were gradually collapsed and covered with randomly distributed cutin materials, finally leading to the *gpat3-2* microspores becoming more and more chaotic-like cotton wools (Figure 4E-II,IV,VI,VIII).

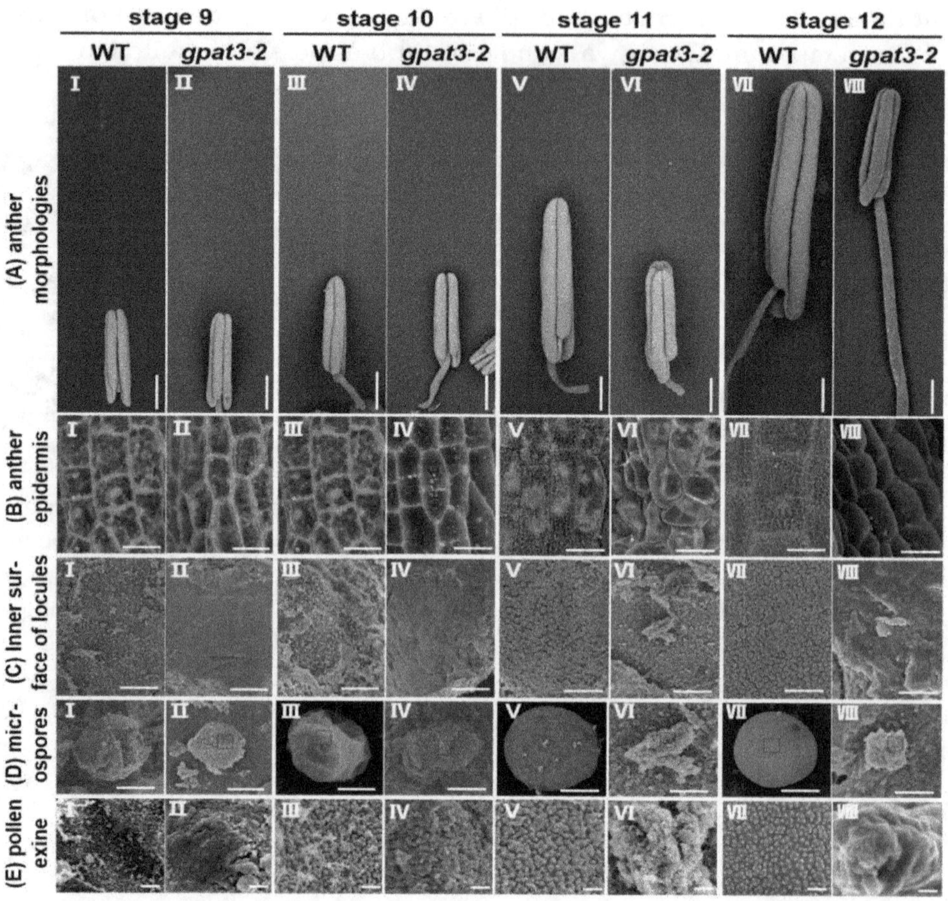

Figure 4. Scanning electron microscopy (SEM) observation of the surface of anther and pollen grains in wild-type (WT) and *gpat3-2* anthers. The anther morphologies (**A**), anther epidermis (**B**), inner surface of anther locules (**C**), microspores (**D**), and pollen exine (**E**) of WT and *gpat3-2* from stages 9 to 12 are shown. (**E**) Macro photograph of the outermost surface of microspores in the areas indicated by red boxes in (**D**). Scale bars = 500 μm in (**A**), 10 μm in (**B**,**C**), 5 μm in (**D**), and 200 nm in (**E**).

Transmission electron microscopy (TEM) observation verified the defects in *gpat3-2* anthers and pollen grains. Obvious differences were first identified at stage 8a; *gpat3-2* anthers showed less tapetum differentiation but more vacuoles and a thicker middle layer (see Figure S1A–F, Supplementary Materials) and pollen mother cells were agglomerated without nucleus compared with the wild-type anther (see Figure S1G,H, Supplementary Materials). At stage 8b, tapetal cells of wild-type anthers began differentiating and became sparse and loose (see Figure S1I,K,M, Supplementary Materials), while the *gpat3-2* tapetum cells were still electron-dense without any signs of degradation (see Figure S1J,L,N, Supplementary Materials). In addition, tetrads of WT were first surrounded by callose that resembled the pollen primexine structure (see Figure S1O, Supplementary Materials); however, tetrads of *gpat3-2* were still naked without a wrapping layer as before (see Figure S1P, Supplementary Materials). These results were almost the same as previously reported by Men et al. [61]. However, at stage 9, the tapetum gradually turned into a small, loose thread resulting from the vigorous PCD process, and only a very small line of residual cells of the middle layer was observed in wild-type anthers (Figure 5A-1,C-1,E-1). Ubisch bodies, which were believed to secrete tapetum-produced sporopollenin precursors for pollen exine formation, began being released from the tapetum and gradually grew into electron-dense orbicules (Figure 5E,F). Pollen exine of microspores with accumulation of particulate sporopollenin in the wild-type locule was visible at this stage (Figure 5G-1,I-1). Although microspores and pre-Ubisch bodies seemed to be formed from the mutant tapetum at stage 9, the tapetal cells and microspores were still electron-dense without any signs of differentiation (Figure 5B-1,D-1,F-1,H-1). In addition, some accumulation of electron-dense materials occurred on the outer surface of mutant microspores, but these structure were unable to grow into mature Ubisch bodies, and formed an abnormal exine structure (Figure 5J-1). From stage 10 to stage 11, the middle layer disappeared completely, and the endothecium layer constantly degraded and thinned accompanied by the collapse of epidermis (Figure 5A-2,C-2,A-3,C-3); the tapetum layer differentiated and degenerated to a lightly stained and thin layer (Figure 5A-2,C-2,E-2,A-3,C-3,E-3). Microspore exine deposition, which underwent vacuolization and two rounds of mitosis, was completed and eventually formed thick exine with distinctive layers of tectum, bacula, and nexine under normal nutrient supply from continuously generated Ubisch bodies (Figure 5G-2,I-2,G-3,I-3). Although the tapetum of the *gpat3-2* mutant appeared to be degraded, it was still dense and deeply stained (Figure 5B-2,D-2,F-2,B-3,D-3,F-3): the precursor of newly formed pre-Ubisch bodies failed to synthesize complete and mature Ubisch bodies (Figure 5F-2,F-3). In addition, the epidermis and endothecium cells of anthers were deformed, but the middle layer was still obviously visible and thick (Figure 5B-2,D-2,B-3,D-3). The microspores formed previously in *gpat3-2* anthers exhibited irregular pollen exine and were also gradually degraded with tapetal cells, resulting in degraded cell remnants in the anther locule (Figure 5B-2,H-2,J-2,B-3,H-3,J-3). At the mature pollen stage (stage 13), a distorted layer of epidermis was also visible in wild-type anther, and the endothecium cells and tapetum were almost degraded with numerous Ubisch bodies attached to the inner side facing the pollen grains (Figure 5A-4,C-4). At stage 13, wild-type microspores grew to spherical pollen grains full of starch, lipids, and other nutrients surrounded by a normal bilayer exine (Figure 5G-4,I-4, white arrowhead). However, the anther wall of the *gpat3-2* mutant was still a three-layer outer structure with a few lipids deposited on the inner side of the middle layer; the tapetum was prominently electron-lucent and no longer a distinct layer at this time, probably due to serious degradation (Figure 5B-4,D-4,F-4). The *gpat3-2* microspores aborted and collapsed, and the framework of the exine distorted and folded without the internal nexine layers; only remnants of abnormal epidermis in the locule were left, resulting from severe degradation (Figure 5H-4,J-4). Ultimately, all the defective tissue burst out; finally, the anther chamber of the *gpat3-2* mutant was filled with degraded fragments and residual abnormal pollen grains.

Together, all these results indicated that the *gpat3-2* mutation affected the differentiation and degradation of the anther wall, the synthesis and supply of Ubisch bodies, and pollen wall formation. The *gpat3-2* defects were different from the *gpat3* mutant reported previously, because *gpat3* microspores could not be released from the tetrads and were still covered with callose at the young microspore stage [63].

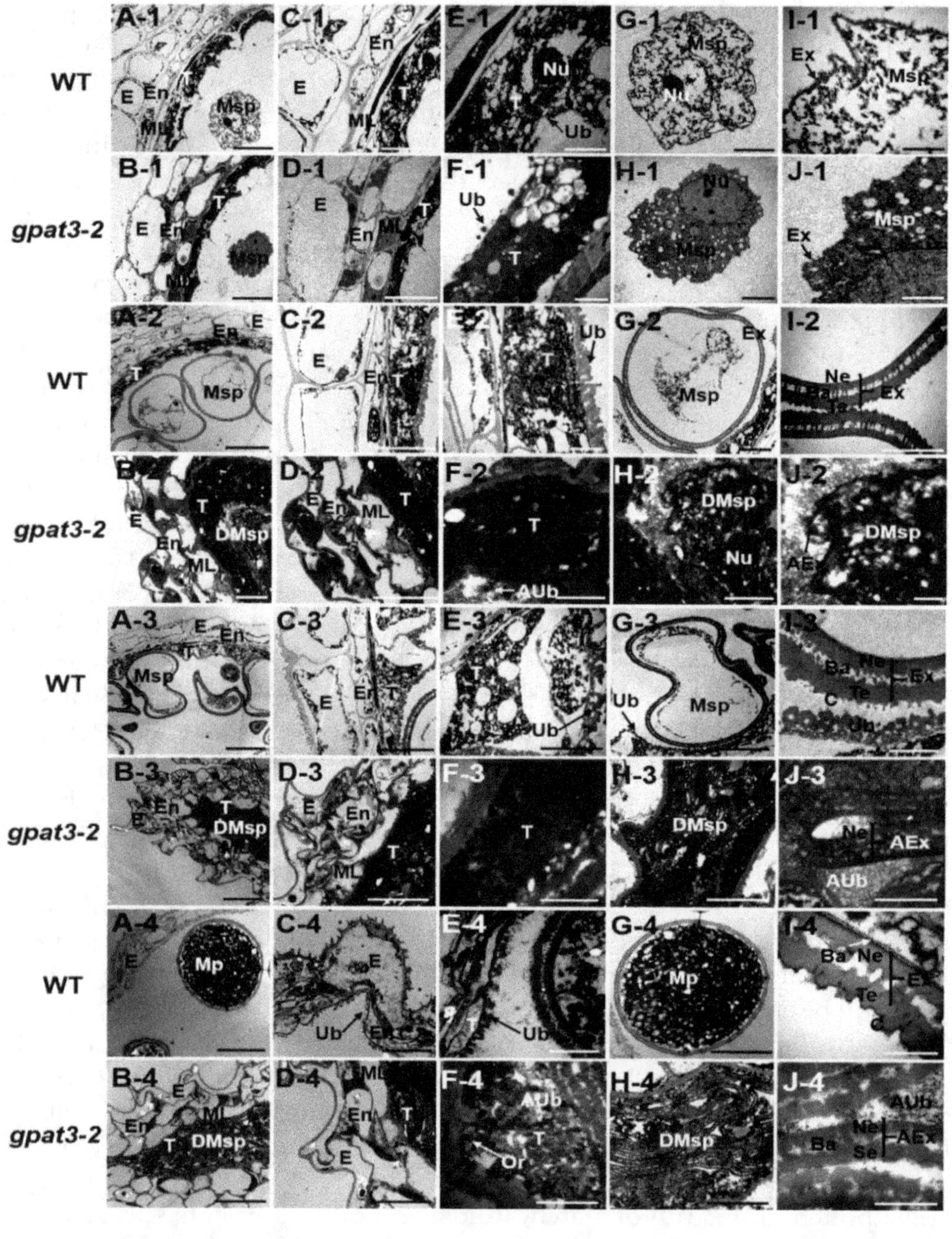

Figure 5. Transmission electron microscopy analysis of anther development in wild-type and *Osgpat3-2* mutant from stages 9–13. The transverse sections of the WT (**A**,**C**,**E**,**G**,**I**) and *gpat3-2* (**B**,**D**,**F**,**H**,**J**) anthers at stage 9 (**A1–J1**), stage 10 (**A2–J2**), stage 11 (**A3–J3**), and stage 13 (**A4–J4**) are compared. Anthers of the wild type (**A**) and *Osgpat3-2* (**B**), showing the anther wall with microspores at different development stages. The layers of anther wall in the wild type (**C**) and *Osgpat3-2* (**D**). Higher magnification of the tapetum cells showing Ubisch body in the wild type (**E**) and *Osgpat3-2* (**F**). The development and morphologies of microspores in the wild type (**G**) and *Osgpat3-2* (**H**). The development and structures of pollen exine in the wild type (**I**) and *Osgpat3-2* (**J**). AEx, abnormal exine; AUb, abnormal Ubisch body; Ba, bacula; C, cuticle; DMsp, degenerated microspores; E, epidermis; En, endothecium; Ex, exine; ML, middle layer; Mp, mature pollen; Msp, microspores; Nu, nucleus; Ne, nexine; Or, orbicule; Se, sexine; T, tapetum; Te, tectum; Ub, Ubisch body. Scale bars = 10 μm in (**A**,**B**), 5 μm in (**C**,**D**), 0.5 μm in (**E**,**F**), 2 μm in (**G**,**H**), and 200 nm in (**I**,**J**).

2.5. Fine Genetic Mapping and Candidate Gene Analysis of the Osgpat3-2 Mutation

To map the *gpat3-2* gene, map-based cloning was performed using the F_2 population from the cross between *gpat3-2* and 02428. Polymorphisms were confirmed as described previously [39,67]. Seven individuals each of the wild-type and *gpat3-2* phenotypes were chosen for linkage analysis using the bulked segregant analysis (BSA) method and the results revealed that *gpat3-2* is located on the long arm of chromosome 11, flanked by simple sequence repeat (SSR) loci RM27172 and RM27326 (Figure 6A). Further primary mapping was conducted using encryption markup with an insertion/deletion (InDel) marker RD1110 and an SSR marker RM27273 in 176 F_2 recessive individuals; the mutation locus was in a 871.4-kb region between markers RM17273 and RM27326 (Figure 6A). For fine-scale mapping of the *gpat3-2* gene, five significantly polymorphic InDel markers were designed based on polymorphisms between *japonica* Nipponbare and *indica* 9311 [68] (Table S1, Supplementary Materials). Using these newly developed markers along with high-resolution genetic linkage analysis with 1354 recessive individuals from F_2 populations of the cross between *gpat3-2* and 02428, the *GPAT3-2* gene was finally delineated to a 26-kb region between ZH-3 and ZH-6 (Figure 6B).

We sequenced the three open reading frames (ORFs) in this region according to the Rice Genome Annotation Project (http://rice.plantbiology.msu.edu/index.shtml) (Figure 6C; Table S2, Supplementary Materials) of both the WT and *gpat3-2* mutant. We found that the *gpat3-2* mutant carried a single nucleotide mutation (G to A) in the first exon of *Loc_Os11g45400* (Figure 6D–F), which resulted in the corresponding amino acids being mutated directly from tryptophan to a stop codon (Figure 6D,G). Two pairs of calcium-dependent activator protein for secretion 1 (CAPS1) enzyme digestion primers were designed to test and verify this alternative splicing site for this mutation using the endonuclease *Nla*IV, and the enzyme digestion results confirmed our prediction (Figure 6H).

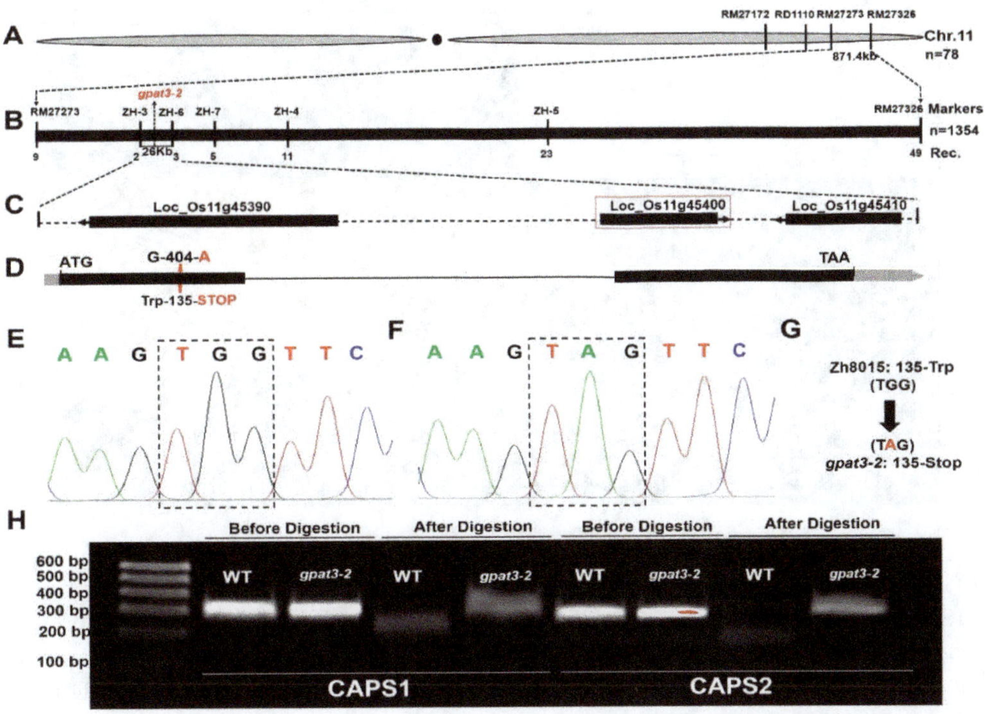

Figure 6. Mapping cloning of the *GPAT3-2* gene. Genetic linkage of the *gpat3-2* locus on chromosome 11 (**A**); Fine mapping of the *GPAT3-2* locus (**B**); The predicted open reading frames (ORFs) in the target region of the rice genome (**C**); Gene structure of the target gene and the *gpat3-2* mutation site (**D**); Sequences of the mutation region in the *Loc_Os11g45400* gene from Zh8015 (**E**) and the *gpat3-2* mutant (**F**); Predicted protein change of the *gpat3-2* mutation (**G**); *Nla*IV electrophoresis before and after enzyme digestion of the mutation site (**H**).

2.6. Function Verification of the *OsGPAT3* Gene

To confirm that male sterility was caused by the mutation in *Loc_Os11g45400*, a 7.9-kb *OsGPAT3* genomic fragment (*gOsGPAT3*) was transformed into calli induced from young panicles of homozygous *gpat3-2* mutant plants to rescue the sterile phenotype of *gpat3-2* (Figure 7A). The *gOsGPAT3* fragment included the 3.35-kb upstream sequence, the full-length ORF of *Loc_Os11g45400* from the wild type, and a 1180-bp region downstream from the termination codon sequence. The transgenic positive plants had a normal seed-setting rate, similar to those of wild type (Figure 7B). The complemented lines exhibited a normal seed-setting rate (Figure 7B) and golden yellow anthers (Figure 7C), and pollen grains accumulated abundant starch granules and could be dyed black by 1.2% I_2/KI solution (Figure 7D–F). In addition, anther transverse sections were examined to further verify the anther developmental process in complementary lines. The microspore mother cells secreted normal microspores at stage 9, the middle layer was invisible, and tapetal cells degraded and had fine lines and irregular shapes (Figure 7G). At stage 11, the microspores underwent mitotic divisions and generated binucleate pollen grains, while the endothecium layer and tapetum cells gradually tapered off and epidermis cells were puffy (Figure 7H). Normal developmental morphologies were also visible at stage 13; fertile pollen grains were produced, the epidermis layer was almost all that was left at this stage, and the anther chamber started shrinking and bursting for pollination (Figure 7I). Therefore, the whole developmental process of the anther in complementation lines was restored. These experiments confirmed that the single-nucleotide mutation in *Loc_Os11g45400* was responsible for the no-pollen phenotype of *gpat3-2*.

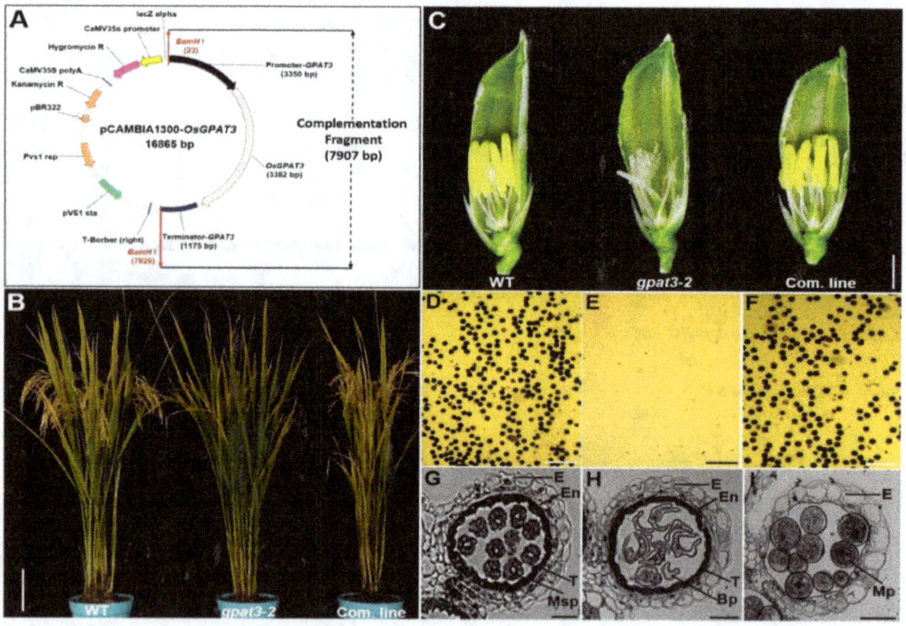

Figure 7. Complementation analysis of the *gpat3-2* mutant using wild-type *OsGPAT3*. The structure of the *gOsGPAT3* plasmid for transformation (**A**). The *gpat3-2* mutant was transformed with the pCAMBIA1300-*OsGPAT3* vector under the 35S promoter for transformation selection and the *OsGPAT3* genomic fragment contained three functional modules, including its native promoter, entire coding sequence, and downstream sequence from the WT for restoration of male fertility. The seed-setting rate of spikelet in wild-type, *gpat3-2*, and complementation plants (**B**). The flower of wild type, *gpat3-2*, and complementation line at stage 12; half of lemma and paleae were removed for clarity (**C**). Pollen grains of wild type (**D**), *gpat3-2* (**E**), and *gOsGPAT3*-complemented mutant (**F**) stained with 1% I_2/KI solution at stage 12. Transverse section analysis of anthers from *gOsGPAT3*-complemented line at stage 9 (**G**), late stage 11 (**H**), and stage 12 (**I**). E, epidermis; En, endothecium; T, tapetum; Msp, microspore; BP, bicellular pollen; Mp, mature pollen. Scale bars = 15 cm in (**B**), 1 mm in (**C**), 5 μm in (**D–F**), and 50 μm in (**G–I**).

To further confirm this result, we designed a target within *Loc_Os11g45400* using CRISPR/Cas9 in the Zh8015 genetic background and obtained three other ideal mutants, *gc-1*, *gc-2*, and *gc-3* (Figure 8A). We also found two allelic mutants, one with a single-base insertion (*gpat3-3*) and the other with a single-nucleotide mutation (*gpat3-4*) (Figure 8A). As expected, these five mutants had small, white anthers without mature pollen grains and with significantly increased transcription levels similar to *gpat3-2* (Figure 8B,C–H,O). A TUNEL assay also uniformly observed disordered anther locule with significantly enhanced PCD signals in almost all layers of anther wall, vascular bundle cells, and cavity of dehiscence compared to wild-type anther (Figure 8I–N). As predicted by the Rice Genome Annotation Project, OsGPAT3 protein includes a typical signal peptide, two transmembrane regions, and a conserved GPAT domain, which contains four acyltransferase motifs, without a phosphatase domain (see Figure S2, Supplementary Materials).

Previous studies showed that OsGPAT3 belongs to the first clade of the conserved land plant sn-2 GPAT family that specifically regulates lipid biosynthesis for anther cuticle and pollen exine formation [60–64]. Further multiple comparisons of the *OsGPAT3* sequence found that the *gpat3-2* mutant and all three CRISPR/Cas9-induced mutants had premature stop codons and produced truncated polypeptides, which resulted in destruction of the conserved domains. While the two allelic mutants were somewhat different, the protein structure of *gpat3-3* changed completely after the single-base insertion and *gpat3-4* carried an amino-acid conversion in the most conservative GPAT domain (see Figure S2, Supplementary Materials). The findings indicated that the six mutant lines carried different mutations which disrupted the function of the conserved GPAT domain, transmembrane region, and acyltransferase motifs (Figure 7; Figure S2, Supplementary Materials). Taken together, the results demonstrated the function of *Loc_Os11g45400* in rice anther wall PCD and pollen development.

2.7. Mutation in OsGPAT3 Affects the Expression of Genes Involved in Both Tapetum PCD and Nutrient Metabolism

OsGPAT3 was reported to regulate the biosynthesis of lipid metabolism for anther cuticle and pollen exine formation [63]. Our results indicated that the mutation of *OsGPAT3* caused severely delayed tapetum PCD and anther wall development, failure of Ubisch body formation, and abnormal degradation of microspores. To understand the role of *OsGPAT3* in lipid metabolism, anther wall development, and tapetum PCD, and to explain the phenotype in *gpat3-2*, we compared the expression level of a series of genes regulating tapetum PCD and synthesis/transportation of sporopollenin precursors during male reproductive development between the wild type and *gpat3-2* mutant using qPCR. The expression pattern of *OsGPAT3* is consistent with the defects shown in *gpat3-2* mutant and is exactly like *PTC1* (Figure 9A, Li et al.) [28].

The *gpat3-2* anthers showed higher expression than the wild type at stage 7, lower expression during meiosis, and tetrad formation (stage 8), but significantly higher expression than the wild type after stage 9, when young microspores were released (Figure 7A). Nearly half of the tapetum PCD-related genes were downregulated in the *gpat3-2* mutant, including *GAMYB*, *DEX1*, and three cysteine proteases, *AP25*, *AP37*, and *OsCP1* (Figure 9A). Eight genes, including two homologous genes of *OsGPAT3* in rice (*Loc_Os05g38350* and *Loc_Os10g42720*), four bHLH transcription factors (*UDT1*, *TDR*, *EAT1*, and *TIP2*) that particularly regulate tapetum degeneration, *OsTGA10*, which is the target transcription factor of *OsMADS8*, and *MTR*, showed coincident upregulation in the *gpat3-2* mutant (Figure 9A). This may further prove that the organelles are still active and the metabolic process was still occurring during late stages of the *gpat3-2* mutant.

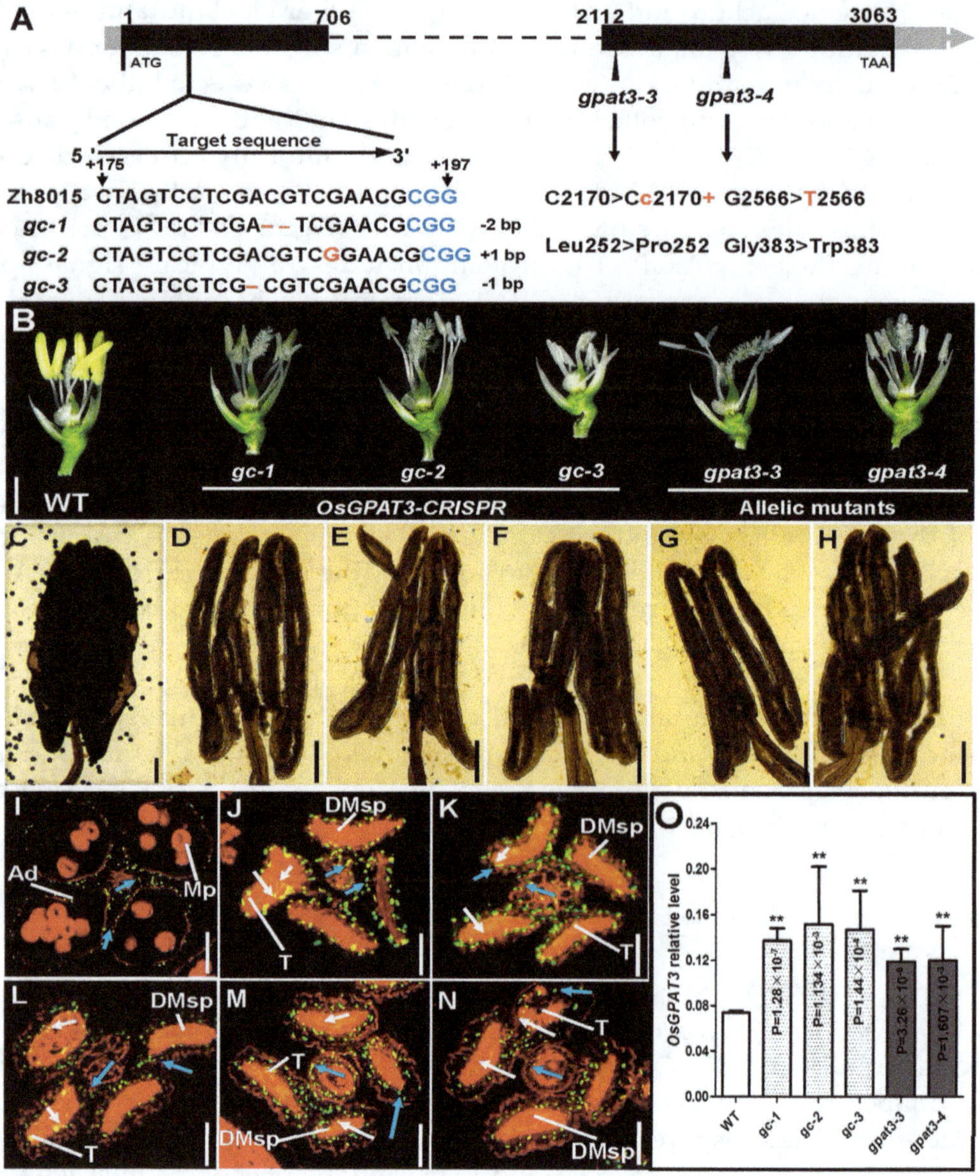

Figure 8. Sequence analysis and phenotypic observation of *OsGPAT3* clustered regularly interspaced short palindromic repeats (CRISPR)/CRISPR-associated 9 (Cas9)-induced mutants and allelic mutants. Gene structure of *OsGPAT3* and mutation analysis of *OsGPAT3* gene in transgenic plants and allelic mutants (**A**). The sequence (5′–CTAGTACTCGACGTCGAAGGCGG–3′) located in the first exon of the *OsGPAT3* gene was selected as the target site of single guide RNA (sgRNA). The black boxes indicate the exons. The blue characters indicate the protospacer adjacent motif (PAM). The red characters indicate the three different types of mutation events generated by CRISPR/Cas9 in the mutants. Phenotypic comparison of the WT and mutant anthers at stage 13; the lemma and paleae were removed for clarity (**B**). Compressed anthers of wild type Zh8015 (**C**) and the mutants (**D**: *gc-1*, **E**: *gc-2*, **F**: *gc-3*, **G**: *gpat3-3*, **H**: *gpat3-4*) after I_2/KI staining. Detection of DNA fragmentation in wild-type (**I**) and mutant (**J**: *gpat3-3*, **K**: *gpat3-4*, **L**: *gc-1*, **M**: *gc-2*, **N**: *gc-3*) anthers using a TUNEL assay at the dehiscence stage (stage 13). White arrows indicate TUNEL-positive signals detected in the tapetum cells of wild-type and *gpat3-2* anthers, while blue arrows indicate TUNEL-positive signal observed in the outer cell layers (including the epidermis, endothecium, and middle layer), and vascular bundle cells. The qPCR analysis of *OsGPAT3* in wild-type, CRISPR/Cas9-induced mutants, and allelic mutants (**O**). Ad, anther dehiscence; DMsp, degenerated microspore; Mp, mature pollen; T, tapetum. Scale bars = 2 mm in (**B**), 500 µm in (**C–H**), and 50 µm in (**I–N**). ** indicates significant differences at $p < 0.01$.

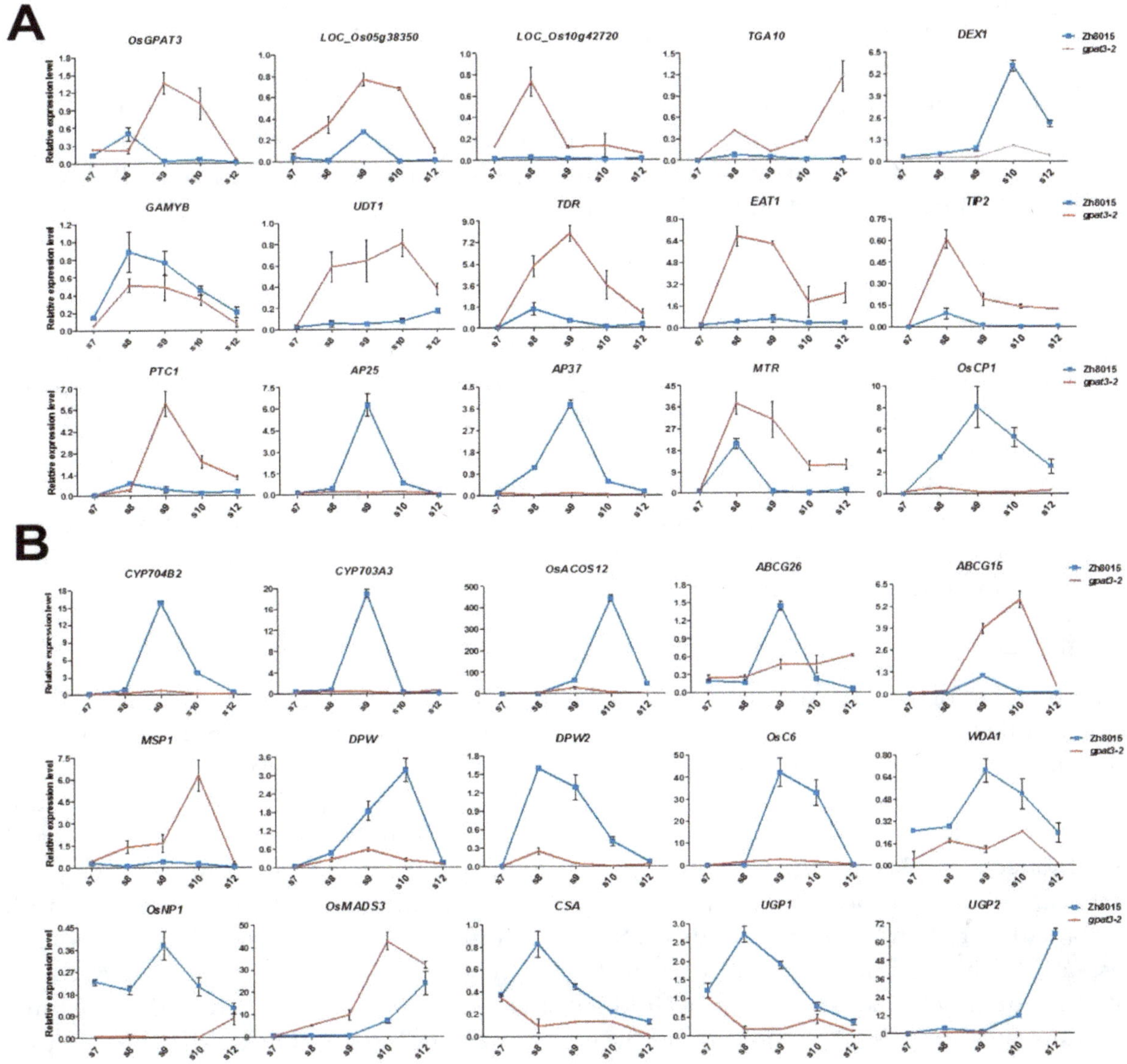

Figure 9. Expression analysis of *GPATs* and male sterility (MS)-involved genes in the WT and *gpat3-2*. The qPCR analysis of genes related to tapetum degeneration retardation in the wild-type and *gpat3-2* anthers at stages 7 to 12 (**A**); The qPCR analysis of genes involved in anther cutin biosynthesis/transport, sugar partitioning, and pollen exine formation in the WT and *gpat3-2* anthers at stages 7 to 12 (**B**). *OsACTIN1* was chosen as a control, and data are shown as means ± SD (n = 3).

The expression pattern of genes involved in lipid/carbohydrate metabolism, synthesis/transportation of sporopollenin precursors and Ubisch bodies, and pollen wall formation also had some interesting features. The majority of genes related to anther cutin biosynthesis and pollen exine formation had significantly reduced expression in the *gpat3-2* mutant, including *CYP704B2*, *CYP703A3*, *OsACOS12*, *DPW*, *DPW2*, *OsC6*, *WDA1*, and *OsNP1*, further confirming the metabolic disorders of these nutrients in the *gpat3-2* mutant (Figure 9B). Three glycometabolism-related regulators, *CSA*, *UGP1*, and *UGP2*, also showed significant downregulation in the *gpat3-2* mutant, indicating that distortion of many other metabolic processes may concomitantly occur in the mutant (Figure 9B). However, the expression levels of two genes were significantly increased in the *gpat3-2* mutant: *MSP1*, a Leu-rich repeat receptor-like protein kinase that affects the number of cells entering into sporogenesis while simultaneously initiating anther wall formation in rice [69], and *OsMADS3*, which regulates late anther development and pollen formation by modulating ROS homeostasis [70]. The expression pattern of two ATP-binding cassette G transporters, *OsABCG15/PDA1* and *OsABCG26*, showed the same upregulation at stage 8 and mature pollen stage (stage 12), but unexpectedly had different expression patterns at stage 8 to stage 10 (Figure 9B). The transcriptional level of *OsABCG15/PDA1* in *gpat3-2* anthers was continuously higher than that of the wild type; however, *OsABCG26* was evidently downregulated at these stages. Together, these results suggest that loss of function of the *OsGPAT3* mutation may affect not only the biosynthesis of lipidic compounds and the anther wall PCD process, but also the formation/transport of Ubisch bodies and sporopollenin precursors, all of which are essential for anther development and pollen formation.

3. Discussion

Rice male reproductive development is determined by a series of specific functional factors, and loss of function of these genes may result in malformed anthers and eventually cause pollen abortion and male sterility, which are essential for hybrid rice breeding. Elucidation of the genetic, molecular, and biochemical mechanisms of male sterility-related genes are of primary importance for both basic research and application of rice breeding [2,15]. In this study, we further characterized the roles of rice *OsGPAT3* in male fertility, PCD-induced anther wall degradation, and pollen maturation on the basis of original research showing that *OsGPAT3* significantly affects anther cuticle biosynthesis and pollen exine formation [63]. This work provides new insights into the role of *OsGPAT3* in anther development and pollen formation.

3.1. OsGPAT3 Controls Ubisch Body Formation and Anther Development in Rice

Sporopollenin precursors are recognized as the most important structural security and nutritional sources for pollen formation [15,71]. Previous studies showed that recessive mutation of *OsGPAT3* can result in metabolism problems such as defective anther cuticle and synthesis failure of Ubisch bodies and pollen exine, which eventually cause complete male sterility [63]. Almost all components of lipid molecules decreased significantly in *osgpat3* mutant anthers [63]. Our study confirmed that *gpat3-2* exhibited a similar phenotype to *gpat3*, i.e., smooth anther surface, abnormal collapse of pollen grains (Figure 4), and amorphic Ubisch bodies (Figure 5F), finally resulting in severe degradation of all cells within the anther locule (Figure 2N,P and Figure 5H). In addition, defects in the anther wall, especially the middle and endothecium layers, were still visible at stage 13. In the *gpat3-2* mutant, this not only blocked transport of lipid substances, such as orbicules from the tapetum layer to the anther wall (Figure 2N,P and Figure 5H), but also hindered anther dehiscence. Expression changes of nutrient metabolism-related genes also confirmed that there are problems in synthesis related pathways. However, the upregulation of *MSP1* and *OsMADS3* may provide a sign that the *gpat3-2* mutant still carried active cell differentiation and degradation for sporogenesis and pollen wall formation. Different expression patterns between *OsABCG26* and OsABCG15/*PDA1* also implied that tapetum-synthesized lipid molecules differ in the process of transport from tapetum cells to anther surface by OsABCG26, and sporopollenin precursor transport to anther locules for anther cuticle formation by OsABCG15/*PDA1* [42–44]. These genes most likely still work to ensure cell activity in the mutant. Thus, our study demonstrated that *OsGPAT3* is required for sporopollenin precursor formation, pollen maturation, and even anther dehiscence in rice. but may functions varied in cell activity and nutrient transportation.

3.2. Loss of Function of OsGPAT3 Causes Abnormal PCD of Both Anther Wall and Pollen Grains

Timely cell differentiation and degradation of the three inner anther somatic layers is of critical importance for the patterning of microspore release, sporopollenin biosynthesis, pollen exine formation, and anther dehiscence. TUNEL assay analysis confirmed that *gpat3-2* exhibited delayed tapetum PCD and abnormal degradation of outer anther layers (Figure 3). The expression pattern of *OsGPAT3* in the *gpat3-2* mutant also suggests a possible feedback regulation of *OsGPAT3* transcription that functions in gradually strengthened degradation of anther wall and microspores in rice. Upregulation of tapetum PCD-related genes such as *bHLH TFs* (*UDT1*, *TDR*, *EAT1*, and *TIP2*) also showed that *gpat3-2* still had vigorous cell activity at stage 13, the appropriate anther dehiscence stage of wild type (Figure 9). However, the expression levels of *OSDEX1*, *GAMYB*, *AP25*, *AP37*, and *OsCP1* decreased significantly resulting from the upregulation of *OsGPAT3*. These results suggest that *OsGPAT3* positively regulates the expression of *OSDEX1*, *GAMYB*, *AP25*, *AP37*, and *OSCP1* in the tapetum, while *bHLH TFS*, *TGA10*, *PTC1*, and *MTR1* may be involved in other parallel regulatory pathways. Unfortunately, the functional network of *OsGPAT3* in rice tapetum development remains unclear; future investigations need to focus on identifying potential upstream genetic regulators using genetic and molecular approaches by expression pattern analysis of *OsGPAT3* in mutants of known PCD or lipid metabolism-associated transcription factors. Researchers can also create more double mutants or even polygene mutations using gene editing.

Our cytological observations showed that the *gpat3-2* mutant can release approximately normal microspores at stage 9 (Figures 1H, 3D and 5), which is significantly different from the *gpat3* mutant reported previously; microspores in *gpat3* were still covered with callose and could not be released from the tetrads at stage 9 [63]. Unfortunately, the newly released micropores in *gpat3-2* mutant gradually shrunk and degraded with the tapetum and other anther tissues at later stages. There were only remnants of abnormal epidermis left until the end in the *gpat3-2* anthers, all CRISPR/Cas9-induced mutants, and allelic mutants (Figure 3H,J,L and Figure 8I–N). This probably resulted from the above defects that hindered synthesis of Ubisch bodies/sporopollenin, but also abnormal PCD of the whole anther wall and pollen grains. In addition, our results also suggested that the middle layer and endothecium layer of the *gpat3-2* mutant had very little degradation (Figures 1 and 5), and the PCD process of all four anther layers were delayed (Figure 3H). Although the anther wall of *gpat3-2* anthers showed accelerated degradation at late development stages, it was too late for anther dehiscence as the pollen grains were already degraded and totally aborted (Figures 2P, 3L and 5). Therefore, our findings demonstrate that *OsGPAT3* is also a key regulator coordinating the differentiation and degradation of different layers for normal pollen exine formation and maturation during late anther development.

3.3. *Proposed Functions of OsGPAT3 in Rice Male Reproductive Development*

Our results further confirm that *OsGPAT3* affects not only metabolic processes such as lipid and carbohydrate metabolism in the early stages of anther development [63], but also differentiation and degradation of the anther wall in rice anthers to control both anther development and pollen formation. Therefore, a putative summary describing various functions of *OsGPAT3* during the male reproductive development process in rice is given below. *OsGPAT3* directly or indirectly affects various lipid and carbohydrate metabolism-related genes to ensure proper biosynthesis of anther cuticle and glycometabolism/sporopollenin precursors for anther development, pollen exine formation, and pollen maturation. *OsGPAT3* does not, however, directly regulate the transportation pathway of these substances. It also directly or indirectly regulates expression of *GAMYB*, *DEX1*, *AP25*, *AP37*, and *OsCP1* required for tapetum development and degeneration. *GAMYB* showed downregulation probably because it also affects pollen exine/Ubisch body formation and gibberellin-inducible nutrient mobilization [26,72]. Furthermore, *OsGPAT3* directly or indirectly affects differentiation and degradation of the anther wall required for pollen maturation and anther dehiscence. In summary, *OsGPAT3* regulates processes required for proper anther development and pollen maturation in different tissues at different stages of rice anther development. Our study also increased the known novel functions of the *OsGPAT3* gene during male reproductive development in rice.

4. Materials and Methods

4.1. *Mutant Material and Plant Growth Conditions*

The *gpat3-2* mutant (sp. *indica*), as the pollen acceptor, was crossed with wild-type Zh8015 and 02428 (sp. *japonica*). The heterozygous F_1 plants were then self-pollinated to generate a BC_1F_2 and an F_2 population for genetic analysis and mapping of the *Osgpat3-2* gene. In the F_2 mapping population, male-sterile plants were selected for gene mapping. All plants were grown in paddy fields of the China National Rice Research Institute during the spring of 2016 and 2017 in Lingshui, Hainan Province, China and the summer of 2016 and 2017 in Hangzhou, Zhejiang Province, China.

4.2. Identification of Mutant Anther and Pollen Development

Plant materials were photographed with a Nikon D800 digital camera and a Carl Zeiss SteREO Lumar V12 stereo fluorescence stereomicroscope (Markku Saari, Jena, Germany). For I_2/KI pollen staining, anthers and pollen grains were immersed in 1.2% I_2/KI solution (30 s) and then photographed using a Leica DM2500 microscope [12,67].

For cross-section observation of anther development, materials were collected and fixed on standard plastic sections as described by Zhang et al. and Li et al. [12,66]. Spikelets of wild type and *gpat3-2* mutant at different developmental stages were collected, based on the length of spikelet, and fixed with FAA (5% *v*/*v* formaldehyde, 5% *v*/*v* glacial acetic acid, and 50% *v*/*v* ethanol) overnight at 4 °C, dehydrated in a graded ethanol series (50–100%), embedded in Technovit 7100 resin (Heraeus Kulzer, Hanau, Germany, and polymerized at 60 °C. Transverse sections of 2.5-μm slices were obtained using a Leica RM2265 Fully Automated Rotary Microtome, stained with 0.2% toluidine blue (Chroma, Solms, Germany) and photographed using a Leica DM2000 biological microscope (Chroma, Solms, Germany).

For transmission electron microscopy (TEM), spikelets at various stages of development were collected and fixed in 2.5% glutaraldehyde in phosphate buffer (pH 7.0) for 16–24 h and then washed with phosphate-buffered saline (PBS; pH 7.2) three times, post-fixed with 1% OsO4 in phosphate buffer (pH 7.0) for 1 h, and washed three more times in phosphate buffer. Following ethanol dehydration, cutting and staining were performed as described previously [73]. After developmental identification, transverse sections were examined with a Model H-7650 TEM (HITACHI, Tokyo, Japan). For scanning electron microscopy (SEM), anthers at various developmental stages were collected and processed as described previously [67].

4.3. TUNEL Assay

The terminal deoxynucleotidyl transferase-mediated dUTP nick-end labeling (TUNEL) assay was performed as in a previous report [29]. Anthers from wild type and *gpat3-2* at different developmental stages were collected and prepared to generate paraffin sections. The selected paraffin sections were dewaxed in xylene and rehydrated in an ethanol series. TUNEL assay was performed using an In Vitro DeadEnd^{TM} Fluorometric TUNEL System, using fluorescein (Promega, Madison, WI 017959, USA) according to the manufacturer's instructions with some modifications. Signals were observed and imaged under a fluorescence confocal scanner microscope (ZEISS LSM 700, Jena, Germany). All pictures were taken in the same setting.

4.4. Molecular Cloning of Osgpat3-2

For mapping the *Osgpat3-2* locus, total DNA was extracted from fresh leaves using the modified cetyl trimethyl ammonium bromide (CTAB) method [74]. InDel (insertion/deletion) markers within the preliminary-mapping region were developed according to sequence differences between the genome sequence of *japonica* Nipponpare and *indica* 9311 (http://www.gramene.org and http://blast.ncbi.nlm.nih.gov), and polymorphisms between the two parents (*gpat3-2* and 02428) were detected. Potential candidate genes for all open reading frames (ORFs) in the fine-mapping interval were identified by referring to the Rice Genome Annotation Project Database (http://rice.plantbiology.msu.edu/index.shtml). The primers for molecular cloning of *Osgpat3-2* are listed in Table S3 (Supplementary Materials). The PCR products were separated by electrophoresis on 8% non-denaturing polyacrylamide gels and visualized by 0.1% $AgNO_3$ staining and NaOH staining with formaldehyde.

For enzyme digestion analysis of the mutation site, the restriction endonuclease *Nla*IV site of the target gene was identified using Primer Premier 5.0 software. Two pairs of primers were designed to generate 298 bp of DNA fragments amplified by KOD-FX (TOYOBO, Osaka, Japan). Purified PCR products were used for *Nla*IV digestion. A total of 100 μL of this reaction system contained 3 μL of HaeII (20 units per 1 μL), 10 μL of NEBuffer (1×), 60 μL of purified DNA template (3 μg), and 27 μL of distilled deionized H_2O (ddH_2O). Two reaction systems were incubated at 37 °C for 3 h followed by 2.5% agarose gel electrophoresis.

4.5. Complementation of the gpat3-2 Mutant

For complementation, a genomic DNA fragment ~7908 bp containing the entire *Osgpat3-2* coding region, a 3350-bp upstream sequence, and a 1180-bp sequence downstream of the termination codon was amplified from wild-type Zhonghui8015 using the primers listed in Table S1 (Supplementary Materials). The amplified fragment was released by *BamHI* digestion and cloned into *BamHI*-digested binary vector pCAMBIA1300 (CAMBIA, Portland, OR, USA, hygromycin resistance) using an In-Fusion Advantage Cloning kit (catalog no. PT4065; Clontech, San Francisco, CA, USA). Then, calli induced using BC_1F_2 seeds and showing the *gpat3-2* genotype were used for *Agrobacterium tumefaciens*-mediated transformation (all calli were selected by sequencing using the primers listed in Table S3, Supplementary Materials).

4.6. Vector Construction for CRISPR/Cas9-Mediated Mutation

Vector construction of CRISPR/Cas9-mediated mutation was processed essentially as described in Wu et al. (2017) [75]. To create the single guide RNA (sgRNA)/Cas9-induced *OsGPAT3* construct, a 23-bp *OsGPAT3*-specific sgRNA/Cas9 target sequence (red marking sequence) was inserted into the *Aar*I site of the pcas9-sgRNA vector, as described in Miao et al. (2013) [76], with some modifications. The primers used are detailed in Table S2 (Supplementary Materials). The aforementioned calli were introduced into wild-type and Zhonghua 11 seeds by *Agrobacterium tumefaciens*-mediated transformation as described previously [77]. The T_0 transgenic mutant plants regenerated from hygromycin-resistant calli were examined for the presence of transgenes using specific Cas-seq primers (Table S3, Supplementary Materials).

4.7. RNA Extraction, First-Strand Synthesis, and qPCR Analysis

Rice anthers at different developmental stages of the *gpat3-2* mutant and wild-type plants were collected for qPCR analysis of gene expression levels. Total RNA was extracted using the TIANGEN RNAprep Pure Plant Kit as described by the supplier. RNA was then reverse-transcribed (RT) from DNase I-treated RNA using Oligo-dT (18) primers in a 20-μL reaction using a SuperScript III Reverse Transcriptase Kit (TOYOBO, Japan). For qPCR, first-strand complementary DNA (cDNA) was diluted three times and then 3 μL of the RT products were used as the template of every PCR reaction using SYBR Premix Ex *Taq* II (TaKaRa) according to the manufacturer's instructions. The qPCR analysis was performed on a Roche LightCycler 480 device using gene-specific primers with the rice *Actin* gene (*Os03g0234200*) as an endogenous control, the relative expression levels were measured using the $2^{-\Delta Ct}$ analysis method, and the results were represented as means ± SD. This analysis examined expression of *OsGPAT3* and two homologs of *OsGPAT3* (Table S4), twelve tapetum PCD-related genes, twelve regulators that participate in lipid metabolism, and three glycometabolism-related regulators [21,31,41,44,50,78].

5. Conclusions

Male reproductive development in rice is important in both the improvement of yield and for an in-depth understanding of the mechanisms of anther development and pollen formation. In this study, we isolated and characterized a candidate recessive gene *OsGPAT3* that regulates anther wall PCD and pollen formation in rice using a typical map cloning method. Complementation analysis and knock-out experiments with the candidate gene further confirmed that the recessive mutation on *OsGPAT3* was responsible for the no-pollen phenotype of *gpat3-2*. Expression patterns of male sterility-related genes also demonstrated that loss of function of *OsGPAT3* caused various alterations in expression levels of both nutrient metabolism-related and tapetum PCD-related regulators, resulting in abnormal anther wall development and pollen formation. Nevertheless, the identification and observation of the *gpat3-2* mutant, together with the five allelic mutants, provided new insights into the function of *OsGPAT3* in regulating anther development and pollen formation in rice.

Author Contributions: L.C., S.C., and L.S. conceived and designed the experiments. X.X. and Z.Y. provided assistance in transection analysis, and TUNEL assay primer design. Y.Z. and P.Y. conducted population construction and phenotypic identification. X.W. and H.W. created the transgenic rice lines and provided assistance in primer design and mapping cloning. A.A. and R.M.K. made significant contributions to manuscript edits. L.S. analyzed the data and drafted the manuscript. Corresponding authors S.C. and L.C. were responsible for the overall concepts and designing the experiments. All authors read and approved the final manuscript.

Acknowledgments: We would like to thank associate professor Jing Yu (Shanghai Jiao Tong University, Shanghai, China) for generously providing many valuable suggestions for the manuscript. In addition, Lianping Sun particularly wishes to thank the god-given arrangements of meeting Siqi Deng last year, and the patience, understanding, and support from her over the past year. Will you marry me and jointly weave our happiness in the rest of our lives?

References

1. Goldberg, R.B.; Beals, T.P.; Sanders, P.M. Anther development: Basic principles and practical applications. *Plant Cell* **1993**, *5*, 1217–1229. [CrossRef] [PubMed]
2. Zhang, D.B.; Liang, W.Q. In Pushing the boundaries of scientific research: 120 years of addressing global issue. *Science* **2016**, *351*, 1223. [CrossRef]
3. Ma, H. Molecular genetic analyses of microsporogenesis and microgametogenesis in flowering plants. *Annu. Rev. Plant Biol.* **2005**, *56*, 393–434. [CrossRef] [PubMed]
4. Zhao, D.Z. Control of anther cell differentiation: A teamwork of receptor-like kinases. *Sex. Plant Reprod.* **2009**, *22*, 221–228. [CrossRef] [PubMed]
5. Wilson, Z.A.; Song, J.; Taylor, B.; Yang, C. The final split: The regulation of anther dehiscence. *J. Exp. Bot.* **2011**, *62*, 1633–1649. [CrossRef] [PubMed]
6. Walbot, V.; Egger, R.L. Pre-meiotic anther development: Cell fate specification and differentiation. *Annu. Rev. Plant Biol.* **2016**, *67*, 365–395. [CrossRef]
7. Scott, R.J.; Spielman, M.; Dickinson, H.G. Stamen structure and function. *Plant Cell* **2004**, *16*, S46–S60. [CrossRef]
8. Matsui, T.; Omasa, K.; Horie, T. Mechanism of anther dehiscence in rice (*Oryza sativa* L.). *Ann. Bot.* **1999**, *84*, 501–506. [CrossRef]
9. Wu, H.M.; Cheun, A.Y. Programmed cell death in plant reproduction. *Plant Mol. Biol.* **2000**, *44*, 267–281. [CrossRef]
10. Parish, R.W.; Li, S.F. Death of a tapetum: A programme of developmental altruism. *Plant Sci.* **2010**, *178*, 73–89. [CrossRef]
11. Jung, K.H.; Han, M.J.; Lee, Y.S.; Kim, Y.W.; Hwang, I.; Kim, M.J.; Kim, Y.K.; Nahm, B.H.; An, G. Rice *Undeveloped Tapetum1* is a major regulator of early tapetum development. *Plant Cell* **2015**, *17*, 2705–2722. [CrossRef] [PubMed]
12. Li, N.; Zhang, D.S.; Liu, H.S.; Yin, C.S.; Li, X.X.; Liang, W.Q.; Yuan, Z.; Xu, B.; Chu, H.W.; Wang, J.; et al. The rice *tapetum degeneration retardation* gene is required for tapetum degradation and anther development. *Plant Cell* **2006**, *18*, 2999–3014. [CrossRef] [PubMed]
13. Liu, J.; Qu, L.J. Meiotic and mitotic cell cycle mutants involved in gametophyte development in *Arabidopsis*. *Mol. Plant.* **2008**, *1*, 564–574. [CrossRef] [PubMed]
14. Wilson, Z.A.; Zhang, D.B. From Arabidopsis to rice: Pathways in pollen development. *J. Exp. Bot.* **2009**, *60*, 1479–1492. [CrossRef] [PubMed]
15. Shi, J.X.; Cui, M.H.; Yang, L.; Kim, Y.J.; Zhang, D.B. Genetic and biochemical mechanisms of pollen wall development. *Trends Plant Sci.* **2015**, *20*, 741–753. [CrossRef] [PubMed]

16. Gavrieli, Y.; Sherma, Y.; Ben-Sasson, S.A. Identification of programmed cell death in situ via specific labeling of nuclear DNA fragmentation. *J. Cell Biol.* **1992**, *119*, 493–501. [CrossRef] [PubMed]
17. Varnier, A.L.; Mazeyrat-Gourbeyre, F.; Sangwan, R.S.; Clément, C. Programmed cell death progressively models the development of anther sporophytic tissues from the tapetum and is triggered in pollen grains during maturation. *J. Struct Biol.* **2005**, *152*, 118–128. [CrossRef] [PubMed]
18. Zhang, D.S.; Liang, W.Q.; Yuan, Z.; Li, N.; Shi, J.; Wang, J.; Liu, Y.M.; Yu, W.J.; Zhang, D.B. *Tapetum Degeneration Retardation* is critical for aliphatic metabolism and gene regulation during rice pollen development. *Mol. Plant* **2008**, *1*, 599–610. [CrossRef] [PubMed]
19. Ji, C.N.; Li, H.Y.; Chen, L.B.; Xie, M.; Wang, F.P.; Chen, Y.L.; Liu, Y.G. A novel rice bHLH transcription factor, *DTD*, acts coordinately with tdr in controlling tapetum function and pollen development. *Mol. Plant* **2013**, *6*, 1715–1718. [CrossRef] [PubMed]
20. Niu, N.N.; Liang, W.Q.; Yang, X.J.; Jin, W.L.; Wilson, Z.A.; Hu, J.P.; Zhang, D.B. *EAT1* promotes tapetal cell death by regulating aspartic proteases during male reproductive development in rice. *Nat. Commun.* **2013**, *4*, 1445. [CrossRef] [PubMed]
21. Fu, Z.Z.; Yu, J.; Cheng, X.W.; Zong, X.; Xu, J.; Chen, M.J.; Li, Z.Y.; Zhang, D.B.; Liang, W.Q. The rice basic helix-loop-helix transcription factor *TDR INTERACTING PROTEIN2* is a central switch in early anther development. *Plant Cell* **2014**, *26*, 1512–1524. [CrossRef] [PubMed]
22. Ko, S.S.; Li, M.J.; Sun-Ben, K.M.; Ho, Y.C.; Lin, Y.J.; Chuang, M.H.; Hsing, H.X.; Lien, Y.C.; Yang, H.T.; Chang, H.C.; et al. The *bHLH142* transcription factor coordinates with *tdr1* to modulate the expression of *eat1* and regulate pollen development in rice. *Plant Cell* **2014**, *26*, 2486–2504. [CrossRef] [PubMed]
23. Ko, S.S.; Li, M.J.; Lin, Y.J.; Hsing, H.X.; Yang, T.T.; Chen, T.K.; Jhong, C.M.; Ku, M.S. Tightly controlled expression of *bHLH142* is essential for timely tapetal programmed cell death and pollen development in rice. *Front. Plant Sci.* **2017**, *8*, 1258. [CrossRef] [PubMed]
24. Ranjan, R.; Khurana, R.; Malik, N.; Badoni, S.; Parida, S.K.; Kapoor, S.; Tyagi, A.K. *bHLH142* regulates various metabolic pathway-related genes to affect pollen development and anther dehiscence in rice. *Sci. Rep.* **2017**, *7*, 43397. [CrossRef] [PubMed]
25. Kaneko, M.; Inukai, Y.; Ueguchi-Tanaka, M.; Itoh, H.; Izawa, T.; Kobayashi, Y.; Hattori, T.; Miyao, A.; Hirochika, H.; Ashikari, M.; et al. Loss-of-function mutations of the rice *GAMYB* gene impair α-amylase expression in aleurone and flower development. *Plant Cell* **2004**, *16*, 33–44. [CrossRef] [PubMed]
26. Aya, K.; Ueguchi-Tanaka, M.; Kondo, M.; Hamada, K.; Yano, K.; Nishimura, M.; Matsuoka, M. Gibberellin modulates anther development in rice via the transcriptional regulation of *GAMYB*. *Plant Cell* **2009**, *21*, 1453–1472. [CrossRef] [PubMed]
27. Liu, Z.H.; Bao, W.J.; Liang, W.Q.; Yin, J.Y.; Zhang, D.B. Identification of *gamyb-4* and analysis of the regulatory role of *GAMYB* in rice anther development. *J. Integr. Plant Biol.* **2010**, *52*, 670–678. [CrossRef]
28. Li, H.; Yuan, Z.; Vizcay-Barrena, G.; Yang, C.Y.; Liang, W.Q.; Zong, J.; Wilson, Z.A.; Zhang, D.B. *PERSISTENT TAPETAL CELL1* encodes a phd-finger protein that is required for tapetal cell death and pollen development in rice. *Plant Physiol.* **2011**, *156*, 615–630. [CrossRef]
29. Chen, Z.S.; Liu, X.F.; Wang, D.H.; Chen, R.; Zhang, X.L.; Xu, Z.H.; Bai, S.N. Transcription factor *OsTGA10* is a target of the MADS protein OsMADS8 and is required for tapetum development. *Plant Physiol.* **2018**, *176*, 819–835. [CrossRef]
30. Yi, J.; Moon, S.; Lee, Y.S.; Zhu, L.; Liang, W.Q.; Zhang, D.B.; Jung, K.H.; An, G. *Defective Tapetum Cell Death 1* (*DTC1*) regulates ROS levels by binding to metallothionein during tapetum degeneration. *Plant Physiol.* **2016**, *170*, 1611–1623. [CrossRef]
31. Yu, J.; Meng, Z.; Liang, W.Q.; Behera, S.; Kudla, J.; Tucker, M.R.; Luo, Z.J.; Chen, M.J.; Xu, D.W.; Zhao, G.C.; et al. A rice Ca^{2+} binding protein is required for tapetum function and pollen formation. *Plant Physiol.* **2016**, *172*, 1772–1786. [CrossRef] [PubMed]
32. Tan, H.X.; Liang, W.Q.; Hu, J.P.; Zhang, D.B. *MTR1* encodes a secretory fasciclin glycoprotein required for male reproductive development in rice. *Dev. Cell* **2012**, *22*, 1127–1137. [CrossRef] [PubMed]
33. Li, L.; Li, Y.X.; Song, S.F.; Deng, H.F.; Li, N.; Fu, X.Q.; Chen, G.H.; Yuan, L.P. An *anther development F-box* (*ADF*) protein regulated by *tapetum degeneration retardation* (*TDR*) controls rice anther development. *Planta* **2015**, *241*, 157–166. [CrossRef] [PubMed]

34. Li, X.W.; Gao, X.Q.; Wei, Y.; Deng, L.; Ouyang, Y.D.; Chen, G.X.; Li, X.H.; Zhang, Q.F.; Wu, C.Y. Rice *APOPTOSIS INHIBITOR5* coupled with two DEAD-Box adenosine 5′-triphosphate-dependent rna helicases regulates tapetum degeneration. *Plant Cell* **2011**, *23*, 1416–1434. [CrossRef] [PubMed]
35. Blackmore, S.; Wortley, A.H.; Skvarla, J.J.; Rowley, J.R. Pollen wall development in flowering plants. *New Phytol.* **2007**, *174*, 483–498. [CrossRef] [PubMed]
36. Zhang, D.B.; Shi, J.X.; Yang, X. Role of lipid metabolism in plant pollen Exine development. *Subcell Biochem.* **2016**, *86*, 315–337. [CrossRef]
37. Li, H.; Pinot, F.; Sauveplane, V.; Werckreichhart, D.; Diehl, P.; Schreiber, L.; Franke, R.; Zhang, P.; Chen, L.; Gao, Y.W.; et al. Cytochrome P450 family member *CYP704B2* catalyzes the v-hydroxylation of fatty acids and is required for anther cutin biosynthesis and pollen exine formation in rice. *Plant Cell* **2010**, *22*, 173–190. [CrossRef]
38. Yang, X.J.; Wu, D.; Shi, J.X.; Yi, H.; Pinot, F.; Grausem, B.; Yin, C.S.; Zhu, L.; Chen, M.J.; Luo, Z.J.; et al. Rice *CYP703A3*, a cytochrome P450 hydroxylase, is essential for development of anther cuticle and pollen exine. *J. Integr. Plant Biol.* **2014**, *56*, 979–994. [CrossRef]
39. Yang, Z.F.; Zhang, Y.X.; Sun, L.P.; Zhang, P.P.; Liu, L.; Yu, P.; Xuan, D.D.; Xiang, X.J.; Wu, W.X.; Cao, L.Y.; et al. Identification of *cyp703a3-3* and analysis of regulatory role of *CYP703A3* in rice anther cuticle and pollen exine development. *Gene* **2018**, *649*, 63–73. [CrossRef]
40. Shi, J.; Tan, H.X.; Yu, X.H.; Liu, Y.Y.; Liang, W.Q.; Ranathunge, K.; Franke, R.B.; Schreiber, L.; Wang, Y.J.; Kai, G.Y.; et al. *Defective pollen wall* is required for anther and microspore development in rice and encodes a fatty acyl carrier protein reductase. *Plant Cell* **2011**, *23*, 2225–2246. [CrossRef]
41. Xu, D.W.; Shi, J.X.; Rautengarten, C.; Yang, L.; Qian, X.L.; Uzair, M.L.; Zhu, L.L.; Luo, Q.L.; An, G.L.; Waßmann, F.L.; et al. *Defective Pollen Wall 2* (*DPW2*) encodes an acyl transferase required for rice pollen development. *Plant Physiol.* **2017**, *173*, 240–255. [CrossRef] [PubMed]
42. Qin, P.; Tu, B.; Wang, Y.P.; Deng, L.C.; Quilichini, T.D.; Li, T.; Wang, H.; Ma, B.T.; Li, S.G. *ABCG15* encodes an ABC transporter protein, and is essential for post-meiotic anther and pollen exine development in rice. *Plant Cell Physiol.* **2013**, *54*, 138–154. [CrossRef] [PubMed]
43. Zhu, L.; Shi, J.X.; Zhao, G.C.; Zhang, D.B.; Liang, W.Q. *Post-meiotic deficient anther1* (*PDA1*) encodes an ABC transporter required for the development of anther cuticle and pollen exine in rice. *J. Plant Biol.* **2013**, *56*, 59–68. [CrossRef]
44. Zhao, G.C.; Shi, J.X.; Liang, W.Q.; Xue, F.Y.; Luo, Q.; Zhu, L.; Qu, G.R.; Chen, M.J.; Schreiber, L.; Zhang, D.B. Two ATP Binding cassette G transporters, rice *ATP binding cassette G26* and *ATP binding cassette G15*, collaboratively regulate rice male reproduction. *Plant Physiol.* **2015**, *169*, 2064–2079. [CrossRef] [PubMed]
45. Chang, Z.Y.; Jin, M.N.; Yan, W.; Chen, H.; Qiu, S.J.; Fu, S.; Xia, J.X.; Liu, Y.C.; Chen, Z.F.; Wu, J.X.; et al. The ATP-binding cassette (ABC) transporter *OsABCG3* is essential for pollen development in rice. *Rice* **2018**, *11*, 58. [CrossRef] [PubMed]
46. Zhang, D.S.; Liang, W.Q.; Yin, C.S.; Zong, J.; Gu, F.W.; Zhang, D.B. *OsC6*, encoding a lipid transfer protein, is required for postmeiotic anther development in rice. *Plant Physiol.* **2010**, *154*, 149–162. [CrossRef]
47. Jung, K.H.; Han, M.J.; Lee, D.Y.; Lee, Y.S.; Schreiber, L.; Franke, R.; Faust, A.; Yephremov, A.; Saedler, H.; Kim, Y.W.; et al. *Wax-deficient anther1* is involved in cuticle and wax production in rice anther walls and is required for pollen development. *Plant Cell* **2006**, *18*, 3015–3032. [CrossRef]
48. Li, Y.L.; Li, D.D.; Guo, Z.L.; Shi, Q.S.; Xiong, S.X.; Zhang, C.; Zhu, J.; Yang, Z.N. *OsACOS12*, an orthologue of Arabidopsis acyl-CoA synthetase5, plays an important role in pollen exine formation and anther development in rice. *BMC Plant Biol.* **2016**, *16*, 256. [CrossRef]
49. Yang, X.J.; Liang, W.Q.; Chen, M.J.; Zhang, D.B.; Zhao, X.X.; Shi, J.X. Rice fatty acyl-CoA synthetase *OsACOS12* is required for tapetum programmed cell death and male fertility. *Planta* **2017**, *246*, 105–122. [CrossRef]
50. Liu, Z.; Lin, S.; Shi, J.X.; Yu, J.; Zhu, L.; Yang, X.J.; Zhang, D.B.; Liang, W.Q. Rice *No Pollen 1* (*NP1*) is required for anther cuticle formation and pollen exine patterning. *Plant J.* **2017**, *91*, 263–277. [CrossRef]
51. Zhu, X.L.; Yu, J.; Shi, J.X.; Tohge, T.; Fernie, A.R.; Meir, S.; Aharoni, A.R.; Xu, D.W.; Zhang, D.B.; Liang, W.Q. The polyketide synthase *OsPKS2* is essential for pollen exine and Ubisch body patterning in rice. *J. Integr. Plant Biol.* **2017**, *59*, 612–628. [CrossRef] [PubMed]

52. Zou, T.; Liu, M.X.; Xiao, Q.; Wang, T.; Chen, D.; Luo, T.; Yuan, G.Q.; Li, Q.; Zhu, J.; Liang, Y.Y.; et al. *OsPKS2* is required for rice male fertility by participating in pollen wall formation. *Plant Cell Rep.* **2018**, *37*, 759–773. [CrossRef] [PubMed]
53. Coleman, R.A.; Lee, D.P. Enzymes of triacylglycerol synthesis and their regulation. *Prog. Lipid Res.* **2004**, *43*, 134–176. [CrossRef]
54. Cao, J.; Li, J.L.; Li, D.; Tobin, J.F.; Gimeno, R.E. Molecular identification of microsomal acyl-CoA:glycerol-3-phosphate acyltransferase, a key enzyme in de novo triacylglycerol synthesis. *Proc. Natl. Acad. Sci. USA* **2006**, *103*, 19695–19700. [CrossRef] [PubMed]
55. Takeuchi, K.; Reue, K. Biochemistry, physiology, and genetics of GPAT, AGPAT, and lipin enzymes in triglyceride synthesis. *Am. J. Physiol. Endocrinol. Metab.* **2009**, *296*, E1195–E1209. [CrossRef] [PubMed]
56. Lung, S.C.; Weselake, R.J. Diacylglycerol acyltransferase: A key mediator of splant triacylglycerol synthesis. *Lipids* **2006**, *41*, 1073–1088. [CrossRef] [PubMed]
57. Li, Y.; Beisson, F.; Koo, A.J.; Molina, I.; Pollard, M.; Ohlrogge, J. Identification of acyltransferases required for cutin biosynthesis and production of cutin with suberin-like monomers. *Proc. Natl. Acad. Sci. USA* **2007**, *104*, 18339–18344. [CrossRef]
58. Pollard, M.; Beisson, F.; Li, Y.; Ohlrogge, J.B. Building lipid barriers: Biosynthesis of cutin and suberin. *Trends Plant Sci.* **2008**, *13*, 236–246. [CrossRef]
59. Beisson, F.; Li-Beisson, Y.; Pollard, M. Solving the puzzles of cutin and suberin polymer biosynthesis. *Curr. Opin. Plant Biol.* **2012**, *15*, 329–337. [CrossRef] [PubMed]
60. Yang, W.; Simpson, J.P.; Li-Beisson, Y.; Beisson, F.; Pollard, M.; Ohlrogge, J.B. A land-plant-specific glycerol-3-phosphate acyltransferase family in Arabidopsis: Substrate specificity, sn-2 preference, and evolution. *Plant Physiol.* **2012**, *160*, 638–652. [CrossRef]
61. Zheng, Z.; Xia, Q.; Dauk, M.; Shen, W.; Selvaraj, G.; Zou, J. Arabidopsis *AtGPAT1*, a member of the membrane-bound glycerol-3-phosphate acyltransferase gene family, is essential for tapetum differentiation and male fertility. *Plant Cell* **2003**, *15*, 1872–1887. [CrossRef] [PubMed]
62. Petit, J.; Bres, C.; Mauxion, J.P.; Tai, F.W.; Martin, L.B.; Fich, E.A.; Joubes, J.; Rose, J.K.; Domergue, F.; Rothan, C. The glycerol-3-phosphate acyltransferase *GPAT6* from tomato plays a central role in fruit cutin biosynthesis. *Plant Physiol.* **2016**, *171*, 894–913. [CrossRef] [PubMed]
63. Men, X.; Shi, J.X.; Liang, W.Q.; Zhang, Q.F.; Lian, G.B.; Quan, S.; Zhu, L.; Luo, Z.J.; Chen, M.J.; Zhang, D.B. *Glycerol-3-Phosphate Acyltransferase 3* (*OsGPAT3*) is required for anther development and male fertility in rice. *J. Exp. Bot.* **2017**, *68*, 513–526. [CrossRef] [PubMed]
64. Xie, K.; Wu, S.W.; Li, Z.W.; Zhou, Y.; Zhang, D.F.; Dong, Z.Y.; An, X.L.; Zhu, T.T.; Zhang, S.M.; Liu, S.S.; et al. Map-based cloning and characterization of *Zea mays male sterility33* (*ZmMs33*) gene, encoding a glycerol-3-phosphate acyltransferase. *Theor. Appl. Genet.* **2018**, *131*, 1363–1378. [CrossRef] [PubMed]
65. Zhang, D.B.; Wilson, Z.A. Stamen specification and anther development in rice. *Chin. Sci. Bull.* **2009**, *54*, 2342–2353. [CrossRef]
66. Zhang, D.B.; Luo, X.; Zhu, L. Cytological analysis and genetic control of rice anther development. *J. Genet. Genom.* **2011**, *38*, 379–390. [CrossRef]
67. Sun, L.P.; Zhang, Y.X.; Zhang, P.P.; Yang, Z.F.; Zhou, X.X.; Xuan, D.D.; Li, Z.H.; Wu, W.X.; Zhan, X.D.; Shen, X.H.; et al. Morphogenesis and Gene Mapping of *deformed interior floral organ 1* (*difo1*), a novel mutant associated with floral organ development in rice. *Plant Mol. Biol. Rep.* **2017**, *35*, 330–344. [CrossRef]
68. Shen, Y.J.; Jiang, H.; Jin, J.P.; Zhang, Z.B.; Xi, B.; He, Y.Y.; Wang, G.; Wang, C.; Qian, L.; Li, X.; et al. Development of genome-wide DNA polymorphism database for map-based cloning of rice genes. *Plant Physiol.* **2004**, *135*, 1198–1205. [CrossRef]
69. Nonomura, K.I.; Miyoshi, K.; Eiguchi, M.; Suzuki, T.; Miyao, A.; Hirochika, H.; Kurata, N. The *MSP1* gene is necessary to restrict the number of cells entering into male and female sporogenesis and to initiate anther wall formation in rice. *Plant Cell* **2003**, *15*, 1728–1739. [CrossRef]
70. Hu, L.F.; Liang, W.Q.; Yin, C.S.; Cui, X.; Zong, J.; Wang, X.; Hu, J.P.; Zhang, D.B. Rice *MADS3* regulates ros homeostasis during late anther development. *Plant Cell* **2011**, *23*, 515–533. [CrossRef]
71. Ariizumi, T.; Toriyama, K. Genetic regulation of sporopollenin synthesis and pollen exine development. *Annu. Rev. Plant Biol.* **2011**, *62*, 437–460. [CrossRef] [PubMed]

72. Hong, Y.F.; Ho, T.H.; Wu, C.F.; Ho, S.L.; Yeh, R.H.; Lu, C.A.; Chen, P.W.; Yu, L.C.; Chao, A.; Yu, S.M. Convergent starvation signals and hormone crosstalk in regulating nutrient mobilization upon germination in cereals. *Plant Cell* **2012**, *24*, 2857–2873. [CrossRef] [PubMed]
73. Li, Z.; Zhang, Y.X.; Liu, L.; Liu, Q.E.; Bi, Z.Z.; Yu, N.; Cheng, S.H.; Cao, L.Y. Fine mapping of the *lesion mimic and early senescence 1* (*lmes1*) in rice (*Oryza sativa.* L). *Plant Physiol. Biochem.* **2014** *80*, 300–307. [CrossRef]
74. Chen, D.H.; Ronald, P.C. A rapid DNA minipreparation method suitable for AFLP and other PCR applications. *Plant Mol. Biol. Rep.* **1999**, *17*, 53–57. [CrossRef]
75. Wu, W.X.; Zheng, X.M.; Chen, D.; Zhang, Y.X.; Ma, W.W.; Zhang, H.; Sun, L.P.; Yang, Z.F.; Zhao, C.D.; Zhan, X.D.; et al. *OsCOL16*, encoding a CONSTANS-like protein, represses flowering by up-regulating *Ghd7* expression in rice. *Plant Sci.* **2017**, *260*, 60–69. [CrossRef]
76. Miao, J.; Guo, D.; Zhang, J.; Huang, Q.; Qin, G.; Zhang, X.; Wan, J.; Gu, H.; Qu, L.J. Targeted mutagenesis in rice using CRISPR-Cas system. *Cell Res.* **2013**, *23*, 1233–1236. [CrossRef]
77. Hiei, Y.; Ohta, S.; Komari, T.; Kumashiro, T. Efficient transformation of rice (*Oryza sativa* L.) mediated by Agrobacterium and sequence analysis of the boundaries of the T-DNA. *Plant J.* **1994**, *6*, 271–282. [CrossRef]
78. Zhang, H.; Liang, W.Q.; Yang, X.J.; Luo, X.; Jiang, N.; Ma, H.; Zhang, D.B. *Carbon Starved Anther* encodes a myb domain protein that regulates sugar partitioning required for rice pollen development. *Plant Cell* **2010**, *22*, 672–689. [CrossRef]

Permissions

All chapters in this book were first published in MDPI; hereby published with permission under the Creative Commons Attribution License or equivalent. Every chapter published in this book has been scrutinized by our experts. Their significance has been extensively debated. The topics covered herein carry significant findings which will fuel the growth of the discipline. They may even be implemented as practical applications or may be referred to as a beginning point for another development.

The contributors of this book come from diverse backgrounds, making this book a truly international effort. This book will bring forth new frontiers with its revolutionizing research information and detailed analysis of the nascent developments around the world.

We would like to thank all the contributing authors for lending their expertise to make the book truly unique. They have played a crucial role in the development of this book. Without their invaluable contributions this book wouldn't have been possible. They have made vital efforts to compile up to date information on the varied aspects of this subject to make this book a valuable addition to the collection of many professionals and students.

This book was conceptualized with the vision of imparting up-to-date information and advanced data in this field. To ensure the same, a matchless editorial board was set up. Every individual on the board went through rigorous rounds of assessment to prove their worth. After which they invested a large part of their time researching and compiling the most relevant data for our readers.

The editorial board has been involved in producing this book since its inception. They have spent rigorous hours researching and exploring the diverse topics which have resulted in the successful publishing of this book. They have passed on their knowledge of decades through this book. To expedite this challenging task, the publisher supported the team at every step. A small team of assistant editors was also appointed to further simplify the editing procedure and attain best results for the readers.

Apart from the editorial board, the designing team has also invested a significant amount of their time in understanding the subject and creating the most relevant covers. They scrutinized every image to scout for the most suitable representation of the subject and create an appropriate cover for the book.

The publishing team has been an ardent support to the editorial, designing and production team. Their endless efforts to recruit the best for this project, has resulted in the accomplishment of this book. They are a veteran in the field of academics and their pool of knowledge is as vast as their experience in printing. Their expertise and guidance has proved useful at every step. Their uncompromising quality standards have made this book an exceptional effort. Their encouragement from time to time has been an inspiration for everyone.

The publisher and the editorial board hope that this book will prove to be a valuable piece of knowledge for researchers, students, practitioners and scholars across the globe.

List of Contributors

Jingyi Liu, Meijing Shi, Jing Wang, Bo Zhang, Yushun Li, Jin Wang and Yan Liang
College of Horticulture, Northwest A&F University, Shaanxi 712100, China
State Agriculture Ministry Laboratory of Northwest Horticultural Plant Germplasm Resources & Genetic Improvement, Northwest A&F University, Shaanxi 712100, China

Ahmed. H. El-Sappah
College of Horticulture, Northwest A&F University, Shaanxi 712100, China
State Agriculture Ministry Laboratory of Northwest Horticultural Plant Germplasm Resources & Genetic Improvement, Northwest A&F University, Shaanxi 712100, China
Genetics Department, Faculty of Agriculture, Zagazig University, Zagazig 44511, Egypt

Esnart Nyirenda Yohane, Hussein Shimelis, Mark Laing, Isack Mathew and Admire Shayanowako
African Centre for Crop Improvement, School of Agricultural, Earth and Environmental Sciences, University of KwaZulu-Natal, Private Bag X01, Scottsville 3201, Pietermaritzburg 3209, South Africa

Yu Pan
Key Laboratory of Horticulture Science for Southern Mountainous Regions, Ministry of Education, Southwest University, No. 2 Tiansheng Road, Beibei, Chongqing 400715, China
Academy of Agricultural Sciences, Southwest University, Chongqing 400715, China

Meichen Zhu, ShuxianWang, Guoqiang Ma, Xiaohu Huang, Cailin Qiao, Rui Wang, Xinfu Xu, Ying Liang, Kun Lu, Jiana Li and Cunmin Qu
Academy of Agricultural Sciences, Southwest University, Chongqing 400715, China
Chongqing Rapeseed Engineering Research Center, College of Agronomy and Biotechnology, Southwest University, No. 2 Tiansheng Road, Beibei, Chongqing 400715, China

Fengqing Han
Institute of Vegetables and Flowers, Chinese Academy of Agricultural Sciences, Key Laboratory of Biology and Genetic Improvement of Horticultural Crops, Ministry of Agriculture, Beijing 100081, China

Xiaoli Zhang
Tianjin Kernel Vegetable Research Institute, The National Key Laboratory of Vegetable Germplasm Innovation, The Enterprise key Laboratory of Tianjin Vegetable Genetics and Breeding, Jinjing Road, Xiqing District, Tianjin 300384, China

Yi Yang, Ming Sun, Cunquan Yuan, Yu Han, Tangchun Zheng, Tangren Cheng and Jia Wang
Beijing Key Laboratory of Ornamental Plants Germplasm Innovation & Molecular Breeding, National Engineering Research Center for Floriculture, Beijing Laboratory of Urban and Rural Ecological Environment, Engineering Research Center of Landscape Environment of Ministry of Education, Key Laboratory of Genetics and Breeding in Forest Trees and Ornamental Plants of Ministry of Education, School of Landscape Architecture, Beijing Forestry University, Beijing 100083, China

Qixiang Zhang
Beijing Key Laboratory of Ornamental Plants Germplasm Innovation & Molecular Breeding, National Engineering Research Center for Floriculture, Beijing Laboratory of Urban and Rural Ecological Environment, Engineering Research Center of Landscape Environment of Ministry of Education, Key Laboratory of Genetics and Breeding in Forest Trees and Ornamental Plants of Ministry of Education, School of Landscape Architecture, Beijing Forestry University, Beijing 100083, China
Beijing Advanced Innovation Center for Tree Breeding by Molecular Design, Beijing Forestry University, Beijing 100083, China

Zilhas Ahmed Jewel, Jauhar Ali and Anumalla Mahender
Rice Breeding Platform, International Rice Research Institute (IRRI), Los Baños, Laguna 4031, Philippines

Jose Hernandez
College of Agriculture, University of the Philippines Los Baños, Laguna 4031, Philippines

Yunlong Pang
Rice Breeding Platform, International Rice Research Institute (IRRI), Los Baños, Laguna 4031, Philippines
College of Agronomy, Shandong Agricultural University, Taian 271018, China

Zhikang Li
Institute of Crop Sciences, Chinese Academy of Agricultural Sciences, Beijing 100081, China

Jinshuai Shu, Lili Zhang, Yumei Liu, Zhansheng Li, Zhiyuan Fang, Limei Yang, Mu Zhuang, Yangyong Zhang and Honghao Lv
Institute of Vegetables and Flowers, Chinese Academy of Agricultural Sciences, Key Laboratory of Biology and Genetic Improvement of Horticultural Crops, Ministry of Agriculture, 12 Zhongguancun Nandajie Street, Beijing 100081, China

Ángela S. Prudencio, Pedro J. Martínez-García, Federico Dicenta and Pedro Martínez-Gómez
Department of Plant Breeding, CEBAS-CSIC, Espinardo, 30100 Murcia, Spain

Olaf Werner and Rosa M. Ros
Department of Plant Biology, Faculty of Biology, University of Murcia, Espinardo, 30100 Murcia, Spain

Madhav Bhatta, Vikas Belamkar and P. Stephen Baenziger
Department of Agronomy and Horticulture, University of Nebraska-Lincoln, Lincoln, NE 68583, USA

Alexey Morgounov
International Maize and Wheat Improvement Center (CIMMYT), 06511 Emek, Ankara, Turkey

Sangrea Shim, Jungmin Ha, Moon Young Kim and Suk-Ha Lee
Department of Plant Science and Research Institute of Agriculture and Life Sciences, Seoul National University, Seoul 08826, Korea
Plant Genomics and Breeding Institute, Seoul National University, Seoul 08826, Korea

Man Soo Choi
National Institute of Crop Sciences, Rural Development Administration, Wanju-gun, Jeollabuk-do 55365, Korea

Sung-Taeg Kang
Department of Crop Science & Biotechnology, Dankook University, Cheonan-si, Chungcheongnam-do 31116, Korea

Soon-Chun Jeong
Bio-Evaluation Center, Korea Research Institute of Bioscience and Biotechnology, Cheongju-si, Chungcheongbuk-do 28116, Korea

Jung-Kyung Moon
National Institute of Agricultural Sciences, Rural Development Administration, Jeonju-si, Jeollabuk-do 54874, Korea

Xiaoming Sui, Mingyuan Zhao, Zongda Xu, Lanyong Zhao and Xu Han
Flower Research Laboratory, College of Forestry, Shandong Agricultural University, Taian 271018, China

Lianping Sun, Xiaojiao Xiang, Zhengfu Yang, Ping Yu, Xiaoxia Wen, Hong Wang, Adil Abbas, Riaz Muhammad Khan, Yingxin Zhang, Shihua Cheng and Liyong Cao
Key Laboratory for Zhejiang Super Rice Research and State Key Laboratory of Rice Biology, China National Rice Research Institute, Hangzhou 310006, China

Index

www.ingramcontent.com/pod-product-compliance
Lightning Source LLC
LaVergne TN
LVHW081018210926
848485LV00004B/888